Mechanisms
of Receptor
Regulation

NEW HORIZONS IN THERAPEUTICS
Smith Kline & French Laboratories Research Symposia Series

Series Editors: George Poste and Stanley T. Crooke
Smith Kline & French Laboratories, Philadelphia, Pennsylvania

---

DOPAMINE RECEPTOR AGONISTS
Edited by George Poste and Stanley T. Crooke

MECHANISMS OF RECEPTOR REGULATION
Edited by George Poste and Stanley T. Crooke

NEW INSIGHTS INTO CELL AND MEMBRANE TRANSPORT
　　PROCESSES
Edited by George Poste and Stanley T. Crooke

Mechanisms of Receptor Regulation

Edited by

GEORGE POSTE and
STANLEY T. CROOKE
Smith Kline & French Laboratories
Philadelphia, Pennsylvania

PLENUM PRESS · NEW YORK AND LONDON

Library of Congress Cataloging in Publication Data

Main entry under title:

Mechanisms of receptor regulation.

(New horizons in therapeutics)
Proceedings of the Second Smith, Kline & French Research Symposium on New Horizons in Therapeutics, held in Philadelphia in 1984.
1. Hormone receptors – Congresses. 2. Cell receptors – Congresses. 3. Metabolic regulation – Congresses. I. Poste, George. II. Crooke, Stanley T. III. Smith, Kline, and French Research Symposium on New Horizons in Therapeutics (2nd: 1984: Philadelphia, Pa.) IV. Series. [DNLM: 1. Binding Sites – congresses. 2. Receptors, Endogenous Substances – congresses. WK 102 M486 1984]

QP571.7.M43   1986   574.87′5   85-28339
ISBN-13: 978-1-4612-9259-3   e-ISBN-13: 978-1-4613-2131-6
DOI: 10.1007/978-1-4613-2131-6

© 1985 Plenum Press, New York
Softcover reprint of the hardcover 1st edition 1985
A Division of Plenum Publishing Corporation
233 Spring Street, New York, N.Y. 10013

Contributors

Menashe Bar-Eli, Department of Immunology, Ben-Gurion University, Beer-Sheva, Israel

Michael J. Berridge, A.R.C. Unit of Insect Neurophysiology and Pharmacology, Department of Zoology, University of Cambridge, Cambridge CB2 3EJ, England

Donald K. Blumenthal, Howard Hughes Medical Institute Laboratory, Department of Pharmacology, University of Washington, Seattle, Washington 98195

Gary M. Bokoch, Department of Pharmacology, Southwestern Graduate School, University of Texas Health Science Center at Dallas, Dallas, Texas 75235

Aaron Ciechanover, Department of Biology, Massachusetts Institute of Technology, Cambridge, Massachusetts 02139. *Present address*: Unit of Biochemistry, Faculty of Medicine, Technion-Israel Institute of Technology, Haifa 31096, Israel

Zanvil A. Cohn, Laboratory of Cellular Physiology and Immunology, The Rockefeller University, New York, New York 10021

Bianca M. Conti-Tronconi, Division of Chemistry, California Institute of Technology, Pasadena, California 91125

Rebecca D. Crawford, Division of Chemistry, California Institute of Technology, Pasadena, California 91125

Pedro Cuatrecasas, Wellcome Research Laboratories, Burroughs Wellcome Company, Research Triangle Park, North Carolina 27709

Michael P. Czech, Department of Biochemistry, University of Massachusetts Medical School, Worcester, Massachusetts 01605

Roger J. Davis, Department of Biochemistry, University of Massachusetts Medical School, Worcester, Massachusetts 01605

S. W. de Laat, Hubrecht Laboratory, 3584 CT Utrecht, The Netherlands

Susan M. J. Dunn, Division of Chemistry, California Institute of Technology, Pasadena, California 91125

Arthur M. Edelman, Howard Hughes Medical Institute Laboratory, Department of Pharmacology, University of Washington, Seattle, Washington 98195

Sabyasachi Ganguly. Department of Molecular Pharmacology, Albert Einstein College of Medicine, Bronx, New York 10461

Shoshana Gill, Department of Chemical Immunology, The Weizmann Institute of Science, Rehovot 76100, Israel

Alfred G. Gilman, Department of Pharmacology, Southwestern Graduate School, University of Texas Health Science Center at Dallas, Dallas, Texas 75235

H. Joseph Goren, Endocrine Research Group, Department of Medical Biochemistry, Faculty of Medicine, University of Calgary, Calgary, Alberta T2N 4N1, Canada

C. Nicholas Hales, Howard Hughes Medical Institute Laboratory, Department of Pharmacology, University of Washington, Seattle, Washington 98195. *Present address*: Department of Clinical Biochemistry, University of Cambridge, Cambridge, England

Ilana Harari, Department of Chemical Immunology, The Weizmann Institute of Science, Rehovot 76100, Israel

Roman Herrera, Department of Molecular Pharmacology, Albert Einstein College of Medicine, Bronx, New York 10461

Morley D. Hollenberg, Endocrine Research Group, Department of Pharmacology and Therapeutics, Faculty of Medicine, University of Calgary, Calgary, Alberta T2N 4N1, Canada

Leonard Jarett, Department of Pathology and Laboratory Medicine, University of Pennsylvania School of Medicine, Philadelphia, Pennsylvania 19104

Jerry Kaplan, Department of Pathology, University of Utah School of Medicine, Salt Lake City, Utah 84132

Edwin G. Krebs, Howard Hughes Medical Institute Laboratory, Department of Pharmacology, University of Washington, Seattle, Washington 98195

Richard Kris, Department of Chemical Immunology, The Weizmann Institute of Science, Rehovot 76100, Israel

Irit Lax, Department of Chemical Immunology, The Weizmann Institute of Science, Rehovot 76100, Israel

Hua Lee, Department of Molecular Pharmacology, Albert Einstein College of Medicine, Bronx, New York 10461

Robert J. Lefkowitz, Howard Hughes Medical Institute and Departments of Medicine and Biochemistry, Duke University Medical Center, Durham, North Carolina 27710

Towia A. Libermann, Department of Chemical Immunology, The Weizmann Institute of Science, Rehovot 76100, Israel

Harvey L. Lodish, Department of Biology, Massachusetts Institute of Technology, Cambridge, Massachusetts 02139

Mary Makowske, Department of Molecular Pharmacology, Albert Einstein College of Medicine, Bronx, New York 10461

R. H. Michell, Department of Biochemistry, University of Birmingham, Birmingham B15 2TT, United Kingdom

David Middlemas, Division of Chemistry, California Institute of Technology, Pasadena, California 91125

W. H. Moolenaar, Hubrecht Laboratory, 3584 CT Utrecht, The Netherlands

Christina Mottola, Department of Biochemistry, University of Massachusetts Medical School, Worcester, Massachusetts 01605

Yoshitomo Oka, Department of Biochemistry, University of Massachusetts Medical School, Worcester, Massachusetts 01605

Yetunde Olowe, Department of Molecular Pharmacology, Albert Einstein College of Medicine, Bronx, New York 10461

Bert W. O'Malley, Department of Cell Biology, Baylor College of Medicine, Houston, Texas 77030

Jeffrey E. Pessin, Department of Biochemistry, University of Massachusetts Medical School, Worcester, Massachusetts 01605

Lilli M. Petruzzelli, Department of Molecular Pharmacology, Albert Einstein College of Medicine, Bronx, New York 10461

Michael A. Raftery, Division of Chemistry, California Institute of Technology, Pasadena, California 91125

Janet D. Robishaw, Department of Pharmacology, Southwestern Graduate School, University of Texas Health Science Center at Dallas, Dallas, Texas 75235

Martin Rodbell, Section on Membrane Regulation, NIADDK, National Institutes of Health, Bethesda, Maryland 20205. *Present address*: National Institute of Environmental Health Sciences, Research Triangle Park, North Carolina 27709

Ora M. Rosen, Department of Molecular Pharmacology, Albert Einstein College of Medicine, Bronx, New York 10461

Joseph Schlessinger, Department of Chemical Immunology, The Weizmann Institute of Science, Rehovot 76100, Israel

Hans Schreiber, Department of Pathology, The University of Chicago, La Rabida–University of Chicago Institute, Chicago, Illinois 60649

Alan L. Schwartz, Division of Pediatric Hematology/Oncology, Children's Hospital Medical Center, Dana-Farber Cancer Institute, and Department of Pediatrics, Harvard Medical School, Boston, Massachusetts 02115

Murray D. Smigel, Department of Pharmacology, Southwestern Graduate School, University of Texas Health Science Center at Dallas, Dallas, Texas 75236

Robert M. Smith, Department of Pathology and Laboratory Medicine, University of Pennsylvania School of Medicine, Philadelphia, Pennsylvania 19104

Jeffrey M. Stadel, Howard Hughes Medical Institute and Departments of Medicine and Biochemistry, Duke University Medical Center, Durham, North Carolina 27710. *Present address*: Smith Kline & French Laboratories, Philadelphia, Pennsylvania 19101

Laurel Stadtmauer, Department of Molecular Pharmacology, Albert Einstein College of Medicine, Bronx, New York 10461

Hans Josef Stauss, Department of Pathology, The University of Chicago, La Rabida–University of Chicago Institute, Chicago, Illinois 60649

Ralph M. Steinman, Laboratory of Cellular Physiology and Immunology, The Rockefeller University, New York, New York 10021

Berta Strulovici, Howard Hughes Medical Institute and Departments of Medicine and Biochemistry, Duke University Medical Center, Durham, North Carolina 27710. *Present address*: Syntex Corporation, Palo Alto, California 94304

Diane Tabarini, Department of Molecular Pharmacology, Albert Einstein College of Medicine, Bronx, New York 10461

Carter Van Waes, Department of Pathology, The University of Chicago, La Rabida–University of Chicago Institute, Chicago, Illinois 60649

Yosef Yarden, Department of Chemical Immunology, The Weizmann Institute of Science, Rehovot 76100, Israel

Preface

It is less than 80 years since John Newport Langley first proposed the role of "receptive substances" as the site of drug action from his observations on the effects of nicotine and curare at the myoneural junction. The many advances in our understanding of receptor biology that have occurred during the intervening period mirror the extraordinary growth of knowledge in the biological sciences and in cell and molecular biology in particular. Receptor biology, in common with many other topics in contemporary biology, is on the threshold of a transition from being a descriptive, phenomenological discipline to one in which underlying mechanisms and regulatory principles can be defined with increasing precision. This change, together with the evolution of powerful analytical techniques and timely convergence of ideas from a number of previously separate fields of inquiry, is generating an increasingly unified theoretical and experimental framework for the study of receptor function.

These themes, and the mood of anticipation that a real understanding of receptor function in health and disease is emerging, are reflected in the papers in this volume, which summarizes the proceedings of the Second Smith Kline & French Research Symposium on New Horizons in Therapeutics held in Philadelphia in 1984.

The growth of knowledge of receptor function can be divided into five periods, each distinguished by discrete conceptual and technical advances. The first originated with the studies of Langley, Ehrlich, and others proposing the existence of different receptor molecules as an explanation for the differing responses of living cells to chemically diverse agents in the extracellular milieu. The second period began in 1926 with publication of the receptor occupancy theory by Gaddum and Clark. This offered a quantitative basis for the analysis of ligand–receptor interactions and the evaluation of differences in cellular responses to specific ligands. This period lasted until approximately 1960 and was characterized by extensive studies on the reactivity of different tissues and organs to drugs and hormones. These studies resulted in the development of many innovative methods in pharmacology that have been of great experimental and clinical importance and have led to the general classification schemes for the major receptor classes in use today. Parallel advances in organic chemistry also afforded researchers with specific ligands for different receptor classes, which have been of inestimable value in the detection and isolation of receptors and in the refinement of receptor taxonomy. These strategies were also responsible for many important therapeutic advances in which the analysis of structure–activity relationships of structurally related ligands created potent receptor agonists or antagonists that continue to provide the core of the modern pharmacopeia.

The next era in receptor research began in the 1960s with improvements in cell culture, electron microscopy, cell fractionation techniques, the introduction of radiolabeled high-affinity ligands, and the development of biochemical and biophysical methods of then unprecedented resolving power. Collectively, these techniques quickly shaped a new conceptual framework in which cell function began to be interpreted in terms of the properties of distinct subcellular compartments. This period also resulted in the recognition that receptors are present not only on the cell surface but also in intracellular organelles. At the same time, the importance of functional integration of subcellular compartments was beginning to be emphasized in the concepts of signal transduction and the role of cAMP as a second messenger in translating the outcome of ligand–receptor interactions into cellular responses. Finally, this period witnessed the first examples of receptor isolation and the ability to study their function in cell-free systems. These achievements, albeit crude in comparison with current studies in receptor purification and reconstitution, were of enormous significance in converting the large body of deductive, though heuristically useful, evidence about receptors obtained from ligand-binding studies studies into a chemical reality.

The next stage in the conceptual evolution of receptor theory, which heralded the start of the current era of research on receptor biology, occurred in the early 1970s with recognition that biological membranes are not static, fixed assemblies but are dynamic structures whose components, including receptor molecules, undergo rapid and reversible topographic rearrangements in response to intracellular and extracellular signals. This concept, embodied in the fluid mosaic model of membrane structure, altered radically our perspective of membrane function and catalyzed an exciting period of fruitful investigation that continues today at a seemingly ever-faster pace. Recognition that membranes and their component receptors are dynamic structures also heightened appreciation of the interrelationships between different membrane systems within the cell and the functional integration of different intracellular compartments as illustrated by transmembrane control of cell-surface receptors by cytoplasmic cytoskeletal elements, the diverse pathways of intracellular organelle traffic in endo- and exocytosis, and the importance of these processes in regulating receptor number, synthesis, turnover, and localization.

In addition to signal transduction arising from the direct interaction of a ligand with its receptor, it is now recognized that the biological actions of many ligands require internalization of ligand–receptor complexes into the cell, followed by a complex sequence of intracellular "processing" reactions in which ligands and their receptors may have very different fates. For some receptors, internalization of ligand–receptor complexes is followed quickly by dissociation of the complexes within the endosome, with receptors being recycled back to the cell surface and the ligands directed either to the lysosome or to the cytosol. In contrast, other receptor–ligand complexes may remain intact and either be exported intact from the cell, providing an efficient method for transcellular ligand transport, or be transferred to lysosomes, where they are degraded with or without recycling of receptors. The mechanisms responsible for these complicated sorting patterns are unknown but are the subject of intense research interest and are discussed at length in this volume.

The already hectic pace of progress in elucidating receptor function promises to accelerate as a result of advances in molecular biology that offer new approaches for the isolation, characterization, and manipulation of receptor molecules. Foremost among these are the ability to clone the genes for receptor molecules using recombinant DNA techniques and to transfer them to defined cell types to engineer functional expression of receptor molecules, new microchemical techniques that permit analysis and synthesis of minute quantities of peptides and nucleic acids, and the use of monoclonal antibodies to map the structure and function of specific domains in both native and genetically modified receptors. It is difficult to overemphasize the potential importance of this powerful trinity. Some may view these techniques as merely supplementing the powerful analytical capabilities of other equally impressive techniques introduced over the past two decades. In our opinion, however, this combination of genetic, immunologic, and microchemical techniques will enhance our understanding of receptor function so dramatically over the next few years that they can be viewed legitimately as marking the start of a new era in receptor research.

The capacity to clone the genes for receptors and associated regulatory subunits and to engineer their overproduction in heterologous cells will enable these molecules to be generated in the quantities needed for detailed chemical and structural analyses. This approach will replace the tedious task of attempting to extract the extremely small amounts of receptor material present in most cells. Analysis of the composition and structure of different receptor classes using these techniques will permit development of a sophisticated molecular taxonomy in which receptor subtypes can be identified reliably without strict reliance on ligand-binding data. Such schemes will also provide insight into the likely evolutionary relationships of different receptors and receptor subtypes. Analysis of the composition and tertiary structure of these molecules will also be invaluable in designing synthetic peptides that mimic specific regions (domains) of the receptor molecule that can be used to generate antibodies that react with these domains, which, in turn, can serve as probes of domain function and facilitate receptor isolation via affinity chromatography. In addition, such peptides, as well as the antibodies formed against them and anti-idiotypic antibodies generated against these antibodies, might be exploited in their own right as receptor agonists and antagonists. The applications of such techniques are limited virtually only by the imagination of the experimenter.

Receptor function can also be explored using transfection techniques in which cloned receptor genes, in native or modified form, are transferred to homologous or heterologous cells. The ability to modify receptor structure and examine the consequences for specific receptor function can be achieved by site-directed mutagenesis of specific nucleotide sequences in receptor genes and by generating chimeric or truncated receptor genes. Transfection and expression of such genes in appropriate target cells will rapidly replace current approaches in which receptors are transferred via fusion of intact cells or the fusion of plasma membrane vesicles, reconstituted viral envelopes, or lipid vesicles containing isolated receptors with intact cells. When these techniques are used in conjunction with monoclonal antibodies to native and genetically modified receptor molecules as probes of receptor identity and function, it becomes possible to examine the effect of subtle changes in receptor structure and composition on receptor biosynthesis, turnover, localization, ligand-binding properties, internalization of ligand–receptor complexes, and receptor recycling behavior.

No single technique or discipline can be preeminent or exist in isolation in facing the many challenging questions in receptor biology that are still to be answered. Successful research in biology and medicine has always required broad intellectual horizons and eclectic skill. Modern biomedical research is no exception and is founded on successful integration of conceptual and technical advances from many different disciplines and an understanding of the pathogenesis of disease at different levels of biological complexity, ranging from molecular biology to events in the intact host.

The rapid pace of discovery in understanding receptor function in health and disease will inevitably lead to improvements in disease treatment in which novel therapeutic agents are developed based on a detailed appreciation of the molecular biology of receptor dysfunction. The strategy for the development of most of the drugs in use today has been to seek agents that act as ligands for target receptors and either stimulate or antagonize biological responses via binding interactions with the receptors. New insights into receptor regulation indicate that receptor function can be modulated in many ways other than affecting ligand binding, and these offer new targets for therapeutic manipulation. These include, but are not limited to, internalization of receptors and receptor–ligand complexes, intracellular uncoupling of ligand–receptor complexes, receptor degradation or recycling, receptor "activation" (e.g., autophosphorylation) or modification (e.g., partial proteolytic degradation), receptor "desensitization" via induced covalent modification(s) and/or uncoupling of receptors and their regulatory subunits, receptor translation and biosynthesis, and perturbation of receptor transport and localization in specific intracellular compartments.

Progress in understanding receptors has advanced from early studies on their distribution in whole animals and isolated organs to the threshold of a detailed dissection of their structure and genetic control. That this has occurred in less than 50 years is a formidable achievement by any scientific standard. Even more exhilarating, however, is the prospect that the mysteries of receptor function and regulation are now close to being answered in a pending era of profound discovery that will result in novel therapies for many important diseases.

George Poste
Stanley T. Crooke

Philadelphia

Contents

Chapter 4

The Insulin Receptor as a Tyrosine-Specific Protein Kinase

*Lilli M. Petruzzelli, Laurel Stadtmauer, Roman Herrera, Mary
Makowske, Sabyasachi Ganguly, Diane Tabarini, Hua Lee, Yetunde
Olowe, and Ora M. Rosen*

Chapter 5

Signal Transduction in Biological Membranes

Martin Rodbell

Chapter 9

Ionic Signal Transduction by Growth Factors

W. H. Moolenaar and S. W. de Laat

Chapter 10

Guanine-Nucleotide-Binding Regulatory Proteins: Membrane-Bound Information Transducers

Alfred G. Gilman, Murray D. Smigel, Gary M. Bokoch, and Janet D. Robishaw

Chapter 11

Role of Cyclic-AMP-Dependent Protein Kinase in the Regulation of Cellular Processes

Edwin G. Krebs, Donald K. Blumenthal, Arthur M. Edelman, and C. Nicholas Hales

Chapter 12

The Homogeneity and Discreteness of Membrane Domains

Zanvil A. Cohn and Ralph M. Steinman

Chapter 13

Internalization and Processing of Peptide Hormone Receptors

Pedro Cuatrecasas

Chapter 14

Sorting and Recycling of Cell Surface Receptors and Endocytosed Ligands: The Asialoglycoprotein and Transferrin Receptors

Aaron Ciechanover, Alan L. Schwartz, and Harvey L. Lodish

Chapter 15

The Nicotinic Acetylcholine Receptor: Its Structure, Multiple Binding Sites, and Cation Transport Properties

Michael A. Raftery, Bianca M. Conti-Tronconi, Susan M. J. Dunn, Rebecca D. Crawford, and David Middlemas

Chapter 16

Adenylate-Cyclase-Coupled β-Adrenergic Receptors: Biochemical Mechanisms of Desensitization

Berta Strulovici, Jeffrey M. Stadel, and Robert J. Lefkowitz

Chapter 17

Control of Receptor Function by Homologous and Heterologous Ligands

Morley D. Hollenberg

Chapter 18

Ligand–Receptor Interactions at the Cell Surface

Morley D. Hollenberg and H. Joseph Goren

Receptor Regulation

Problems and Perspectives

LEONARD JARETT and ROBERT M. SMITH

1. Introduction

It is now well accepted that the interaction of ligands, especially peptide hormones, with their surface receptor is the initial step in a chain of events that leads to subsequent biological responses. Therefore, the regulation of surface receptors is crucial as a control point for these responses. Current investigation of hormone action must focus, in part, on regulation of receptors, which goes beyond the simple binding studies of the past decade or more that opened the field to further exploration.

Figure 1 demonstrates the important role of the ligand–receptor complex in the control of subsequent biological responses. Multiple effects can be elicited by an interaction of a ligand with its receptor. This is especially true of a hormone such as insulin, whose pleiotropic responses are probably caused by more than one mechanism. The multiple effector systems that can be activated by a ligand receptor interaction are represented as E-1, E-2, E-3, etc. The jagged lines are unknown coupling systems linking the receptor to the effector after occupancy of the receptor.

E-1 could be an effector system that generates intracellular messengers or mediators, represented by S-1, S-2, etc., which bring about alterations of various intracellular metabolic pathways. The insulin mediator described by our laboratory (Jarett, *et al.* 1983) would fall into this category, as would the more classic second messenger, cAMP. It is pos-

LEONARD JARETT and ROBERT M. SMITH • Department of Pathology and Laboratory Medicine, University of Pennsylvania School of Medicine, Philadelphia, Pennsylvania 19104.

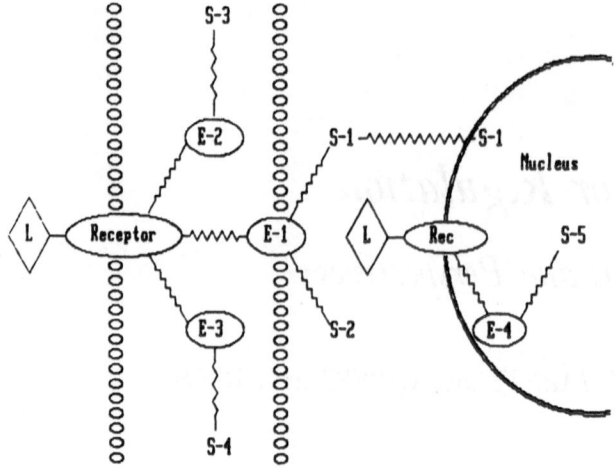

Figure 1. A schematic representation of the multiple effects of ligand–receptor interaction.

sible that some of the cytoplasmic signals, such as S-1, can affect nuclear responses as well. Alternatively, the ligand could interact directly with receptors on the nuclear membrane, as has been suggested for insulin (Goldfine and Smith, 1976; Vigneri *et al.*, 1978), and activate a separate effector system (E-4), which generates a different or similar signal or mediator (S-5) to that generated in the cytoplasm.

Various membrane responses can be elicited by the binding of a ligand to its receptor. These responses could involve the activation of potential effector mechanisms such as phospholipid methylation and receptor phosphorylation (E-2, E-3). The signals (S-3, S-4) generated by these effector systems have not been defined but may regulate membrane processes such as transport of molecules across the membrane, receptor down-regulation, etc. Studies from our laboratory have shown that insulin, within seconds, increased phospholipid methylation in adipocyte membranes (Kelly and Jarett, 1984). Our laboratory has recently found that the addition of insulin mediator to intact adipocytes had no effect on glucose transport but activated pyruvate dehydrogenase (unpublished observations). Therefore, an alternative mechanism other than the insulin mediator must be found to explain the stimulation of glucose transport by insulin. Such findings support the concept expressed by this schematic that hormones such as insulin have multiple pathways for carrying out their pleiotropic responses.

2. Receptor Regulation

In general, receptors can be regulated in three ways. These include alterations in (1) number, (2) affinity, or (3) location. Receptor regulation can result from several processes. The first process relates to receptor-specific ligands such as insulin, epidermal growth factor, or glucagon interacting with their own receptors. Secondly, receptors can be regulated by ligands other than their natural ligands, such as receptor antibodies, lectins, or other hormones. For instance, insulin occupancy of its receptor has been shown to regulate the appearance of IgF II receptors (Czech *et al.*, 1983), and β-adrenergic agents are reported to regulate insulin receptors (Pessin *et al.*, 1983). Finally, chemical agents, such as dithiothreitol, can affect receptors by alteration of the receptor *per se* or of surrounding membrane components (Ozawa *et al.*, 1979; Vauquelin *et al.*, 1979; Suen *et al.*, 1980: Schweitzer *et al.*, 1980; Jacobs and Cuatrecasas, 1980; Walker *et al.*, 1981; Moore *et al.*, 1983).

Receptor number appears to be regulated by the following general mechanisms. The first mechanism involves down-regulation or loss of receptors from the cell surface after occupancy by the ligand. This phenomenon can involve internalization, degradation, and/or altered recycling of the receptor and, in some cases, shedding of the receptor. Down-regulation has been shown to occur on a number of cell types with a variety of hormones (Gavin *et al.*, 1974; Lesniak and Roth, 1976; Marshall and Olefsky, 1980; Galbraith *et al.*, 1980; Balkin and Sonenberg, 1981; Amatruda *et al.*, 1982; Lloyd and Ascoli, 1983). A second mechanism is receptor desensitization, which indicates that the receptor is still present on the surface membrane but not in an accessible or active form for the ligand. This process has been studied extensively for β-adrenergic receptors (David and Lefkowitz, 1981) and may occur with the insulin receptor (Karlsson *et al.*, 1979; Grunfeld *et al.*, 1980). In contrast to down-regulation, up-regulation or increase of receptors after occupancy by their ligand has been shown to occur with prolactin (Shiu and Friesen, 1981) and, more recently, with insulin receptors on chondrosarcoma cells (Stevens *et al.*, 1983). Finally, the number of receptors can be altered by changes in the rate of synthesis of the receptor.

Receptor affinity can be regulated by receptor-specific ligands, other ligands, and various chemical agents. In some cases, occupancy of a receptor by its ligand results in either positive or negative cooperative effects, which change the affinity of the receptor (DeMeyts, 1976). Membrane components have been identified by several groups (Krupp and Livingston, 1978; Maturo and Hollenberg, 1978; Harmon *et al.*, 1983) that interact with the receptor and change its affinity. This putative affinity regulator has yet to be purified and chemically characterized.

Receptor organization and distribution can be regulated by at least three general mechanisms. The first mechanism involves the actual cross linking of receptors by multivalent receptor-specific or other ligands such as immunoglobulins or lectins (Schlessinger et al., 1977; Perelson, 1980, Perelson and DeLisi, 1980). The second mechanism involves induction of biochemical membrane or cellular responses by occupancy of the receptor with monovalent ligands such as insulin. Induced responses could regulate such processes as receptor migration, aggregation, internalization, and recycling. Finally, various agents that alter membrane and/or cellular constituents, such as disulfide bonds, phospholipids, and cytoskeletal elements, can alter receptor organization and distribution (Domnina et al., 1977; Bourguignon and Singer, 1977; Klausner et al., 1980; Jarett and Smith, 1983).

There are several classes of biochemical membrane reactions that could be involved in receptor regulation. One of these processes could certainly be phosphorylation, since several different receptors have been shown to have tyrosine kinases that are capable of autophosphorylation (Carpenter et al., 1978; Zick et al., 1983; Rosen et al., 1983; Pike et al., 1983). Another hormone-sensitive membrane process that could be involved in receptor regulation is phospholipid methylation (Hirata and Axelrod, 1980; Nieto and Catt, 1983; Kelly and Jarett, 1984). Finally, oxidation/reduction changes of membrane components containing disulfide bonds and sulfhydryl groups could result in receptor regulation (Jacobs and Cuatrecasas, 1980; Jarett and Smith, 1983).

The above discussion has focused on general pathways and mechanisms by which receptors can be regulated. However, the uniqueness of ultrastructure and function exhibited by different cell types clearly distinguishes one cell from another. This cellular uniqueness will probably prevent the development of unitarian models of receptor regulation that can be applied from one cell type to another or from one type of receptor to another. There may even be distinct mechanisms for regulating theoretically identical receptors for the same ligand on different cell types.

The following are specific examples of some of these points. The elegant studies of Anderson, Brown, and Goldstein established the important role of coated structures in the uptake of low-density lipoprotein in normal human fibroblasts and their role in disease states (Anderson et al., 1976, 1977; Brown and Goldstein, 1979). Subsequently, coated pits have been implicated in the uptake of various ligands by different cells (Maxfield et al., 1978; Willingham et al., 1979). At the same time it has been clearly documented that certain ligand-receptor complexes are excluded from coated structures (Carney and Bergmann, 1982; Bergmann and Carney, 1982; Montesano *et al.*, 1982; Ackerman *et al.*, 1983; Smith and Jarett, 1983, 1984). In addition, the same ligand–receptor complex may be internalized via coated structures on one cell type and not on another (Gorden *et al.*, 1978; Haigler *et al.*, 1979; McKanna *et al.*, 1979). As a second example, agents such as dithiothreitol (DTT) that affect binding of ligands to their receptors have been shown to have different effects on the same ligand–receptor interaction on different cells: DTT increased the binding of insulin to adipocyte membranes (Schweitzer *et al.*,, 1980), decreased the binding to placental membranes (Jacobs and Cuatrecasas, 1980), and had no effect on binding to liver membranes (Schweitzer *et al.*, 1980; Jacobs and Cuatrecasas, 1980).

The organizational pattern of the insulin receptor on these three cell types differs markedly as well, as determined by electron microscopy (Smith and Jarett, 1984). Finally, insulin has been shown to down-regulate its receptor on a variety of cell types (Galbraith *et al.*, 1980; Gavin *et al.*, 1974; Marshall and Olefsky, 1980), but recently the hormone has been shown to up-regulate its receptor on cultured chondrosarcoma chondrocytes (Stevens *et al.*, 1983). As a result of these variabilities, investigators must be cautious in proposing and adopting models that can become the accepted dogma for all cells and all ligands. These dogma can stifle acceptance of valid new but differing data.

3. Insulin Receptor Studies

This section describes studies on the insulin receptor that have illustrated cell-to-cell variability in the initial organizational pattern of the occupied insulin receptor and in the subsequent behavior of the occupied receptor. These studies have utilized both biochemical and morphological approaches, using monomeric ferritin–insulin as an ultrastructural marker for the occupied insulin receptor. This electron-dense insulin conjugate is indistinguishable from native insulin as determined by biological and immunologic activity and by its ability to compete with [^{125}I]-insulin in binding assays (Smith and Jarett, 1982).

Studies were designed to determine the original native distribution of the occupied insulin receptor on a variety of cells, and the data from those studies are summarized in Table I. The initial distribution was determined by incubating glutaraldehyde-prefixed cells with monomeric ferritin–insulin until steady-state binding has been attained. The fixation process was gentle enough not to alter insulin binding but did prevent postoccupancy movement of the receptors (Jarett and Smith, 1977). The

Table I. Initial Distribution of Occupied Insulin Receptors on Various Tissues

Cell types	Distribution	Organization
Rat adipocytes	Random, entire cell	Groups $\gg$ singles
Placenta (syncytial trophoblast)	Nonrandom; microvilli only	Groups $>$ singles
3T3-L1 adipocytes	Nonrandom; primarily microvilli	Singles $\gg$ groups
IM9 lymphocytes	Nonrandom; primarily microvilli	Singles $>$ groups
H4(IIEC3) hepatoma cells	Nonrandom; primarily microvilli	Singles $>$ groups

rat adipocyte has been studied extensively. The occupied insulin receptors occurred as either single receptors or in groups of up to six receptor molecules, randomly distributed over the entire cell surface. Over two-thirds of the receptors were in groups, and fewer than one-third were found as single receptors (Jarett and Smith, 1977). Insulin receptors on the syncytial trophoblast of human placenta were found to be nonrandomly distributed, restricted to the distal portion of the microvilli. These receptors occurred singly or as small groups (Nelson et al., 1978). Recent studies with 3T3-L1 adipocytes revealed the insulin receptors to be found initially in a nonrandom pattern predominantly on the microvilli, occurring as single or paired receptors. Patterns similar to the 3T3-L1 adipocytes were found with IM-9 lymphocytes and H4(IIEC3) hepatoma cells. However, on these latter two cells, the receptors were more commonly seen as groups of two or three (Smith and Jarett, 1984).

Table II summarizes the distribution of occupied insulin receptors on the same tissues after steady-state binding was attained with nonfixed cells. No difference was found between the initial and steady-state distributions of the occupied insulin receptors on the rat adipocyte. Data from a number of studies have proven that the insulin receptor on the adipocyte occurs naturally in groups prior to and independent of occupancy of the receptor by insulin (Jarett and Smith, 1977) and that once occupied, these receptors do not migrate and cannot be further aggregated (Lyen et al., 1983). Similarly, the occupied insulin receptors on the placental membranes remained restricted to the distal portion of the microvilli, possibly relating to studies showing marked differences in the glycoprotein composition of the microvilli from the proximal to the distal portion (Nelson et al., 1978). In contrast to these two tissues, the insulin receptors on 3T3-L1 adipocytes, IM-9 lymphocytes, and the H4 hepatoma cells all showed substantial migration and aggregation of receptors after occupancy, primarily onto the intervillous cell surface. Whereas single receptors were more frequently seen in the initial state, large clusters of

Table II. Steady-State Distribution of Occupied Insulin Receptors on Various Tissues

Cell types	Distribution	Organization
Rat adipocytes	No migration/aggregation; (HR) on entire cell surface	Groups ⩾ singles
Placenta (syncytial trophoblast)	No migration/aggregation; (HR) on microvilli only	Groups > singles
3T3-L1 adipocytes	Substantial migration/aggregation; (HR) primarily on intervillous cell surface	Clusters > groups > singles
IM9 lymphocytes	Substantial migration/aggregation and shedding; cell-associated (HR) primarily on intervillous cell surface	Clusters > groups > singles
H4(IIEC3) hepatoma cells	Substantial migration/aggregation; (HR) primarily on intervillous cell surface	Clusters > groups > singles

receptors were found at steady state in greater quantity than even the small groups of receptors. Single receptor sites were still found on these cells, but they were predominantly on the microvilli. The IM-9 lymphocytes differed in one major respect from the other two cells on which aggregating clusters of receptors were found. The clusters of receptors were frequently seen sloughed or shed from the surface of the cell. This observation was consistent with biochemical data concerning insulin binding to this cell and the shedding of receptors from its surface (Berhanu and Olefsky, 1982).

These ultrastructural studies clearly show cell-to-cell variability in the initial organization and distribution of the insulin receptor and in the subsequent behavior of the occupied receptor.

Our laboratory has documented a number of other differences regarding the organization and regulation of the insulin receptor in various tissues. Space does not permit detailed discussion of these data, but they have been presented in a recent review (Smith and Jarett, 1984). In summary, these other differences included the following. (1) On the rat adipocyte, insulin and presumably its receptor were internalized in non-coated pinocytotic invaginations and not in coated pits. In fact, insulin receptors were excluded from the coated pits (Smith and Jarett, 1983). On other cells, both noncoated and coated invaginations seem to play a role in insulin internalization. (2) The groups of insulin receptors on adipocytes are immobile and seem to be held together by membrane disulfide bonds that are not the intrareceptor disulfide bonds (Jarett and Smith, 1983). Insulin receptors on other cell types, i.e., 3T3-L1 adipocytes, IM-

9 lymphocytes, and H4(IIEC3) hepatoma cells, appear to be mobile and not restricted as in the rat adipocyte. (3) Insulin receptor migration has been shown not to occur by ligand cross linking of the receptors (Lyen et al., 1983); Jarett and Smith, 1977). Instead, it appears that the interaction of insulin with its receptor on specific cell types initiates a biochemical membrane or cellular response that causes migration and aggregation to occur.

4. Conclusion

Future biochemical and morphological studies of different receptors on different cells should provide further insight into the complex mechanisms by which receptors are regulated and the role this regulation plays in the biological responses initiated by the ligand–receptor interaction. Subsequent chapters in this book provide current information on the problems and mechanisms of receptor regulation; this will provide the basis for future studies.

References

Ackerman, G. A., Yang, J., and Wolken, K.W., 1983, Differential surface labeling and internalization of glucagon by peripheral leukocytes, J. Histochem. Cytochem. 31:433–440.

Amatruda, J. M., Newmeyer, H. W., and Chang, C. L., 1982, Insulin-induced alterations in insulin binding and insulin action in primary cultures of rat hepatocytes, Diabetes 31:145–148.

Anderson, R. G. W., Goldstein, J. L., and Brown, M. S., 1976, Localization of low density lipoprotein receptors on plasma membrane of normal human fibroblasts and their absence in cells from a familial hypercholesterolemia homozygote, Proc. Natl. Acad. Sci., U.S.A. 73:2434–2438.

Anderson, R. G. W., Goldstein, J. L., and Brown, M. S., 1977, A mutation that impairs the ability of lipoprotein receptors to localise in coated pits on the cell surface of human fibroblasts, Nature 270:695–699.

Balkin, M. S., and Sonenberg, M., 1981, Hormone-induced homologous and heterologous desensitization in the rat adipocyte, Endocrinology 109:1176–1183.

Bergmann, J. S., and Carney, D. H., 1982, Receptor-bound thombin is not internalized through coated pits in mouse embryo cells, J. Cell. Biochem. 20:247–258.

Berhanu, P., and Olefsky, J. M., 1982, Photoaffinity labeling of insulin receptors in viable cultured human lymphocytes, demonstration of receptor shedding and degradation, Diabetes 31:410–417.

Bourguignon, L. Y. W., and Singer, S. J., 1977, Transmembrane interactions and the mechanism of capping of surface receptors by their specific ligands, Proc. Natl. Acad. Sci. U.S.A. 74:5031–5035.

Brown, M. S., and Goldstein, J. L., 1979, Receptor-mediated endocytosis: insights from the lipoprotein receptor system, *Proc. Natl. Acad. Sci. U.S.A.* **76:**3330–3337.

Carney, D. H., and Bergmann, J. S., 1982, ^{125}I-Thrombin binds to clustered receptors on noncoated regions of mouse embryo cell surfaces, *J. Cell Biol.* **95:**697–703.

Carpenter, G., King, L., Jr., and Cohen, S., 1978, Epidermal growth factor stimulates phosphorylation in membrane preparations *in vitro, Nature* **276:**409–410.

Czech, M. P., Oppenheimer, C. L., and Massague, J., 1983, Interrelationships among receptor structures for insulin and peptide growth factors, *Fed. Proc.* **42:**2598–2601.

Davies, A. O., and Lefkowitz, R. J., 1981, Regulation of adrenergic receptors, in: *Receptors and Recognition,* series B, Vol. 13 (R. J. Lefkowitz, ed.), Chapman and Hall, New York, pp. 83–121.

DeMeyts, P., 1976, Cooperative properties of hormone receptors in cell membranes, *J. Supramol. Struct.* **4:**241–258.

Domnina, L. V., Pletyushkina, O. Y., Vasiliev, J. M., and Gelfand, I. M., 1977, Effects of antitubulins on the redistribution of crosslinked receptors on the surface of fibroblasts and epithelial cells, *Proc. Natl. Acad. Sci., U.S.A.* **74:**2865–2868.

Galbraith, R. A., Tucker, S., Wise, C., and Buse, M. G., 1980, Insulin binding to erythroblastic leukemic cells is decreased by insulin and increased by amino acids, *Biochem. Biophys, Res. Commun.* **96:**1434–1440.

Gavin, J. R. III, Roth, J., Neville, D. M., Jr., DeMeyts, P., and Buell, D. N., 1974, Insulin-dependent regulation of insulin receptor concentrations: A direct demonstration in cell culture, *Proc. Natl. Acad. Sci. U.S.A.* **71:**84–88.

Goldfine, I. D., and Smith, G. J., 1976, Binding of insulin to isolated nuclei, *Proc. Natl. Acad. Sci. U.S.A.* **73:**1427–1431.

Gorden, P., Carpentier, J.-L, Cohen, S., and Orci, L., 1978, Epidermal growth factor: Morphological demonstration of binding, internalization, and lysosomal association in human fibroblasts, *Proc. Natl. Acad. Sci. U.S.A.* **75:** 5025–5029.

Grunfeld, C., Van Obberghen, E., Karlsson, F. A., and Kahn, C. R., 1980, Antibody-induced desensitization of the insulin receptor, *J. Clin. Invest.* **66:**1124–1134.

Haigler, H. T., McKanna, J. A., and Cohen, S., 1979, Direct visualization of the binding and internalization of a ferritin conjugate of epidermal growth factor in human carcinoma cells A-431, *J. Cell Biol.* **81:**382–395.

Harmon, J. T., Hedo, J. A., and Kahn, C. R., 1983, Characterization of a membrane regulator of insulin receptor affinity, *J. Biol. Chem.* **258:**6875–6881.

Hirata, F., and Axelrod, J., 1980, Phospholipid methylation and biological signal transmission, *Science* **209:**1082–1090.

Jacobs, S., and Cuatrecasas, P., 1980, Disulfide reduction converts the insulin receptor of human placenta to a low affinity form, *J. Clin. Invest.* **66:**1424–1427.

Jarett, L., and Smith, R. M., 1977, The natural occurrence of insulin receptors in groups on adipocyte plasma membranes as demonstrated with monomeric ferritin-insulin, *J. Supramol. Struct.* **6:**45–59.

Jarett, L., and Smith, R. M., 1983, Partial disruption of naturally ocurring groups of insulin receptors on adipocyte plasma membranes by dithiothreitol and N-ethylmaleimide: The role of disulfide bonds, *Proc. Natl's Acad. Sci., U.S.A.* **80:**1023–1027.

Jarett, L., Kiechle, F. L., Parker, J. C., and Macaulay, S. L., 1983, The chemical mediators of insulin action: Possible targets for postreceptor defects, *A. J. Med.* **74:**31–37.

Karlsson, F. A., Van Obberghen, E., Grunfeld, C., and Kahn, C. R., 1979, Desensitization of the insulin receptor at an early postreceptor step by prolonged exposure to antireceptor antibody, *Proc. Natl. Acad. Sci. U.S.A.* **76:**809–813.

Kelly, K. L., and Jarett, L., 1984, Insulin stimulation of phospholipid methylation in isolated rat adipocyte plasma membranes, *Proc. Natl. Acad. Sci. U.S.A.* **81:**1089–1092.

Klausner, R. D., Bhalla, D. K., Dragsten, P., Hoover, R. L., and Karnovsky, M. J., 1980, Model for capping derived from inhibition of surface receptor capping by free fatty acids, *Proc. Natl. Acad. Sci. U.S.A.* **77**:437–441.

Krupp, M. N., and Livingston, J. N., 1978, Insulin binding to solubilized material from fat cell membranes: Evidence for two binding species, *Proc. Natl. Acad. Sci. U.S.A.* **75**:2593–2597.

Lesniak, M. A., and Roth, J., 1976, Regulation of receptor concentration by homologous hormone: Effect of human growth hormone on its receptor in IM-9 lymphocytes, *J. Biol. Chem.* **251**:3720–3729.

Lloyd, C. E., and Ascoli, M., 1983, On the mechanisms involved in the regulation of the cell-surface receptors for human choriogonadotropin and mouse epidermal growth factor in cultured Leydig tumor cells, *J. Cell Biol.* **96**:521–526.

Lyen, K. R., Smith, R. M., and Jarett, L., 1983, Differences in the ability of anti-insulin antibody to aggregate monomeric ferritin-insulin occupied receptor sites on liver and adipocyte plasma membranes, *Diabetes* **32**:648–653.

Marshall, S., and Olefsky, J. M., 1980, Effects of insulin incubation on insulin binding, glucose transport, and insulin degradation by isolated rat adipocytes, *J. Clin. Invest.* **66**:763–772.

Maturo, J. M. III, and Hollenberg, M. D., 1978, Insulin receptor: Interaction with nonreceptor glycoprotein from liver cell membranes, *Proc. Natl. Acad. Sci. U.S.A.* **75**:3070–3074.

Maxfield, F. R., Schlessinger, J., Shechter, Y., Pastan, I., and Willingham, M. C., 1978, Collection of insulin, EGF, and alpha-2-macroglobulin in the same patches on the surface of cultured fibroblasts and common internalization, *Cell* **14**:805–810.

McKanna, J. A., Haigler, H. T., and Cohen, S., 1979, Hormone receptor topology and dynamics: Morphological analysis using ferritin-labeled epidermal growth factor, *Proc. Natl. Acad. Sci. U.S.A.* **76**:5689–5693.

Montesano, R., Roth, J., Robert, A., and Orci, L., 1982, Non-coated membrane invaginations are involved in binding and internalization of cholera and tetanus toxins, *Nature* **296**:651–653.

Moore, W. V., Wohnlich, L. P., and Fix, J. A., 1983, Role of disulfide bonds in human growth hormone binding and dissociation in isolated rat hepatocytes and liver plasma membranes, *Endocrinology* **112**:2152–2158.

Nelson, D. M., Smith, R. M., and Jarett, L., 1978, Nonuniform distribution and grouping of insulin receptors on the surface of human placental syncytial trophoblast, *Diabetes* **27**:530–538.

Nieto, A., and Catt, K. J., 1983, Hormonal activation of phospholipid methyltransferase in the Leydig cell, *Endocrinology* **113**:758–762.

Ozawa, Y., Chopra, I. J., Solomon, D. H., and Smith, F., 1979, The role of sulfhydryl groups in thyrotropin binding and adenylate cyclase activities of thyroid plasma membranes, *Endocrinology* **105**:1221–1225.

Perelson, A. S., 1980, Receptor clustering on a cell surface. II. Theory of receptor cross-linking by ligands bearing two chemically distinct functional groups, *Math Biosci.* **49**:87–110.

Perelson, A. S., and DeLisi, C., 1980, Receptor clustering on a cell surface. I. Theory of receptor cross-linking by ligands bearing two chemically identical functional groups, *Math Biosci.* **48**:71–110.

Pessin, J. E., Gitomer, W., Oka, Y., Oppenheimer, C. L., and Czech, M. P., 1983, β-Adrenergic regulation of insulin and epidermal growth factor receptors in rat adipocytes, *J. Biol. Chem.* **258**:7386–7394.

Pike, L. J., Bowen-Pope, D. F., Ross, R., and Krebs, E. G., 1983, Characterization of platelet-derived growth factor-stimulated phosphorylation in cell membranes, *J. Biol. Chem.* **258:**9383–9390.

Rosen, O. M., Herrera, R., Olowe, Y., Petruzzelli, L. M., and Cobb, M. H., 1983, Phosphorylation activates the insulin receptor tyrosine protein kinase, *Proc. Natl. Acad. Sci. U.S.A.* **80:**3237–3240.

Schlessinger, J., Elson, E. L., Webb, W. W., Yahara, I., Rutishauser, U., and Edelman, G. M., 1977, Receptor diffusion on cell surfaces modulated by locally bound concanavalin A, *Proc. Natl. Acad. Sci. U.S.A.* **74:**1110–1114.

Schweitzer, J. B., Smith, R. M., and Jarett, L., 1980, Differences in organizational structure of insulin receptor on rat adipocyte and liver plasma membranes: Role of disulfide bonds, *Proc. Natl. Acad. Sci. U.S.A.* **77:**4692–4696.

Shiu, R. P. C., and Friesen H. G., 1981, Regulation of prolactin receptors in target cells, in: *Receptors and Recognition*, series B. Vol. 13 (R. J. Lefkowitz, ed.), Chapman and Hall, New York, pp. 67–81.

Smith, R. M., and Jarett, L., 1982, A simplified method of producing biologically active monomeric ferritin-insulin for use as a high resolution ultrastructural marker for occupied insulin receptors, *J. Histochem. Cytochem.* **30:**651–656.

Smith, R. M., and Jarett, L., 1983, Quantitative ultrastructural analysis of receptor-mediated insulin uptake into adipocytes, *J. Cell. Physiol.* **115:**199–207.

Smith, R. M., and Jarett, L., 1985, Tissue specific variations in insulin receptor dynamics: A high resolution ultra-structural and biochemical approach, in: *Insulin, Its Receptor and Diabetes* (M. D. Hollenberg, ed.), Marcel Dekker, New York (in press).

Stevens, R. L., Austen, K. F., and Nissley, S. P., 1983, Insulin-induced increase in insulin binding to cultured chondrosarcoma chondrocytes, *J. Biol. Chem.* **258:**2940–2944.

Suen, E. T., Stefanini, E., and Clement-Cormier, Y. C., 1980, Evidence for essential thiol groups and disulfide bonds in agonist and antagonist binding to the dopamine receptor, *Biochem. Biophys, Res. Commun.* **96:**953–960.

Vauquelin, G., Bottari, S., Kanarek, L., and Strosberg, A. D., 1979, Evidence for essential disulfide bonds in β_1-adrenergic receptors of turkey erythrocyte membranes, *J. Biol. Chem.* **254:**4462–4469.

Vigneri, R., Goldfine, I. D., Wong, K. Y., Smith, G. J., and Pezzino, V., 1978, The nuclear envelope: The major site of insulin binding in rat liver nuclei, *J. Biol. Chem.* **253:**2098–2103.

Walker, J. W., Lukas, R. J., and McNamee, M. G., 1981, Effects of thio-group modifications on the ion permeability control and ligand binding properties of *torpedo californica* acetylcholine receptor, *Biochemistry* **20:**2191–2199.

Willingham, M. C., Maxfield, F. R., and Pastan, I. H., 1979, Alpha-2-macroglobulin binding to the plasma membrane of cultured fibroblasts, *J. Cell Biol.* **82:**614–625.

Zick, Y., Wittaker, J., and Roth, J., 1983, Insulin stimulated phosphorylation of its own receptor, *J. Biol. Chem.* **258:**3431–3434.

Patterns in Receptor Behavior and Function

JERRY KAPLAN

1. Introduction

Scientific principles are derived by observing the behavior of individual members of a set and, from a collection of such observations, deducing the rules that govern that set's behavior. For example, the properties of molecules are defined by physicochemical principles that limit their behavior. By defining the "rules" that dictate a molecule's behavior, predictions can be made regarding the behavior of other members in a similar set; i.e., patterns lead to expectations (Judson, 1980). The ability to make testable predictions is the only gauge of a model's validity.

Just as the rules that govern the behavior of molecules can be defined, so can the rules that govern the behavior of membrane components. The behavior of membrane components is defined not only by their physicochemical properties but by limitations imposed by evolutionary and functional considerations. Previously I defined a classification of membrane receptors that grouped membrane receptors according to function (Kaplan, 1981; Anderson and Kaplan, 1983). Members of a given class demonstrate similar behavioral features, allowing for the deduction of rules that govern the behavior of these molecules. In this work I wish to elaborate on some of the rules that appear to govern receptor function as well as to define a new effector mechanism for hormone action.

Membrane receptors can be divided into two groups based on whether the major function of the receptor is to transmit information that modifies cell behavior or metabolism (class I receptors, Table I) or to

JERRY KAPLAN • Department of Pathology, University of Utah School of Medicine, Salt Lake City, Utah 84132.

Table I. Properties of Class I Receptors

1. Formation of a surface receptor–ligand complex leads to changes in cell behavior.
2. Receptors are randomly distributed on cell surface.
3. Formation of a ligand–receptor complex is independent of divalent cations.
4. Exposure of cells to a ligand results in a decrease in the number of receptors for that ligand and a decrease in response to that ligand.
5. For most class I receptors internalization results in receptor degradation. There are some class I receptors capable of recycling.

clear ligand from the extracellular space (class II receptors, Table II). The latter receptors function to remove injurious ligands from bodily fluids or to provide some metabolite or nutrient required for cell growth.

2. Class I Receptors

Binding of ligand to class I receptors results in changes in cell behavior or metabolism. The spectrum of changes is broad, ranging from secretion of a specific hormone (chorionic gonadotrophin, thyroid-releasing hormone) to directed cell movement or chemotaxis (fifth component of complement) to mitogenesis (epidermal growth factor). The available data indicate that almost all changes in cell behavior are a consequence of the formation of a surface receptor–ligand complex. For many receptors, the mechanism by which surface receptor–ligand complexes generate signals can be explained in precise molecular terms. For example, formation of the appropriate hormone–receptor complex can activate adenylate cyclase catalytic units by promoting the binding of GTP

Table II. Properties of Class II Receptors

1. Formation of a surface receptor–ligand complex results in the clearance of that ligand from the extracellular space. Ligands that are taken up either provide a required nutrient (e.g., vitamin B_{12}) or are potentially injurious (e.g., proteases).
2. Formation of a receptor–ligand complex is dependent on pH and/or Ca^{2+}.
3. Receptors may be distributed randomly on the cell surface and/or prelocalized in coated pits.
4. All class II receptors recycle. Internalization of the ligand–receptor complex is accompanied by the intracellular dissociation of the complex and the reappearance of receptors on the cell surface.
5. The activity or concentration of almost all class II receptors is unaffected by ligand binding and internalization.

to coupling factor: all three components are present in the plasma membrane, and cAMP can be produced in a cell-free system. Thus, there is no need to postulate ligand internalization as a prerequisite for the action of hormones that function through adenylate cyclase (glucagon, vasoactive intestinal polypeptide, β-adrenergic agents).

Along a similar vein, agents that function by modulating intracellular Ca^{2+} do so as a consequence of formation of a surface receptor–ligand complex. Thus, binding of antigen to receptor-bound IgE on mast cells effects an increase in intracellular Ca^{2+} without requiring ligand internalization. It is much harder to prove that agents such as epidermal growth factor, insulin, or somatomedin exert their mitogenic effect solely by the formation of a surface receptor–ligand complex, since hours may elapse between formation of a surface receptor–ligand complex and DNA synthesis or mitosis. In the intervening time period, extensive changes occur in cellular metabolism, making it difficult to define the causal factor. Indeed, evidence exists suggesting that mitogenesis results from internalized ligand (Schreiber *et al.*, 1981). All of the agents mentioned above also cause a series of rapid changes in cell metabolism. Among such alterations are increases in glucose uptake (Karnieli *et al.*, 1981), K^+ transport (Boonstra *et al.*, 1983), Ca^{2+} transport (Villereal, 1981), increased intracellular pH (Rothenberg *et al.*, 1983), and increased pinocytic activity (Chinkers *et al.*, 1979; Connolly *et al.*, 1981). Data indicate that these events directly result from formation of a surface receptor–ligand complex (Wiley and Cunningham, 1982; Wiley and Kaplan, 1984).

If a surface receptor–ligand complex is the molecular species responsible for initiating changes in cell behavior, then dissolution of that complex would terminate the signal. There are two possible mechanisms by which a surface receptor–ligand complex can be terminated: (1) $(R \cdot L)_{surface} \rightarrow R \cdot L_{internal}$ by internalizing the complex or (2) physical dissociation of the complex $(R \cdot L)_s \rightarrow R_s + L$. In the first case, internalized ligand–receptor complexes may be physically separated from an effector system and therefore would be unable to activate it (Mukhergee *et al.*, 1975). Hormones that activate adenylate cyclase may provide an example of such a case. Internalization of a specific ligand–receptor complex may occur independently of adenylate cyclase catalytic unit or of the coupling protein (Segaloff and Ascoli, 1981; Strulovici *et al.*, 1983). The removal of the hormone–receptor complex would not necessarily deplete the membrane of catalytic subunits or prevent increases in cAMP in response to a second agent.

An equally efficacious route to terminate a signal transmission evoked by a ligand-receptor complex would be dissociation of the ligand from the receptor. An example of a situation in which ligand dissociation constitutes the major route of terminating a receptor-initiated event is the acetylcholine-induced alteration of ion transport. Acetylcholine receptors are not internalized, and acetylcholine has a rapid rate of dissociation from the receptor (Gardner and Fambrough, 1979). The neuromuscular junction contains a high level of acetylcholinesterase, which hydrolyzes acetylcholine, further lowering the extracellular concentration (Landau, 1978).

The factor that determines whether internalization or dissociation is the major route for signal termination is the ratio between the dissociation rate (K_{-1}) and the endocytic rate (K_e). If the rate of dissociation is greater than the rate of internalization, then the former would constitute the major mechanism of signal termination. For example, epidermal growth factor receptor complexes on fibroblasts exhibit a rate constant for internalization of 0.14 min^{-1} and an apparent rate constant for dissociation of 0.74 min^{-1} (Knauer *et al.*, 1984). Consequently, at a given time, less than half of the cell-bound ligand is internalized. If we consider a $T_{1/2}$ for internalization of 5 min as an average value, then dissociation constants of 0.14 or greater must exist in order for dissociation to be a significant mechanism in terminating signals evoked by a surface receptor-ligand complex.

One important consequence of internalization is that the concentration of surface receptors is lowered. Reduction in surface receptors results in a loss of sensitivity of cells to subsequent ligand challenge. Loss of receptors and subsequent loss of ligand sensitivity is referred to as desensitization or down-regulation. Within the past 2 years it has become clear that internalized class I receptor complexes may have different fates. In the cases of epidermal growth factor (Carpenter and Cohen, 1976), chorionic gonadotrophin (Ascoli and Puett, 1978), interferon (Branca and Baglioni, 1982), or platelet-derived growth factor (Huang *et al.*, 1982; Bowen-Pope and Ross, 1982), internalization of a ligand-receptor complex results in the degradation of that complex. The subsequent reappearance of receptor activity occurs as a result of biosynthesis. However, for a number of receptors, it appears that internalization does not immediately signal the destruction of the receptor. Studies demonstrate that receptors for insulin (Marshall *et al.*,, 1981), bombesin (Pandol *et al.*, 1982), enkephalin (Blanchard *et al.*, 1983), f-Met-Leu-Phe (Zigmond *et al.*, 1982), and β-adrenergic agents (Strulovici *et al.*, 1983) may recycle back to the cell surface and be capable of further signal transmission.

However, internalization of the aforementioned receptors does not appear to be totally benign. Binding of insulin to its receptor increases the rate of degradation of the receptor such that after extended incubation in the presence of the hormone, receptor recovery is dependent on *de novo* synthesis (Kasuga *et al.*, 1981). Studies on the f-Met-Leu-Phe receptor of polymorphonuclear leukocytes reveal a progressive irreversible loss of receptors as a function of incubation time when cells are incubated with high concentrations of ligand (Sullivan and Zigmond, 1980). The molecular properties that determine whether an internalized class I receptor is degraded or recycled are not known.

3. Class II Receptors

The major function of these receptors is to mediate ligand accumulation, and all class II receptors recycle. Internalization of a surface receptor–ligand complex is followed by the dissociation of the intracellular complex. The internalized ligand is either stored, degraded, or in some instances passed through the cell. The receptors that mediate internalization are spared from degradation and are returned to the cell surface, capable of further rounds of ligand uptake. As a consequence of receptor recycling, cells accumulate ligand without requiring the *de novo* synthesis of receptors for each round of ligand internalization. For class I receptors that recycle, it appears that the rate constant for recycling of the internalized receptor is relatively long with respect to internalization. In the case of the insulin receptor on 3T3 adipocytes, the $T_{1/2}$ for internalization is 5 min, and the $T_{1/2}$ for receptor recovery is 20 min (Knutson *et al.*, 1983). This recovery rate is substantially slower than that for class II receptors [i.e., $T_{1/2}$ 4.2 min asialoglycoprotein receptor (Schwartz *et al.*, 1982) and 7.2 min transferrin receptor on hepatoma cells (Ciechanover *et al.*, 1983)].

In every case studied so far, internalization of receptor-ligand complexes occurs via coated pits. These are areas of the plasma membrane in which the cytoplasmic surface contains high concentrations of the protein clathrin, giving these membranes a coated or bristled aspect. The role of clathrin in promoting internalization or providing an environment capable of "capturing" surface receptor–ligand complexes has been recently reviewed (Anderson and Kaplan, 1983).

Both class I and class II receptor–ligand complexes internalize via coated pits. A major difference between the two classes is that whereas class I receptors are uniformly dispersed on cell surfaces, some class II receptors are preclustered in coated pits. In the absence of ligand, receptors for low-density lipoproteins on human fibroblasts (Anderson *et al.*, 1977), ferritin on reticulocytes (Fawcett, 1965), transferrin on human adenocarcinoma cells (Hopkins and Trowbridge, 1983), α_2-macroglobulin–protease complexes and mannose-terminal glycoproteins in macrophages (Kaplan *et al.*, 1984), and α_2-macroglobulin–protease complexes on granulosa cells (Hopkins, 1982) appear to be localized in coated pits. Whether asialoglycoprotein receptors in liver parenchymal cells are localized in coated pits is unresolved (Geuze *et al.*, 1982; Wall and Hubbard, 1981). There are, however, class II receptors that are not prelocalized in coated pits, for example, the Fc receptor for IgG on the intestinal mucosa of suckling rats (Abrahamson and Rodewald, 1981).

Finally, on some cell types, receptors may exhibit a heterogeneous distribution and be distributed both randomly and in coated pits. An example of this is the low-density-lipoprotein receptor on tumor cells (Anderson *et al.*, 1981). Kinetic analysis of low-density-lipoprotein uptake suggests that receptors that are not localized in coated pits are less efficient in internalizing ligand than those in the coated pit. The elements of receptor structure that specify whether or not a receptor is located in coated pits are not yet known. The fact that mutations exist that affect low density lipoprotein receptor distribution (Anderson *et al.*, 1977) suggests that site-directed mutagenesis of receptor genes may provide an approach to answering this question.

Another feature that distinguishes class II from class I receptors is that formation of class II receptor-ligand complexes is pH and/or Ca^{2+} dependent. With only a few exceptions, formation of class II receptor-ligand complexes requires Ca^{2+}. Chelation of Ca^{2+} results in the dissociation of the receptor–ligand complex. The rate of dissociation of the ligand–receptor complex is markedly increased by reducing the pH from neutrality (7.0) to an acidic pH of 5.5–6.0. The ability to regulate ligand–receptor affinity is critical for class II receptor function. If receptor and ligand are to have different fates, then some mechanism must exist to promote dissociation and separate the two. One mechanism that may accomplish this function is hydrogen ion concentration. Receptor–ligand complexes are internalized into an intracellular compartment of low pH. Studies have demonstrated that the pH surrounding internalized α_2-macroglobulin (Tycho and Mayfield, 1981) and transferrin (vanRenswoude *et al.*, 1982) is on the order of 5.5–6.0. At this concentration, most class II receptors exhibit marked decreases in ligand affinity. Agents that increase the pH of acidic compartments prevent ligand–receptor dissociation and also prevent receptor recycling (Tietze *et al.*, 1980; Kaplan and Keogh, 1981; Ciechanover *et al.*, 1983). *In vitro* studies as well as genetic approaches have demonstrated the existence of an ATP-dependent hydrogen pump in coated vesicles and intermediate vesicles (Merion *et al.*, 1983; Robbins *et al.*, 1983; Xie *et al.*, 1983). However, the metabolic features and biochemical nature of the pump have not yet been defined.

Although increased hydrogen ion concentration can reduce ligand–

receptor affinity, other factors must play a role in mediating receptor–ligand dissociation. There are two considerations that lead to this conclusion. (1) A number of class I receptor–ligand complexes dissociate under low-pH conditions, and yet the unoccupied receptor does not recycle (i.e., platelet-derived growth factor, Bowen-Pope and Ross, 1982). (2) Even if the affinity of the ligand–receptor complex is reduced by a factor of ten, the increase in ligand and/or receptor concentration in the endosome should insure that receptors remain occupied.

One feature that can regulate ligand–receptor affinity is a reduced calcium concentration in the endosome. In at least one case, the effect of pH and Ca^{2+} chelation on dissociation rate is synergistic (Harford *et al.*, 1984). That is, the rate constant for ligand–receptor dissociation is markedly increased by simultaneously reducing pH and Ca^{2+} concentration. Although data demonstrating that the endocytic apparatus also transports calcium are yet forthcoming, two lines of evidence suggest that possibility. Cells that are taking up fluid by pinocytosis would also take up Ca^{2+}, and since the fluid-phase markers accumulate in lysosomes, one would suspect that lysosomes would accumulate Ca^{2+} to high concentrations. Examination of thin sections of cells by X-ray probe analysis has not revealed any accumulation of Ca^{2+} by lysosomes or endosomes, suggesting that some transport system must exist.

Further evidence for a Ca^{2+} transport system in the endocytic apparatus is suggested by studies on the uptake of vitellogenin by oocytes. Vitellogenin is a large (460,000 molecular weight) lipoglycophosphoprotein found in the plasma of female oviparous vertebrates. Via specific receptors, amphibian oocytes can internalize large quantities of vitellogenin, up to 19 μg/cell per day. Internalized vitellogenin is stored in yolk platelets and is the precursor of the major yolk proteins of the oocyte. Vitellogenin is a highly phosphorylated protein containing by mass 1.2–1.6% of phosphorus. It also contains a large amount of calcium, ~200 molecules Ca^{2+}/molecule of vitellogenin, present as a calcium phosphate complex (Wallace, 1970). The yolk platelet accumulates vitellogenin to concentrations high enough for the protein to form paracrystalline arrays. However, measurement of calcium levels in isolated yolk platelets by atomic absorption spectroscopy has provided no evidence for calcium accumulation (H. S. Wiley, personal communication, 1984). This result strongly suggests that somewhere in the endocytic apparatus Ca^{2+} is removed from vitellogenin prior to its storage in yolk granules. These observations suggest that the endocytic apparatus must have a calcium transport system.

There are at least three exceptions to the rule that class II receptors are calcium dependent. Analysis of the physiology of these receptors provides insight into how functional considerations can alter receptor behavior. Receptors for mannose-phosphate-terminal glycoproteins mediate the accumulation of hydrolytic enzymes in lysosomes. Binding of mannose-phosphate-terminal glycoproteins to receptors is calcium independent but is sensitive to acid pH (Rome *et al.*, 1979). Although this receptor can mediate lysosomal accumulation of extracellular ligand, its major function appears to be transport of newly synthesized lysosomal enzymes from Golgi apparatus to lysosomes. If the endosomal apparatus contains a Ca^{2+} transport system, then this receptor must have evolved to function in a low-calcium environment. Once the receptor–ligand complex reaches the lysosome, there is a rapid hydrolysis of mannose-phosphate, the recognition marker for the receptor (Gabel *et al.*, 1982). Binding of ligand to receptor would be prevented by loss of the recognition marker and would result in unoccupied receptors.

Another modification of receptor behavior is seen in the IgG receptor in neonatal intestinal epithelial cells. Binding of ligand to receptor is enhanced by low pH, and dissociation of ligand from receptor is promoted by high or neutral pH (Rodewald, 1976). Binding of IgG to receptor occurs in the acidic environment of the intestinal lumen. The receptor–ligand complex is passed through the cell, most probably transversing through acidic compartments. The ligand–receptor complex is ultimately transported to the basal–lateral surface of the epithelium where, at a neutral pH, the ligand dissociates from the receptor (Abrahamson and Rodewald, 1981). Dissociation may occur because the concentration of ligand on the basal–lateral surface is low relative to the concentration in an intracellular vesicle.

The most interesting modification of the ionic behavior of class II receptors is seen in the transferrin-mediated delivery of iron. Transferrin is the major iron-binding protein in plasma, and the cellular uptake of iron is mediated by the binding of diferric transferrin to high-affinity receptors. The surface receptor–diferric-transferrin complex is internalized into the endosome where, in an acidic environment, iron dissociates from transferrin. At the acidic pH of the lumen, apotransferrin remains receptor bound, and the complex is recycled to the cell surface. At neutral pH the rate of dissociation of apotransferrin from the cell surface is increased by an order of magnitude, and released apotransferrin is capable of binding iron and repeating the cycle (Iacopetta and Morgan, 1983; Dautry-Varsat *et al.*, 1983; Klausner *et al.*, 1983). It is of interest to note that iron is a multivalent cation, and some mechanism must exist to get it across the endosome membrane. These observations suggest that when receptors are shown to be calcium independent they mediate either transcellular or intracellular ligand traffic.

The concentration of class II receptors in cell membranes can be controlled by a number of factors. For example, the number of mannose-terminal glycoprotein receptors on macrophages depends on the stage of activation of the macrophage (Imber *et al.*, 1982). Within a few days of birth, receptors for IgG in intestinal cells disappear (Rodewald, 1976). Although a variety of developmental cues can regulate class II receptor number, receptor levels are unaffected by either ligand binding or internalization. Receptor-mediated internalization does not affect receptor turnover (Tanabe *et al.*, 1979). This lack of ligand-induced regulation is understandable in the context of receptor function. Class II receptors mediate the removal of noxious agents (i.e., lysosomal hydrolases, proteases) or the uptake of a desired metabolite or macromolecule (IgG, vitellogenin). In either case, a ligand-induced loss of receptor activity would be deleterious for either the cell or the organism.

There are two instances in which class II receptor number is regulated as a consequence of ligand internalization. As a result of interaction of cells with ligand, receptor number for either low-density lipoproteins (Brown and Goldstein, 1975) or transferrin (Ward *et al.*, 1982) is subject to regulation. These receptors are regulated not by ligand binding or internalization but by metabolic products derived from the internalized ligand. Biosynthesis of the low density lipoprotein receptor is regulated by cholesterol liberated from the lysosomal hydrolysis of low density lipoproteins (Brown and Goldstein, 1975). Biosynthesis of the transferrin receptor is affected by the intracellular accumulation of iron (Ward *et al.*, 1982).

Recent evidence suggests that iron exerts its effect as a substrate for heme biosynthesis and that increased intracellular heme effects a decrease in receptor biosynthesis (Ward *et al.*, 1982). This mode of regulation classically fits the tenants of feedback regulation in which the product of a metabolic pathway (heme, cholesterol) regulates the concentration of the first step in the pathway (receptor number), thereby controlling substrate flux. There are a number of other features in common between the low density lipoprotein and transferrin receptor that would result in evolutionary pressure to regulate ligand accumulation. They are the only two receptors whose ligand is present continuously in a high steady-state concentration. For example, the K_D for diferric transferrin is $\sim 10^{-8}$, whereas the concentration of diferric transferrin in plasma is 2×10^{-5} M. Consequently, surface receptors would always be occupied by ligand. Similar calculations pertain to the low density lipoprotein receptor. Although both iron and cholesterol are required compounds, cellular accumulation in excess of biosynthetic requirements results in pathophysiological sequelae.

Defects occur in the regulation of the low density lipoprotein receptor resulting from defective catabolism of internalized low density lipoprotein, and as a consequence cells accumulate low density lipoprotein to great excess (Goldstein et al., 1975). Analogous pathophysiological consequences occur when the ligand concentrations for other class II receptors are abnormally increased. A normally unregulated receptor, when faced with a high concentration of ligand, will continue to internalize that ligand, resulting in unregulated ligand accumulation and cellular "engorgement." An example of this phenomenon is the unregulated accumulation of apoprotein-E-containing lipoproteins by the apoprotein-E receptor on smooth muscle cells and macrophages (Mahley and Innerarity, 1983). When exposed to high ligand concentration, cells accumulate ligand to the extent that they develop large lipoid inclusions resembling those seen in foam cells present in an athrosclerotic plaque. Indeed, unregulated low density lipoprotein uptake is thought to be involved in the etiology of atherosclerosis (Mahley, 1981; Brown et al., 1980).

4. Recycling and the Endosome

The fate of most of the ligands internalized by class I and class II receptors is lysosomal hydrolysis. However, many of the receptors that mediate ligand accumulation are spared from lysosomal catabolism although these receptors are not inherently resistant to hydrolase action. Exposure to proteases of receptors for transferrin (Lamb et al., 1983), low-density lipoprotein (Goldstein and Brown, 1974), and mannose-terminal glycoproteins (Stahl et al., 1980) results in their inactivation. There is no a priori reason to think that these receptors are selectively resistant to lysosomal enzymes. The mechanism that seems to protect these receptors from being catabolized is that during endocytosis ligand–receptor dissociation occurs in a nonlysosomal compartment, and receptors may never be exposed to lysosomal hydrolases.

A number of independent lines of evidence demonstrate that internalized ligand–receptor complexes are initially directed to a nonlysosomal compartment. Among such studies are (1) the examination of the cellular distribution of internalized receptors and ligand by electron microscopy (Wall and Hubbard, 1981; Geuze et al., 1982) and (2) the examination of the intracellular distribution of radiolabeled ligand using fractionation techniques (Merion and Sly, 1983). In these studies analysis of the distribution of [^{125}I]-transferrin has been particularly informative, since this ligand does not accumulate in lysosomes and is only found in the endosome (Lamb et al., 1983; van Renswoude et al., 1982). (3) Demonstration of a time lag between the internalization of a ligand and its appearance in lysosomes (Harford *et al.*, 1983), (4) the demonstration that low temperature can prevent transfer of components from the intermediate vesicle to the lysosome (Dunn *et al.*, 1980), (5) the isolation of mutants that exhibit a defective endosomal pH and altered patterns of ligand accumulation (Robbins *et al.*, 1983; Merion *et al.*, 1983).

These studies indicate the existence of a compartment between the internalization of a ligand and its deposition in lysosomes. Evidence indicates that this compartment is a distinct cellular organelle. In the absence of ligand, a large number of class II receptors are present in this organelle. Studies on the subcellular distribution of transferrin receptors revealed that ~70% of total cell receptor activity was localized in this compartment (Lamb *et al.*, 1983). Studies have demonstrated that large intracellular receptor pools exist for α-macroglobulin–protease complexes (Kaplan and Keogh, 1981), mannose-terminal glycoproteins (Tietze *et al.*, 1982), galactose-terminal glycoproteins (Harford *et al.*, 1983; Weigle and Oka, 1983; 1984), and insulin (Deutsch *et al.*, 1982). The available evidence indicates that unoccupied receptors from this pool can be recruited to the cell surface, thus maintaining a constant number of surface receptors.

The factors that control receptor recycling are not known. Recent studies by Weigle and co-workers suggest that receptor recycling can be divorced from receptor internalization. Their studies indicate that the rates of these processes exhibit different temperature coefficients and that by altering incubation temperature one can alter steady-state receptor number (Weigle, 1981). Receptor recycling also appears more sensitive than internalization to reductions in cellular ATP levels (Clarke and Weigle, 1982). Incubation of cells with agents that alter endosomal pH can inhibit recycling without affecting internalization (Kaplan and Keogh, 1981; Tietze *et al.*, 1980; Gonzalez-Noriega *et al.*, 1980).

5. Role of the Endosome in Regulating Cell Surface Area

As mentioned, one role of the endosome may be to act as a reservoir of receptors and membrane. Recruitment of unoccupied receptors to the cell surface compensates for receptors internalized during ligand uptake. The result of this process is that cells can maintain a relatively constant surface receptor density and cell surface area in the face of high levels of endocytic activity. Examination of the literature reveals circumstancesin which there may be radical changes in cell surface area. During phagocytosis, macrophages are capable of internalizing prodigous

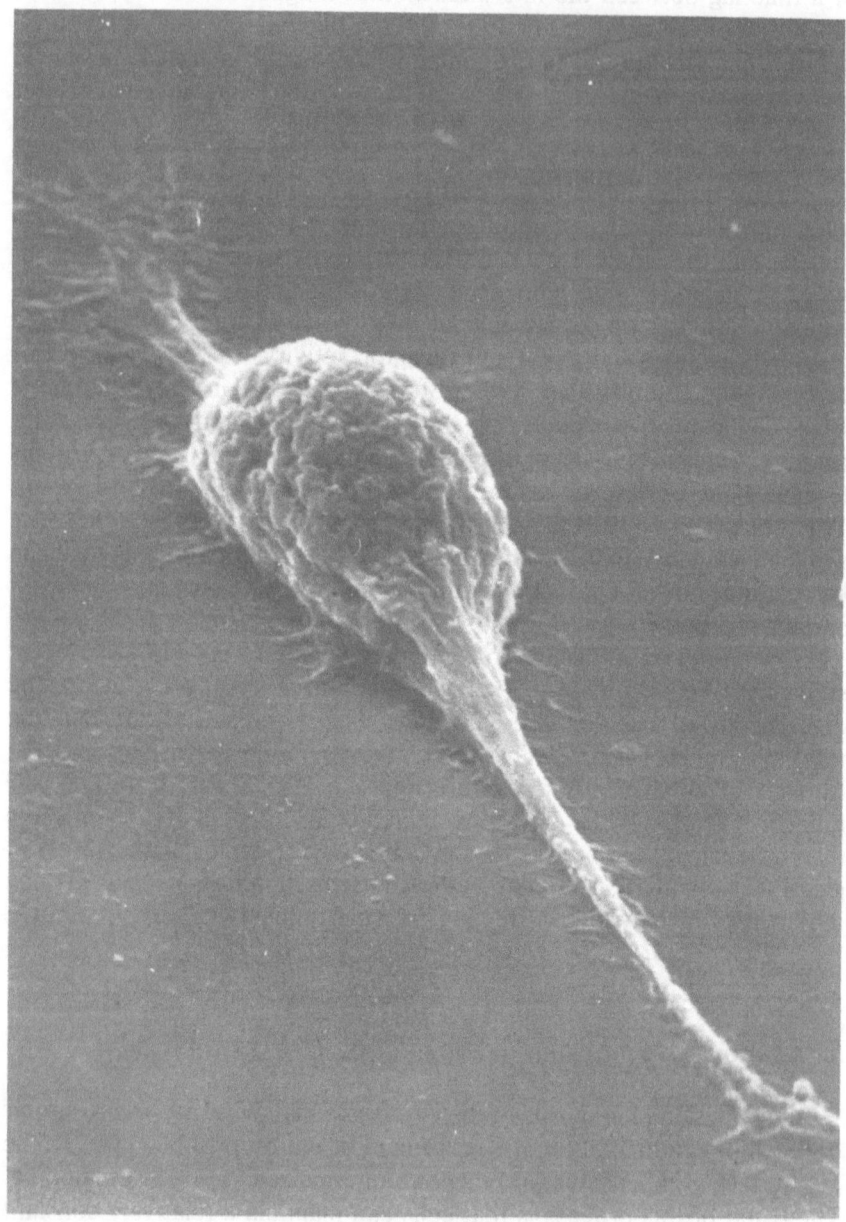

Figure 1. Effect of phorbol myristate acetate on the morphology of J774 macrophages. Cultured J774 tumor macrophages were incubated in the presence or absence of phorbol myristic acetate (0.1 µg/ml) for 20 min. The cells were then fixed with glutaraldehyde, dehydrated in a graded series of ethanol, shadowed with gold, and examined by scanning electron microscopy. A: Control cells. B: Phorbol myristate acetate-treated cells.

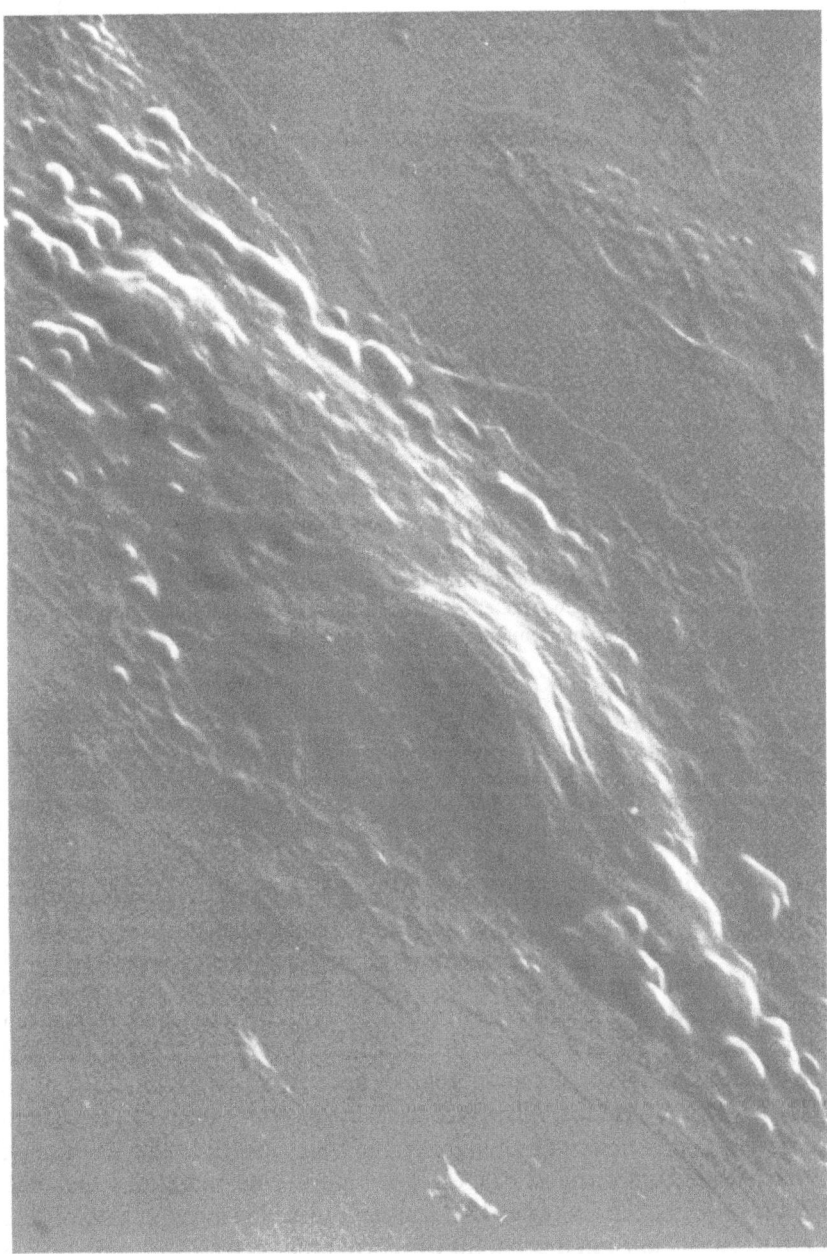

Figure 1. (continued)

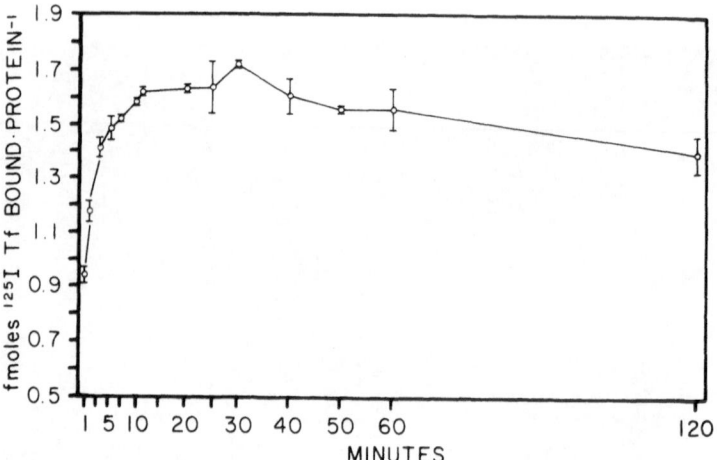

Figure 2. The kinetics of phorbol-myristate-acetate-induced increase in the binding of trans-ferrin to surface receptors. J774 cells attached to plastic dishes were incubated at 37°C in Eagle's minimal media. Phorbol myristic acetate (0.1 μg/ml) was added to the cultures, and at specified times the cells were placed at 0°C to measure cell surface transferrin binding activity. Binding of [^{125}I]-transferrin to cells was performed as described by Ward *et al.* (1982). Nonspecific binding was determined and subtracted from total binding, and what is presented is specific binding (femtomoles [^{125}I]-transferrin per microgram cell protein).

amounts of surface membrane. Petty *et al.*, (1981) determined that phag-ocytosis of IgG-coated liposomes by guinea pig peritoneal macrophages results in the internalization of ~100% of the cell's surface. However, direct measurement of cell surface area before and after phagocytosis revealed that surface area decreased by only 25%. They suggested that the internalized membrane was replaced by membrane from an intra-cellular reservoir.

Another instance in which intracellular stores of membrane appear to be recruited to the cell surface is suggested by the study of Phaire-Washington *et al.*, (1980). These workers demonstrated that exposure of mouse peritoneal macrophages to the tumor promoter phorbol myristate acetate (PMA) resulted in an increased rate of fluid-phase pinocytosis and a threefold increase in cell surface area. Buys *et al.*, (1984) have recently demonstrated that similar changes occur on exposure to PMA of the cul-tured mouse tumor macrophage J774. Figure 1 demonstrates that PMA effects radical changes in cell morphology, resulting in cells that are much flatter. This change in morphology is correlated with an apparent increase in surface area. Buys *et al.*, demonstrated that the increase in surface area may result from fusion of the endosome with the cell surface. This conclusion was based on the observation that addition of PMA to cells

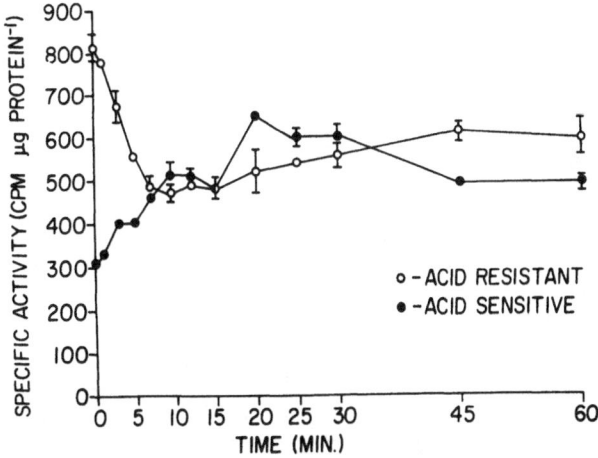

Figure 3. Effect of phorbol myristic acetate on the cellular distribution of [^{125}I]-transferrin (Tf). Cells wre incubated at 37°C for 90 min with 50 nM [^{125}I]-Tf until steady-state binding had been reached. At this point, phorbol myristate acetate (0.1 μg/ml) was added to the plates, and at specified times plates were placed on ice. The plates were washed three times with cold phosphate-buffered saline and then treated to remove unbound ligand with 0.2 M acetic acid in 0.5 M NaCl to distinguish between cell surface (acid-strippable) and internalized (acid-resistant) ligand (Hargler *et al.*, 1979). Zero time represents the time at which the phorbol myristate acetate was added to the cells.

resulted in a rapid increase in surface transferrin receptors (Fig. 2) and a concomitant decrease in intracellular transferrin content (Fig. 3). Since [^{125}I]-transferrin only binds to receptors on the surface and in the endosome, increased surface receptor number would result from the fusion of the endosome with the cell surface. Similar increases in cell surface area are seen in alveolar macrophages exposed to the calcium ionophore A23187. Measurement of the numbers of four different surface receptors revealed coordinate increases after ionophore treatment (Fig. 4).

The effects of PMA and the calcium ionophore can be separated by a number of criteria. Although alveolar macrophages respond to the Ca^{2+} ionophore, they do not respond to PMA. The cultured macrophage cell line J774 responds to both PMA and the Ca^{2+} ionophore, but the effects are not additive (Table III). The ionophore-induced increase in surface transferrin receptor number required the presence of extracellular Ca^{2+}, whereas the effect of PMA was independent of extracellular calcium. We conclude from these observations that PMA and A23187 are separate signals acting on a common effector mechanism.

The above studies indicate that the endosome can act as a reservoir of membrane capable of being recruited to the cell surface. As a conse-

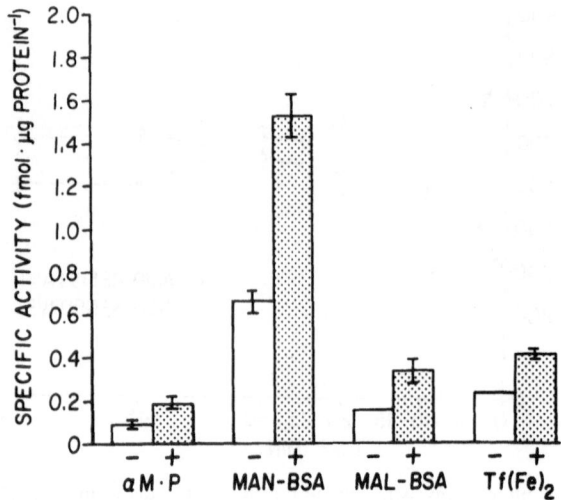

Figure 4. The effect of calcium ionophore A23187 on alveolar macrophage surface receptor number. Rabbit alveolar acrophages were obtained by bronchial lavage and incubated at 37°C in Eagle's minimal media. The cells were incubated at 37°C in the presence or absence of the calcium ionophore A23187 (1.3 μM) for 6 min. The cells were placed at 0°C, and the binding of [^{125}I]-transferrin, [^{125}I]-mannose bovine serum albumin, [^{125}I]-maleated bovine serum albumin, and α_2-macroglobulin·[^{125}I]-trypsin complexes was determined. For details of ligand preparation see Buys *et al.* (1984). In all instances, what is presented is specific binding (mean ± standard error of the mean).

quence of fusion of this compartment with the plasma membrane, surface area increases, as does the absolute number of surface receptors per cell. One consequence of an increase in surface area is an increase in fluid-phase pinocytosis. Holtzman and colleagues suggested that an increase in pinocytic activity may reflect the attempt of cells to reduce surface area increased as a result of secretory phenomena (Holtzman, 1977), a process they referred to as "compensatory endocytosis" (Schacher *et al.*, 1976). One of the early effects on fibroblasts of epidermal growth factor is increased fluid-phase pinocytosis. Wiley and Cunningham (1982) presented kinetic data indicating that this effect was caused by formation of a surface receptor–ligand complex. They further suggested that increased pinocytic activity was the result of a secretory process.

In a collaborative effort, Wiley and I demonstrated that addition of epidermal growth factor to cultured human skin fibroblasts resulted in a rapid, but transient, increase in surface transferrin receptor number (Wiley and Kaplan, 1984). Cells increased their surface receptors within 1 min of addition of epidermal growth factor; receptor number reached

Table III. Effect of PMA and the Ca^{2+} Ionophore A23187 on Transferrin Receptor Number[a]

	Specific activity (fmole/μg protein) (mean ± S.E.M.)	Percentage change
Ca^{2+}-containing media		
Control	1.03 ± 0.08	—
+ PMA	1.70 ± 0.05	+66
+ A23187	1.57 ± 0.06	+53
+ PMA + A23187	1.75 ± 0.07	+70
Ca^{2+}-free media		
Control	1.17 ± 0.04	—
+ PMA	1.81 ± 0.005	+54
+ A23187	1.26 ± 0.05	+8
+ PMA + A23187	1.53 ± 0.15	+30

[a] Cells (J774) were incubated in Hank's Buffered Salt Solution (HBSS) that either contained 1 mM Ca^{2+} or was Ca^{2+}-free. Either PMA (0.1 μg·ml^{-1}) or A23187 (1.3 μM) was added to the cultures, and the cells were allowed to incubate at 37°C for 25 min or 6 min, respectively. The cells were then placed at 0°C and washed, and the specific binding was determined. To determine if the effects of PMA and A23187 were additive, cells were incubated with PMA for 19 min and then exposed to A23187 for a further 6 min prior to assay.

a maximum by 4 min and by 45 min had declined to essentially base-line levels. The kinetics of this response and its dose dependency match those of the increased fluid-phase pinocytosis.

Examination of the literature reveals that a number of the rapid cellular responses to epidermal growth factor exhibit similar kinetics. Among such responses are increased metabolite transport, increased electrolyte transport (Boonstra *et al.*, 1983; Moolenar *et al.*, 1982; Rothberg *et al.*, 1982), increased calcium uptake, and increased intracellular pH (Rothenberg *et al.*, 1983). These observations lead us to speculate that the short-term pleiotropic effects of epidermal growth factor are caused by translocation of the endosome to the plasma membrane. This hypothesis has been advanced by others (Leinhard, 1983) to explain the effects of insulin on glucose transport (Cushman and Wardzala, 1980; Karnieli *et al.*, 1981; Kono *et al.*, 1982; Suzuki and Kono, 1980). In a preliminary study Wiley has found that insulin also increases surface transferrin receptors in human fibroblasts. We extend this hypothesis and suggest that translocation of intracellular vesicles to cell surfaces may be a widespread mechanism by which many class I receptors exert their effects.

Table IV represents a survey of a number of different systems and ligands in which the common factor is vesicle translocation. The nature

Table IV. Examples of Class I Receptor–Ligand Complexes That Affect Vesicle–Cell Surface Fusion

Ligand	Cell	Effect	"Messenger"[a]	Selected references
Sperm	Egg	Granule release	Ca^{2+}	Steinhardt and Epel (1974) Fisher and Rebhun (1981)
Insulin Epidermal growth factor Platelet-derived growth factor	Fibroblasts	Increased metabolite, electrolyte transport, and intracellular pH	?	Moore (1981) Boonstra et al. (1983) Cushman and Wardzala (1980) Kono et al. (1982) Moolenaar et al. (1982) Rothenberg et al. (1983)
Vasopressin	Epithelial cell, kidney, bladder	Increased H_2O permeability	Ca^{2+}	Harmanci et al. (1980)
Gastrin Histamine Acetylcholine	Acinar cell, stomach	H^+ secretion	Ca^{2+}	Masur et al. (1972) Forte et al. (1981)
f-Met-Leu-Phe C_{5a}	Polymorphonuclear neutrophil	Superoxide anion production	Ca^{2+}	Hoffstein et al. (1982)
Phorbol ester		Intracellular pH, granule release	Protein kinase C	White and Estensen (1974)
Cholecystokinin	Acinar cell, pancreas	Enzyme secretion	Ca^{2+}	Gardner (1979)
IgE	Mast cell	Granule release	Ca^{2+}	Metzger (1977) Ishizaka and Ishizaka (1975)
Thrombin Collagen	Platelet	Granule release	Ca^{2+}	Knight et al. (1982) Rink et al. (1982)
Phorbol ester			Protein kinase C	
Phorbol ester Ionophore	Macrophage	Increased surface area	Ca^{2+} Protein kinase C	Phaire-Washington et al. (1980) Buys et al. (1984)

[a] The determination that intracellular free calcium is involved is derived principally from the fact that addition of Ca^{2+} ionophores mimic the secretagogue.

and biochemical properties of the vesicles differ in each cell type, but in each instance fusion of the vesicle with cell surface occurs in response to formation of a surface receptor–ligand complex. Analysis of the table reveals an interesting feature in that in those instances in which the "second messenger" is known, that messenger is either calcium or protein kinase C. Ameba, when exposed to calcium ionophores, increase the rate of fluid phase pinocytosis (Prusch, 1980). If, as we suspect, this is another example of compensatory pinocytosis, then vesicle translocation may be a primitive response to external stimuli. Vesicle fusion would then be another example of the conservative process of evolution and represent a common cellular mechanism, the specifics of which have been altered by evolutionary pressure.

ACKNOWLEDGMENTS. I would like to express my appreciation to Elissa A. Keogh, Diane McVey, John Ward, Jamie Lamb, Sandra Buys, Ina Jordan, and Scott Rogers for doing most (all) of the work. I would like to express my appreciation to Dr. Steve Wiley for his useful comments and to Dr. James P. Kushner for his thoughts on evolution. I particularly wish to express my gratitude to Jana Lawton, Eleanor Hart, and Rich Ajioka for their help in preparing this manuscript. The studies presented in this manuscript were supported by grants from the National Institutes of Health (AM30534, HL26922) and from the R. J. Reynolds Foundation. J. Kaplan is a recipient of a National Institutes of Health Research Career Development Award (HL 00598).

References

Abrahamson, D. R., and Rodewald, R., 1981, Evidence for the sorting of endocytic vesicle contents during the receptor mediated transport of IgG across the newborn rat intestine, *J. Cell Biol.* 91:270–280.

Anderson, R. G. W., and Kaplan, J., 1983, Receptor-mediated endocytosis, in: Current Topics in Cell Biology (B. Satir, ed.), Academic Press, pp. 1–51.

Anderson, R. G. W., Goldstein, J. L., and Brown, M. S., 1977, A mutation that impairs the ability of lipoprotein receptors to localize in coated pits on the cell surface of human fibroblasts, *Nature* 270:695–699.

Anderson, R. G. W., Brown, M. S., and Goldstein, J. L., 1981, Inefficient internalization of receptor bound low density lipoprotein in human carcinoma A-431 cells, *J. Cell Biol.* 88:441–452.

Ascoli, M., and Puett, D., 1978, Degradation of receptor bound human choriogonadotropin by murine Leydig tumor cells, *J. Biol. Chem.* 253:4892–4899.

Blanchard, S. G., Chang, K. J., and Cuatrecasas, P., 1983, Characterization of the association of tritiated enkephalin with neuroblastoma cells under conditions optimal for receptor down regulation, *J. Biol. Chem.* 258:1092–1097.

Boonstra, J., Moolenaar, W. H., Harrison, P. H., Moed, P., van der Saag, P. T., and deLatt, W. S., 1983, Ionic responses and growth stimulation induced by nerve growth factor and epidermal growth factor in rat pheochromacytoma (pc 12) cells, *J. Cell Biol.* **97**:92–98.

Bowen-Pope, D. F., and Ross, R., 1982, Platelet derived growth factor. VI. specific binding to cultured cells, *J. Biol. Chem.* **257**:5161–5171.

Branca, A. A., and Baglioni, C., 1982, Down-regulation of the interferon receptor, *J. Biol. Chem.* **257**:13197–13200.

Bridges, K., Harford, J., Ashwell, G., and Klausner, R. D., 1982, Fate of receptor and ligand during endocytosis of asialoglycoproteins by isolated hepatocytes, *Proc. Natl. Acad. Sci. U.S.A.* **79**:350–354.

Brown, M. S., and Goldstein, J. L., 1975, Regulation of the activity of the low density lipoprotein receptor in human fibroblasts, *Cell* **6**:307–316.

Brown, M. S., Basu, S. K., Falck, J. R., Ho, Y. K., and Goldstein, J. L., 1980, The scavenger cell pathway for lipoprotein degradation: Specificity of the binding site that mediates the uptake of negatively charged LDL by macrophages, *J. Supramol. Struct.* **13**:67–81.

Buys, S. H., Keogh, E. A., and Kaplan, J., 1984, Fusion of intracellular membrane pools with cell surfaces of macrophages stimulated with phorbol esters with calcium ionophores, *Cell* **38**:569–576.

Carpenter, G., and Cohen, S., 1976, ^{125}I-labeled epidermal growth factor. Binding, internalization and degradation in human fibroblasts, *J. Cell Biol.* **71**:159–171.

Chinkers, M., McKanna, J. A., and Cohen, S., 1979, Rapid induction of morphological changes in human carcinoma cells A-431 by epidermal growth factor, *J. Cell Biol.* **83**: 260–265.

Ciechanover, A., Schwartz, A. L., Dautry-Varsat, A., and Lodish, H. F., 1983, Kinetics of internalization and recycling of transferrin and transferrin receptor in a human hepatoma cell line, *J. Biol. Chem.* **258**:9681–9689.

Clarke, B. L., and Weigel, P. H., 1982, Effect of metabolic energy poisons on endocytosis mediated by the asialoglycoprotein receptor, *J. Cell Biol.* **95**:A442.

Connolly, J. L., Green, S. A., and Greene, L. A., 1981, Pit formation and rapid changes in surface morphology of sympathetic neurons in response to nerve growth factor, *J. Cell Biol.* **90**:176–180.

Cushman, S. M., and Wardzala, L. J., 1980, Potential mechanism of insulin action on glucose transport in the isolated rat adipose cell, *J. Biol. Chem.* **255**:4758–4762.

Dautry-Varsat, A., Ciechanover, A., and Lodish, H. F., 1983, pH and the recycling of transferrin during receptor-mediated endocytosis, *Proc. Natl. Acad. Sci. USA* **80**:2258–2262.

Deutsch, P. J., Rosen, O. M., and Rubin, C. S., 1982, Identification and characterization of a latent pool of insulin receptors in 3T3-Li adipocytes, *J. Biol. Chem.* **257**:5350–5358.

Dunn, W. A., Hubbard, A. L., and Aronson, N. N., Jr., 1980, Low temperature selectively inhibits fusion between pinocytic vesicles and lysosomes during heterophagy of ^{125}I-asialofetuin by the perfused rat liver, *J. Biol, Chem.* **255**:5971–5978.

Fawcett, D. W., 1965, Surface specializations of absorbing cells, *J. Histochem. Cytochem.* **13**:75–91.

Fisher, G. W., and Rebhun, L. I., 1981, Turn on of endocytotic processes in response to sea urchin egg activation accompanies restructuring of the egg surface, *J. Cell Biol.* **91**:185a.

Forte, J. G., Black, J. A., Forte, T. M., Machen, T. E., and Wolosin, J. M., 1981, Ultrastructural changes related to functional activity in gastric oxyntic cells, *Am J. Physiol.* **241**:G349–358.

Gabel, C. A., Goldberg, D. E., and Kornfelds, S., 1982, Lysosomal enzyme oligosaccharide phosphorylation in mouse lymphoma cells: Specificity and kinetics of binding to the mannose-6-phosphate receptor *in vivo, J. Cell Biol.* **95**:536–543.

Gardner, J. D., 1979, Regulation of pancreatic exocrine function *in vitro*: Initial steps in the actions of secretagogues, *Annu. Rev. Physiol.* **41**:55–66.

Gardner, J. M., and Fambrough, D. M., 1979, Acetylcholine receptor degradation measured by density labeling: Effects of cholinergic ligands and evidence against recycling, *Cell* **16**:661–674.

Geuze, N. J., Slot, J. W., Strous, G. J., Lodish, H. F., and Schwartz, A. L., 1982, Immunocytochemical localization of the receptor for asialoglycoproteins in rat liver cells, *J. Cell Biol.* **92**:865–870.

Goldstein, J. L., and Brown, M. S. 1974, Binding and degradation of low density lipoproteins by cultured human fibroblasts, *J. Biol. Chem.* **249**:5153–5162.

Goldstein, J. L., Dana, S. E., Faust, J. R., Beaudet, A. L., and Brown, M. S., 1975, Role of lysosomal acid lipase in the metabolism of plasma low density lipoprotein: Observations in cultured fibroblasts from a patient with cholesteryl ester storage disease, *J. Biol. Chem.* **250**:8487–8795.

Gonzalez-Noriega, A., Grubb, J. W., Talkad, V., and Sly, W. B., 1980, Chloroquine inhibits lysosomal enzyme pinocytosis and enhances lysosomal enzyme secretide by impairing receptor recycling, *J. Cell Biol.* **85**:839–852.

Haigler, H. T., McKanna, J. A., and Cohen, S., 1979, Rapid stimulation of pinocytosis in human carcinoma cells A-431 by epidermal growth factor, *J. Cell Biol.* **83**:82–90.

Harford, J., Bridges, K., Ashwell, G., and Klausner, R. D., 1983, Intracellular dissociation of receptor bound asialoglycoproteins in cultured hepatocytes: A pH mediated nonlysosomal event, *J. Biol. Chem.* **258**:3191–3197.

Harford, J., Klausner, R. D., Wolkoff, A. W., Bridges, K. R., and Ashwell, G., 1984, Asialoglycoprotein catabolism by hepatocytes: Insights from perturbation of the endocytic pathway, in: *The Molecular Basis of Lysosomal Storage Disorders* (R. O. Brady and J. A. Barranger, eds.), Academic Press, Orlando, 1984, pp. 149–162.

Harmanci, M. C., Stern, P., Kachadorian, W. A., Vatin, H., and DiScala, V. A., 1980, Vasopressin and collecting duct intramembranous particle clusters: A dose–response relationship, *Am. J. Physiol.* **239**:F560–F564.

Hoffstein, S. T., Friedman, R. S., and Weissmann, G., 1982, Degranulation, membrane addition and shape change during chemotactic factor-induced aggregation of human netrophils, *J. Cell Biol.* **95**:234–241.

Holtzman, E., 1977, The origin and fate of secretory packages, especially synaptic vesicles, *Neuroscience* **2**:327–355.

Hopkins, C. R., 1982, Early events in the receptor mediated endocytosis of E.G.F. and α_2M.P complex in membrane recycling, *Ciba Found. Symp.* **92**:235–242.

Hopkins, C. R., and Trowbridge, I. S., 1983, Internalization and processing of transferrin and the transferrin receptor in human carcinoma A431 cells, *J. Cell Biol.* **97**:508–521.

Huang, S. J., Huang, S. S., Kennedy, B., and Deuel, T. F., 1982, Platelet derived growth factor. Specific binding to target cells, *J. Biol. Chem.* **257**:8130–8136.

Iacopetta, B. J., and Morgan, E. H., 1983, The kinetics of transferrin endocytosis and iron uptake from transferrin in rabbit reticulocytes, *J. Biol. Chem.* **258**:9108–9115.

Imber, M. J., Pizzo, S. V., Johnson, W. J., and Adams, D. O., 1982, Selective diminution of the binding of mannose by murine macrophages in the late stage of activation, *J. Biol. Chem.* **257**:5129–5135.

Ishizaka, T., and Ishizaka, K., 1975, Biology of immunoglobin E: Molecular basis of reaginic hypersensitivity, *Prog. Allergy* **19**:60–121.

Judson, H. F., 1980, *The Search for Solutions*, Holt, Rinehart, Winston, New York.

Kaplan, J., 1981, Polypeptide-binding membrane receptors: Analysis and classification, *Science* **212**:14–20.

Kaplan, J., and Keogh, E. A., 1981, Analysis of the effect of amines on inhibition of receptor-mediated and fluid phase pinocytosis in rabbit alveolar macrophages, *Cell* **24**:925–932.

Kaplan, J., McVey, D., and Wiley, H. S., 1984, Phenylarsine oxide induced increase in alveolar macrophage surface receptors: Evidence for fusion of internal receptor pools with the cell surface, *J. Cell Biol.* **101**:121–129.

Karnieli, E., Zarnowski, M. J., Hissin, P. J., Simpson, I. A., Salans, L. B., and Cushman, S. W., 1981, Insulin-stimulated translocation of glucose transport systems in the isolated rad adipose cell, *J. Biol. Chem.* **256**:4772–4777.

Kasuga, M., Kahn, C. R., Hedo, J. A., Van Obberghen, E., and Yamada, K. M., 1981, Insulin inducted receptor loss in cultured human lymphocytes is due to accelerated receptor degradation, *Proc. Natl. Acad. Sci. USA* **78**:6917–6921.

Klausner, R. P., Ashwell, G., vanRenswoude, J., Harford, J. B., and Bridges, K. R., 1983, Binding of apotransferrin to K562 cells: Explanation of the transferrin cycle, *Proc. Natl. Acad, Sci. U.S.A.* **80**:2263–2266.

Knauer, D. J., Wiley, H. S., and Cunningham, D. D., 1984, Relationship between epidermal growth factor receptor occupancy and mitogenic response: Quantitative analysis using a steady state model system, *J. Biol. Chem.* **259**:5623–5631.

Knight, D. E., Hallam, T. J., and Scrutton, M. C., 1982, Agonist selectively and second messenger concentration in Ca^{2+} mediated secretion, *Nature* **296**:256–257.

Knutson, V. P., Ronnett, G. V., and Lane, M. D., 1983, Rapid, reversible internalization of cell surface insulin receptors. Correlation with insulin-induced down regulation, *J. Biol. Chem.* **258**:12139–12142.

Kono, T., Robinson, F. W., Blevins, T. L., and Ezaki, O., 1982, Evidence that translocation of the glucose transport activity is the major mechanism of insulin action on glucose transport in fat cells, *J. Biol. Chem.* **257**:10962–10967.

Lamb, J. E., Ray, F., Ward, J. H., Kushner, J. D., and Kaplan, J. 1983, Internalization and subcellular localization of transferrin and transferrin receptors in HeLa cells, *J. Biol. Chem.* **285**:8751–8758.

Landau, E. M., 1978, Function and structure of the acetylcholine receptor at the muscle end-plate. *Prog. Neurobiol.* **10**:253–288.

Mahley, R. W., 1981, Cellular and molecular biology of lipoprotein metabolism in atherosclerosis, *Diabetes* **30**(Suppl 2):60–65.

Mahley, R. W., and Innerarity, T. L., 1983, Lipoprotein receptors and cholesterol hemostasis, *Biochim. Biophys. Acta* **737**:197–222.

Marshall, S., Greena, A., and Olefsky, J. M., 1981, Evidence for recycling of insulin receptors in isolated adipocytes, *J. Biol. Chem.* **256**:11464–11470.

Masur, S. K., Holtzman, E., and Walter, R., 1972, Hormone-stimulated exocytosis in the toad urinary bladder (some possible implications for turnover of surface membranes), *J. Cell Biol.* **52**:211–219.

Merion, M., and Sly, W. S., 1983, The role of intermediate vesicles in the adsorptive endocytosis and transport of ligand to lysosomes by human fibroblasts, *J. Cell Biol.* **96**:644–650.

Metzger, H., 1977, The cellular receptor for IgE, in: *Receptors and Recognition*, Vol. 4 (P. Cuatrecacas and M. F. Greaves, eds.), Chapman and Hall, London, pp. 74–102.

Moolenaar, W. H., Yarden, Y., deLaat, S. W., and Schlessinger, J., 1982, Epidermal growth factor induces electrically silent Na^+ influx in human fibroblasts, *J. Biol. Chem.* **257**:8502–8506.

Moore, R. D., 1981, Stimulation of $Na^+:H^+$ by insulin, *Biophys, J.* **33**:203–210.

Mukherjee, C., Caron, M. G., and Lefkowitz, R. J., 1975, Catecholamine-induced subsensitivity of adenylate cyclase associated with loss of β-adrenergic receptor binding sites, *Proc. Natl. Acad. Sci. U.S.A.* **72**:1945–1949.

Pandol, S. J., Jensen, R. T., and Gardner, J. D., 1982, Mechanism of (Tyr⁴) bombesin induced desensitization in dispersed acini from guinea pig pancreas, *J. Biol. Chem.* **257**:12024–12029.

Petty, H. R., Hafeman, D. G., and McConnell, H. M., 1981, Disappearance of macrophage surface folds after antibody-dependent phagocytosis, *J. Cell Biol.* **89**:223–229.

Phaire-Washington, L., Wang, E., and Silverstein, S. C., 1980, Phorbol myristate acetate stimulates pinocytosis and membrane spreading in mouse peritoneal macrophages, *J. Cell Biol.* **86**:634–640.

Prusch, R. D., 1980, Endocytic sucrose uptake in amoeba proteins induced with the calcium ionophore A23187, *Science* **209**:691–692.

Rink, T. J., Smith, S. W., and Tsien, R. Y., 1982, Cytoplasmic free Ca^{2+} in human platelets: Ca^{2+} thresholds and Ca-indpendent activation for shape-change and secretion, *FEBS Lett.* **148**:21–26.

Robbins, A. R., Peng, S. S., and Marshall, J. L., 1983, Mutant Chinese hamster ovary pleiotropically defective in receptor-mediated endocytosis. **96**:1064–1071.

Rodewald, R., 1976, pH dependent binding of immunoglobulins to intestinal cells of the neonatal rat, *J. Cell Biol.* **71**:666–670.

Rome, L. N., Weissman, B., and Neufeld, E. F., 1979, Direct demonstration of a lysosomal enzyme α-L-iduronidase to receptors on cultured fibroblasts, *Proc. Natl. Acad. Sci. U.S.A.* **76**:2331–2334.

Rothenberg, P., Reuss, L., and Glaser, L., 1982, Serum and epidermal growth factor transiently depolarize quiescent BSC-1 epithelial cells, *Proc. Natl. Acad. Sci. U.S.A.* **79**:7783–7787.

Rothenberg, P. Glaser, L., Schlessinger, P., and Cassel, D., 1983, Epidermal growth factor stimulates amiloride-sensitive ²²Na⁺ uptake in A-431 cells, *J. Biol. Chem.* **258**:4883–4889.

Schacher, S., Holtzman, E., and Hodd, D. C., 1976, Synaptic activity of frog retinal photoreceptors (A peroxidase uptake study), *J. Cell Biol.* **70**:178–192.

Schreiber, A. B., Yarden, Y., and Schlessinger, J., 1981, A non-mitogenic analogue of epidermal growth factor enhances the phosphorylation of endogenous membrane proteins, *Biochem. Biophys. Res. Commun.* **101**:517–523.

Schwartz, A. L., Fridovich, S. E., and Lodish, H. F., 1982, Kinetics of internalization and recycling of the asialoglycoprotein receptor in a hepatoma cell line, *J. Biol. Chem.* **257**:4230–4237.

Segaloff, D. L., and Ascoli, M., 1981, Removal of the surface bound human choriogonadotropin results in the cessation of hormonal responses in cultured Leydig tumor cells, *J. Biol. Chem.* **256**:11420–11423.

Stahl, P., Schlessinger, P. H., Sigardson, E., Rodman, J. S., and Lee, Y. C., 1980, Receptor mediated pinocytosis of mannose glycoconjugates by macrophages: Characterization and evidence for receptor recycling, *Cell* **19**:207–215.

Steinhardt, R. A., and Epel, D., 1974, Activation of sea-urchin eggs by a calcium ionophore, *Proc. Natl. Acad. Sci. U.S.A.* **71**:1915–1919.

Strulovici, B., Stadel, J. M., and Lefkowitz, R. J., 1983, Functional integrity of desensitized β-adrenergic receptors: Internalized receptors reconstitute catecholamine-stimulated adenylate cyclase activity, *J. Biol. Chem.* **258**:6410–6414.

Sullivan, S. J., and Zigmond, S. H., 1980, Chemotactic receptor modulation in polymorphonuclear leukocytes, *J. Cell Biol.* **85**:703–711.

Suzuki, K., and Kono, T., 1980, Evidence that insulin causes translocation of glucose transport activity to the plasma membrane from an intracellular storage site, *Proc. Natl. Acad. Sci. U.S.A.* **77:**2542–2545.

Tanabe, T., Pricer, E. J., and Ashwell, G., 1979, Subcellular membrane topology and turnover of a rat hepatic binding protein specific for asialoglycoproteins, *J. Biol. Chem.* **254:**1038–1043.

Tietze, C., Schlessinger, P., and Stahl, P., 1980, Chloroquinine and ammonium ion inhibit receptor mediated endocytosis of mannose conjugates by macrophages: Apparent inhibition of receptor recycling, *Biochem. Biophys. Res. Commun.* **93:**1–8.

Tietze, C., Schlesinger, P., and Stahl, P., 1982. Mannose specific endocytosis receptor in alveolar macrophage: Demonstration of two functionally distinct intracellular pools of receptor and their role in ligand recycling, *J. Cell Biol.* **92:**417–424.

Tycho, B., and Mayfield, F. R., 1982, Rapid acidification of endocytic vesicles containing α_2-macroglobulin, *Cell* **28:**643–651.

vanRenswoude, J., Bridges, K. R., Harford, J. B., and Klausner, R. D., 1982, Receptor-mediated endocytosis of transferrin and the uptake of Fe in K562 cells: Identification of a nonlysosomal acidic compartment, *Proc. Natl. Acad. Sci. U.S.A.* **79:**6186–6190.

Villereal, M. L., 1981, Sodium fluxes in human fibroblasts: Effect of serum, Ca^{2+} and amiloride, *J. Cell. Physiol.* **107:**359–369.

Wall, D. A., and Hubbard, A. L., 1981, Galactose specific recognition system of mammalian liver: Receptor distribution on the hepatocyte cell surface, *J. Cell Biol.* **90:**687–696.

Wallace, R. A., 1970, Studies on amphibian yolk. IX. *Xenopus* vitellogenin, *Biochim. Biophys. Acta* **215:**176–183.

Ward, J. H., Kushner, J. P., and Kaplan, J., 1982, Regulation of HeLa cell transferin receptor number, *J. Biol. Chem.* **257:**10317–10323.

Weigle, P. H., 1981, Evidence that the hepatic asialoglycoprotein receptor is internalized during endocytosis and that receptor recycling can be uncoupled from endocytosis at low temperature, *Biochem. Biophys. Res. Commun.* **101:**1419–1425.

Weigle, P. H., and Oka, J. A., 1983, The large intracellular pool of asialoglycoprotein receptors functions during the endocytosis of asialoglycoproteins by isolated rat hepatocytes, *J. Biol. Chem.* **258:**5095–5102.

Weigle, P. H., and Oka, J. A., 1984, Recycling of the hepatic asialoglycoprotein receptor in rat hepatocytes. Receptor ligand complexes in an intracellular slowly dissociating pool return to the cell surface prior to dissociation, *J. Biol. Chem.* **259:**1150–1154.

White, J. G., and Estensen, R. D., 1974, Selective labelization of specific granules in polymorphonuclear leukocytes by phorbol myristate acetate, *Am. J. Pathol.* **75:**45–60.

Wiley, H. S., and Cunningham, D. D., 1982, Epidermal growth factor stimulates fluid phase endocytosis in human fibroblasts through a signal generated at the cell surface, *J. Cell. Biochem.* **19:**383–396.

Wiley, H. S., and Kaplan, J., 1984, Epidermal growth factor rapidly redistributes pools of transferrin receptors, *Proc. Natl. Acad. Sci.* **81:**7456–7460.

Xie, X. S., Stone, D. K., and Racker, G., 1983, Determinants of clathrin-coated vesicle acidification, *J. Biol. Chem.* **258:**14834–14838.

Zigmond, S. H., Sullivan, S. J., and Lauffenburger, D. A., 1982, Kinetic analysis of chemotactic peptide modulation, *J. Cell Biol.* **92:**34–43.

The Membrane Receptor for Epidermal Growth Factor

Structural and Functional Studies

JOSEPH SCHLESSINGER, IRIT LAX, SHOSHANA GILL,
RICHARD KRIS, TOWIA A. LIBERMANN,
MENASHE BAR-ELI, ILANA HARARI, and YOSEF YARDEN

1. Introduction

A useful model system for exploring the molecular mechanisms underlying the proliferation of eucaryotic cells is the mode of action of the cellular mitogen epidermal growth factor (EGF). Epidermal growth factor is a small protein containing 53 amino acid residues (Carpenter and Cohen, 1980). It binds tightly to a specific membrane receptor. The occupation of the receptor molecule leads to activation of the pleiotropic response culminating in DNA synthesis and cell proliferation (Carpenter and Cohen, 1980; Schlessinger 1983).

The membrane receptor for EGF is a 170,000-dalton glycoprotein that possesses several domains: (1) an extracellular portion, which contains the EGF binding site, (2) a transmembrane region, (3) a tyrosine-specific kinase activity, and (4) autophosphorylation site(s).

The binding of EGF to the external portion enhances the activity of the tyrosine-specific, cyclic-nucleotide-independent, protein kinase domain, which in turn phosphorylates various cellular proteins and auto-

*JOSEPH SCHLESSINGER, IRIT LAX, SHOSHANA GILL, RICHARD KRIS, TOWIA A. LIB
ERMANN, ILANA HARARI, and YOSEF YARDEN* • Department of Chemical Immunology, The Weizmann Institute of Science, Rehovot 76100, Israel. *MENASHE BAR-
ELI* • Department of Immunology, Ben-Gurion University, Beer-Sheva, Israel.

phosphorylates specific site(s) on the receptor molecule (Carpenter *et al.*, 1978, 1979; Cohen *et al.*, 1980). The EGF receptor kinase activity is similar to the tyrosine-specific kinase associated with the transforming protein encoded by oncogenes belonging to the *src* gene family (Collett *et al.*, 1979, 1980; Hunter and Sefton, 1980; Levinson *et al.*, 1980). Antibodies generated against the transforming protein of Rous sarcoma virus, $PP60_{src}$, are phosphorylated on tyrosine residues by membrane preparations containing EGF receptor from A-431 cells (Chinkers and Cohen, 1981; Kudlow *et al.*, 1981). Interestingly, these antibodies do not immunoprecipitate the EGF receptor kinase.

In this chapter we describe the purification of EGF receptor and the analysis of its structural domains using specific immunologic probes. Several forms of EGF receptor are purified by an efficient purification procedure utilizing immunoaffinity chromatography with monoclonal antibodies against the receptor molecule. Based on these analyses, we propose a model for the structure of the EGF receptor.

2. Results

Complete understanding of the mechanism of action of EGF requires the characterization of its membrane receptor in molecular terms. Necessary conditions required for the characterization of the membrane receptor for EGF require an efficient procedure for the purification of an active EGF receptor kinase. This would enable the analysis of its functional domains and the interaction between them. It would also afford the establishment of the amino acid sequence of the receptor molecule. To date, the most efficient approach for the determination of the amino acid sequence of a protein is to analyze it from the nucleotide sequence of cDNAs corresponding to the encoded receptor. This is most efficiently achieved by utilizing oligonucleotide primers synthesized according to partial amino acid sequence of the receptor molecule for the identification of the appropriate cDNA clone from a cDNA library.

2.1. Purification of EGF Receptor by Immunoaffinity Chromatography

We have previously described the generation of various monoclonal antibodies against the EGF receptor (Schreiber *et al.*, 1981, 1983; Schlessinger *et al.*, 1984). One hybridoma cell line, denoted 29.1, secretes IgG_1 antibodies. These antibodies are coupled to Sepharose beads and are used for the purification of EGF receptor by immunoaffinity chromatography. Yarden *et al.* (1984) developed a rapid two-step procedure for the purification of EGF receptor from cultured human A-431 cells. In the first step of purification, detergent-solubilized A-431 cells are immobilized on an immunoaffinity column conjugated with the 29.1-IgG₁ antibody. Subsequently, the immobilized EGF receptor is eluted from the antibody column and then absorbed and specifically eluted from either lectin agarose or Cibachrome blue–Sepharose column. Different elution protocols yield different forms of the EGF receptor.

2.1.1. Purification of an Active Receptor Kinase

Active EGF receptor was recovered from the 29.1-IgG₁ column by rapid elution with an eluant composed of 2 M urea in 50 mM HEPES buffer, pH 7.5, containing 150 mM NaCl, 10% glycerol, 2 mM EGTA, 15 mM $MgCl_2$, 1% aprotinin, 5 μg/ml leupeptin, and 1% Triton X-100. The elution was repeated several times, and the eluants were combined and subsequently dialyzed against 10 mM HEPES buffer, pH 7.5, containing 100 mM NaCl and 0.1% Triton X-100 to remove the urea.

In order to remove minor protein contaminants, the eluted receptor was further purified by lectin chromatography. The eluted receptor was stirred for 1 hr at 4°C with wheat germ agglutinin (WGA) agarose beads. After several washes, the receptor was eluted with buffer containing 0.5 M N-acetylglucosamine, 1 mM $CaCl_2$, and 1 mM $MnCl_2$; SDS-PAGE analysis of the pure receptor revealed only the 170,000-dalton receptor molecule (Fig. 1). The total yield of the two-step purification is approximately 8%. The major loss occurs during the elution from the immunoaffinity column (Yarden *et al.*, 1984).

The eluted pure EGF receptor possesses several activities. It specifically binds [^{125}I]-labeled EGF. Moreover, the binding of EGF activates the tyrosine-specific kinase, leading to autophosphorylation of the receptor molecule. The binding constant of [^{125}I]-EGF to the pure receptor is 2×10^{-8} M^{-1}. This value is five- to tenfold lower than the value measured for the binding constant of [^{125}I]-EGF to intact cells. Similar reduced affinity was also measured for the binding of EGF to membranes prepared from A-431 cells. Like the native EGF receptor (Linsley and Fox, 1980), the pure receptor retains the capacity to covalently cross link [^{125}I]-labeled EGF. The addition of excess EGF inhibits the covalent linkage of the radiolabeled EGF to the receptor molecule (Yarden *et al.*, 1984).

The pure EGF receptor also retains the EGF-induced protein kinase activity that leads to autophosphorylation of the receptor molecule. Figure 2 depicts phosphorylation experiments with the pure receptor and with Triton extracts of A-431 cells. It is well established that the addition of EGF to Triton extracts of A-431 cells results in the phosphorylation of

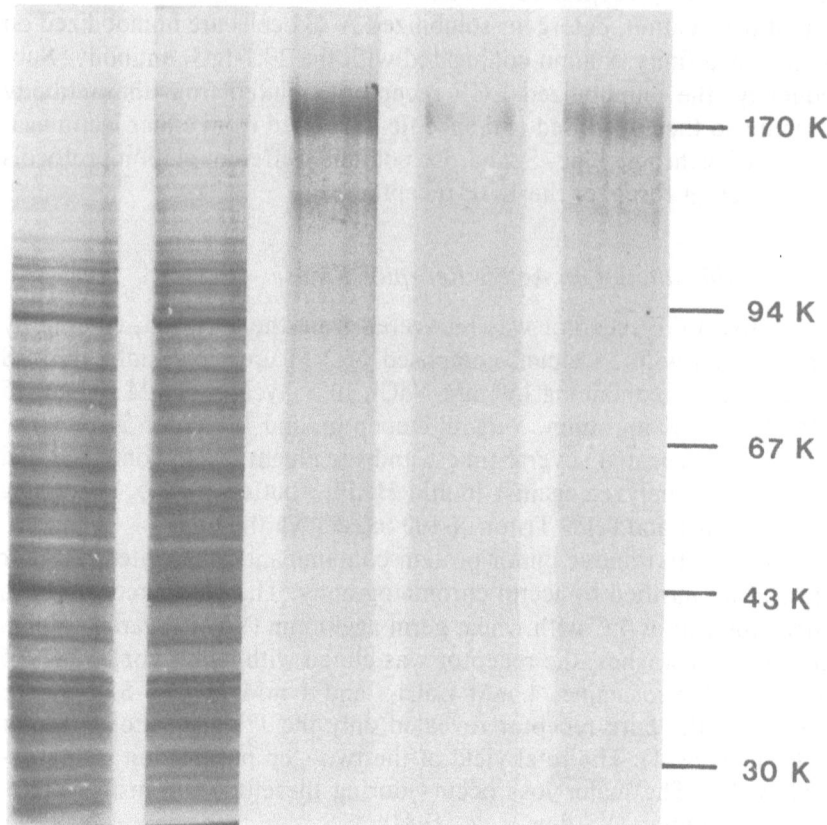

Figure 1. SDS-PAGE analysis of the EGF receptor at different stages of purification. The EGF receptor from A-431 cells was purified by immunoaffinity chromatography with monoclonal antibody, 29.1-IgG$_1$ and then by Cibachrome-blue–Sepharose as described in the text. Coomassie blue stained polyacrylamide gels of different purification stages of EGF receptor. Lane a: Triton extract of A-431 cells. Lane b: A fraction of the extract that was not absorbed on the 29.1-IgG$_1$ Sepharose column. Lane c: Fraction eluted from the column at pH 2.5. Lane d: Fraction eluted from Cibachrome-blue–Sepharose and immediately processed for electrophoresis. Lane e: Cibachrome-blue–Sepharose-eluted material after fourfold concentration by Amicon.

various membrane proteins as well as the autophosphorylation of the 170,000-dalton EGF receptor (Fig. 2). This activity is also retained in the pure receptor molecule: EGF induces, in a dose-dependent manner, the autophosphorylation of the 170,000-dalton receptor molecule (Fig. 2, lanes c and d). Phosphoamino acid analysis of the [^{32}P]-labeled 170,000-dalton protein showed that the phosphorylation occurs exclusively on

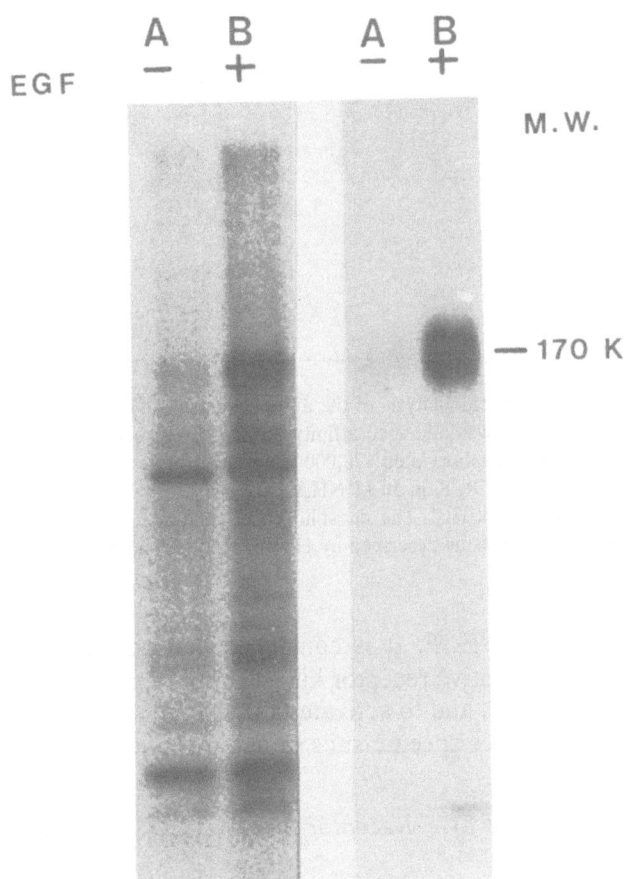

Figure 2. EGF-stimulated protein kinase activity of the affinity-purified EGF receptor. Epidermal growth factor (100 ng/ml) was added to the affinity-purified receptor for 10 min at 4°C together with 3 mM MnCl₂ and 10 μM of [γ-³²P]-ATP (5 μCi). The reaction was terminated by the addition of sample buffer and then analyzed by SDS-PAGE. Left: Autoradiogram of SDS gel of Triton extract of A-431 cells in the absence (lane a) or presence (lane b) of 50 nM EGF. Right: Autoradiogram of SDS gel of pure receptor in the absence (lane c) or presence (lane d) of 50 nM EGF.

tyrosine residues (Fig. 3). Moreover, comparison of the phosphoamino map of the 170,000-dalton phosphoprotein before (basal) and after the addition of EGF indicates that the elevated phosphorylation occurs on the same peptide fragments. A similar phosphopeptide map of the receptor molecule was also observed when a native form of EGF receptor from shed membrane vesicles prepared from A-431 cells was autophosphory-

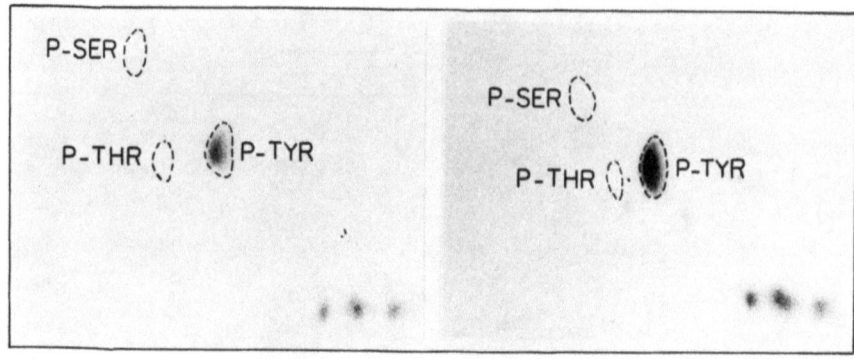

Figure 3. Phosphoamino acid analysis of the autophosphorylated pure receptor. Epidermal growth factor (50 ng/ml) was added to affinity-purified EGF receptor as described in the legend to Fig. 2. The phosphorylated 170,000-dalton band was excised from the gel, trypsinized (50 ng/ml trypsin-TPCK in 50 M NH_4HCO_3, pH 8) for 24 hr and then hydrolyzed by boiling for 2 min in 6 N HCl. The phosphorylated amino acids were analyzed by two-dimensional electrophoresis as described by Hunter and Sefton (1980).

lated in response to EGF. It is concluded that purified EGF receptor kinase is similar to native receptor kinase in its capacity to bind EGF, to cross link [^{125}I]-EGF, and to activate the tyrosine-specific kinase, which in turn phosphorylates specific sites on the receptor molecule.

2.1.2. Purification of EGF Receptor for the Analysis of Its Primary Structure

For the analysis of the primary structure of EGF receptor, it is possible to use an inactive form of the EGF receptor kinase. For this purpose we used an elution protocol that substantially increases the yield of the purification but causes inactivation of the EGF-sensitive kinase. As before, detergent-solubilized A-431 cells were immobilized on a 29.1-IgG$_1$–Sepharose column. The column was then washed with buffer, and the immunoaffinity-purified receptor was phosphorylated by the addition of [γ-^{32}P]-ATP (20 μCi, 5000 Ci/mmole) and 3 mM $MnCl_2$. After 10 min at 4°C, the receptor was eluted with a solution composed of 50 mM glycine-HCl buffer at pH 2.5, 0.1% Triton X-100, 150 mM NaCl, and 10% glycerol. The peak fractions were pooled according to radioactive content.

The eluted receptor appeared as a 170,000-dalton polypeptide on polyacrylamide gels. It retained the capacity to bind various monoclonal antibodies against EGF receptor, and it bound specifically [^{125}I]-labeled EGF. However, it lost both basal and the EGF-induced kinase activity.

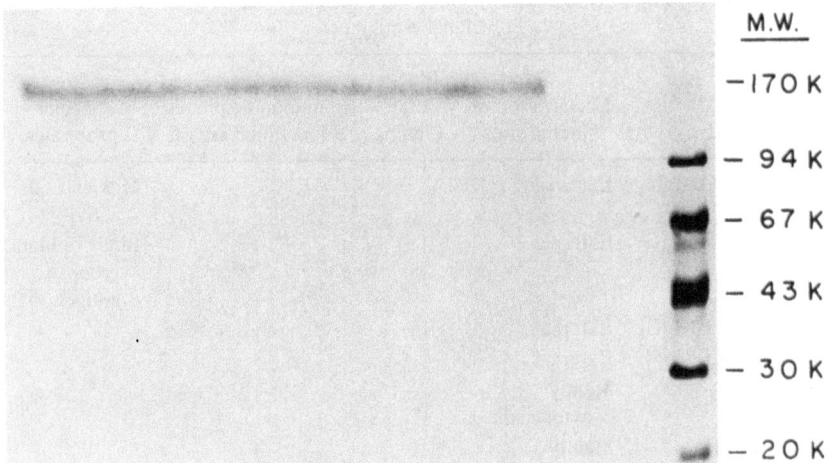

Figure 4. Autoradiogram of preparative gel electrophoresis of immunoaffinity-purified autophosphorylated EGF receptor. The EGF receptor from a Triton extract of A-431 cells was absorbed on 29.1-IgG–Sepharose beads, exposed to $[\gamma\text{-}^{32}P]$-ATP and $McCl_2$, and then eluted as described in the text. Autoradiogram of the unfixed preparative gel is shown, as is a Coomassie-stained lane of molecular weight standard proteins that also contains 2% of the material loaded on the preparative gel.

For the analysis of the primary structure of the EGF receptor molecule (Downward *et al.*, 1984), the eluted receptor was first heated for 3 min at 95°C in sample buffer and then applied on a preparative polyacrylamide gel (3 × 120 × 150 mm). The $[^{32}P]$-labeled receptor was easily located on the unfixed gel (Fig. 4). The receptor was finally electroeluted from the gel and used for the analysis of the primary structure of EGF receptor and as an immunogen (Downward *et al.*, 1984; Yarden *et al.*, 1984).

Using this procedure we can purify approximately 150 µg of receptor from 2×10^9 cultured A-431 cells.

2.2. *Structural Domains on EGF Receptor: Analysis with Immunologic Probes*

Immunologic reagents based on the various monoclonal and polyclonal antibodies generated against the EGF receptor afford the assignment of various structural and functional domains on the receptor molecule.

A repertoire of monoclonal antibodies against the EGF receptor of human epidermoid carcinoma A-431 cells were generated. These anti-

Table I. The Mapping of Antigenic Determinants on EGF Receptor with Monoclonal and Polyclonal Antibodies

Antibody	Localization of binding sites	Blocks EGF binding	Immunoprecipitates a functional kinase	Special properties
Monoclonal 2G2-IgM	External	+	−	Mimics EGF activity
Monoclonal TL5-IgG$_3$	External	−	+	Binds to blood group A antigen
Monoclonal 29.1-IgG$_1$	External	−	+	
Monoclonal I.L. IgM	External	−	+	
Polyclonal R1 antibody	Mainly cytoplasmic	−	+	
Polyclonal R2 antibody	Mainly external	−	+	
Polyclonal anti-*src*-1	Cytoplasmic	−	+	
Polyclonal anti-*src*-2	Cytoplasmic	−	+	Blocks kinase activity

bodies together with various polyclonal antibodies against EGF receptor allow the assignment of various structural and functional domains on the receptor. Table I summarizes the properties of the various antibodies we have generated.

2.2.1. Monoclonal Antibodies against EGF Receptor

We have previously described various monoclonal antibodies against EGF receptor (Schlessinger *et al.*, 1984). The first antibody we developed, denoted 2G2-IgM, binds to EGF receptor of human and mouse cells. This antibody blocked the binding of [^{125}I]-EGF to the receptor and induced various EGF-like effects (Schreiber *et al.*, 1981, 1983; Hapgood *et al.*, 1983). Hence, it acts as a full agonist of EGF, mimicking both its early and delayed effects (Schreiber *et al.*, 1981; Hapgood *et al.*, 1983; Schlessinger *et al.*, 1984). Monovalent Fab' fragments of the 2G2-IgM antibody did not induce DNA synthesis in human fibroblasts (Schreiber *et al.*, 1983). However, the Fab' fragments blocked the binding of [^{125}I]-EGF and activated the EGF-sensitive kinase (Schreiber *et al.*, 1983). The inhibition of EGF binding by the monoclonal antibody 2G2-IgM could result from direct competition of the two ligands for the same binding site. Alternatively, 2G2-IgM and its Fab' fragment could block the binding of

EGF by occupying a domain on the receptor close to binding site for EGF and thereby sterically hinder the binding of the growth factor to the receptor.

Another antibody, TL-5-IgG$_3$ (Table I) reveals another structural feature of the EGF receptor from A-431 cells. TL-5-IgG$_3$ was found to react strongly with the blood group A active glycoprotein derived from an ovarian cyst and from sheep gastric mucus (Gooi *et al.*, 1984). Structurally defined oligosaccharides were used to inhibit the binding of radiolabeled blood group A active glycoprotein to the TL-5-IgG$_3$ antibody (Gooi *et al.*, 1984). Hence, we concluded that the EGF receptor of A-431 cells has the following blood-group-A-like structure:

$$\text{Gal-Nac}\alpha_{1-3}\text{Gal (or Gal-NAc).}$$
$$\underset{\text{Fuc}\alpha}{\pm\; 1,2}$$

The monoclonal antibody TL5-IgG$_3$ does not interfere with the binding of EGF to its receptor, nor does it possess any EGF-like activity. This antibody can be used for the immunoprecipitation of the EGF receptor of A-431 cells (Schlessinger *et al.*, 1984).

Another monoclonal antibody, 29.1-IgG$_1$, also binds to an external domain on the EGF receptor. Like TL5-IgG$_3$, it does not interfere with the binding of EGF to the receptor molecule. Moreover, the binding of 29.1-IgG$_1$ to EGF receptor does not affect the binding of TL5-IgG$_3$ to the receptor, indicating that the two antibodies recognize distinct antigenic determinants. We are currently characterizing the antigenic determinants recognized by this antibody as well as its biological properties. In any event, the 29.1-IgG$_1$ antibody is our best reagent for large-scale purification of EGF receptor using immunoaffinity chromatography. We have recently generated eight more monoclonal antibodies that recognize the external portion of EGF receptor. All of these new antibodies bind to intact A-431 cells. They do not interfere with the binding of EGF to the receptor, and they immunoprecipitate a functional EGF receptor kinase. These new monoclonal antibodies all belong to the IgM class. Further characterization of their properties is currently in progress. The properties of one member of these group of reagents denoted I.L. IgM are described in Table I.

2.2.2. Polyclonal Antibodies against EGF Receptor

We are using two different types of polyclonal antibodies that recognize the EGF receptor. The first antibodies were generated against pure denatured EGF receptor. The EGF receptor from A-431 cells was purified by immunoaffinity chromatography using the monoclonal 29.1-IgG$_1$-Sepharose column. The pure denatured 170,000-dalton polypeptide electroeluted from a preparative polyacrylamide gel was used as an immunogen. Two rabbits were immunized, and two different antisera, respectively denoted anti-EGF-receptor R1 and anti-EGF-receptor R2, were raised (Table I). These two antisera have different properties, although similar reagents and protocols were used for the immunization of the two rabbits. The R1 antiserum does not bind essentially to intact A-431 cells at a dilution of 1:40 of the antiserum (5 mg/ml). However, at this dilution the antiserum precipitates various forms of solubilized receptor and recognizes the receptor molecule in different assays (Lax *et al.*, 1984). Hence, we propose that the R1 antibodies are directed against the cytoplasmic portion of EGF receptor, which became exposed after solubilization with detergents. The properties of R2 antiserum are different. This antiserum binds to intact A-431 cells without interfering with the binding of EGF to the receptor. The R2 antiserum also precipitates a functional EGF receptor kinase (Table I).

The second approach for obtaining antibodies against EGF receptor involves utilization of antibodies generated against synthetic peptides from other proteins that possess functions that are similar to those of EGF receptor and thus may be structurally similar in a local sense. For example, the EGF receptor kinase appears to be similar in many respects to the kinase activity of the transforming protein of Rous sarcoma virus PP60^{src}. It was shown that a synthetic peptide derived from the phosphorylation site of PP60^{src} is phosphorylated by the EGF-sensitive kinase (Pike *et al.*, 1982). Moreover, antibodies against PP60^{src} are phosphorylated by preparations containing EGF receptor from A-431 cells (Chinkers and Cohen, 1981; Kudlow *et al.*, 1981). However, the antibodies against PP60^{src} fail to precipitate the EGF receptor kinase from A-431 and other cells.

In order to further explore this possible homology, we have prepared polyclonal rabbit antibodies against two peptides from PP60^{src} (Fig. 5). The antibodies were prepared against the peptide Arg-Leu-Ile-Glu-Asp-Asn-Glu-Tyr-Thr-Ala-Arg-Gln-Gly-Ala-Lys-Phe-Pro, designated *src* 1, and against the peptide Asn-Arg-Glu-Val-Leu-Asp-Gln-Val-Glu-Arg-Gly-Tyr-Arg-Met-Pro, designated as *src* 2. The *src* 1 contains residues 412–428 of PP60^{src} and corresponds to the phosphorylation site of PP60^{src}. The *src* 2 contains residues 488–502 of PP60^{src}. It was previously reported that this antibody binds to a region that is close to the kinase domain of PP60^{src} (Gentry *et al.*, 1983; Tamura *et al.*, 1983).

We have used several approaches to determine possible specific in-

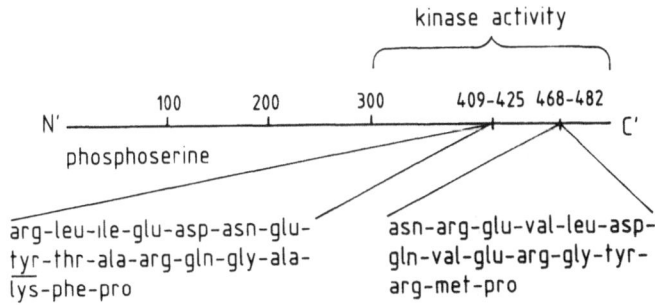

Figure 5. Diagram describing the structure of PP60^{src} and the two synthetic peptides. Polyclonal rabbit antibodies were generated against the *src* 1 and *src* 2 peptides (Lax *et al.*, 1984).

teractions between the EGF receptor kinase and the two antibodies against *src* 1 and *src* 2. Both antibodies recognize the 170,000-dalton receptor polypeptide in "Western" blotting experiments (Lax *et al.*, 1984). Furthermore, both anti-*src*-1 and anti-*src*-2 bind to immobilized EGF receptor and interact specifically with EGF receptor in two sensitive radioimmunoassays (Lax *et al.*, 1984). Whereas anti-*src*-1 immunoprecipitates a funcational EGF receptor kinase (Hortsch *et al.*, 1983; Libermann *et al.*, 1983; Schlessinger *et al.*, 1984), anti-*src*-2 blocks the autophosphorylation of EGF receptor (Lax *et al.*, 1984).

Neither antibody binds to intact cells, but they do recognize various forms of the solubilized receptor. It is concluded that at least two cytoplasmic domains of the EGF receptor are antigenically and presumably also structurally related to specific domains on PP60^{src}.

3. Discussion

A model describing the EGF receptor as an "allosteric receptor" composed of various functional domains was proposed by Schlessinger *et al.* (1983). According to this model, EGF receptor involves various functional domains or sites: (1) the combining site for EGF, which serves as an allosteric regulator of the receptor molecule; (2) the tyrosine-specific, cyclic-nucleotide-independent protein kinase; (3) the locus of phosphorylation sites at tyrosine and possibly at serine residues, which most likely resides in the cytoplasmic portion of the receptor; (4) an aggregation site, which facilitate the receptor–receptor and receptor–coated-pit interaction; (5) a modulation site for the interaction with other molecules known to modulate the binding and activity of EGF [e.g., the tumor pro-

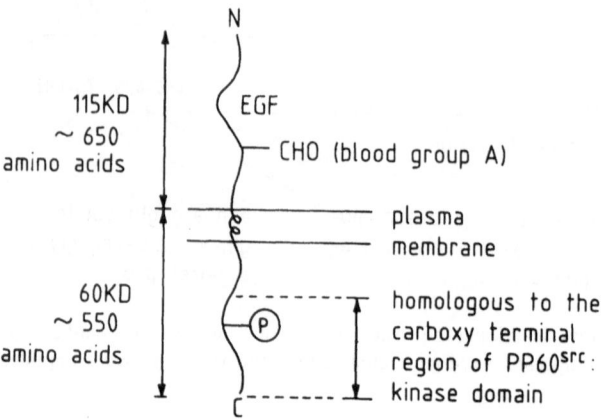

Figure 6. A model for the membrane receptor of EGF of A-431 cells. The external portion of EGF receptor (~115,000 kd) possesses the EGF binding site and a blood group A oligosaccharide. The cytoplasmic portion of EGF receptor possesses the kinase activity and the phosphorylation site(s). It is postulated that the carboxy-terminal tail of the receptor faces the cytoplasm and that a part of this region is homologous to the carboxy-terminal region of the transforming protein encoded by the *src* gene family.

moter phorbol ester (TPA), the growth factor PDGF]; and (6) the site of attachment of EGF receptor-associated carbohydrates.

The immunologic analysis described in this chapter provides novel information concerning the location of the various antigenic determinants on the 170,000-dalton EGF receptor polypeptide. Figure 6 depicts a model for the structure of EGF receptor based on the studies presented in this chapter and studies described by others.

All the monoclonal antibodies described in Table I bind to the external portion of EGF receptor. One of them binds to a blood group A carbohydrate that is attached to EGF receptor. However, the antigenic determinants of EGF receptor recognized by the polyclonal antibody R1 and by the two antibodies generated against the peptides from PP60[src] are located at the cytoplasmic portion of the EGF receptor. Furthermore, the kinase activity stimulated by EGF in intact cells exposed to [γ-^{32}P]-ATP is only observed when the cells are permeabilized. Therefore, we support the view that the kinase domain is part of the cytoplasmic portion of the receptor. Proteolytic cleavage of EGF receptor suggests that the external domain, which contains the EGF binding site and oligosaccharides, has a molecular weight of-115,000 (Mayes and Waterfield, 1984). Thus, the cytoplasmic and transmembrane domains of EGF receptor have an approximate total molecular weight of 60,000. Our results suggest that at least two domains within the cytoplasmic portion of EGF receptor are antigenically related to two domains of the PP60^{src} molecule. One domain is related to the major phosphotyrosine site on PP60^{src}, and the second domain ins either related to the kinase domain of PP60^{src} or can somehow modulate the kinase activity. The inhibition of the autophosphorylation of EGF receptor by an antibody that binds to the kinase domain of PP60^{src} (Gentry *et al.*, 1983; Tamura *et al.*, 1983) suggests that this antibody binds to a domain on the EGF receptor with a similar function. However, the mechanism by which the anti-*src*-2 antibody neutralizes the kinase activity of EGF receptor is not clear.

The two synthetic peptides from PP60^{src} belong to the carboxy-terminal region of this transforming protein. This region is homologous to the transforming proteins encoded by other members of the *src* gene family, *mos*, *erb*-B, *abl*, *yes*, and *fps* (Bishop, 1983). The size of this region is approximately 280 amino acids. On the basis of the analysis presented in this chapter, we would like to propose that the cytoplasmic portion of EGF receptor is the carboxy-terminal region of the 170,000-dalton polypeptide. Moreover, we propose that the EGF receptor also possesses a region that is homologous to the carboxy-terminal domain of the *src* gene family. Furthermore, since other receptors for growth factors such as PDGF receptor (Ek *et al.*, 1982), insulin receptor (Kasuga *et al.*, 1982), and IGF$_1$ receptor (Rubin *et al.*, 1983) are also tyrosine-specific kinases, it is possible that their kinase activity is also located at their cytoplasmic domain (carboxy terminal?) and that it is related to the kinase portion of EGF receptor and to the other kinases encoded by the *src* gene family.

4. Concluding Remarks

We have recently sequenced several peptides that were derived from the EGF receptor from human cells in collaboration with Dr. M. Waterfield. We have shown that each of six peptides derived from the human EGF receptor very closely matched a part of the deduced sequence of the V-*erb*-B transforming protein of avian erythroblastosis virus (AEV) (Downward *et al.*, 1984).

The V-*erb*-B oncogene is a member of the *src* gene family. Our results support the view that the transforming protein encoded by V-*erb*-B acquired the cellular gene sequence of a truncated EGF receptor lacking the external binding domain but retaining the transmembranal domain and a domain involved in stimulating cell proliferation (Downward *et al.*, 1984).

ACKNOWLEDGMENTS. We acknowledge useful discussions with M. Waterfield and A. Ulrich. This work was supported by grants from the National Institutes of Health CA-25820 (J.S.), from the U.S.-Israel Binational Science Foundation (J.S.), and from the Stiftung Volkswagenwerke (J.S.). M.B.E. is a recipient of the Igal Alon Award.

References

Bishop, M. J., 1983, Cell oncogenes and retroviruses, *Annu. Rev. Biochem.* **52**:301–354.

Carpenter, G., and Cohen, S., 1979, Epidermal growth factor, *Annu. Rev. Biochem.* **48**:193–216.

Carpenter, G., King, L., Jr., and Cohen, S., 1978, Epidermal growth factor stimulates phosphorylation in membrane preparation *in vitro*, *Nature* **276**:409–410.

Carpenter, G., King, L., Jr., and Cohen, S., 1979, Rapid enhancement of protein phosphorylation in A-431 cell membrane preparations by epidermal growth factor, *J. Biol. Chem.* **254**:4884–4891.

Chinkers, M., and Cohen, M. S., 1981, Purified EGF receptor-kinase interacts specifically with antibodies to Rous sarcoma virus transforming protein, *Nature* **290**:516–519.

Cohen, S., Carpenter, G., and King, L., Jr., 1980, Epidermal growth factor–receptor protein kinase interactions. Co-purification of receptor and epidermal growth factor-enhanced phosphorylation activity, *J. Biol. Chem.* **255**:4834–4842.

Cohen, S., Hiroshi, U., Christa, S., and Michael, C., 1982, A native 170,000 epidermal growth factor receptor-kinase complex from shed plasma membrane vesicles, *J. Biol. Chem.* **257**:1523–1531.

Collett, M. S., Erikson, E., Purchio, A. F., Brugge, J. S., and Erikson, R. L., 1979, A normal cell protein similar in structure and function to avian sarcoma virus transforming gene product, *Proc. Natl. Acad. Sci. U.S.A.* **76**:3159–3163.

Collett, M. S., Purchio, A. F., and Erikson, R. L., 1980, Avian sarcoma virus transforming protein, PP60^{src}, shows protein kinase activity specific for tyrosine, *Nature* **285**:167–169.

Downward, J., Yarden, Y., Mayes, E., Scrace, G., Totty, N., Stockwell, P., Ullrich, A., Schlessinger, J., and Waterfield, M., 1984, Close similarity of epidermal growth factor receptor and V-*erb*-B oncogene protein sequences, *Nature* **307**:521–526.

Ek, B., Westermark, B., Wasteson, A., and Heldin, C.-H., 1982, Stimulation of tyrosine-specific phosphorylation by plateled-derived growth factor, *Nature* **295**:419–420, 1982.

Gentry, L. E., Rohrschneider, L. R., Casnellie, J. E., and Krebs, E. G., 1983, Antibodies to defined region of PP60^{src} neutralize the tyrosine specific kinase activity, *J. Biol. Chem.* **258**:11219–11228.

Gooi, H. C., Schlessinger, J., Lax, I., Yarden, Y., Libermann, T. A., and Feizi, T., 1983, Monoclonal antibody reactive with the human epidermal growth factor receptor recognizes the blood group A antigen, *Biosci. Rep.* **3**:1045–1052.

Hapgood, J., Liberman, T. A., Lax, I., Yarden, Y., Schreiber, A. B., Naor, Z., and Schlessinger, J., 1983, Monoclonal antibodies against EGF receptor induce prolactin synthesis in cultured rat pituitary cells (GH$_3$), *Proc. Natl. Acad. Sci. U.S.A.* **80**:6451–6455.

Hortsch, M., Schlessinger, J., Gootwine, E., and Webb, C. G., 1983, appearance of a functional EGF-receptor kinase during rodent embryogenesis, *EMBO J.* **2**:1937–1941.

Hunter, T., and Sefton, B. M., 1980, The transforming gene product of Rous sarcoma virus phosphorylates tyrosine, *Proc. Natl. Acad. Sci. U.S.A.* **77**:1311–1315.

Kasuga, M., Zick, Y., Blithe, D. L., Crettaz, M., and Kahan, C. R., 1982, Insulin stimulates tyrosine phosphorylation of insulin receptor in a cell free system, *Nature* **298**:667–669.

Kudlow, J. E., Buss, J. E., and Gill, G. N., 1981, Anti-PP60^{src} antibodies are substrates for EGF-stimulated protein kinase, *Nature* **290**:519–521.

Laemmli, U. K., 1970, Cleavage of structural proteins during the assembly of the head of bacteriophage T4, *Nature* **227**:680–683.

Lax, I., Bar-Eli, M., Yarden, Y., Libermann, T. A., and Schlessinger, J., 1984, Antibodies to two defined regions on PP60^{src} interact specifically with the EGF receptor kinase system, *Proc. Natl. Acad. Sci. U.S.A.* **81**:5911–5915.

Levinson, A. D., Oppermann, H., Varmus, H. E., and Bishop, J. M., 1980, The purified product of the transforming gene of avian sarcoma virus phosphorylates tyrosine *J. Biol. Chem.* **255**:11973–11980.

Libermann, T. A., Razon, N., Bartal, A. D., Yarden, Y., Schlessinger, J., and Soreq, M., 1984, Expression of epidermal growth factor receptors in human brain tumors, *Cancer Res.* **44**:753–760.

Linsley, P. S., and Fox, C. F., 1980, Direct linkage of EGF to its receptor. Characterization and biological activity, *J. Supramol. Struct.* **14**:441–.

Mayes, E. L. V., and Waterfield, M. D., 1984, Biosynthesis of epidermal growth factor receptors in A-431 cells, *EMBO J.* **3**:531–537.

Rubin, J. B., Shia, M. A., and Pilch, P. F., 1983, Stimulation of tyrosine specific phosphorylation *in vitro* by insulin like growth factor 1, *Nature* **305**:438–440.

Savage, C. R., Jr., and Cohen, S., 1972, Epidermal growth factor and a new derivative, *J. Biol. Chem.* **247**:7609–7611.

Schlessinger, J., Schreiber, A. B., Levi, A., Lax, I., Libermann, T., and Yarden, Y., 1983, Regulation of cell proliferation by epidermal growth factor, *CRC Crit. Rev. Biochem.* **14**:93–111.

Schlessinger, J., Lax, I., Yarden, Y., Kanety, H., and Libermann, T. A., 1984, Monoclonal antibodies against the membrane receptor for epidermal growth factor: A versatile for structural and mechanistic studies, in: *Receptors and Recognition* (M. Greaves, ed.), Chapman and Hall, London **17**:279–303.

Schreiber, A. B., Yarden, Y., and Schlessinger, J., 1981, A non-mitogenic analogue of epidermal growth factor enhances the phosphorylation of endogenous membrane proteins, *Biochem. Biophys. Res. Commun.* **101**:517–523.

Schreiber, A. B., Liberman, T. A., Lax, I., Yarden, Y., and Schlessinger, J., 1983, Biological role of EGF receptor clustering: Investigation with monoclonal anti-EGF receptor antibodies *J. Biol. Chem.* **258**:846–853.

Snyder, M. A., Bishop, J. M., Colby, W. W., and Levinson, A. D., 1983, Phosphorylation of tyrosine 416 is not required for the transforming properties and kinase activity of PP60^{src}, *Cell* **32**:891–901.

Tamura, T., Bauer, H., Birr, C., and Pipkorn, R., 1983, Antibodies against synthetic peptides as a tool for functional analysis of the transforming protein PP60^{src}, *Cell* **34**:587–596.

Ushiro, H., and Cohen, S., 1980, Identification of phosphotyrosine as a product of epidermal growth factor-activated protein kinase in A-431 cell membranes, *J. Biol. Chem.* **255**:8363–8365.

Yarden, Y., Schreiber, A. B., and Schlessinger, J., 1982, A non-mitogenic analogue of EGF induces the early responses mediated by EGF, *J. Cell Biol.* **92**:687–693.

Yarden, Y., Harari, I., and Schlessinger, J., 1985, Purification of EGF receptor from A-431 cells by immunoaffinity chromatography with monoclonal antibody against EGF receptor, *J. Biol. Chem.* **260**:315–319.

4

The Insulin Receptor as a Tyrosine-Specific Protein Kinase

LILLI M. PETRUZZELLI, LAUREL STADTMAUER,
ROMAN HERRERA, MARY MAKOWSKE,
SABYASACHI GANGULY, DIANE TABARINI, HUA LEE,
YETUNDE OLOWE, and ORA M. ROSEN

1. Introduction

Despite detailed understanding of the structure, biosynthesis, and metabolic effects of insulin, it is not known how the interaction of insulin with its receptor generates a biological signal (Cobb and Rosen, 1984). Evidence for mediation of insulin action by a small peptide second messenger has been presented (Larner *et al.*, 1978; Seals and Jarrett, 1980; Seals and Czech, 1980). Recently, however, the discovery that the insulin receptor is associated with an insulin-dependent protein kinase activity (Kasuga *et al.*, 1982a,b,c; Petruzzelli *et al.*, 1982; Van Obberghen and Kowalski, 1982; Haring *et al.*, 1982; Zick *et al.*, 1983a,b; Avruch *et al.*, 1982; Shia and Pilch, 1983; Van Obberghen *et al.*, 1983; Kasuga *et al.*, 1983; Rosen *et al.*, 1983) has suggested additional or alternative models for transduction of the insulin signal (for review, see Cobb and Rosen, 1984). In this chapter we present (1) our evidence that the insulin receptor is itself an insulin-dependent tyrosine protein kinase; (2) the properties of the enzyme derived from an insulin-sensitive cell line, 3T3-L1, and from human placenta; and (3) possible roles of the insulin-dependent protein kinase in bioregulation.

LILLI M. PETRUZZELLI, LAUREL STADTMAUER, ROMAN HERRERA, MARY MAKOWSKE, SABYASACHI GANGULY, DIANE TABARINI, HUA LEE, YETUNDE OLOWE, and ORA M. ROSEN • Department of Molecular Pharmacology, Albert Einstein College of Medicine, Bronx, New York 10461.

2. Development of a Monoclonal Antibody to the Human Placental Insulin Receptor

Antibodies to the insulin receptor have been invaluable reagents for identifying and characterizing this molecule. To facilitate our studies, we developed monoclonal antibodies by immunizing mice with partially purified preparations of human placental receptor. SJL/J mice were immunized with wheat germ agglutinin–agarose eluates of placental extracts, and titers were assessed by ability to immunoprecipitate $[^{125}I]$-insulin receptor complexes in the presence of antimouse IgG (see Section 2). The spleen from the mouse with the highest titer was fused with the NSO myeloma cell line (obtained through the courtesy of Dr. Cesar Millstein), and a small portion of the splenocytes was used to repopulate the spleens of three irradiated SJL/J mice (Fox *et al.*, 1984). The latter were boosted immediately and again 72 hr later with the wheat germ agglutinin eluate, and 7 days after the repopulation their spleens were removed and fused with NSO cells. The hybridomas from the initial and repopulation fusions were grown initially in medium containing hypoxanthine aminopterm, and thymidine (HAT) with 10% fetal calf serum (FCS) and then transferred to regular Dulbecco's minimum essential (DME) medium with 10% FCS.

The culture media from 800 clones were assayed by direct immunoprecipitation as follows. A 2% Triton X-100 extract of placental membranes (approximately 100 fmole insulin-binding activity per assay) was incubated with $[^{125}I]$-insulin in HEPES buffer, pH 7.9, for 1 hr at 24°C. Tubes were then chilled on ice, and 100-μl aliquots of hybridoma supernatants were added and incubated with the receptor–insulin complex for 16 hr at 4°C. Normal mouse serum (1 μl) was then added to each tube followed by 60 μl sheep antimouse antibody. After a 2-hr incubation at 4°C, the tubes were centrifuged at 3000 rpm for 10 min at 4°C, and the immunoprecipitates were washed once with 1 ml of 100 mM Tris-HCl buffer, pH 7.4, containing 0.1% bovine serum albumin (BSA) and 0.1% Triton X-100. The pellets were then assayed for ^{125}I. Six positive clones survived all of the necessary manipulations. Of these, five interreacted with a number of placental membrane glycoproteins including the insulin receptor and may have been directed against a common carbohydrate moiety.

One clone, CII 25.3, derived from the repopulation fusion, immunoprecipitated only the insulin receptor. It was subcloned twice and then used to induce ascites in Balb/C × SJL/J F_1 hybrids. The CII 25.3 IgG_1 was purified by chromatography on protein A Sepharose. It exhibited the following properties: (1) At concentrations as high as 25 μg/ml, it had no effect on either insulin binding or the insulin-dependent protein kinase

Table I. Specificity of the Anti-insulin Receptor Monoclonal Antibody CII 25.3

Source of receptor	Total > ^{125}I < insulin-receptor complex (cpm)	Immunoprecipitated complex (cpm)	Specific immuno-precipitation (%)
Human placenta	37,196	39,555	106
Human IM 9 lymphocytes	14,872	13,047	88
Monkey (*Macaca fascicularis*) liver	10,531	31,075	295
Mouse 3T3-L1 adipocytes	17,177	275	2
Rat liver	11,900	110	1

The initial incubation mixture (200 μl) contained 20 μl of a 2% Triton X-100 extract of membranes from each source plus > ^{125}I < insulin (100 fmoles, 100,000 cpm). After 1 h at 24°, antibody (6 μg) was added in the presence or absence of 10 μg unlabelled insulin and incubated at 4° for 16 h. Sheep antimouse IgG (60 μl) was then added to each tube (along with 1 μl normal mouse serum) and the incubation was continued for 2 h at 4°. The immunoprecipitates were collected by centrifugation at 3000 × g for 10 min at 4°, washed with 1 ml of 100 mM Tris HCl buffer pH 7.5 containing 0.1% BSA and 0.1% Triton X-100 and assayed for radioactivity. The values presented were obtained after subtracting the radioactivity found in the control tubes to which excess insulin had been added. Total insulin receptor complex was determined by polyethylene glycol precipitation as described by Siegel *et al.* (1981).

activity of the human placental receptor. In fact, insulin-dependent tyrosine protein kinase activity could be demonstrated directly in immunoprecipitates or on columns composed of antibody coupled to the succinylated derivative of diaminodipropylamino agarose. (2) It was species specific, reacting with human and Old World monkey insulin receptor but not rodent receptor (See Table I). Preliminary data derived from direct immunoprecipitation and from a competitive displacement radioimmunoassay with [^{32}P]-labeled human placental insulin receptor indicate significant tissue-specific differences in the antigenic properties of the insulin receptor of the Old World monkey *Macaca fascicularis* (unpublished data). The potential of correlating insulin-binding activity, protein kinase activity, and antigenic properties of the insulin receptor of primate tissues with the structural information that can be gained from immunoprecipitation of the receptors of these tissues may provide new insights into tissue-specific insulin responsiveness.

3. Purification of the Insulin-Dependent Protein Kinase from Human Placenta

Central to the study of insulin-stimulated protein phosphorylation is the question of the relationship between the kinase and the insulin-binding activity of the receptor. Several lines of evidence suggest that both activities reside in the same molecule (Roth and Cassell, 1983; Machicao *et al.*, 1982, Zick *et al.*, 1983; Shia and Pilch, 1983; Van Obberghen *et al.*, 1983; Kasuga *et al.*, 1983), but the most direct approach to the answer, short of molecular cloning, is to assay the protein kinase activity of the purified receptor. We have previously reported a procedure for purifying the insulin-binding activity of the human insulin receptor (Siegel *et al.*, 1981). The final step involved elution of the receptor from an insulin–sepharose resin with 4.5 M urea at pH 6.3. This eluate had no protein kinase activity. The affinity chromatography step was then reevaluated in an effort to preserve the receptor's putative kinase activity, and the following critical changes were made. Elution was performed with 0.15 M sodium acetate buffer, pH 5.5, containing 0.5 M NaCl, 10% glycerol, 1 mM dithiothreitol (DTT), and 0.1% Triton X-100 (Petruzzelli *et al.*, 1984). Under these conditions, the insulin-dependent tyrosine protein kinase activity copurified precisely with the insulin-binding activity of the receptor. The preparation appeared homogeneous after SDS-PAGE. The silver-stained gel showed only two bands (after reduction) of M_r 95,000 and M_r 135,000, and both the binding and kinase activities of the intact receptor were completely immunoprecipitated by the monoclonal antibody CII 25.3 The overall recovery of these activities of the receptor after the purification procedure was 20–25%.

The essential features of the revised procedure are the elimination of urea and elution from the affinity resin at pH 5.5 in the presence of 1 mM DTT and 0.5 M NaCl. Elution in the buffer outlined above but in the absence of DTT led to equivalent recovery of the insulin-binding activity but only 10% of the expected protein kinase activity. The DTT, under these conditions, did not result in dissociation of either the insulin or the receptor. Copurification to apparent homogeneity, immunoprecipitation of both activities of the homogeneous oligomer by a monoclonal antibody, and evidence that the self-phosphorylation of the receptor occurs by an intramolecular reaction mechanism (see Section 6) support the proposition that the insulin-dependent tyrosine protein kinase activity is intrinsic to the receptor molecule or, to put it otherwise, that the insulin receptor is an insulin-dependent protein kinase.

4. Properties of the Purified Human Placental Insulin-Dependent Protein Kinase

The physical properties of the insulin receptor purified to preserve the protein kinase activity (Petruzzelli *et al.*, 1982) appear to be the same as those previously ascribed to the purified insulin-binding activity from this tissue (Siegel *et al.*, 1981). A minimum oligomeric molecular weight was estimated to be 350,000, and two types of subunits, M_r 95,000 and M_r 135,000 were evident following disulfide reduction. As previously established for this and other tissue sources, insulin binds to the 135,000-dalton subunit, and the 95,000-dalton β subunit is a substrate for the insulin-dependent protein kinase (see Cobb and Rosen, 1984). The homogeneous receptor–kinase has the following characteristics. (1) It has one high-affinity insulin binding site per molecule (Petruzzelli *et al.*, 1984). (2) The K_d for the insulin receptor complex and the K_a for insulin activation of the insulin-dependent protein kinase are both 5 nM (Petruzzelli *et al.*, 1984). (3) The optimum conditions for assaying the protein kinase activity include the presence of 2 mM Mn^{2+}, 12 mM Mg^{2+}, and 100 μM ATP. (4) The protein kinase domain of the oligomer is more sensitive to heat and to N-ethylmaleimide (NEM) than the insulin-binding domain (L. Stadtmauer and R. Herrera, unpublished observations). The molecule heated at 45°C for 2 min retains more than 80% of its insulin-binding activity and only 50% of its protein kinase activity. Similarly, incubation with 10 mM NEM for 30 min at 23°C inactivates the kinase but has no effect on insulin-binding capacity.

An important property of the purified kinase is that it phosphorylates the β subunit of the receptor as well as exogenous substrates (see Section 6) exlusively on tyrosine. This is a feature predictable from studies on the impure insulin-dependent protein kinase (Petruzzelli *et al.*, 1981), which also exhibits this specificity. It contrasts with the observations in intact cells (Kasuga *et al.*, 1982a) that insulin promotes phosphorylation of the β subunit of the receptor predominantly on serine residues and only to a minor extent on tyrosine. This potentially interesting inconsistency suggests either that the insulin-dependent protein kinase assayed *in vitro* is missing a component(s) of the reaction, that the relative abundances of phosphotyrosine and phosphoserine on the isolated receptor reflect an artifact introduced during cell breakage, or, perhaps most likely, that a serine protein kinase distinct from the receptor recognizes either the insulin–receptor complex of the self-phosphorylated receptor.

5. Substrates for the Insulin-Dependent Protein Kinase

Apart from the β subunit of the receptor itself, the endogenous substrates for the insulin-dependent protein kinase have not yet been identified. Studies using exogenous substrates with partially purified insulin-dependent protein kinase isolated from cultured 3T3-L1 adipocytes and the kinase purified from human placenta are summarized below. In all cases, phosphorylation is exclusively on tyrosine residues.

5.1. Peptides

Four Tyr-containing peptides can serve as substrates (Stadtmauer and Rosen, 1983): angiotensin II, a synthetic peptide containing the autophosphorylation site of pp60[src], angiotensin III inhibitor, and gastrin. In each case, the principal effect of insulin is to increase the V_{max} of the reaction. The V_{max} of the purified placental kinase for angiotensin II is 80 nmole/min per mg protein, assayed in the absence of DTT and at 23°C (Petruzzelli et al., 1984). The K_m of the kinase for the first three peptide substrates is between 1 and 8 mM; for gastrin it is between 100 and 200 μM. Although no critical substrate sequence can be identified from such a limited analysis, the presence of acidic amino acid residues to the N-terminal side of the tyrosine appears to improve the affinity for the substrate.

5.2. Proteins

Anti-src IgG (Stadtmauer and Rosen, 1983), casein αS, histone H2b (Petruzzelli et al., 1982; Stadtmauer and Rosen, 1983), and phosphoglyceromutase can serve as substrates. The K_ms for casein and histone are in the micromolar range, whereas the K_ms for the peptides cited above are millimolar. This may reflect a requirement for protein structure beyond the sequence immediately surrounding the phosphotyrosine.

It is evident that the amino acid sequence of a "physiological" substrate for the insulin-dependent protein kinase, e.g., the amino acid sequence of the phosphopeptides of the β subunit of the receptor, may be more illuminating than an investigation of model substrates for which the enzyme has poor affinity.

6. Properties of the Phosphoreceptor: Consequences of Reversible Autophosphorylation

The purified insulin-dependent protein kinase incorporates approximatley 2 moles of phosphate per mole β subunit (Petruzzelli et al., 1984). Two principal [^{32}P]-labeled tryptic phosphopeptides have been purified by high-performance liquid chromatography (HPLC). The peptides phosphorylated in the presence or absence of insulin behave identically on

HPLC, providing evidence that the activation of β subunit phosphorylation by insulin is not attributable to the phosphorylation of unique sites (Petruzzelli *et al.*, 1984).

In dilute solution, the initial rate of phosphorylation of the β subunit by the insulin-dependent protein kinase is independent of receptor concentration, indicating that self-phosphorylation occurs by an intramolecular mechanism (Petruzzelli *et al.*, 1984). The proximity of the two tyrosine sites on the β subunit is not known, so it is premature to conjecture about the difficulties inherent in a multisite intramolecular reaction. Whether or not the phosphorylation of the β subunit can occur by an intermolecular route remains unanswered. This is similar in concept to asking whether one type of receptor can phosphorylate another, a subject of considerable biological interest.

Since the insulin receptor can be phosphorylated stoichiometrically *in vitro*, one can compare the properties of the phosphorylated and nonphosphorylated oligomers. Insulin binds to both phospho- and nonphosphoreceptor with the same association constant and dissociates at the same rate from both types of receptor. The dissociation constant calculated from a Scatchard analysis is about 5 nM for both (R. Herrera, unpublished observations). Phosphorylation does, however, influence the properties of the protein kinase activity of the receptor. It both activates the kinase and renders it insulin independent (Rosen *et al.*, 1983). Dependence on insulin is restored by dephosphorylation with bacterial alkaline phosphatase. It is not known whether phosphorylation alters the properties of this kinase in intact cells. If it does, there are two potentially important implications. First, activated phosphorylated receptor could phosphorylate and thereby activate adjacent unoccupied receptors, initiating a chain of kinase activation with only limited intial hormone occupancy. Second, reversible kinase inactivation might be regulated by a tyrosine-specific (or receptor-specific) phosphoprotein phosphatase rather than, or in addition to, dissociation of hormone. In fact, a phosphatase has been identified in placental membranes that dephosphorylates the tyrosine residues of the epidermal growth factor receptor of A431 cells, the insulin receptor, and histone phosphorylated on tyrosine residues by the insulin-dependent protein kinase. The activity is severalfold more active on phosphotyrosine than on phosphoserine substrates. Studies are now in progress to fully purify this activity and establish whether or not it is identical to the well-studied membrane-associated placental alkaline phosphatase. A distinct tyrosine-specific phosphoprotein phosphatase has also been detected in placental cytosol. Since it is likely that the phosphorylation sites on the β subunit are accessible to the cytosol of the cell, physiologically relevant phosphatases might be present in either or both cellular compartments.

The potential significance of autophosphorylation in receptor clustering, internalization, and turnover remains to be explored.

7. Insulin-Promoted Phosphorylation in 3T3-L1 Adipocytes

The availability of a cultured insulin-sensitive cell line such as the 3T3-L1 adipocytes (Green and Kehunde, 1974) permits one to consider physiological questions that are more difficult to study in tissues. Petinent to this discussion are the approaches that can be taken to find out which proteins are phosphorylated when intact cells are treated with insulin and whether these proteins are phosphorylated directly by the insulin receptor kinase or indirectly by other protein kinases that are themselves activated directly or indirectly by the insulin-activated kinase. When differentiated 3T3-L1 adipocytes are incubated with ^{32}p for 1 hr, treated with insulin for 2–15 min, and finally homogenized and subjected to SDS-PAGE, at least six proteins become phosphorylated (relative to control's receiving no insulin).

The first is the insulin receptor itself. Two points about this phosphorylation should be reiterated: (1) most of the ^{32}P found in the β subunit labeled in intact cells is on seryl residues and not tyrosine; (2) this protein is not sufficiently prevalent to be detectable unless specifically immunoprecipitated from detergent extracts of membranes by antireceptor antibody. Both points suggest that critical phosphoproteins related to insulin action may be present in low concentration, difficult to resolve, phosphorylated on amino acid residues in addition to tyrosine, or impossible to detect without preliminary purification and/or specific immunoprecipitation. In the future, it may be more profitable to make some educated guesses about likely substrates for insulin-dependent phosphorylation and proceed with specific quests than to scan whole-cell lysates for [^{32}P]-labeled proteins using resolving techniques such as two-dimensional PAGE.

The second protein, ATP citrate lyase, is phosphorylated on serine in response to either insulin or isoproterenol (Swergold et al., 1982). Phosphorylation has no reported effects on enzyme activity and may reflect the fortuitous availability of an amino acid sequence that can be recognized by the cAMP-dependent protein kinase as well as one or more cAMP-independent protein kinases.

By far the most dramatic phosphorylation induced by insulin is on serine residues of ribosomal protein S6 (Smith et al., 1979, 1980). The effect occurs rapidly with nanomolar concentrations of insulin, is exclusively on seryl residues, and is mimicked by polyclonal antireceptor antibody. An effect of S6 phosphorylation on ribosomal function has been observed (Burkhard and Traugh, 1983), and an S6 protein kinase activity that is enhanced by exposing intact cells to insulin has now been detected by two laboratories (Cobb and Rosen, 1983; Perisic and Traugh, 1983); for review see Cobb and Rosen, 1984).

Three additional soluble proteins have recently been detected. one has a M_r of 15,000, and although its identity is unknown, it is clear that it is not the fatty acid binding protein of similar molecular weight that is induced during differentiation of 3T3-L1 cells (Spiegelman *et al.*, 1983). Another protein has a M_r of about 68,000. The phosphate incorporated into the latter two proteins is alkali labile and therefore probably not on tyrosyl residues. The other soluble protein has a M_r of 23,000 and may be the same as the protein reported by Blackshear *et al.*, (1982, 1983) and by Ramakrishna and Benjamin (1983). Interestingly, this protein remains soluble after heating at 80°C for 10 min, and its phosphate is alkali stable. Phosphoamino acid analysis, however, shows both phosphoserine and phosphothreonine but no phosphotyrosine. Phosphorylation of the insulin receptor, ATP citrate lyase, and S6 in response to insulin is not peculiar to 3T3-L1 cells; it is not known whether this is also the case for the phosphoproteins of M_r 15,000, M_r 23,000, and M_r 68,000. Purification of these proteins is under way in an effort to identify them and the protein kinases that catalyze their phosphorylation. With the exception of the β subunit of the insulin receptor, it is likely that none of the proteins cited is a direct substrate for the insulin-dependent tyrosine protein kinase, although the protein kinases that phosphorylate them might be.

8. Concluding Remarks

The discovery that insulin promotes the phosphorylation of its own receptor in intact cells and that some of this phosphorylation is on tyrosine residues placed insulin in the company of another, structurally dissimilar, growth factor, epidermal growth factor, and also a family of retroviruses whose transforming proteins are tyrosine protein kinases. The list of growth factors able to trigger tyrosine protein kinase activity now includes platelet-derived growth factor, insulinlike growth factor I, and transforming growth factors as well as epidermal growth factor, and insulin.

Much remains to be understood about the relationship between insulin and protein phosphorylation. We know the following:

1. The insulin receptor is a bifunctional oligomer that binds insulin and possesses insulin-dependent tyrosine protein kinase activity.
2. The purified kinase can phosphorylate tyrosine residues on exogenous peptide and protein substrates.
3. Phosphorylation of the β subunit of the receptor occurs by an intramolecular reaction with a stoichiometry of two phosphates per β chain.
4. Insulin promotes the phosphorylation of at least six proteins in insulin-sensitive 3T3-L1 cells. With the exception of the receptor itself, phosphorylation of these proteins is probably not a simple, direct consequence of activation of the cell surface insulin-dependent tyrosine protein kinase.
5. Phosphorylation of the β subunit of the insulin receptor *in vitro* has no effect on insulin binding but activates the kinase in the absence of insulin.

The following critical issues remain unanswered:

1. The nature of the enzyme that catalyzes the phosphorylation of the insulin receptor on serine residues *in vivo*.
2. The functional consequences of receptor phosphorylation *in vivo*.
3. The basis for substrate specificity of the insulin-dependent protein kinase.
4. The identities of the endogenous substrates for the insulin dependent protein kinase and the relationship between these targets and the indirect effects of insulin on serine and threonine phosphorylation *in vivo*.
5. The relationship between the various growth factor and viral tyrosine protein kinases.

Although it is our working hypothesis that the tyrosine protein kinase activity of the insulin receptor is biologically meaningful, an unambiguous role for tyrosine protein kinase activity in receptor function or insulin action has yet to be documented.

ACKNOWLEDGMENTS. We are greatly indebted to Ms. Ellyn Fischberg and Dr. M. D. Scharff for teaching us how to make monoclonal antibodies. This research has been supported in part by Grants NIH 2R01-GM-29042-04, 5R01-AM-09038, T32-GM7288, and ACS BC-12N.

References

Avruch, J., Nemenoff, R. A., Blackshear, P. J., Pierce, M. W., and Osathanondh, R., 1982, Insulin-stimulated tyrosine phosphorylation of the insulin receptor in detergent extracts of human placental membranes, *J. Biol. Chem.* **257**:15162–15166.

Blackshear, P. J., Nemenoff, R. A., and Avruch, J., 1982, Preliminary characterization of a heat-stable protein from rat adipose tissue whose phosphorylation is stimulatd by insulin, *Biochem. J.* **204:**817-824.

Blackshear, P. J., Nemenoff, R. A., and Avruch, J., 1983, Insulin and growth factors stimulate the phosphorylation of a M_r-22,000 protein in 3T3-L1 adipocytes, *Biochem. J.* **214:**11-19.

Burkhard, S. J., and Traugh, J. A., 1983, Changes in ribosome function by cAMP-dependent and cAMP-independent phosphorylation of ribosomal protein S6, *J. Biol. Chem.* **258:**14003-14008.

Cobb, M. H., and Rosen, O. M., 1983, Description of a protein kinase derived from insulin-treated 3T3-L1 cells that catalyzes the phosphorylation of ribosomal protein S6 and casein, *J. Biol. Chem.* **258:**12472-12481.

Cobb, M. H., and Rosen, O. M., 1984, The insulin receptor tyrosine protein kinase, *Biochim. Biophys. Acta.* **728:**1-8.

Fox, P. C., Berenstein, E. H., and Siraganian, R. P., 1984, Enhancing the frequency of antigen-specific hybridomas, *Eur. J. Immunol.* **11:**431-434.

Green, H., and Kehinde, O., 1974, Sublines of mouse 3T3 cells that accumulate lipid, *Cell* **1:**113-116.

Haring, H. U., Kasuga, M., and Kahn, C. R., 1982, Insulin receptor phosphorylation in initial adipocytes and in a cell-free system, *Biochem. Biophys. Res. Comm.* **108:**1538-1545.

Kasuga, M., Karlsson, F. A., and Kahn, C. R., 1982a, Insulin stimulates the phosphorylation of the 95,000-dalton subunit of its own receptor, *Science* **215:**185-187.

Kasuga, M., Zick, Y., Blithe, D. L., Karlsson, F. A., Haring, H. U., and Kahn, C. R., 1982b, Insulin stimulation of phosphorylation of the β subunit of the insulin receptor, *J. Biol. Chem.* **257:**9891-9894.

Kasuga, M., Zick, Y., Blithe, D. L., Crettaz, M., and Kahn, C. R., 1982c, Insulin stimulates tyrosine phosphorylation of the insulin receptor in a cell-free system, *Nature* **298:**667-669.

Kasuga, M., Fujita-Yamaguchi, Y., Blithe, D. L., and Kahn, C. R., 1983, Tyrosine-specific protein kinase activity is associated with the purified insulin receptor, *Proc. Natl. Acad. Sci. U.S.A.* **80:**2137-2141.

Larner, J., Lawrence, J. C., Walkenbach, R. J., Roach, P. J., Hazen, R. J., and Huange, L. C., 1978, Insulin control of glycogen synthesis, *Adv. Cyclic Nucleotide Res.* **9:**425-439.

Machicao, F. Urumow, T., and Wieland, O. H., 1982, Phosphorylation–dephosphorylation of purified insulin receptor from human placenta, *FEBS Lett.* **149:**96-100.

Peristic, O., and Traugh, J. A., 1983, Protease-activated kinase II mediates multiple phosphorylation of ribosomal protein S6 in reticulocytes, *J. Biol. Chem.* **258:**13998-14002.

Petruzzelli, L., Herrera, R., and Rosen, O. M., 1984, The insulin receptor is an insulin-dependent tyrosine protein kinase: Copurification of insulin binding and protein kinase activities to homogeneity from human placenta, *Proc. Natl. Acad, Sci. U.S.A.* **81:**3327-3331.

Petruzzelli, L. M., Ganguly, S., Smith, C. J., Cobb, M. H., Rubin, C. S., and Rosen, O. M., 1982, Insulin activates a tyrosine-specific protein kinase in extracts of 3T3-L1 adipocytes and human placenta, *Proc. Natl. Acad. Sci. U.S.A.* **79:**6792-6796.

Ramakrishna, S., and Benjamin, W. B., 1983, Insulin stimulates phosphorylation of a heat-stable protein in rat adipose tissue, *Biochem. Biophys. Res. Commun.* **117:**758-764.

Rosen, O. M., Herrera, R., Olowe, Y., Petruzzelli, L. M., and Cobb, M. H., 1983, Phosphorylation activates the insulin receptor tyrosine protein kinase, *Proc. Natl. Acad. Sci. U.S.A.* **80:**3237-3240.

Roth, R. A., and Cassell, D. J., 1983, Insulin receptor: Evidence that it is a protein kinase, *Science* **219**:299–301.

Seals, J. R., and Jarrett, L., 1980, Activation of pyruvate dehydrogenase by direct addition of insulin to an isolated plasma membrane/mitochondria mixture: Evidence for generation of insulin's second messenger in a subcellular system, *Proc. Natl. Acad. Sci. U.S.A.* **77**:77–81.

Seals, J. R., and Czech, M., 1980, Evidence that insulin activates an intrinsic plasma membrane protease in generating a secondary chemical mediator, *J. Biol. Chem.* **255**:6529–6531.

Shia, M. A., and Pilch, P. F., 1983, The β subunit of the insulin receptor is an insulin-activated protein kinase, *Biochemistry* **22**:717–721.

Siegel, T. W., Ganguly, S., Jacobs, S., Rosen, O. M. and Rubin, C. S., 1981, Purification and properties of the human placental insulin receptor, *J. Biol. Chem.* **256**:9266–9273.

Smith: C. J., Wejksnora, P. J., Warner, J. B., Rubin, C. S., and Rosen, O. M., 1979, Insulin-stimulated protein phosphorylation in 3T3-L1 cells, *Proc. Natl. Acad. Sci. U.S.A.* **76**:2725–2729.

Smith, C. J., Rubin, C. S., and Rosen, O. M., 1980, Insulin-treated 3T3-L1 adipocytes and cell-free extracts derived from them incorporate ^{32}P into ribosomal protein S6, *Proc. Natl. Acad. Sci. U.S.A.* **77**:2641–2645.

Spiegelman, B. M., Frank, M., and Green, H., 1983, Molecular cloning of mRNA from 3T3-L1 adipocytes, *J. Biol. Chem.* **258**:10083–10089.

Stadtmauer, L., and Rosen, O. M., 1983, Phosphorylation of exogenous substrates by the insulin receptor-associated protein kinase, *J. Biol. Chem.* **258**:6682–6685.

Swergold, G. D., Rosen, O. M., Rubin, C. S., 1982, Hormonal regulation of the phosphorylation of ATP citrate lyase in 3T3-L1 adipocytes. Effects of insulin and isoproterenol, *J. Biol. Chem.* **257**:4207–4215.

Van Obberghen, E., and Kowalski, A., 1982, Phosphorylation of the hepatic insulin receptor: Stimulatory effects of insulin on intact cells in a cell-free system, *FEBS Lett.* **143**:179–182.

Van Obberghen, E., Rossi, B., Kowalski, A., Gazzano, H., and Ponzio, G., 1983, Receptor-mediated phosphorylation of the hepatic insulin receptor: Evidence that the M_r 95,000 receptor subunit is its own kinase, *Proc. Natl. Acad. Sci. U.S.A.* **80**:945–949.

Zick, Y., Kasuga, M., Kahn, C. R., and Roth, J., 1983a, Characterization of insulin-mediated phosphorylation of the insulin receptor in a cell-free system, *J. Biol. Chem.* **258**:75–80.

Zick, Y., Whittaker, J., and Roth, J., 1983b, Insulin-stimulated phosphorylation of its own receptor: Activation of a tyrosine-specific protein kinase that is tightly associated with the receptor, *J. Biol. Chem.* **258**:3431–3434.

Signal Transduction in Biological Membranes

MARTIN RODBELL

1. Information Processing: Some Generalities

Information processing is the key to the biology of learning and adaptation. Receptors are essential components of such cognitive systems. However, it should be emphasized that receptors *per se* have no intrinsic function; i.e., as with facts, they cannot speak for themselves. To be functional, receptors must be incorporated into systems that process or transduce the incoming signals. On the other hand, to the extent that their removal results in loss of information processing, receptors are the key to the phenomenon known as habituation or desensitization, i.e., the waning of information processing observed on repeated signal input.

Two general types of information-processing systems are commonly observed. One is short term in that signal processing is rapid in both onset and offset. The other is long term and expresses memory of the input system long after the initial signal has been withdrawn. For experimental purposes, the former systems are more amenable for detailed study and are the subject of most of the discussion in this volume.

Over the past 20 years, receptors have passed from the near-mythical quality with which they were held to that of the macromolecules they have proven to be. For example, the nicotinic cholinergic receptor has been purified and shown to consist of five separate proteins that are tightly integrated and contain all of the elements necessary for information processing including signal recognition, transduction, and conductance of ions. Hence, information processing with this system involves multiple

MARTIN RODBELL • Section on Membrane Regulation, NIADDK, National Institutes of Health, Bethesda, Maryland 20205. *Present address*: National Institute of Environmental Health Sciences, Research Triangle Park, North Carolina 27709.

components integrated and incorporated into a membrane as a unit. Thus far, this is the only purified receptor system that displays all of the qualities of an information-processing unit and is clearly responsible for responses of the cell to the incoming neurotransmitter.

As is discussed in detail in this volume, hormone-sensitive adenylate cyclase systems are also multicomponent systems. However, unlike the cholinergic system, the components are not bonded in a manner that allows isolation of the complete system as a unit. Moreover, in the case of certain steroid receptors, the recognition component is compartmentalized from the processes in the nucleus where signal transduction takes place. Thus, information processing can be composed not only of distinct molecules but also of molecules that are spread over different parts of the cell.

Because of our ability to tag receptors with radioactive markers, these elements have proven to be the easiest to isolate and characterize. The molecules responsible for transduction have proven to be most difficult to detect and isolate. A major problem is knowing the nature of the signal processing that takes place subsequent to ligand interaction with the receptor. As is emphasized in this and subsequent chapters, it is essential to understand the nature of the signals arising initially from the stimulus of the external signal before one can begin to undertake the task of unraveling the nature of transduction. The enormity of the problem is underscored by the actions of such hormones as insulin, which induces a variety of different responses in its target cells, including enhancement of growth, stimulation of synthesis of proteins, fat, and carbohydrates, and alterations of ion, amino acid and sugar transport. Are these various responses the result of a unique chemical signal produced by a single information-processing system, or are there multiple information-processing systems intercalated with a common pool of receptors?

2. Transduction and the Adenylate Cyclase System

As a biological term, transduction has classically referred to the transfer of information between viruses and bacteriophages and their host cells. Put into a somewhat different context, several years ago transduction was used as a term for describing the transfer of information between the receptors for a variety of hormones and adenylate cyclase, the enzyme responsible for the production of cAMP (Rodbell *et al.*, 1969). At that time, hormone-sensitive adenylate cyclase systems were the only information-processing systems that could be investigated at the level of isolated plasma membranes. As a model for information transfer, it was suggested that the system can be described in abstract terms as a tripartite system composed of recognition (R) or discriminator units, transducer, and an effector (E) or amplifying component that produces signals at a higher level than the incoming signal. After a decade of research, it became clear that, in fact, recognition, transduction, and amplification are carried out by distinct macromolecular components. Although not fully understood, transduction involves GTP-binding proteins (abbreviated N). Here I discuss briefly the possible role of receptors in the function of the N units, the possible organization of receptors and N units in the membranes, and the growing evidence that N proteins may be responsible for signal transduction in other membrane information transfer systems.

The production of cAMP in animal cells is one of the most highly regulated processes known in biology. A large and ever growing number of hormones and neurotransmitters, each acting through distinctive receptors, regulate the production of cAMP by acting on adenylate cyclase systems in the outer cell membrane. Because of the multiplicity of receptor types, it is clear that receptors are distinct molecules from the enzyme. The discovery that hormone action invariably requires the presence of micromolar concentrations of GTP, whether the hormones act by stimulating or inhibiting adenylate cyclase, led to the concept (schematically represented in Fig. 1) of distinct GTP-binding proteins, designated N_s and N_i, that control, respectively, stimulation and inhibition of adenylate cyclase. This concept has been verified recently by the isolation of two GTP-binding proteins that have the properties of N_s and N_i (Sternweiss *et al.*, 1981; Bokoch *et al.*, 1983).

Still controversial is how the receptors are linked to the N units. One hypothesis suggests that R units only become coupled to N units on liganding of hormones to their receptors (Stadel *et al.*, 1982). In this theory, the act of coupling induced by hormones is the mechanism by which the N units become "activated." An alternative theory (Rodbell, 1980) suggests that receptors and N units may exist in free and coupled forms but that only the latter are responsible, when occupied by hormone and GTP, for activation of N. Indirect evidence that R and N may be complexed as large aggregates or complexes was obtained from target analysis of two systems (Schlegel *et al.*, 1979, 1980). In these studies it was found that the combined actions of GTP and hormones converted the large targets to smaller targets. This finding led to the theory that the concerted actions of hormone and GTP cause disaggregation of oligomers of RN complexes into forms ("monomers") that were capable of interacting with adenylate cyclase. In contrast to the coupling theory of hormone action, this theory suggests that receptors complexed to the N units prevent the latter from interacting with the enzyme; the small ligands (hormones and

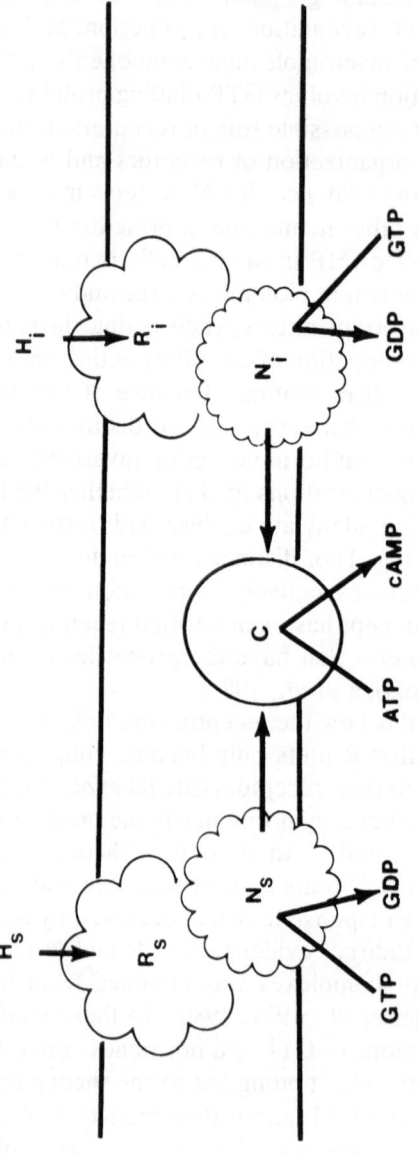

Figure 1. A schematic representation of GTP binding proteins that stimulate (N_s) or inhibit (N_i) adenylate cyclase and also have GTPase activity.

GTP) serve to release these structural constraints, bringing about "activation" of the N units.

Consistent with this theory are recent studies of the reconstitution of purified N units and β-adrenergic receptors in lipid vesicles (Brandt *et al.*, 1983). Hormone action was followed by measuring the production of inorganic phosphate and GDP by a GTPase activity associated with the N_s unit. In the absence of receptor (and detergents that inhibit GTPase activity), the N_s unit displayed appreciable GTPase activity. However, when N_s units were reconstituted in lipid vesicles with receptor, no GTPase activity was detected. Following the addition of hormone, GTPase activity was restored to levels observed with the uncoupled N_s units.

Based on mathematical modeling of the action of a nonhydrolyzable analogue of GTP (guanylylimidodiphosphate), it has been suggested that the enzyme system can take distinct transition states and that there is a slow transition to the "activated" state in the absence of hormone. The role of the hormone is to accelerate the rate of this transition, leading to activation of both adenylate cyclase and the GTPase associated with the N units (Rendell *et al.*, 1975, 1977). According to the "disaggregation" theory of hormone action, the hormones accelerate the rate of release of the N unit from its oligomeric association with the RN complex; in the absence of hormone, nonhydrolyzable analogues induce slow release of the N units. Presumably GTP is relatively inactive because the rate of hydrolysis by the "activated" GTPase on the released N unit is faster than the rate of activation (or coupling) with adenylate cyclase.

Although the disaggregation theory can explain the kinetic behavior of adenylate cyclase systems and has the merit of relating structure with function, the actual process of hormone transduction is still poorly understood. The recent findings that N_s and N_i are heterodimers and that guanine nucleotides or fluoride ions induce dissociation of the heterodimers into "active" forms of the GTP-binding protein (see Gilman *et al.*, Chapter 10) raise the possibility that hormones and GTP act by inducing dissociation of the GTP-binding subunit from the complexes between R and N units.

3. GTP Binding Proteins: A Family of Membrane Regulatory Proteins

In addition to N_s and N_i, it was also postulated a few years ago (Rodbell, 1980) that there may be types of N units (N_x) that regulate membrane processes other than adenylate cyclase. The basis for this suggestion was the finding that, as with receptors coupled to N_s or N_i, agonist binding to receptors that were not involved in the regulation of adenylate cyclase activity produced a marked decrease in binding affinities in the presence of guanine nucleotides. Since such changes in receptor affinity states are associated with the transduction process in the case of adenylate cyclase regulation, it seemed reasonable to suggest that other receptor types were similarly linked to GTP-dependent processes involving N units. Since that suggestion, there have been several reports that guanine nucleotides are involved in the regulation of membrane-bound protein kinases (Walaas *et al.*, 1981), cAMP phosphodiesterase (Heyworth *et al.*, 1983), and calcium gating in mast cells (Gomperts, 1983). However, the best evidence thus far that N units regulate processes other than adenylate cyclase is the isolation of an N unit, termed transducin, that regulates a cGMP phosphodiesterase in rod outer segments. Transducin is activated by light activation of rhodopsin. Recent papers have shown that transducin has a remarkably similar structure to that of both N_s and N_i in that not only do they have identical β subunits but their α subunits (the GTP-binding proteins) share structural homologies (Manning and Gilman, 1983; Abood *et al.*, 1982).

Although there is still only indirect evidence for the participation of N units in other membrane functions, the reasons are now even more compelling to invoke the general thesis that GTP acts, through a family of N units, on a variety of membrane-associated processes, each of which has specific types of receptors associated with the N units.

Clearly needed for testing this hypothesis are other types of signal-generating systems similar to adenylate cyclase and cGMP phosphodiesterases that can be examined with cell-free membrane preparations. Since many hormones appear to affect the metabolism of phosphatidylinositol, we may learn that there are, indeed, other signal-generating systems with which the thesis can be examined. These may involve both calcium gating and the production of inositol triphosphate.

4. Organization of Receptors and Transduction Elements in Membranes

A popular theory of hormone action is that receptors are distributed uniformly over the surface of the membrane and that the action of the hormone is to induce the receptor to react with an effector system. Consistent with this idea are numerous studies showing that some hormones induce receptor aggregation and that aggregation correlates with transduction. A notable example of ligand-induced aggregation of receptors that demonstrates this point is the immunoglobulin E(IgE) receptors on mast cells, which, when liganded by IgE, initiate exocytosis (Perez-Montfort *et al.*, 1983). The receptor in this case is a composite of at least three subunits. However, it is not known whether this complex contains all of the elements necessary for signal transduction or what the nature of the molecular consequences of receptor aggregation is.

Another model system for receptor organization and transduction is the rhodopsin-stimulated phosphodiesterase system in rod outer sgements discussed briefly in Section 3. In the absence of light, the N units (transducin) are associated with the inner or cytoplasmic face of the rhodopsin membrane in the form of large (9- to 12-nm) particles (Roof and Heuser, 1982). These particles are one-tenth the concentration of rhodopsin, in accord with the ratio of rhodopsin to transducin in these membranes. When exposed to light and GTP, the particles dissociate from the membrane coincident with activation of phosphodiesterase (Roof *et al.*, 1982). Only a quantum of light is necessary to discharge hundreds of transducin molecules (Pober and Bitensky, 1979). The structural basis of such amplification is not understood. As a possibility, one might consider that rhodopsin molecules in the membrane interact in a concerted fashion such that one light-activated molecule is sufficient to "energize" the release of all associated transducin molecules. In this case, the efficiency of transducin is dependent on the organization of a large number of receptor molecules to which relatively few transducer molecules are attached.

As discussed in Section 2, the receptors and N units involved in adenylate cyclase regulation may be complexed in the form of large oligomeric structures. In view of the concentrations of receptors and N units in most membranes, however, they can only cover small patches of the membrane. Less certain than the rhodopsin–transducin relationship is the relative concentration of R and N units involved in the regulation of adenylate cyclase. In many cyclase systems, occupation of only a few percent of the total population of receptors leads to essentially maximal activation of adenylate cyclase. The simplest explanation is that the concentration of the enzyme is far less than that of the receptors and the N units. However, the situation is more complex, since the relative concentrations of N_s and N_i have to be considered in the equation.

5. Summary

To summarize this brief introduction to the subject of transduction in biological membranes, two systems—the light-activated phosphodiesterase system in rod outer segments and hormone-sensitive adenylate cyclase systems in the cell membrane—have proven to be excellent model systems for investigating signal transduction. Perhaps the most revealing aspect of the transduction process thus far ascertained is that they are extraordinarily complex systems. Each system contains a minimum of five units if only the receptor, N units, and enzyme are counted. More, certainly, are required for the overall transduction processes. The second notable point is that transduction in these two systems involves the participation of not only the initial signal input at the receptor level but also regulatory ligands (GTP and metal ions) that act in concert to bring about the large changes in structure accompanying the transduction processes. The third point is that the transduction processes involve changes in the distribution of the components within the membrane or the association of the components with the membrane. Finally, and perhaps most importantly, it appears that the receptors and their associated N units are organized in fashions poised for amplification of the incoming signals. The precise nature of the organization and the events taking place immediately following alterations in the receptor are still largley unresolved issues. One point that is clear from the studies reported thus far with these model systems is that the cascade of events among signal receptor, transduction, and amplification involves dissociations and associations between the various macromolcules comprising these systems. Thus, the actual messengers are the proteins that shuttle back and forth both within and out of the framework of the membranes.

References

Abood, M. E., Hurley, J. B., Pappone, M.-C., Bourne, H. R., and Stryer, L., 1982, Functional homology between signal-coupling proteins, *J. Biol. Chem.* **257**:10540–10543.

Bokoch, G. M., Katada, T., Northup, J. K., Hewlett, E. L., and Gilman, A. G., 1983, Identification of the predominant substrate for ADP-ribosylation by islet activating protein, *J. Biol. Chem.* **258**:2072–2075.

Brandt, D. R., Asano, T., Pedersen, S. E., and Ross, E. M., 1983, Reconstitution of catecholamine-stimulated guanosinetriphosphatase activity, *Biochemistry* **22**:4357–4362.

Gomperts, B. D., 1983, Involvement of guanine nucleotide-binding protein in the gating of Ca^{2+} by receptors, *Nature* **306**:64–66.

Heyworth, C. M., Rawal, S., and Houslay, M. D., 1983, Guanine nucleotides can activate the insulin-stimulated phosphodiesterase in liver plasma membranes, *FEBS Lett.* **154**:87–91.

Manning, D. R., and Gilman, A. G., 1983, The regulatory components of adenylate cyclase and transducin, *J. Biol. Chem.* **258**:7059–7063.

Perez-Montfort, R., Fewtrell, C., and Metzger, H., 1983, Changes in the receptor for immunoglobulin E coincident with receptor-mediated stimulation of basophilic leukemia cells, *Biochemistry* **22**:5733–5737.

Pober, J. S., and Bitensky, M. W., 1979, Light-regulated enzymes of vertebrate retinal rods, *Adv. Cyclic Nucleotide Res.* **11:**265–301.

Rendell, M. S., Rodbell, M., and Berman, M., 1977, Activation of hepatic adenylate cyclase by guanyl nucleotides, *J. Biol. Chem.* **252:**7909–7912.

Rendell, M., Salomon, Y., Lin, M. C., Rodbell, M., and Berman, M., 1975, The hepatic adenylate cyclase system: A mathematical model for the steady state kinetics of catalysis and nucleotide regulation, *J. Biol. Chem.* **250:**4253–4260.

Rodbell, M., 1980, The role of hormone receptors and GTP-regulatory proteins in membrane transduction, *Nature* **284:**17–22.

Rodbell, M., Birnbaumer, L., and Pohl, S. L., 1969, Hormones, receptors, and adenyl cyclase activity in mammalian cells, in: *The Role of Adenyl Cyclase and Cyclic 3'5'-AMP in Biological Systems* (T. W. Rall, M. Rodbell, and P. Condliffe, eds.), Fogarty International Center Washington, pp. 59–76.

Roof, D. J., and Heuser, J. E., 1982, Surfaces of rod photoreceptor disk membranes: Integral membrane components, *J. Cell Biol.* **95:**487–500.

Roof, D. J., Korenbrot, J. I., and Heuser, J. E., 1982, Surfaces of rod photoreceptor disk membranes: Light activated enzymes, *J. Cell Biol.* **95:**501–509.

Schlegel, W., Kempner, E. S., and Rodbell, M., 1979, Activation of adenylate cyclase in hepatic membranes involves interactions of the catalytic unit with multimeric complexes of regulatory proteins, *J. Biol. Chem.* **254:**5168–5176.

Schlegel, W., Cooper, D. M. F., and Rodbell, M., 1980, Inhibition and activation of fat cell adenylate cyclase by GTP is mediated by structures of different size, *Arch. Biochem. Biophys.* **201:**678–682.

Stadel, J. M., De Lean, A., and Lefkowitz, R. J., 1982, Molecular mechanisms of coupling in hormone receptor–adenylate cyclase systems, *Adv. Enzymol.* **53:**1–43.

Sternweis, P. C., Northup, J. K., Smigel, M. D., and Gilman, A. G., 1981, The regulatory component of adenylate cyclase: Purification and properties, *J. Biol. Chem.* **256:**11517–11526.

Walaas, O., Horn, R. S., Lystad, E., and Adler, L., 1981, Insulin dependent protein phosphorylation in membranes. Isolation and characterization of a phosphorylated proteolipid from sarcolemma. *FEBS Lett.* **128:**133–136.

6

Receptor-Controlled Phosphatidylinositol 4,5-Bisphosphate Hydrolysis in the Control of Rapid Receptor-Mediated Cellular Responses and of Cellular Proliferation

R. H. MICHELL

1. Introduction

Calcium ions were assigned an important role in the control of cell functions such as contraction and secretion many years ago, but this early physiological and pharmacological work was largely ignored by most biochemists until quite recently (e.g., Rasmussen, 1969; Berridge, 1975). Moreover, the discovery of cAMP in the 1950s established a pattern of thought, particularly dominant among biochemists in the 1960s, in which the search for cellular signaling mechanisms was concentrated on attempts to identify enzymically synthesized and water-soluble intracellular "second messengers."

The discovery by Hokin and Hokin (1953) that receptor-directed agonists often stimulate the metabolism of inositol phospholipids predates the discovery of cAMP by a couple of years, but it has taken 30 years for us even to come close to an understanding of the mechanism and function of this response. During the first 20 of these 30 years, the major characteristics of receptor-stimulated inositol lipid metabolism were de-

R. H. MICHELL • Department of Biochemistry, University of Birmingham, Birmingham B15 2TT, England.

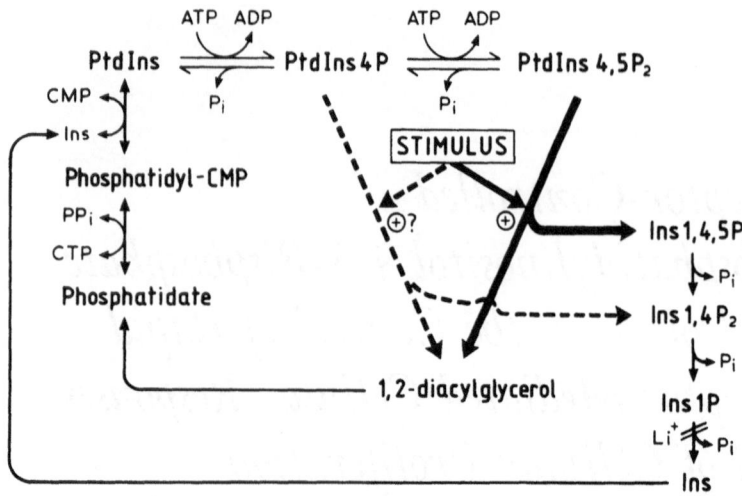

Figure 1. The pathways of inositol lipid metabolism that are activated by receptors that mobilize calcium in the cytosol and activate protein kinase C. Note that the appearance of InsP₃ in stimulated cells proves that PtdIns4,5P₂ is hydrolyzed in stimulated cells, but it is still not clear whether receptors control the hydrolysis of PtdIns4P (see the text).

fined. These foundations of the field were laid almost entirely by Mabel and Lowell Hokin, working for most of the time in Madison.

A major step forward came about 10 years ago, when my colleagues and I realized that activation of certain types of cell-surface receptors always causes stimulation of inositol lipid breakdown, irrespective of the final physiological response evoked in the stimulated cells, and that this reaction might somehow be responsible for a rise in cytosolic Ca^{2+} ion concentration in the stimulated cells (Lapetina and Michell, 1973; Michell, 1975; Michell *et al.*, 1977). This was the first clear indication that the increase in inositol lipid metabolism is always associated with a single class of receptor-linked signaling processes rather than with a particular style of cellular response (e.g., secretion or ion pumping), and it opened the way for the rapid progress of the last few years.

When we first proposed that receptor-stimulated inositol lipid metabolism is an essential step in Ca^{2+} mobilization, it appeared that receptors activated the breakdown of phosphatidylinositol (PtdIns) and that Ca^{2+} was mobilized into the cytoplasm as a result of the opening of ion "gates" in the plasma membrane (Michell, 1975; Michell *et al.*, 1977). We now know that these proposals were correct in essence but wrong in detail (Fig. 1). It appears that receptors activate the breakdown of phosphatidylinositol 4,5-bisphosphate (PtdIns4,5P₂) and possibly also of phosphatidylinositol 4-phosphate (PtdIns4P) (Michell *et al.*, 1981; Michell, 1982, 1983; Weiss *et al.*, 1982; Berridge, 1984) and that the result is release of Ca^{2+} into the cytoplasm from a cell-associated store (Putney, 1978, 1981; Jones and Michell, 1978; Michell, 1979, 1982). The inositol 1,4,5-trisphosphate (Insl,4,5P₃) released during hydrolysis of PtdIns4,5P₂ appears to be a water-soluble intracellular "second messenger" that is responsible for the release of cell-associated Ca^{2+} into the cytosol (Streb *et al.*, 1983; see Berridge, Chapter 8). The location of the Insl,4,5P₃-sensitive Ca^{2+} pool is uncertain, but it may be in some element of the endoplasmic reticulum membrane system. 1,2-Diacylglycerol (1,2-DG), the other product of PtdIns4,5P₂ breakdown, is the first conclusively identified membrane-associated "second messenger" molecule: it activates a novel phospholipid- and Ca^{2+}-dependent protein kinase known as protein kinase C (Nishizuka, 1983).

The remainder of this chapter is devoted to brief discussions of three topics that relate to the signals that use PtdIns4,5P₂ hydrolysis as a enzymic amplification step and to the cellular responses that they evoke. First, I summarize some recent evidence that suggests that some at least of the Ca^{2+}-mobilizing V₁-vasopressin receptors of a variety of cells may be receptors for a vasopressinlike peptide (VLP) that is released at sympathetic nerve terminals rather than for vasopressin. Secondly, I briefly recapitulate the recent evidence that leads to the conclusion that receptors activate phosphodiesterase-catalyzed hydrolysis of PtdIns4,5P₂ and that this reaction has the characteristics to be expected of an essential coupling step in receptor-stimulated Ca^{2+} mobilization. Finally, I offer some speculative considerations on the possible role of inositol lipid metabolism in controlling cell proliferation in both normal and malignant cells.

2. "V₁ Vasopressin Receptors": Receptors for Hormone or Neurotransmitter

Renal diuresis is restrained by circulating vasopressin, and the Brattleboro rat, which lacks the vasopressin of the hypothalamoneurohypophyseal tract, suffers from hereditary diabetes insipidus. Control of renal diuresis is through vasopressin receptors of the V₂ subclass: these receptors effectively control diuresis at circulating vasopressin concentrations around 10^{-11} M by activating renal adenylate cyclase (Jard, 1981).

The circulating vasopressin concentration needed to appreciably raise blood pressure in intact animals is much higher, and there has been doubt as to whether circulating vasopressin normally has any important role in controlling the dynamics of the peripheral circulation. However,

Table I. Cells and Tissues in Which Vasopressin Stimulates the Metabolism of Inositol
Phospholipids

Tissue	References
Hepatocytes	Kirk *et al.*, 1979, 1982a,b; Michell *et al.*, 1979, 1981; Creba *et al.*, 1983; Thomas *et al.*, 1983
Aorta smooth muscle	Takhar and Kirk, 1981
Renal mesangial cells	Troyer *et al.*, 1983
WRK1 mammary tumor cells	Monaco and Woods, 1983
Platelets	E. MacIntyre, personal communication
Adipocytes	R. Rubio and P. Newsholme, unpublished data
Hippocampus	L. Stephens and S. Logan, personal communication
Sympathetic ganglia	Hanley *et al.*, 1984; Bone *et al.*, 1984

recent studies suggest that the lack of effect of small quantities of infused vasopressin on blood pressure may arise because a vasopressin-induced fall in blood flow through peripheral vascular beds is balanced by compensatory feedback inhibition of cardiac output. Spinally transected animals lack this feedback control and show an increase in blood pressure even after quite modest infusions of vasopressin (Cowley, 1983).

The vasopressin receptors of the peripheral vasculature are of the V_1 type, as are the vasopressin receptors that control glycogenolysis in the liver: these receptors act by stimulating the hydrolysis of PtdIns4,5P_2 and hence raising the cytosolic Ca^{2+} concentration in the stimulated cells (Kirk *et al.*, 1979, 1981; Michell *et al.*, 1979; Takhar and Kirk, 1981; Creba *et al.*, 1983). Stimulation of inositol lipid metabolism by vasopressin has now been observed in a variety of cells and tissues (see Table I), suggesting that V_1-vasopressin receptors are widely distributed in the body rather than confined to liver and the vasculature.

We recently discovered that vasopressin stimulates PtdIns4,5P_2 breakdown in isolated superior cervical sympathetic ganglia from rats (Hanley *et al.*, 1984; Michell *et al.*, 1984; Bone *et al.*, 1984). This was unexpected, since we are aware of no previous studies that suggest any direct action of vasopressin on these ganglia. However, this response was sufficiently striking for it to provoke an investigation of whether sympathetic ganglia might contain any vasopressinlike immunoreactive material. In these studies, we observed that the principal noradrenergic neurons of sympathetic ganglia from several mammalian species stain intensely with antibodies both to vasopressin and to oxytocin (Hanley *et al.*, 1984). Moreover, extracts of ganglia contain a material that is larger than either oxytocin or vasopressin and is immunologically reactive with several high-specificity antibodies raised against the neurohypophyseal

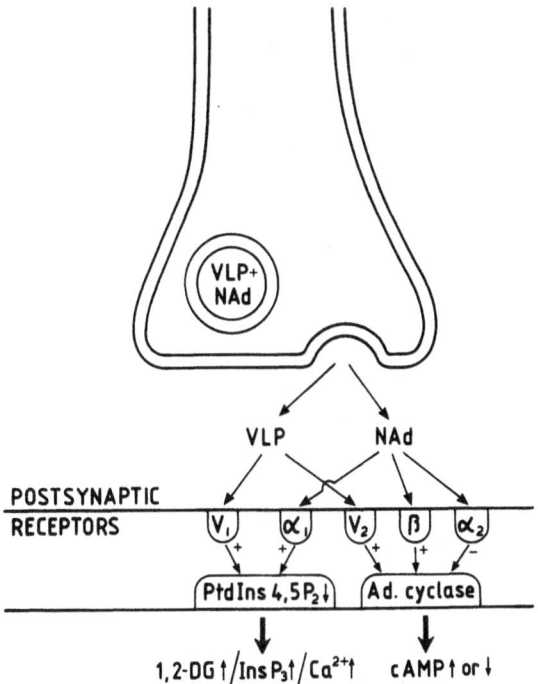

Figure 2. A speculative synthesis of what might happen at the surface of a cell innervated by sympathetic nerves that release both norepinephrine and VLP. Note that we do not yet know whether or not these two neurotransmitters are stored in the same presynaptic vesicles and released together. Thus far, all of the known receptors for norepinephrine and for peptides of the vasopressin/oxytocin family act through control of either inositol lipid hydrolysis or adenylate cyclase activity, but other receptors for these ligands may remain to be discovered. (N.B. The pre- and postsynaptic membranes are diagrammatically represented on quite different scales simply to facilitate a clear representation of postsynaptic events.)

peptide hormones. Staining of peripheral tissues (artery, liver, salivary glands) suggests that the same (or very similar) material is present in the sympathetic nerve fibers that innervate these tissues. On the basis of these studies, we have suggested that the postganglionic noradrenergic neurons of the mammalian sympathetic nervous system may normally contain a vasopressinlike peptide (VLP) that is released as a cotransmitter with norepinephrine (Hanley *et al.*, 1984; Michell *et al.*, 1984).

Extensive further studies will be needed to define the characteristics and function of VLP in detail, and here the Brattleboro rat may prove particularly valuable. Although this mutant rat lacks the hormonal vasopressin that is derived from the neurohypophysis, its sympathetic nervous system retains substantial quantities of VLP (Hanley *et al.*, 1984).

It may therefore offer an opportunity for the function of VLP to be analyzed *in situ* without interference from hormonal vasopressin.

As mentioned above, there is doubt as to whether the widespread
V_1 receptors of the tissues are sufficiently sensitive to circulating vasopressin to play any substantial role in its actions. The discovery of VLP
in the sympathetic nervous system (and possibly also in central neurons,
where epinephrine and vasopressinlike immunoreactivity coexist) may
offer a quite different role for many "vasopressin receptors": some at
least are likely to function as receptors for VLP, a neurotransmitter released close to the responsive cells. In this context, it is interesting to
note that norepinephrine influences cells through activation of at least
four different types of receptors (α_1, α_2, β_1, β_2) and that there are two
types of "vasopressin receptors" (V_1 and V_2). However, it seems likely
that most or all of the information flowing through these six types of
receptors is channeled into only two major signaling systems (Fig. 2), so
that the extra flexibility of response conferred by corelease of norepinephrine and VLP is likely to be quantitative rather than qualitative.

3. Phosphatidylinositol 4,5-Bisphosphate Hydrolysis as a Coupling Reaction in Receptor-Mediated Signaling

As mentioned in the Section 1, it was thought for many years that
receptors stimulated the phosphodiesterase-catalyzed hydrolysis of
PtdIns. This conclusion was based on a large number of studies in which
stimuli evoked one or more of the following changes in the cellular status
of PtdIns: (1) a decrease in its concentration, (2) a rapid reduction in the
labeling of PtdIns (with [^{32}P]- or [^{3}H]-inositol or [^{3}H]-glycerol) in cells
that had been prelabled before stimulation, (3) a sustained increase in the
turnover of PtdIns in cells stimulated for long periods. Although there
may be some situations in which stimulation of cells provokes some direct
enzyme-catalyzed breakdown of PtdIns (Farese, 1984), an increasing
number of workers believe that the stimulated consumption of PtdIns
defined by the above observations is largely a reflection of the use of
PtdIns as a substrate for the synthesis of PtdIns4P and PtdIns4,5P$_2$. Although previous studies had hinted at the possibility that receptors initially
stimulated the hydrolysis of PtdIns4,5P$_2$, the first explicit proposal that
this is the case, and that the previously observed cellular depletion of
PtdIns occurs because of its role as a precursor for newly synthesized
PtdIns4,5P$_2$, did not come until 1981 (Michell *et al.*, 1981; Kirk *et al.*,
1981).

Our proposal that receptors activate a PtdIns4,5P$_2$ phosphodiesterase rather than a PtdIns phosphodiesterase arose from studies in which we observed a very rapid decline in the labeling of PtdIns4,5P$_2$ and PtdIns4P in hepatocytes exposed to high concentrations of vasopressin (Michell *et al.*, 1981; Kirk *et al.*, 1981; Creba *et al.*, 1983); these results have since been confirmed by others (Rhodes *et al.*, 1983; Thomas *et al.*, 1983). This depletion of PtdIns4,5P$_2$ in vasopressin-stimulated hepatocytes starts immediately on hormone addition, can produce a depletion of cellular PtdIns4,5P$_2$ at a rate of at least 1% per second, and leads to a new steady-state concentration of PtdIns4,5P$_2$ about 40–75% of the original value within 1 min (Creba *et al.*, 1983; Thomas *et al.*, 1983). We had previously shown that the stimulation of PtdIns labeling and PtdIns depletion in hepatocytes is mediated by vasopressin receptors (or VLP receptors?) of the V$_1$ type, and this is also true of the vasopressin-stimulated decrease in the steady-state concentration of PtdIns4,5P$_2$ (Kirk *et al.*, 1978, 1981; Creba *et al.*, 1983).

Rat hepatocytes are responsive to a variety of different stimuli, some of whose actions are Ca^{2+}-mediated, whereas others control cells through other mechanisms (e.g., activation or inhibition of adenylate cyclase). Only those stimuli that act by mobilizing Ca^{2+} within a second or two (V$_1$ vasopressin, angiotensin, ATP, α_1-adrenergic, platelet-activating factor) stimulate a decrease in the steady-state concentration of PtdIns4,5P$_2$ and sometimes also of PtdIns4P (Creba *et al.*, 1983; Thomas *et al.*, 1983; Shukla *et al.*, 1983).

These results suggested to us that activation of Ca^{2+}-mobilizing receptors in hepatocytes was probably causing a phosphodiesterase-catalyzed breakdown of PtdIns4,5P$_2$ to 1,2-DG and Insl,4,5,P$_2$ (see Fig. 1), and it seemed likely that a similar reaction sequence would underlie receptor-stimulated inositol lipid metabolism in other cells (Michell *et al.*, 1981). Confirmation of this view has come rapidly from many quarters. First, there have been a large number of reports of very rapid receptor-stimulated PtdIns4,5P$_2$ depletion in cells stimulated by Ca^{2+}-mobilizing stimuli (see Michell, 1983; Michell *et al.*, 1984; Berridge, 1984 for lists). In addition to the stimuli listed above for the hepatocyte, effective agonists include muscarinic cholinergic stimuli in several tissues, substance P, pancreozymin, 5-hydroxytryptamine, TRH, *f*-Met-Leu-Phe, thrombin, ADP, TSH, and concanavalin A.

Secondly, the products of inositol lipid breakdown in the stimulated cells include an inositol trisphosphate (InsP$_3$). This result, first obtained by Akhtar and Abdel-Latif (1980) and later defined in detail by Berridge and his colleagues (Berridge *et al.*, 1983; Berridge, 1983, 1984, Chapter 8, this volume), is particularly important since PtdIns4,5P$_2$ is the only known component of mammalian cells that can rapidly give rise to an inositol trisphosphate. (It should be noted, however, that none of the studies reported to date have included evidence that the accumulated inositol trisphosphate is Ins:1,4,5P$_3$, the isomer to be expected from the hydrolysis of PtdIns4,5P$_2$.) Once again, this result has been rapidly confirmed and extended in other laboratories (see Michell *et al.*, 1984; Berridge, 1984 for refs.): in our laboratory, we have seen receptor-stimulated accumulation of an InsP$_3$ in hepatocytes, adipocytes, and sympathetic ganglia exposed to vasopressin (Hanley *et al.*, 1984, Michell *et al.*, 1984a,b; Bone *et al.*, 1984; R. Rubio and P. Newsholme, unpublished data), in oxytocin-stimulated uterine strips (C. J. Kirk, unpublished data), and in HL60 promyelocytes stimulated with a macrophage-derived colony-stimulating activity (G. Guy, C. Bunce, and G. Brown, unpublished data).

One of the remaining unresolved questions about stimulated inositol lipid breakdown is whether receptors control the hydrolysis only of PtdIns4,5P$_2$ or of both PtdIns4P and PtdIns4,5P$_2$. It is clear from the studies in many laboratories that both InsP$_2$ aand InsP$_3$ accumulate in stimulated cells before there is any appreciable accumulation of inositol monophosphate, but only in few studies is there any clear evidence that InsP$_3$ accumulates before InsP$_2$. Perhaps the only convincing data of this type are in a study that suggested that only InsP$_3$ accumulates in GH3 pituitary cells that are briefly stimulated with TRH below their normal growth temperature (McPhee *et al.*, 1984). Two studies of the stimulated parotid have also addressed this question directly by stimulating cells until new steady-state, levels of InsP$_2$ and InsP$_3$ were established and then removing the stimulus by application of an antagonist (Downes and Wuseman, 1983; D. Aub and J. W. Putney, personal communication). In both of these studies, it appeared that the rate of accumulation of InsP$_2$ was too great for it to be produced solely by the action of phosphatase on InsP$_3$, leading to the tentative conclusion that the receptor-activated phosphodiesterase must attack both PtdIns4P and PtdIns4,5P$_2$.

The apparent contradiction between GH3 cell and parotid data can be resolved in two possible ways. One solution would be for the interpretation of the parotid data to prove wrong for some reason, so leaving PtdIns4,5P$_2$ hydrolysis as the only receptor-activated reaction. Alternatively, the phosphodiesterase might be somewhat "flexible" in its substrate specificity, acting on one or both of the two polyphosphoinositides as dictated by the precise conditions of cellular stimulation. Although this latter alternative lacks the simple elegance of control of a single reaction by receptors, such flexible substrate choice by the receptor-controlled phosphodiesterase would have the interesting effect of varying the relative rates of production of InsP$_3$ and 1,2-DG, the two messenger molecules produced during inositol lipid breakdown.

4. Stimulated Inositol Lipid Metabolism and Cell Proliferation

In 1968, Fisher and Mueller showed that phytohemagglutinin (PHA) from *Phaseolus vulgarus* causes a very rapid increase in the inositol lipid metabolism of lymphocytes from human blood. They noted that the response appeared essentially identical to the stimulation of inositol lipid metabolism that is evoked in many other cells by hormones and neurotransmitters. This was a particularly striking observation, since the main effect of PHA that was known at that time was a stimulation of lymphocyte proliferation that occurred 2–3 days after its addition. Early in the 1970s, Heino Diringer and his colleagues embarked on a series of studies that allowed them, by 1977, to make the general statement that all conditions that favor the proliferation of fibroblasts in culture, including viral transformation, also cause a rapid and sustained turnover of PtdIns in these cells (Diringer and Friis, 1977). It seemed, from these studies and others that followed, that rapid inositol lipid metabolism was somehow an invariable accompaniment of, and maybe a prerequisite for, rapid cell proliferation. In particular, the speed with which inositol lipid turnover was provoked on addition of a proliferogenic stimulus suggested that this response might be an important early step in the initiation of cell proliferation.

A second striking characteristic of a subgroup of rapidly proliferating cells, those transformed to a malignant phenotype by viruses, is that they will grow at much lower Ca^{2+} concentrations than will untransformed cells (see Balk *et al.*, 1979; Whitfield *et al.*, 1979; Durham and Walton, 1982): this change in phenotype occurs soon after viral transformation (Ribiero and Armelin, 1984). Such observations led Balk and his colleagues to suggest in 1976 that ''cell replication is initiated by cytosol calcium in excess of a critical level'' and that neoplastic cells proliferate autonomously because their cytoplasmic calcium levels are abnormally high. There is also substantial evidence to suggest that an elevated calcium concentration constitutes one of the normal signals that initiate cell proliferation (see Whitfield *et al.*, 1979; Durham and Walton, 1982). It thus seems likely that the function of at least some of the transforming gene products coded by viral oncogenes may be to substantially enhance the activity of a normal calcium-mobilizing mechanism that usually mobilizes cytosolic calcium only within the concentration range characteristic of nonproliferating cells.

Table II. The Situations in Which There Is a Positive Correlation between the Rate of
Turnover of Inositol Lipids and the Proliferation of Cells[a]

T-lymphocytes (from 1968 onwards)
    Rapid inositol lipid turnover is initiated rapidly by all polyclonal mitogens so far tested
    but not by nonmitogenic lectins. This response, like the proliferative response, is inhibited
    by low-density lipoprotein (LDL). Hagesawa-Sasaki and Sasaki (1983) have shown that
    this response can be observed in a PHA-stimulated clonal T-cell line (CCRF-CEM) and
    that it involves a rapid (<1 min) breakdown of PtdIns4,5P$_2$, which precedes changes
    in PtdIns metabolism. Whether similar changes occur during the activation of B-lym-
    phocytes is still uncertain.
Fibroblasts in culture (from 1972)
    All conditions leading to rapid cell proliferation that have so far been tested lead to rapid
    and sustained inositol lipid turnover: inositol phosphate headgroups are renewed, but
    the diacylglycerol backbone of the lipids is conserved. Effective stimuli include low
    cell density, serum (probably through the action of PDGF; Habenicht *et al.*, 1981), and
    transformation by either SV40 or Rous sarcoma viruses. Some mitogens that act only
    in concert with others, such as vasopressin, bombesin, and prostaglandin F$_{2\alpha}$, are ef-
    fective stimuli (Rosengurt and Sinnet-Smith, 1983; McPhee *et al.*, 1984). (The reported
    effects of EGF are somewhat contradictory: see the text.) Inhibition of fibroblast pro-
    liferation by dexamethazone slows inositol lipid turnover (Grove *et al.*, 1983).
Cells growing and differentiating normally *in vivo* (from 1980)
    Eye lens and cardiac muscle cells proliferate rapidly early in embryogenesis; they also
    show rapid inositol lipid turnover. As they stop dividing and undergo terminal differ-
    entiation, there a selective decrease in the rate of turnover of PtdIns as compared with
    other phospholipids.

[a] References for most of these observations can be found in Michell (1982b). More recent references are
given in detail.

   In 1979, I recognized that there may be a link between these two sets
of observations. Rapid inositol lipid turnover of the type seen in proliferating cells, if it is initiated by phosphodiesterase-catalyzed reaction of
the type normally controlled by receptor activation, is a reaction that
would naturally cause a continuous mobilization of Ca^{2+} into the cytosol
(Michell, 1979b). It thus appeared that the control of cell proliferation in
both normal and malignant cells might be mediated at least in part by the
same type of inositol lipid breakdown mechanism as is employed by those
receptors that exert short-term control over cell functions by elevation
of cytosol Ca^{2+} concentration and activation of protein kinase C. In a
fuller review of the relationship between inositol lipid metabolism and
cell proliferation a few years later, I was able to point to further extensions
of the correlation between rapid cell proliferation and rapid turnover of
inositol lipids (Michell, 1982b). Table II briefly summarizes this information and brings it up to date. The studies that most clearly illustrate
the probable importance of this process to normal cell proliferation *in vivo*

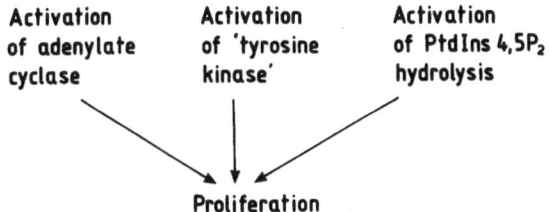

Figure 3. A diagrammatic representation of the fact that cell proliferation is controlled by a variety of extracellular agents whose receptors activate different effector systems, with simultaneous activation of different effector systems causing greater than additive effects on cell proliferation.

are those of Zelenka and her colleagues (Zelenka, 1980; Vu *et al.*, 1983), who have shown that during the normal differentiation of lens epithelial cells to lens fibers the half-life of PtdIns increases at least tenfold, from a few hours to more than 2 days.

Despite the striking correlation between inositol lipid turnover and cell proliferation, there is still a dearth of ideas on the basis of this relationship. In particular, it appears paradoxical that receptor-stimulated inositol lipid breakdown can sometimes be a short-term control signal in nondividing cells but at other times be implicated in longer-term control of cell proliferation. Two recent sets of observations may provide clues to the resolution of this puzzle. The first new factor is the growing awareness that cell proliferation is normally under the control of a family of synergistically acting extracellular signals; each signal falls into one of several groups, with each group having a different mechanism of action. For example, the studies of Enrique Rozengurt and others have revealed that effective initiation of cell proliferation in Swiss 3T3 cells (an "immortalized" fibroblast cell line) can be achieved by a combination of two signals even when neither is an effective mitogen when acting alone (e.g., Rozengurt and Sinnett-Smith, 1983; McPhee *et al.*, 1984). Effective proliferative stimuli can be generated by combining one signal that acts through a receptor that appears to activate a "tyrosine kinase" (e.g., EGF, IGF, insulin) with a second signal that activates inositol lipid hydrolysis, thus mobilizing Ca^{2+} and activating protein kinase C (e.g., bombesin, vasopressin, prostaglandin $F_{2\alpha}$). A third group of receptors that appear to synergize with these two in provoking cell proliferation comprises those receptors that activate adenylate cyclase and hence increase cellular cAMP levels (Rozengurt, 1983). This general situation is summarized in Fig. 3.

An entirely new twist is added to this story by observations recently made by two groups at Harvard and Rochester (Sugimoto *et al.*, 1984; Macara *et al.*, 1984). They have shown that the proteins coded by the viral oncogenes *src* and *ros*, which had previously been regarded as tyrosine-directed protein kinases, can act as inositol lipid kinases. Moreover, when cells are infected with viruses bearing temperature-sensitive mutants of these oncogenes, there are marked increases in the cellular concentrations of PtdIns4P and PtdIns4,5P$_2$: these occur only at the permissive temperature. The studies mentioned above suggest that there is likely to be a rapid metabolic flux through PtdIns, PtdIns4P, and PtdIns4,5P$_2$ in transformed and proliferating cells, so it is unlikely that the accumulation of polyphosphoinositides in these cells is caused by a decrease in their consumption. Indeed, the Rochester group presented preliminary evidence of an increase in the cellular concentrations of InsP$_2$ and InsP$_3$ in *ros*-transformed cells, suggesting the occurrence of rapid polyphosphoinositide turnover in these cells (Macara *et al.*, 1984). It therefore appears that the presence in transformed cells of oncogene-coded proteins with tyrosine kinase activity causes an increase in the net synthesis of PtdIns4P and PtdIns4,5P$_2$. This might occur because the activities of the intrinsic cellular kinases responsible for synthesis of these lipids are increased by their phosphorylation on tyrosine residues. If this were to prove correct, it would provide the first convincing example of control of important cellular enzyme reactions by tyrosine phosphorylation and would identify PtdIns kinase and PtdIns4P kinase as possible key participants in the generation of the transformed phenotype.

An alternative explanation suggested by the results from Harvard and Rochester is that the "tyrosine kinases" encoded by the viral oncogenes are really inositol lipid kinases. However, the assays of the lipid kinase activities expressed by these proteins have all been made under conditions quite unlike those that the proteins will encounter in the plasma membrane of an infected cell, so it is not yet possible to assess the true significance of the observed slow phosphorylation of PtdIns by these proteins. A key point is that these oncogene-encoded kinases might achieve substantial control over cell function by a very low rate of phosphorylation of the tyrosine residues of sensitive proteins (if proteins whose activity is controlled in this way can be identified), but they would have to express much higher activities if they were to catalyze a substantial fraction of the rapid metabolic flux through PtdIns4,5P$_2$. Moreover, it seems a little surprising that phosphorylation of two substrates as different as PtdIns and PtdIns4P might be catalyzed by the same protein. Obviously these new data raise many more questions than they answer.

If tyrosine kinases encoded by oncogenes stimulate the synthesis of PtdIns4P and PtdIns4,5P$_2$, then do the intrinsic tyrosine kinases of normal cells do the same thing? We do not know but may find a few clues in the limited published work. First, Sawyer and Cohen (1981) have observed that EGF, whose receptors express a hormone-stimulated tyrosine kinase activity, stimulates inositol lipid turnover in A431 cells. However, these cells are not stimulated to divide by EGF despite possession of a very large number of EGF receptors, and McPhee *et al.*, (1984) failed to observe any stimulation of inositol lipid metabolism by mitogenic concentrations of EGF in Swiss 3T3 cells. Studies with insulin, another hormone whose receptor expresses hormone-stimulated tyrosine kinase activity, have also given mixed results. In some cells it seems to have little or no effect on inositol lipid metabolism (e.g., Kirk *et al.*, 1981b), but Farese *et al.*, (1982) have reported insulin-stimulated changes in the inositol lipid concentrations of adipose tissue that look remarkably similar to those observed in virally transformed fibroblasts.

The fact that emerges most clearly from an examination of many of the studies of inositol lipid metabolism in cells exposed to proliferogenic stimuli is that inositol lipid metabolism is somehow implicated in cellular responses to these stimuli. However, only in a small number of studies is there any clear clue as to the identity of the controlled metabolic steps or the mechanisms of the control. Systematic studies are now needed to remedy these deficiencies.

Finally, I can offer some very speculative thoughts on possible relationships among the activities of oncogene-coded proteins, the metabolism of inositol lipids, and the control of cellular proliferation. In the last couple of years some order has been emerging from studies of the transforming properties of individual oncogenes in various types of cell, notably fibroblasts, both as primary cultures and as "immortal" cell lines (e.g., Swiss 3T3 or NIH 3T3 cells). The cell lines, which show a relatively high frequency of spontaneous transformation, can also be converted to the transformed state by infection either with *ras* or *sis* oncogenes (Bishop, 1983; Newbold and Overall, 1983; Clarke *et al.*, 1984) or with any of the oncogenes that encode plasma membrane-associated tyrosine kinases (e.g., *src, ros, abl*). The *ras* gene encodes a plasma-membrane-associated protein (p21) that binds GTP, and the *sis* gene product is one polypeptide chain of secreted platelet-derived growth factor (PDGF), which is probably the most effective known fibroblast mitogen, possibly by virtue of the fact that it both activates a receptor with tyrosine kinase activity (Hunter, 1982) and potently stimulates inositol lipid breakdown (Habenicht *et al.*, 1981). By contrast, primary cultures of normal fibroblasts (as compared with "immortalized" lines such as 3T3) can only be converted to the transformed phenotype by simultaneous infection with two oncogenes. One of these can be *ras*, but it seems likely that the other (e.g., *myc*, E1a, or large T) has to encode a protein with a nuclear location (Land *et al.*, 1983; Ruley, 1983; Newbold and Overall, 1983) and that activation of this type of gene somehow "primes" cells for division; permanent activation of *myc* may render cells "immortal". Synthesis of nuclear protooncogene products such as that encoded by c-*myc* can also be turned on in appropriate target cells by potent mitogens such as PHA and PDGF (Kelly *et al.*, 1983): presumably the diffusible signals generated at the cell surface by the receptors for these agents initiate a sequence of events that culminates in activation of the transcription of these genes.

One way of integrating most of this information into a coherent whole would be to propose that the major intracellular signals that control cell proliferation are the inositol trisphosphate and diacylglycerol that arise from receptor-controlled PtdIns4,5P$_2$ hydrolysis and that proliferation occurs in competent cells when the sustained concentrations of these two combined signals exceed some critical threshold value. If this is the case, then it is reasonable to suppose that perturbation of the quantity or intrinsic activity of any of the cellular proteins essential to signaling through this pathway, whether involved in synthesis of PtdIns4,5P$_2$, control of PtdIns4,5P$_2$ hydrolysis, or response to the released InsP$_3$ and diacylglycerol, would modify the growth status of cells: i.e., any of these proteins might be the products of normal cellular protooncogenes or of transforming viral oncogenes.

A speculative synthesis of such ideas is presented in Fig. 4. This scheme bears some resemblance to the scheme offered by Macara *et al.*, (1983), and the idea that the p21 protein encoded by *ras* may be a GTP-dependent transducing protein involved in control of PtdIns4,5P$_2$ breakdown has occurred independently to L. Cantley (personal communication). The key proposals are the following. (1) The rate of generation of signals by the PtdIns4,5P$_2$ breakdown pathway is controlled both by the rate of synthesis of PtdIns4,5P$_2$ and by the degree to which PtdIns4,5P$_2$ phosphodiesterase is activated by receptors. (2) Tyrosine kinases, whether controlled by growth factors or encoded by viral oncogenes, control the rate of PtdIns4,5P$_2$ synthesis. (3) The contribution to proliferation of signals that activate adenylate cyclase is not indicated in Fig. 4 but might involve the ability of cAMP to stimulate the kinases that synthesize PtdIns4,5P$_2$ (Enyedi *et al.*, 1983; Sarkadi *et al.*, 1984). (4) Increased quantities of a receptor that responds to a ubiquitous growth stimulus that activates PtdIns4,5P$_2$ hydrolysis might be mitogenic. The glycoprotein product of the *fms* gene might be such a receptor. (5) There is limited evidence to suggest that control of PtdIns4,5P$_2$ hydrolysis by receptors may involve a GTP-dependent regulatory protein (Gomperts, 1983): the products of the *ras* genes are obvious candidates for this role.

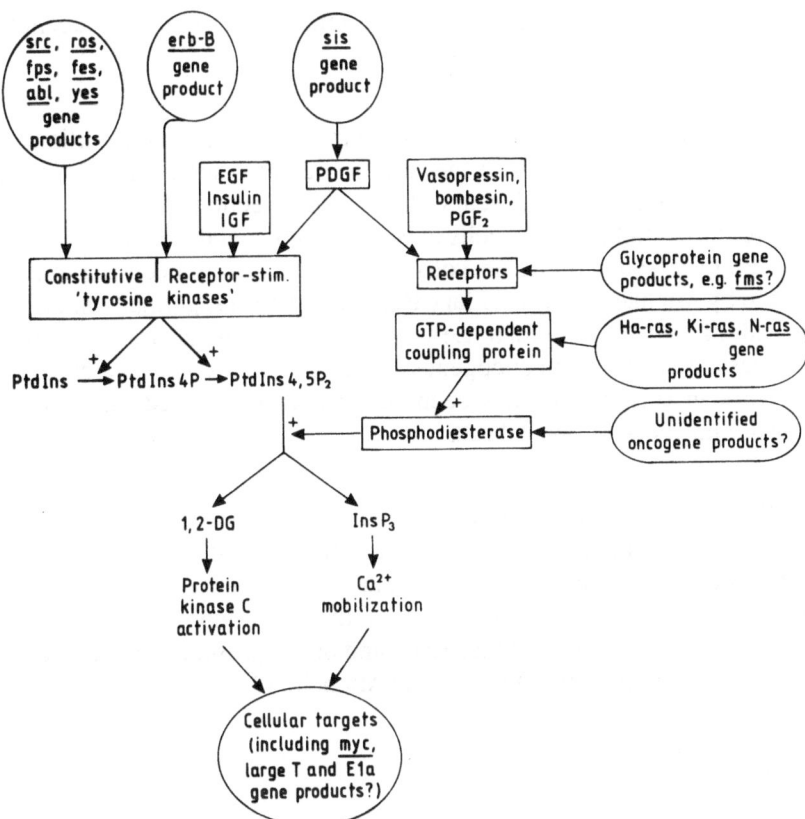

Figure 4. A highly speculative synthesis of possible ways in which the products of some viral oncogenes and of the corresponding cellular protooncogenes might be implicated in the synthesis of inositol lipids, in the receptor-controlled breakdown of these lipids, and in cellular responses to the signals thus generated.

Moreover, point mutations of normal c-*ras* genes make them tumorigenic (Capon *et al.*, 1983; Shimizu *et al.*, 1983; Yuasa *et al.*, 1981). Possibly such mutations convert these GTP-binding proteins from a receptor-activatable form to a constitutively active state. (6) An obvious candidate for an oncogene product is the phosphodiesterase that hydolyzes PtdIns4,5P$_2$, but we so far have no information on the molecular characteristics of this enzyme, and there have been no comparisons of its activity in transformed and nontransformed cells. (7) An excess of any of the direct or indirect cellular targets of diacylglycerol or InsP$_3$ might lead to uncontrolled proliferation. The only primary target so far identified is protein kinase C (Nishizuka, 1983), but others must include the receptor of the

InsP₃-sensitive Ca^{2+} store and at least some of the many intracellular targets of protein kinase C and Ca^{2+}.

ACKNOWLEDGMENTS. I would like to acknowledge the contributions of colleagues past and present to the results and ideas herein. In particular, Chris Kirk has for several years shared the direction of the studies of vasopressin-stimulated liver cells in Birmingham, and the characterization of VLP in the sympathetic nervous system has mainly been achieved in London by Michael Hanley, Hilary Benton, Stafford Lightman, and Kathryn Todd. I am very grateful to Lewis Cantley and Ian Macara for allowing me access to their data on oncogenes and inositol lipids before publication. Our work in Birmingham has been supported by the Medical Research Council and the Wellcome Trust.

Note Added in Proof

This chapter was written in the midst of very rapid advances in our understanding of the functions of inositol lipids in cellular signalling. Much of this recent information has been gathered in subsequent reviews of protein kinase C (Nishizuka, 1984), of the Ca^{2+}-mobilizing actions of Ins1,4,5P3 (Berridge and Irvine, 1984) and of the present status of inositol lipid-mediated signaling (Downes and Michell, 1985).

References

Akhtar, R. A., and Abdel-Latif, A. A., 1980, Requirement for calcium ions in acetylcholine-stimulated phosphodiesteratic cleavage of phosphatidyl-*myo*-inositol 4,5-bisphosphate in rabbit iris smooth muscle, *Biochem. J.* **192**:783–791.

Balk, S. D., Polimeni, P. I., Hoon, B. S., LeSturgeon, D. N., and Michell, R. S., 1979, Proliferation of Rous sarcoma virus-infected, but not of normal, chicken fibroblasts in a medium of reduced calcium and magnesium, *Proc. Natl. Acad. Sci. U.S.A.* **76**:3913–3916.

Berridge, M. J., 1975, The interaction of cyclic nucleotides and calcium in the control of cellular activity, *Adv. Cyclic Nucleotide Res.* **6**:1–98.

Berridge, M. J., 1983, Rapid accumulation of inositol trisphosphate reveals that agonists hydrolyse polyphosphoinositides instead of phosphatidylinositol, *Biochem. J.* **212**:849–858.

Berridge, M. J., 1984, Inositol trisphosphate and diacylglycerol as intracellular messengers, *Biochem. J.* **220**(2):345–360.

Berridge, M. J., and Irvine, R. F., 1984, Inositol triphosphatate, a novel second messenger in cellular signal transduction, *Nature* **312**:315–321.

Berridge, M. J., Dawson, R. M. C., Downes, C. P., Heslop, J. P., and Irvine, R. F., 1983, Changes in the levels of inositol phosphates after agonist-dependent hydrolysis of membrane phosphoinositides, *Biochem. J.* **212**:473–482.

Bishop, J. M., 1983, Cellular oncogenes and retroviruses, *Annu. Rev. Biochem.* **52**:301–354.

Bone, E. A., Fretten, P., Palmer, S., Kirk, C. J., and Michell, R. H., 1984, Rapid accumulation of inositol phosphates in isolated rat superior cervicial sympathetic ganglia exposed to V_1-vasopressin and muscarinic cholingergic stimuli, *Biochem. J.* 221:803–811.

Capon, D. J., Seburg, P. H., McGrath, J. P., Hayflick, J. S., Edman, U., Levinson, A. D., and Goeddel, D. V., 1983, Activation of Ki-*ras*2 gene in human colon and lung carcinomas by two different point mutations, *Nature* 304:507–513.

Clarke, M. F., Westin, E., Schmidt, D., Josephs, S. F., Ratner, L., Wong-Staal, F., Gallo, R. C., and Reitz, M. S., 1984, Transformation of NIH 3T3 cells by a human c-*sis* cDNA clone, *Nature* 308:464–467.

Cowley, A. W., 1982, Vasopressin and cardiovascular regulation, in: *International Review of Physiology*, Vol. 26 (A. C. Guyton and J. E. Hall, eds.), University Park Press, Baltimore, pp. 189–242.

Creba, J. A., Downes, C. P., Hawkins, P. T., Brewster, G., Michell, R. H., and Kirk, C. J., 1983, Rapid breakdown of phosphatidylinositol 4-phosphate and phosphatidylinositol 4,5-bisphosphate in rat hepatocytes stimulated by vasopressin and other calcium-mobilizing hormones, *Biochem. J.* 212:733–747.

Diringer, H., and Friis, R. R., 1977, Changes in phosphatidylinositol metabolism correlated with growth state of normal and Rous sarcoma virus-transformed Japanese quail cells, *Cancer Res.* 37:2978–2984.

Drummond, A. H., Bushfield, M., and McPhee, C. H., 1984, Thyrotropin-releasing hormone-stimulated [^{3}H]-inositol metabolism in GH3 pituitary turmour cells; studies with lithium, *Mol. Pharmacol.* 25:201–208.

Downes, C. P., and Wusteman, M. M., 1983, Breakdown of polyphosphoinositides and not phosphatidylinositol accounts for muscarinic agonist-stimulated inositol phospholipid metabolism in rat parotid glands, *Biochem. J.* 216:633–640.

Downes, C. P., and Michell, R. H., 1985, Inositol phospholipid breakdown as a receptor-controlled generator of second messengers, in: *Molecular Mechanisms of Transmembrane Signalling* (P. Cohen and M. D. Houslay, eds.) pp. 3–56, Elsevier, New York.

Downward, J., Yarden, Y., Mayes, E., Scrace, G., Totty, N., Stockwell, P., Ullrich, A., Schlessinger, J., and Waterfield, M. D., 1984, Close similarity of epidermal growth factor receptor and v-*erb*-B oncogene protein sequences, *Nature* 307:521–527.

Durham, A. C. H., and Walton, J. M., 1982, Calcium ions and the control of proliferation in normal and cancer cells, *Biosci. Rep.* 2:15–30.

Enyedi, A., Faragó, A., Sarkadi, B., Szász, I., and Gardos, G., 1983, Cyclic AMP-dependent protein kinase stimulates the formation of polyphosphoinositides in the plasma membranes of different blood cells, *FEBS Lett.* 161:158–162.

Farese, R. V., 1984, Phospholipids as intermediates in hormone action, *Mol. Cell. Endocrinol.* 35:1–14.

Farese, R. V., Larson, R. E., and Sabir, M. A., 1982, Insulin acutely increases phospholipids in the phosphatidate–inositide cycle in rat adipose tissue, *J. Biol. Chem.* 257:4042–4045.

Fisher, D. B., and Mueller, G. C., 1968, An early alteration in the phospholipid metabolism of lymphocytes by PHA, *Proc. Natl. Acad. Sci. U.S.A.* 60:1396–1402.

Gomperts, B. D., 1983, Involvement of guanine nucleotide-binding protein in the gating of calcium by receptors, *Nature* 306:64–66.

Grove, R. I., Willis, W. D., and Pratt, R. M., 1983, Dexamethazone affects phosphatidylinositol synthesis and degradation in culture human embryonic tumour cells, *Biochem. Biophys. Res. Commun.* 110:200–207.

Habenicht, A. J. R., Glomset, J. A., King, W. C., Nist, C., Mitchell, C. D., and Ross, R.,

1981, Early changes in phosphatidylinositol and arachidonic acid metabolism in quiescent Swiss 3T3 cells stimulated to divide by platelet-derived growth factor, *J. Biol. Chem.* **256**:12329–12335.

Hagesawa-Sasaki, H., and Sasaki, T., 1983, Phytohaemagglutinin induces rapid degradation of phosphatidylinositol 4,5-bisphosphate and transient accumulation of phosphatidic acid and diaclylglycerol in a human T lymphoblastoid line (CCRF-CEM), *Biochim. Biophys. Acta* **754**:305–314.

Hanley, M. R., Benton, H. P., Lightman, S. L., Todd, K., Bone, E. A., Fretten, P., Palmer, S., Kirk, C. J., and Michell, R. H., 1984, A vasopressin-like peptide in the mammalian sympathetic nervous system, *Nature* **309**:258–261.

Hokin, M. R., and Hokin, L. E., 1953, Enzyme secretion and the incorporation of ^{32}Pi into phospholipids of pancreas slices, *J. Biol. Chem.* **203**:967–977.

Hunter, T., 1982, Phosphotyrosine, a new protein modification, *Trends Biochem. Sci.* **8**: 246–249.

Jard, S., 1980, Oxytocin and vasopressin receptors, in: *Cellular Receptors for Hormones and Neurotransmitters* (D. Schulster and A. Levitzki, eds.), John Wiley & Sons, London, pp. 253–266.

Jones, L. M., and Michell, R. H., 1978, Stimulus–response coupling at alpha-adrenergic receptors, *Biochem. Soc. Trans.* **6**:673–688.

Kelly, K., Cochran, B. H., Stiles, C. D., and Leder, P., 1983, Cell-specific regulation of the c-*myc* gene by lymphocyte mitogens and platelet-derived growth factor, *Cell* **35**:603–610.

Kirk, C. J., Rodriques, L. M., and Hems, D. A., 1979, The influence of vasopressin and related peptides on glycogen phosphorylase activity and phosphatidylinositol metabolism in hepatocytes, *Biochem. J.* **178**:493–496.

Kirk, C. J., Creba, J. A., Downes, C. P., and Michell, R. H., 1981a, Hormone-stimulated metabolism of inositol lipids and its relationship to hepatic receptor function, *Biochem. Soc. Trans.* **9**:377–379.

Kirk, C. J., Michell, R. H., and Hems, D. A., 1981b, Phosphatidylinositol metabolism in rat hepatocytes stimulated by vasopressin, *Biochem. J.* **194**:155–165.

Land, H., Parada, L. F., and Weinberg, R. A., 1983, Tumorigenic conversion of primary embryo fibroblasts requires at least two cooperating oncogenes, *Nature* **304**:596–602.

Lapetina, E. G., and Michell, R. H., 1973, Phosphatidylinositol metabolism in cells receiving extracellular stimulation, *FEBS Lett.* **31**:1–10.

Macara, I. G., Marinetti, G. V., and Balduzzi, P. C., 1984, Transforming protein of avian sarcoma virus UR2 is associated with phosphatidylinositol kinase activity: Possible role in tumorigenesis, *Proc. Natl. Acad. Sci. U.S.A.* **81**:2728–2732.

Macphee, C. H., Drummond, A. H., Otto, A. M., and Jiminez de Asua, L., 1984, Prostaglandin F2alpha stimulates phosphatidylinositol turnover and increases the cellular content of 1,2-diacylglycerol in confluent resting Swiss 3T3 cells, *J. Cell. Physiol.* **119**:35–40.

Michell, R. H., 1975, Inositol phospholipids and cell surface receptor function, *Biochim. Biophys. Acta* **415**:81–147.

Michell, R. H., 1979a, Mechanisms of cell-surface receptors for hormones and neurotransmitters, in: *Companion to Biochemistry*, Vol. 2 (A. T. Bull, J. R. Lagnado, J. O. Thomas, and K. F. Tipton, eds.), Longman, London, pp. 205–228.

Michell, R. H., 1979b, Inositol phospholipids in membrane function, *Trends Biochem. Sci.* **4**:128–131.

Michell, R. H. (ed.), 1982a, Inositol phospholipids and cell calcium, *Cell Calcium* **3**:285–502.

Michell, R. H., 1982b, Inositol lipid metabolism in dividing and differentiating cells, *Cell Calcium* 3:429–440.

Michell, R. H., 1983, Polyphosphoinoside breakdown as the initiating reaction in receptor-stimulated inositol lipid metabolism, *Life Sci.* 32:2083–2085.

Michell, R. H., 1984, Oncogenes and inositol lipids, *Nature* 308:770.

Michell, R. H., Jafferji, S. S., and Jones, L. M., 1977, The possible involvement of phosphatidylinositol breakdown in the mechanism of stimulus-response coupling at receptors which control cell-surface calcium gates, in: *Function and Biosynthesis of Lipids* (N. G. Bazan, R. R. Brenner, and N. M. Giusto, eds.), Plenum Press, New York, pp. 447–464.

Michell, R. H., Kirk, C. J., and Billah, M. M., 1979, Hormonal stimulation of phosphatidylinositol breakdown, with particular reference to the hepatic effects of vasopressin, *Biochem. Soc. Trans.* 7:861–865.

Michell, R. H., Kirk, C. J., Jones, L. M., Downes, C. P., and Creba, J. A., 1981, The stimulation of inositol lipid metabolism that accompanies calcium mobilization in stimulated cells: Defined characteristics and unanswered questions, *Phil. Trans. R. Soc. Lond.* [*Biol.*] 296:123–137.

Michell, R. H., Bone, E. A., Fretten, P., Palmer, S., Kirk, C. J., Hanley, M. R., Benton, H., Lightman, S. L., and Todd, K., 1984a, Inositol lipid breakdown in receptor-mediated responses of sympathetic ganglia and sympathetically innervated tissues, in: *Inositol and Phosphoinositides* (J. E. Bleasdale, J. Eichberg, and G. Hauser, eds.), pp. 221–236, Humana Press, New York.

Michell, R. H., Hawkins, P. T., Palmer, S., and Kirk, C. J., 1984b, Phosphodiesterase-catalysed breakdown of phosphatidylinositol 4,5-bisphosphate initiates receptor-stimulated inositol lipid metabolism, in: *Calcium Regulation in Biological Systems* (S. Ebarhi, M. Endo, K. Imahori, S. Kalciuchi, and Y. Nishizulca, eds.), pp. 85–103, Takeda Foundation, Kyoto.

Mitchell, R. S., Elgas, R. J., and Balk, S. D., 1976, Proliferation of Rous sarcoma virus-infected, but not of normal chicken fibroblasts in oxygen-enriched environment: Preliminary report, *Proc. Natl. Acad. Sci. U.S.A.* 73:1265–1268.

Newbold, R. F., and Overall, R. W., 1983, Fibroblast immortality is a prerequisite for transformation by EJ c-Ha-*ras* oncogene, *Nature* 304:648–651.

Nishizuka, Y., 1983, Calcium, phospholipid turnover and transmembrane signalling, *Phil. Trans. R. Soc., Lond.* [*Biol.*] 302:101–112.

Nishizuka, Y., 1984, The role of protein kinase C in cell surface signal transduction and tumour promotion, *Nature* 308:693–698.

Putney, J. W., 1979, Stimulus–permeability coupling: Role of calcium in the receptor regulation of membrane permeability, *Pharmacol. Rev.* 30:209–245.

Putney, J. W., 1981, Recent hypotheses regarding the phosphatidylinositol effect, *Life Sci.* 29:1183–1194.

Rasmussen, H., 1970, Cell communication, calcium ion and cyclic adenosine monophosphate, *Science* 170:404–412.

Rhodes, D., Prpic, V., Exton, J. H., and Blackmore, P. F., 1983, Stimulation of phosphatidylinositol 4,5-bisphosphate hydrolysis in hepatocytes by vasopressin, *J. Biol. Chem.* 258:2770–2773.

Ribiero, S. M. F., and Armelin, H. A., 1984, Calcium and magnesium requirements for growth are not concomitantly reduced during cell transformation, *Mol. Cell. Biochem.* 59:173–181.

Rozengurt, E., 1983, Growth factors, cell proliferation and cancer: An overview, *Mol. Biol. Med.* 1:169–181.

Rozengurt, E., and Sinnett-Smith, J., 1983, Bombesin stimulation of DNA synthesis and cell division in cultures of Swiss 3T3 cells, *Proc. Natl. Acad. Sci. U.S.A.* **80:**2936–2940.

Ruley, H. E., 1983, Adenovirus early region 1A enables viral and cellular transforming genes to transform primary cells in culture, *Nature* **304:**602–606.

Sarkadi, B., Enyedi, A., Farago, A., Meszaros, G., Kremmer, T., and Gardos, G., 1983, Cyclic AMP-dependent protein kinase stimulates the formation of polyphosphoinositides in lymphocyte plasma membrane, *FEBS Lett.* **152:**195–198.

Sawyer, S. T., and Cohen, S., 1981, Enhancement of calcium uptake and phosphatidylinositol turnover by epidermal growth factor in A431 cells, *Biochemistry* **20:**6280–6286.

Shukla, S. D., Buxton, D. B., Olson, M. S., and Hanahan, D. J., 1983, Acetylglyceryl ether phosphorylcholine: A potent activator of hepatic phosphoinositide metabolism and glycogenolysis, *J. Biol. Chem.* **258:**10212–10214.

Shimizu, K., Birnbaum, D., Ruley, M. A., Fasano, O., Suard, Y., Edlund, L., Taparowsky, E., Goldfarb, M., and Wigler, M., 1983, Structure of the Ki-*ras* gene of the human lung carcinoma cell line Calu-1 *Nature* **304:**497–500.

Streb, H., Irvine, R. F., Berridge, M. J., and Shulz, I., 1983, Release of Ca^{2+} from a nonmitochondrial intracellular store in pancreatic acinar cells by inositol 1,4,5-trisphosphate, *Nature* **306:**67–69.

Sugimoto, Y., Whitman, M., Cantley, L. C., and Erikson, R. L., 1984, Evidence that the Rous sarcoma transforming gene product phosphorylates phosphatidylinositol and diacylglycerol, *Proc. Natl. Acad. Sci. U.S.A.* **81:**2117–2121.

Sukumar, S., Notario, V., Martin-Zanca, D., and Barbacid, M., 1983, Induction of mammary carcinomas in rats by nitrosomethylurea involves malignant activation of H-*ras*-1 locus by single point mutations, *Nature* **306:**658–661.

Takhar, A.P.S., and Kirk, C. J., 1981, Stimulation of inorganic-phosphate incorporation into phosphatidylinositol in rat thoracic aorta mediated through V_1-vasopressin receptors, *Biochem. J.* **194:**167–172.

Thomas, A. P., Marks, J. S., Coll, K. E., and Williamson, J. R., 1983, Quantitation and early kinetics of inositol lipid changes induced by vasopressin in isolated and cultured hepatocytes, *J. Biol. Chem.* **258:**5716–5725.

Troyer, D. A., Kreisberg, J. I., Schwertz, D. W., and Venkatachalem, M. A., 1983, Phosphatidyinositol turnover in cultured rat glomerular mesangial cells exposed to vasopressin, *Fed. Proc.* **42:**1259.

Vu, N.-D., Chepko, G., and Zelenka, P., 1983, Decreased turnover of phosphatidylinositol accompanies *in vitro* differentiation of embryonic chicken lens epithelial cells into lens fibres, *Biochim. Biophys. Acta* **750:**105–111.

Weiss, S. J., McKinney, J. S., and Putney, J. W., 1982, Receptor-mediated net breakdown of phosphatidylinositol 4,5-bisphosphate in parotid acinar cells, *Biochem. J.* **206:**555–560.

Whitfield, J. F., Boynton, A. L., Macmanus, J. P., Sikorska, M., and Tsang, B. K., 1979, The regulation of cell proliferation by calcium and cyclic AMP, *Mol. Cell. Biochem.* **27:**155–179.

Yuasa, Y., Srivastava, S. K., Dunn, C. Y., Rhim, J. S., Reddy, E. P., and Aaronson, S. A., Acquisition of transforming properties by alternative point mutations within c-*bas*/ *has* human protooncogene, *Nature* **303:**775–779.

Zelenka, P. S., 1980, Changes in phosphatidylinositol metabolism during differentiation of lens epithelial cells into lens fibres in the embryonic chick, *J. Biol. Chem.* **255:**1296–1300.

Requirements for Steroid Hormone Action in Eucaryotic Cells

BERT W. O'MALLEY

1. Introduction

Over the past decade, work from many laboratories has led to a generally accepted hypothesis for the primary synthetic events involved in steroid hormone action. Although no universal agreement exists for the specific molecular details, the overall steps in the pathway have been elucidated.

Receptors for steroid hormones, initially described by Jensen and co-workers (1962), are tissue-specific binding proteins for steroidal ligands that have high affinity ($K_d \sim 10^{-10}$ M) for hormone and exist in low concentration ($2-6 \times 10^4$ molecules) in target cells. On binding the entering hormone, the receptor appears to undergo an ill-defined "activation" reaction, which enhances the affinity of receptor for nuclear interphase chromosomes. The activated hormone–receptor complex accumulates in the nucleus and is found bound to chromosomal DNA. Within 30 min, synthesis of new high-molecular-weight precursor to messenger RNA is initiated. The aggregate experimental evidence from many model systems and laboratories indicates that the level of DNA transcription is the primary focal point of steroid action in target cells. Following accumulation of precursor mRNA, these molecules are processed in a complex splicing reaction so that all intervening sequence (intron) RNA is removed. At this point, the mature mRNA can now relocate to the cytoplasm, attach to ribosomes, and code for the hormone-mediated synthesis of enzymes and structural or secretory proteins (Ringold *et al.*, 1977; Harris *et al.*, 1975; Chan *et al.*, 1973, Rhoads *et al.*, 1971).

BERT W. O'MALLEY • Department of Cell Biology, Baylor College of Medicine, Houston, Texas 77030.

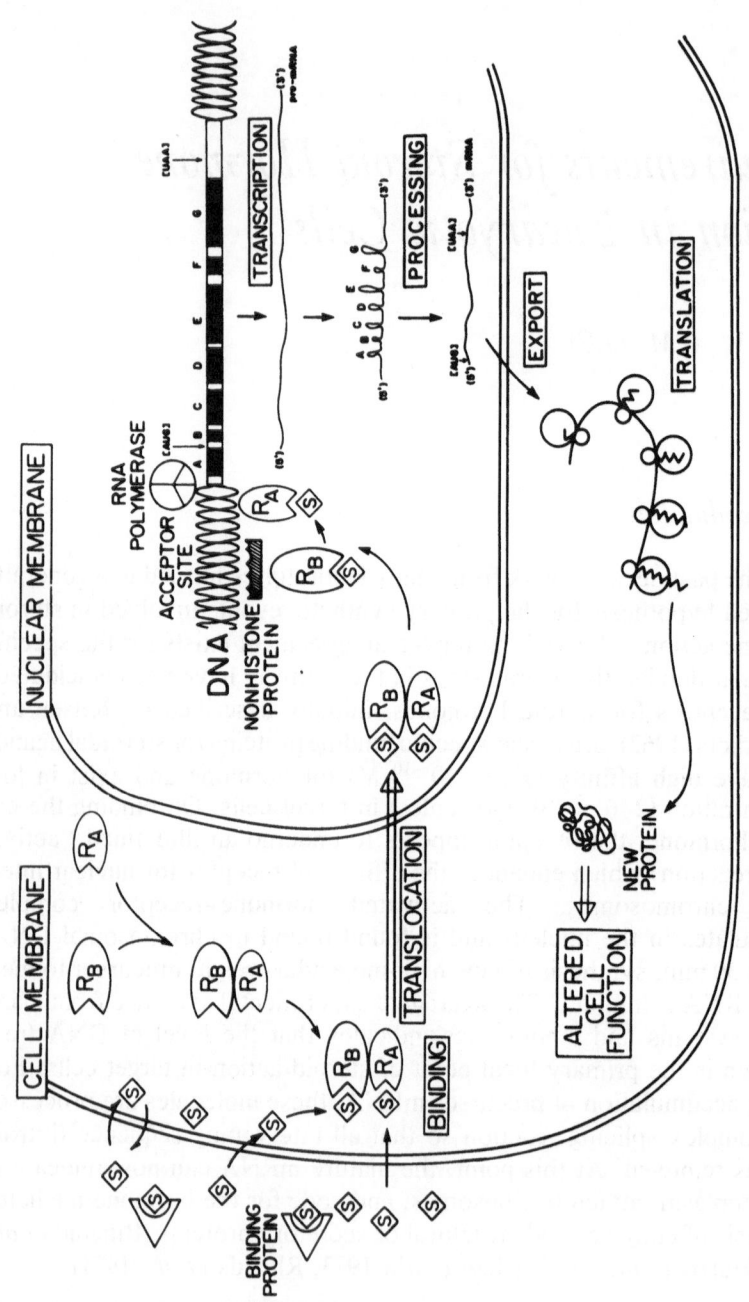

Figure 1. Pathway for progesterone-evoked synthesis of proteins.

In the chick oviduct, a series of investigations designed to define the pathway for progesterone (and estrogen) effects on egg-white protein (e.g., ovalbumin, ovomucoid, avidin) synthesis has led us to postulate the specific sequence of events shown in Fig. 1 (O'Malley *et al.*, 1969, 1979; Schrader *et al.*, 1981).

It has been postulated that hormones act primarily at the level of DNA transcription from the following lines of evidence. Hormone–receptor complexes accumulate in the nuclear compartment and bind to chromosomal DNA with high affinity. Steroid-regulatable genes contain sequences in their 5'-flanking regions that preferentially bind receptors with an affinity greater than average DNA (Payvar *et al.*, 1981, Compton *et al.*, 1983). Removal of these 5'-flanking sequences prevents induction of gene expression by steroid hormones (Hynes *et al.*, 1983; Dean *et al.*, 1983; Renkowitz *et al.*, 1982). Precursor mRNA and mature mRNA both accumulate in response to steroid hormone action. Following accumulation of steroid hormone receptors on the nuclear chromosomes, radiolabeling experiments demonstrate that synthesis of nascent pre-mRNA is stimulated (Ringold *et al.*, 1975; Swaneck *et al.*, 1979; McKnight *et al.*, 1975; Chan *et al.*, 1973). Thus, although steroids can have an effect on mRNA half-life under certain conditions, their primary action appears to occur at the DNA or transcription level.

2. Receptors for Progesterone

In chick oviduct cells, we have reported the existence of a receptor complex composed of two hormone-binding subunits (Schrader *et al.*, 1981; Schrader and O'Malley, 1972; Sherman *et al.*, 1970) (Fig. 1). Subunit A (mol. wt. 79,000), which binds to deproteinized DNA with high affinity ($K_d \sim 10^{-10}$ M), is considered to be the "effector" subunit (O'Malley *et al.*, 1972; Spelsberg *et al.*, 1972). Subunit B (mol wt. 105,000), which binds to interphase chromosomes ($K_d \sim 5 \times 10^{-9}$ M), has been postulated to play a "specifier" role in chromosomal localization of hormone–receptor complex (McKnight *et al.*, 1975). The structural proof and structure–function speculations have been published in detail previously (O'Malley *et al.*, 1972; Grody *et al.*, 1982). It is noteworthy that a similar subunit structure has been reported recently for the progesterone receptor in human cultured cells (Lessey *et al.*, 1983).

3. Postreceptor Specificity

In the present chapter, we wish to emphasize the complexity of steroid hormone action at the level of the intact cell. We suggest that steroid

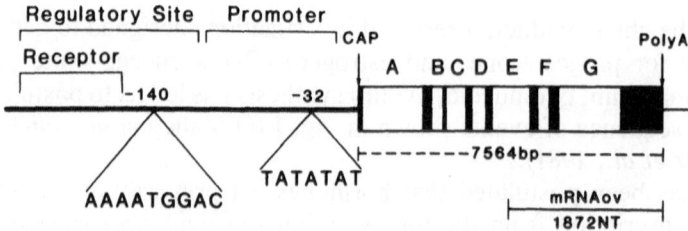

Figure 2. Primary structure of the ovalbumin gene.

hormone regulation of gene expression may require a series of coordinate structural interactions occurring at least at four separate levels of cellular organization.

Although receptors for steroid hormones appear to be critical molecules necessary for implementing intracellular hormone action, they are obviously not sufficient of themselves. Since cells with apparently identical receptors can activate different genes on stimulation by the same hormone, it is clear that other structural determinants must exist that either limit or facilitate the access of the hormone–receptor complexes to their appropriate nuclear regulatory sites. These determinants cannot exist entirely within the primary sequence of DNA because it appears that the structural and regulatory sequences within and around eucaryotic genes are essentially the same in all cell types of a given individual. Thus, if steroid hormones regulate protein synthesis and cell growth primarily by regulating mRNA synthesis and not RNA processing, but each regulatable cell has the identical complement of genes and the same receptor, one must then look to explanations that involve higher-order chromosomal or nuclear structure. In fact, it appears that to adequately explain the molecular basis of steroid hormone action, we must consider at least the structural relationships and interactions of receptors, structural and flanking sequences of genes, chromosomal differentiation-specific domains, and the nuclear matrix or skeleton.

In the chick oviduct, we have characterized four genes (ovalbumin, X, Y, and ovomucoid) in terms of their primary structure and inducibility in response to hormone administration. The ovalbumin gene (Fig. 2) has been completely sequenced, together with its surrounding genomic regions (Woo *et al.*, 1981; Breathnach *et al.*, 1977). The mature mRNA of 1872 nucleotides is generated by a series of complex splicing reactions from a high-molecular-weight precursor of 7564 bp transcribed intact from the gene itself. The split nature of the gene, containing eight exons and seven introns, has been described in detail elsewhere (Spelsberg *et al.*, 1972; Grody *et al.*, 1982). Our more recent evidence suggests that the ovalbumin gene contains certain sequences in its adjacent 5'-flanking region that are considered important for both accurate and efficient expression.

4. Induction of Transcription

Accurate initiation of transcription of the ovalbumin gene appears to be influenced primarily by the TATATAT box located at -32 base pairs upstream from the first nucleotide $(+1)$ of the structural gene (Fig. 2). This AT-rich heptamer acts as a specifier sequence in that it directs RNA polymerase II to initiate transcription at the proper site located 32 bases downstream. If as little as a single base pair is changed to a G–C within the TATA box, accurate transcription is abolished (O'Malley *et al.*, 1972, 1979). This result is supported by similar observations in a number of other laboratories for a variety of genes (Corden *et al.*, 1980).

Although the TATA box may be considered a part of the eucaryotic promoter region, it can be no more than one of the sequence participants because it does not have a great capacity to modulate the rate of gene transcription. In fact, sequences located slightly further upstream from the TATA box (-95 to -48) form an important part of the basal promoter. The "hormone control" region, however, appears to be structurally separate and is located further upstream (-222 to -95). These regions are shown schematically in Fig. 2, and the experimental evidence for this hypothesis is discussed in more detail below. Finally, we have attempted to ascertain the importance of steroid receptor–DNA interactions in the hormone-regulatable induction of ovalbumin gene expression. Toward this end, we have searched for DNA sequences that might display a capacity to preferentially attract steroid receptors. In fact, such a region has been identified within this hormone control region (-200 to -150) for the ovalbumin gene (Compton *et al.*, 1983) and is also shown in Fig. 2. This region of the genome binds the A subunit of the progesterone receptor of chick oviduct with an order of magnitude higher affinity compared to other nonspecific DNA sequences.

Of major interest have been the results of our recent experiments designed to determine the region of 5'-flanking sequence near the ovalbumin gene that is required for hormone-mediated induction of transcription of this gene. We have employed a tissue culture transfection system in which an ovalbumin globin fusion gene (ovalglobin) cloned together with SV40 and plasmid sequences is used to transiently transfect cells in culture (Dean *et al.*, 1983; Knoll *et al.*, 1983). This "fusion" gene consists of the 5' region of the chicken ovalbumin gene (-753 to $+41$) and the structural region of the chicken β-globin gene ($+115$ to $+1479$). We found this experimental model satisfactory because it utilized a homologous hybrid gene of small size that should produce a globin transcript, easily identifiable in untransformed oviduct cells after *in vitro* transfer experiments. Since it still retains the putative ovalbumin promoter, we would systematically alter the 5'-flanking sequences and monitor their regulatory potential after reintroduction into oviduct cells. Finally, the same recombinant contains the SV40 early region genes for T antigen, which are not hormonally regulated and serve as an internal control for quantification in acute gene transfer experiments.

When this gene was transfected into HeLa S_3 cells, ovalglobin gene transcription initiated at the ovalbumin gene cap site, as measured by S1 nuclease and primer extension analysis. However, no regulation of transcription by steroid hormones (progesterone, estradiol, etc.) was observed in these cells or in a number of heterologous cells (TC7, MCF7, T47D, etc.). Nevertheless, since the basal promoter allowed for efficient and accurate transcription, we utilized the heterologous transfer systems to define the hormone-independent promoter for the ovalbumin gene. We then constructed a series of 5'-flanking deletions using a construction method that insures that each deletion end point is joined to exactly the same pBR322 sequences.

The results of these studies indicate that the basal promoter sequence lies between -95 and the cap site. In fact, the rate of transcription appears most dependent on the 29-nucleotide sequence between -77 and -48 (Knoll *et al.*, 1983).

In a separate series of experiments, we were surprised to find that transfection of the ovalglobin gene into untransformed, primary monolayer cultures of oviduct tubular gland cells led to a significant (five- to 20-fold) induction of ovalglobin RNA when progesterone was added to the culture. This observation suggested that some tissue-specific factor required for the induction and this factor was present only in the homologous cell type. Deletion of critical 5'-flanking sequences near the gene abolished the induction. The results have been described in detail elsewhere (Dean *et al.*, 1983; Knoll *et al.*, 1983). Oviduct cells in primary monolayer culture were exposed to either intact ovalglobin (-753) or deletion constructs in which all but 323, 222, or 95 nucleotides of 5'-flanking sequence of ovalbumin gene had been removed. Half of the cultures were then exposed to progesterone. At the end of 36–48 hr, the RNA was extracted, electrophoresed on a denaturing gel, and hybridized directly to β-globin and SV40 early gene probes. The presence of progesterone stimulated the accumulation of transcripts of the intact ovalglobin gene. Deletion of sequences 5' of -95 eliminated most, if not all, of the progesterone-mediated induction of transcripts. Deletion of sequences located 5' to -222 still allowed regulation of transcription by progesterone, as did another deletion to -323. Neither the deletions themselves nor the presence of progesterone had any effect on the level of SV40 early gene transcripts.

These results, taken together, demonstrate that the hormone regulatory region and the basal promoter for the ovalbumin gene are distinct sequences with no obvious overlap. Why these two transcriptional control regions within the proximal 5'-flanking sequence of the same gene have evolved as separate structural entities can serve only as a speculative question at present. Nevertheless, it is clear that DNA rearrangement or transpositions that result in the relocation of the hormone regulatory region to regions in proximity with the promoters of unrelated structural genes can bring these genes under control of steroid hormones. In such a way, evolutionary forces can create a network of structural genes that are under the control of a single hormone.

These experiments indicated that the removal of upstream sequences in the region of -95 to -222 leads to the elimination of the capacity to respond to hormone. It was of considerable interest that the sequences containing both the capacity to respond to hormone and the preferential DNA binding site for receptor were located within this same region of the genome (Compton *et al.*, 1983, 1985; Davidson *et al.*, 1983) (Fig. 2).

We have raised an additional question as to the mechanism by which more than one hormone can activate the same structural gene. Again, the ovalbumin gene provides a suitable model to define such a phenomenon, since the ovalbumin gene can be regulated by three hormones, namely, progesterone, estrogen, and glucocorticoid. Using the identical set of deletions described above, we have found that removal of sequences in the region of -95 to -222 leads to a simultaneous loss of response to all three hormones. Although the precise site of interaction of each hormone–receptor complex remains to be determined in order to claim sequence identity, it appears that the control sequences for separate receptors are at least overlapping in nature.

We and others (Renkowitz *et al.*, 1982) have searched for sequences in this region of the ovalbumin gene that might be common to the 5'-flanking regions of other chicken genes responsive to the same hormone. Surprisingly, six genes that are regulated by progesterone (or estrogen) have a common sequence located at a similar position (-140) upstream from the gene. Considering these sequences as a group, we can derive a nine-base-pair consensus sequence of AAAAT(G/T)G(G/A)C. The statistical likelihood of this occurring by chance is low ($\sim 10^{-6}$) (Davidson *et al.*, 1983). Moreover, three other genes of the chicken (globin, α-actin, and collagen), which are not regulated by steroid hormones, do not contain this sequence in the same location. The requirement for this intriguing sequence in regulatory control must now be investigated by *in vitro* mutagenesis experiments in which all or part of the nomeric sequence is removed or altered while the adjacent sequences are left intact.

5. Tissue-Specific Factors

As mentioned above, the entire picture relating to steroid hormone regulation of gene expression cannot be explained by even the most explicit definition of receptors and primary DNA structure. We need to understand why the same receptors do not regulate the same gene sequences in different cells. To clarify this "nuclear capacity" to respond to hormones, we are required to use a biochemical probe that could reliably distinguish various higher-order structural states of eucaryotic interphase chromosomes. To this end, we utilized the DNase I digestion assay (Weintraub and Grondine, 1976) to determine structural differences in the genomic area containing the ovalbumin gene family relative to other cells in which these same genes are not expressed. This work has been published in detail elsewhere (Lawson *et al.*, 1982; Stumph *et al.*, 1983) and is only summarized here to illustrate conceptual points.

Approximately 100 kb of DNA containing and surrounding the X, Y, and ovalbumin genes exhibit a preferential sensitivity to DNase I in the chromatin of oviduct cells. In other tissues, where these genes are not expressed, no such preferential sensitivity is observed. We interpret this to mean that in cells where these genes are to be expressed, the surrounding chromatin DNA in this 100-kb domain is packaged differently from the majority of the DNA in the bulk of the chromatin. In chicken, the ovomucoid, GAPDH, U1, vitellogenin, and globin genes all exist in DNase-1-sensitive chromosomal domains of varying sizes in the tissues in which they are expressed. Presumably, all active genes share a similar altered structure in regard to their chromosomal configuration.

However, the fact that a gene is in the DNase-I-sensitive state is not entirely sufficient for it to be actively transcribed. This is demonstrated by the fact that the DNase I sensitivity of the entire domain persists in nuclei isolated from hormonally withdrawn chicks in spite of the shutdown of ovalbumin gene transcription. An analogous result was obtained for the globin gene itself in transcriptionally inactive erythrocytes. Therefore, DNase I sensitivity appears to reflect a more accessible chromatin structure, which in turn relates to the development capacity of a cell to express the gene in question. It can be viewed as a necessary but not wholly

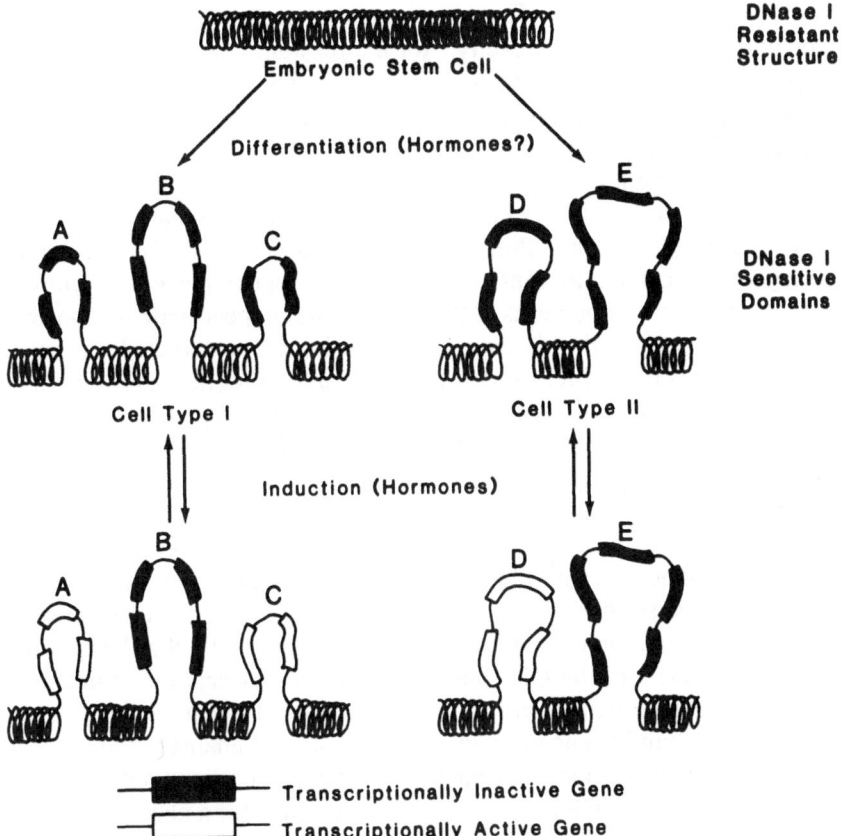

Figure 3. Relationship between cell differentiation and DNase I sensitivity of tissue-specific genes—a working model.

sufficient step in the prior commitment of a cell to allow a certain gene to be expressed. Such a mechanism would make it possible for distinct cell types to respond to a single inducer, each in its own individual and distinctive manner (Fig. 3).

In other words, all genes that are ever to be expressed in a given cell must be contained within these unraveled or accessible regions of chromatin at the time of terminal differentiation. The chromosomal domains appear to be related to molecular differentiation since they are tissue-specific and "irreversible." The DNA that is not contained in these domains could be passively packaged into a more complex higher-order structure by histones. The DNA in such higher-order structure, the majority of DNA in each cell type, would be unavailable for interactions with regulatory molecules. The containment of such genes in these structures only provides the "capacity" for expression. Once included in this "expressible" domain, it is now accessible to regulatory factors or influences such as hormone–receptor complexes (Fig. 3).

Thus, it is important to understand what mechanisms are involved in the establishment of DNase-I-sensitive domains during cellular differentiation. Initially, we have chosen to look for possible signals in the DNA itself. Sequences either within or near the regions where the chromatin undergoes the change in configuration seem a logical place to begin the search. Gross rearrangements of the DNA do not appear to be responsible for establishment of the domain, since we have not detected any such rearrangements in oviduct tissue as compared to other tissues (Stumph *et al.*, 1983). Logically, it seems quite possible that repetitive DNA sequences could play a role in defining domains. We have observed that a certain subset of these sequences belongs to a single family of repeats termed CR1. These CR1 family members are preferentially located near the transition regions at each end of the domain. We have also assigned a nucleotide sequence polarity to the CR1 sequences to find that they are always oriented so that their 3' ends are pointing to the included gene set in the domain (i.e., inverted repeats). This "polarity rule" has been noted in seven members of the CR1 family located in four separate domains.

If the CR1 sequences are potentially involved in establishing or defining the ovalbumin gene domain as well as other domains in the chicken cell chromatin, what kinds of mechanisms might be involved? A likely possibility is that the CR1s may act as binding sites for a class of proteins that induce conformational changes in the chromatin. Although other mechanisms not involving protein–DNA interactions are certainly possible, recent experiments performed in our laboratory have provided encouragement for us to search for the existence of CR1 binding proteins. We have devised a new method to purify cellular sequence-specific binding proteins, which will be published elsewhere (Weintraub and Grondine, 1976). From total mixtures of nuclear DNA binding proteins, we can selectively purify a protein that has a preferential affinity for members of the CR1 middle-repeat family. The biological function of these fascinating proteins has not yet been elucidated.

6. Role of the Nuclear Matrix

Finally, it would be appropriate to conclude by discussing an even more complex structural interaction of cellular genes and genomic domains with the nuclear matrix (Berezney and Coffey, 1974; Capco *et al.*,

1982). In a series of recently published experiments, we have investigated the possibility that selected regions of genomic DNA might be attached to the nuclear matrix or nucleoskeleton. Although the nuclear matrix has not been rigorously defined in biochemical or cell-biological terms, its potential importance in DNA synthesis (Berezney and Coffey, 1974; Robinson *et al.*, 1982) and processing of mRNA precursors (Ciejek *et al.*, 1982) has been documented. In recent experiments, we questioned whether the differential structural attachment of genomic sequences might cause these sequences to be made more or less available to the transcription apparatus.

For these studies, we isolated interphase chromosomes that had been treated previously with 2 M NaCl to remove histones and loosely adherent nuclear proteins. This preparation has been analyzed by others using electron microscopy and sedimentation analysis; the uncoiled DNA attached to the residual protein matrix exists in "loop" form with an average length of 30–100 kb (Paulson and Laemmb, 1977).

Our approach was to release the dehistonized unattached DNA present in the loops by digestion with a site-specific restriction endonuclease (Ciejek *et al.*, 1983). Under the conditions used here, 80–90% of the total DNA bound to the protein matrix was released by the enzymatic digestion. After separation and purification of the DNA in the bound (10% of total) and released (90% of total) fractions, the presence of specific sequences was assayed by the Southern blot–hybridization technique using defined cloned chicken DNA probes. Since equal amounts of DNA from the released and bound fractions are applied to the gel, preferential association of a specific sequence in the bound fraction should be detectable because that fraction represents only 10% of the total DNA. A nonspecific association would result in a random distribution of sequences between bound and released fractions, and in that case no enrichment would be seen. The pattern of hybridization in the released fraction was found to be substantially enriched in ovalbumin DNA sequences. In contrast, vitellogenin sequences were depleted in the matrix-bound fraction. Since the ovalbumin gene is transcribed in the oviduct and vitellogenin gene is not transcribed, this suggests that transcriptional activity of a gene may be associated with the nuclear matrix.

The results revealed that neither DNase-I-resistant DNA outside the ovalbumin family domain nor DNase-I-sensitive, but not expressed, DNA within the domain was preferenitally bound to the nuclear matrix (Ciejek *et al.*, 1983). In contrast, the entire ovalbumin gene was markedly enriched in the matrix-bound DNA fraction. Moreover, the X and Y genes, expressed at markedly different rates but in response to the same hormone, were also preferentially attached to nuclear matrix. Finally, other genes such as ovomucoid, present in separate DNase-I-sensitive domains, were also found attached to the oviduct matrix, but genes not expressed in this tissue (e.g., globin and vitellogenin) were not bound to the matrix. In contrast, the ovalbumin gene was not associated with the nuclear matrix in hen or rooster liver, a tissue in which it is not expressed. Perhaps more importantly, our results showed that if steroid hormone was withdrawn from the oviduct, the ovalbumin gene was found to be no longer associated with the matrix (Ciejek *et al.*, 1983).

In fact, we found that all the actively transcribed gene sequences that we tested were associated with the nuclear matrix, whereas nontranscribed sequences were localized in the unattached "loop" structures and could be released by restriction nuclease treatment. This conclusion is consistent with the idea that the nucleus is a highly organized organelle and that transcription may occur on a matrix structure rather than free in solution. This attachment to the matrix could either facilitate transcription of DNA by RNA polymerase or be a concomitant of transcription. Cessation of transcription in the presence of actinomycin D does not itself lead to release of genes from the matrix.

It is interesting to note that steroid hormone receptors also have been found associated with salt-insoluble nuclear subfractions and the nuclear matrix (Clark and Peck, 1976; Barrack and Coffey, 1980). On hormonal withdrawal, the receptors were no longer associated with the nuclear matrix. Although receptors could play some role in the attachment of the inducible genes to the nuclear matrix, it is unlikely that the receptor is the sole protein component binding the active gene to the matrix structure.

7. Conclusions

In conclusion, it is fair to speculate that the cellular forces involved in steroid hormone induction of gene expression are complex indeed. These parameters are summarized in Fig. 4. At the present time, our best guess on the major structural determinants for induction are as follows: (1) steroid receptor is the obligatory and active intermediate to transduce the informational signal inherent in the hormone to the regulatable gene; (2) the linear sequence of the gene itself is of obvious importance since it not only contains the inherited structural code for the protein but appears to contain structurally distinct "promoter" and "regulatory" sequences, the latter of which both binds receptor and determines the maximal rate of hormone-induced gene expression; (3) inducible genes are contained within large structurally distinct (DNase-I-sensitive) domains, which are an index of molecular differentiation, and which are likely to

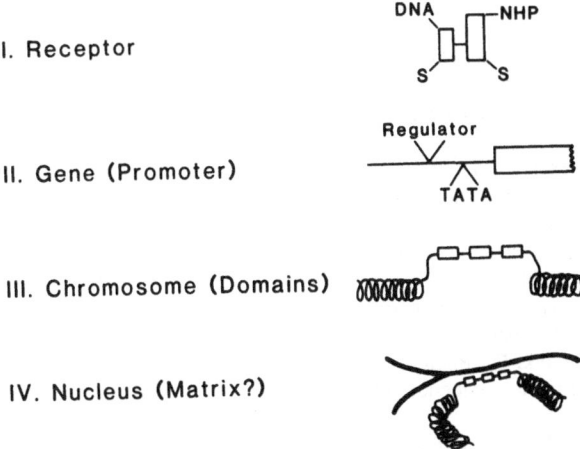

I. Receptor

II. Gene (Promoter)

III. Chromosome (Domains)

IV. Nucleus (Matrix?)

Figure 4. Determinants of hormonal induction of gene expression.

maintain the capacity of genes to respond to inductive influences; and (4) the chromatin itself undergoes a specific attachment to the nuclear matrix so that actively expressed regions of these domains appear to be more firmly bound and, perhaps, more easily transcribed by the nuclear tran-scriptive appartus. This picture is only complicated further by consid-eration of other potentially important levels of substructure such as mod-ification of primary DNA sequence (e.g., methylation and z-DNA) and chromatin fine structure (DNase hypersentitivity).

It appears safe to speculate that only by obtaining more precise struc-tural and functional information on each of these levels of regulation can we understand completely the molecular mechanism of steroid hormone action. To accomplish this task, a continued application of the combined technologies of molecular biology and cell biology will most certainly be required.

References

Barrack, E. R., and Coffey, D. S., 1980, The specific binding of the estrogens and androgens to the nuclear matrix of sex hormone responsive tissues, *J. Biol. Chem.* **255**:7265–7275.

Berezney, R., and Coffey, D. S., 1974, Identification of a nuclear protein matrix, *Biochem. Biophys. Res. Commun.* **60**:1410–1417.

Breathnach, R., Mandel, J. L., and Chambon, P., 1977, Ovalbumin gene is split in chicken DNA, *Nature* **270**:314–319.

Capco, D. G., Wan, K. M., and Penman, S., 1982, The nuclear matrix: Three dimensional architecture and protein composition, *Cell* **29**:847–858.

Chan, L., Means, A. R., and O'Malley, B. W., 1973, Rates of induction of specific translatable messenger RNAs for ovalbumin and avidin by steroid hormones, *Proc. Natl. Acad. Sci. U.S.A.* **70**:1870–1874.

Ciejek, E. M., Nordstorm, J. L., Tsai, M. J., and O'Malley, B. W., 1982, Ribonucleic acid precursors are associated with the chick oviduct nuclear matrix, *Biochemistry* **21**:4945–4953.

Ciejek, E. M., Tsai, M. J., and O'Malley, B. W., 1982, Actively transcribed genes are associated with the nuclear matrix, *Nature* **306**:607–609.

Clark, J. H., and Peck, E. J., Jr., 1976, Nuclear retention of receptor–oestrogen complex and nuclear acceptor sites, *Nature* **260**:635–637.

Compton, J. G., Schrader, W. T., and O'Malley, B. W., 1983, DNA sequence preference of the progesterone receptor, *Proc. Natl, Acad. Sci. U.S.A.* **80**:16–20.

Compton, J. G., Schrader, W. T., and O'Malley, B. W., 1985, (in preparation).

Corden, J., Waslylyk, B., Buchalder, A., Sassone-Corsi, P., Kedinger, C., and Chambon, P., 1980, Promoter sequences of eukaryotic protein-coding genes, *Science* **209**:1406–1414.

Davidson, E. H., Jacobs, H. T., and Britten, R. J., 1983, Very short repeats and coorinate induction of genes, *Nature* **301**:468–470.

Dean, D. C., Knoll, B. J., Riser, M. E., and O'Malley, B. W., 1983, A 5' flanking sequence essential for progesterone regulation of an ovalbumin fusion gene, *Nature* **305**:551–558.

Dugaiczyk, A., Woo, S. L. C., Lai, E. C., Mace, M. L., Jr., McReynolds, L., and O'Malley, B. W., 1978, The natural ovalbumin gene contains seven intervening sequences, *Nature* **274**:328–333.

Grody, W., Schrader, W. T., and O'Malley, B. W., 1983, Activation, transformation and subunit structure of steroid-hormone receptors, *Endocrine Rev.* **3**:141–163.

Harris, S. E., Rosen, J. M., Means, A. R., and O'Malley, B. W., 1975, Use of a specific probe for ovalbumin messenger RNA to quantitate estrogen-induced gene transcripts, *Biochemistry* **14**:2072–2081.

Hynes, N., van Ooyen, A. J. J., Kennedy, N., Herrlich, P., Ponta, H., and Groner, B., 1983, Subfragments of the large terminal repeat cause glucocorticoid-responsive expression of mouse mammary tumor virus and of an adjacent gene, *Proc. Natl. Acad. Sci. U.S.A.* **80**:3637.

Jensen, E. V., and Jacobson, H. I., 1962, *Recent Prog. Horm. Res.* **18**:387–414.

Knoll, B. J., Zarucki-Schulz, T., Dean, D. C., and O'Malley, B. W., 1983, Definition of the ovalbumin gene promoter by transfer of an ovalglobin fusion gene into cultured cells, *Nucl. Acid Res.* **11**:6733–6754.

Lawson, G. M., Knoll, B. J., March, C. J., Woo, S. L. C., Tsai, M. J., and O'Malley, B. W., 1982, Definition of 5' and 3' structural boundaries of the chromatin domain containing the ovalbumin multigene family, *J. Biol. Chem.* **257**:1501–1507.

Lessey, B. A., Alexander, P. S., and Horwitz, K. B., 1983, The subunit structure of human breast cancer progesterone receptors: Characterization by chromatography and photoaffinty labeling, *Endocrinology* **112**:1267–1274.

McKnight, G. S., Pennequin, P., and Schimke, R. T., 1975, Induction of ovalbumin mRNA sequences by estrogen and progesterone in chick oviduct as measured by hybridization to complementary DNA, *J. Biol. Chem.* **250**:8105–8110.

O'Malley, B. W., McGuire, W. L., Kohler, P. O., and Korenman, S. G., 1969, Studies on the mechanism of steroid hormone regulation of the synthesis of specific proteins, *Recent Prog. Horm. Res.* **25**:105–160.

O'Malley, B. W., Spelsberg, T. C., Schrader, W. T., Chytil, F., and Steggles, A. W., 1972, Mechanisms of interaction of a hormone–receptor complex with a genome of a eukaryotic target cell, *Nature* **235**:141–144.

O'Malley, B. W., Roop, D. R., Lai, E. C., Nordstrom, J. L., Catterall, J. F., Swaneck, G. E., Colbert, D. A., Tsai, M. J., Dugaiczyk, A., and Woo, S. L. C., 1979, The ovalbumin gene: Organization, structure, transcription and regulation, *Recent Prog. Horm. Res.* **35**:1–46.

Paulson, J. R., and Laemmli, U. K., 1977, The structure of histone-depleted metaphase chromosomes, *Cell* **12**:817–828.

Payvar, F., Wrange, O., Carlstedt-Duke, J., Okret, S., Gustafasson, J. A., and Yamamoto, K. R., 1981, Purified glucocorticoid receptors bind selectively *in vitro* to a cloned DNA fragment whose transcription is regulated by glucocorticoids *in vivo*, *Proc. Natl. Acad. Sci. U.S.A.* **78**:6628–6637.

Renkawitz, R., Bueg, H., Graf, T., Mathias, P., Grez, M., and Schutz, 1982, Expression of a chicken lysozyme recombinant gene is regulated by progesterone and dexamethasone after microinjection into oviduct cells, *Cell* **31**:167–176.

Rhoads, R. E., McKnight, G. S., and Schimke, R. T., 1971, Synthesis of ovalbumin in a rabbit reticulocyte cell-free system programmed with hen oviduct ribonucleic acid, *J. Biol. Chem* **246**:7407–7410.

Ringold, G., Yamamoto, K. R., Bishop, J. M., and Varmus, H. E., 1977, Glucocorticoid-stimulated accumulation of mouse mammary tumor virus RNA increased rate of synthesis of viral RNA, *Proc. Natl. Acad. Sci. U.S.A.* **74**:2879–2883.

Ringold, G. M., Yamamoto, K. R., Tomkins, G. M., Bishop, J. M., and Varmus, H. E., 1975, Dexanetgasibe-mediated induction of mouse mammary tumor virus RNA: A system for studying glucocorticoid action, *Cell* **6**:299–305.

Robinson, S. I., Nolkin, B. D., and Vogelstein, B., 1982, The ovalbumin gene is associated with the nuclear matrix of chicken oviduct cells, *Cell* **28**:99–106.

Sanzo, M. A., Stevens, B., Tsai, M. J., and O'Malley, B. W., 1985, Isolation of a protein fraction that binds preferenitally to chicken middle repetitive DNA, *Biochemistry* (in press).

Schrader, W. T., and O'Malley, B. W., 1972, Progesterone-binding components of chick oviduct—IV. Characterization of purified subunits, *J. Biol. Chem.* **247**:51–59.

Schrader, W. T., Birnbaumer, M. E., Hughes, M. R., Weigel, N. L., Grody, W. W., and O'Malley, B. W., 1981, Studies on the structure and function of the chicken progesterone receptor, *Recent Prog. Horm. Res.* **37**:583–633.

Sherman, M. R., Corvol, P. L., and O'Malley, B. W., 1970, Progesterone-binding components of a chick oviduct, *J. Biol. Chem.* **245**:6085–6095.

Spelsberg, T. C., Steggles, A. W., Chytil, F., and O'Malley, B. W., 1972, Progesterone-binding components of chick oviduct—V. Exchange of progesterone-binding capacity from target to nontarget tissue chromatins, *J. Biol. Chem.* **247**:1368–1374.

Stumph, W. E., Baez, M., Beattie, W. G., Tsai, M. J., and O'Malley, B. W., 1983a, Characterization of deoxyribonucleic acid sequences at the 5' and 3' borders of the 100 kilobase pair ovalbumin gene domain, *Biochemistry* **22**:306–315.

Swaneck, G. E., Nordstrom, J. L., Kreutzaler, F., Tsai, M. J., and O'Malley, B. W., 1979, Effect of estrogen on gene expression in chicken oviduct: Evidence for transcriptional control of ovalbumin gene, *Proc. Natl. Acad. Sci. U.S.A.* **76**:1049–1053.

Weintraub, H., and Groudine, M., 1976, Chromosomal subunits in active genes have an altered conformation, *Science* **193**:848–856.

Woo, S. L. C., Beattie, W. G., Catterall, J. F., Dugaiczyk, A., Staden, R., Brownlee, G. G., and O'Malley, B. W., 1981, Complete nucleotide sequence of the chicken chromosomal ovalbumin gene and its biological significance, *Biochemistry* **20**:6437–6446.

Zarucki-Schulz, T., Tsai, S. Y., Itakura, K., Soberon, X., Wallance, R. B., Tsai, M. J., Woo, S. L. C., and O'Malley, B. W., 1982, Point mutagenesis of the ovalbumin gene promoter sequence and its effect on *in vitro* transcription, *J. Biol. Chem.* **257**:11070–11077.

Inositol Trisphosphate and Diacylglycerol as Intracellular Second Messengers

MICHAEL J. BERRIDGE

1. Introduction

Inositol lipids play a central role in the signal transduction mechanism used by a large number of hormones and neurotransmitters. The primary action of these agonists is to stimulate the hydrolysis of phosphatidylinositol 4,5-bisphosphate (PIP_2) to produce diacylglycerol (DG), which remains in the plane of the membrane, and inositol 1,4,5-trisphosphate (IP_3), which is released to the cytoplasm (for details see Michell, Chapter 6). The fascinating aspect of this receptor mechanism is that both products resulting from the agonist-dependent cleavage of PIP_2 appear to function as second messengers to activate separate cellular processes. The neutral DG activates a protein kinase (C-kinase) that phosphorylates specific proteins (Takai *et al.*, 1979; Kishimoto *et al.*, 1980; Kuo *et al.*, 1980; Nishizuka, 1983), whereas IP_3 appears to function as a second messenger to mobilize intracellular calcium (Berridge, 1983; Streb *et al.*, 1983; Berridge *et al.*, 1984b; Burgess *et al.*, 1984; Joseph *et al.*, 1984). In order for a molecule to be considered as a second messenger, the following criteria must be satisfied:

1. The concentration or turnover of the putative second messenger must increase rapidly during stimulation with hormones or neurotransmitters.

MICHAEL J. BERRIDGE • A.R.C. Unit of Insect Neurophysiology and Pharmacology, Department of Zoology, University of Cambridge, Cambridge CB2 3EJ, England.

2. Specific degradative mechanisms must be present to remove the internal signal once the external agonist has been withdrawn.
3. The second messenger must be capable of activating some internal effector system responsible for altering cellular activity.

In the following sections we consider how these criteria have been satisfied for both DG and IP_3, which can thus be considered as the putative second messengers responsible for mediating the effects of those agonists that act through the inositol lipids.

2. Formation of Diacylglycerol and Inositol Trisphosphate

The beauty of this receptor mechanism lies in the fact that both second messengers are produced by a single transduction event, namely, the hydrolysis of PIP_2. The latter is one of the inositol-containing lipids that is formed by the stepwise phosphorylation of phosphatidylinositol (PI). Phosphate groups are added to the 4 and 5 positions of the inositol ring to produce the PIP_2 that is used by the receptor mechanism. When such receptors are occupied, there is a rapid cleavage of PIP_2 by a phospholipase C enzyme to give DG and IP_3 (Fig. 1). The first indication that agonists hydrolyze PIP_2 was obtained from studies on iris smooth muscle, where acetylcholine stimulated the breakdown of PIP_2 (Abdel-Latif et al., 1977) leading to the accumulation of IP_3 (Akhtar and Abdel-Latif, 1980). Such agonist-dependent hydrolysis of PIP_2 has also been described in liver (Michell et al., 1981; Creba et al., 1983; Rhodes et al., 1983; Thomas et al., 1983), parotid gland (Weiss et al., 1982; Downes and Wusteman, 1983), blood platelets (Billah and Lapetina, 1982a; Agranoff et al., 1983), cloned rat pituitary cells (Martin, 1983; Rebecchi and Gershengorn, 1983), and insect salivary gland (Berridge, 1983; Litosch et al., 1984). The most convenient and sensitive method of monitoring this hydrolysis of PIP_2 is to measure the formation of the two products.

The first study on inositol phosphates was conducted by Durell et al. (1968), who showed that synaptosomes responded to acetylcholine with an increase in the level of inositol 1,4-bisphosphate (IP_2). Of more interest was the observation that iris smooth muscle responded to acetylcholine with an increase in the level of IP_3 (Akhtar and Abdel-Latif, 1980). Similar increases in the level of IP_3 following cell activation have now been recorded in blood platelets (Agranoff et al., 1983), parotid gland (Berridge et al., 1983; Downes and Wusteman, 1983), the insect salivary gland (Berridge, 1983; Berridge et al., 1983, 1984a), cloned rat pituitary cells (Martin, 1983; Rebecchi and Gershengorn, 1983), and Swiss 3T3 cells (Berridge et al., 1984b).

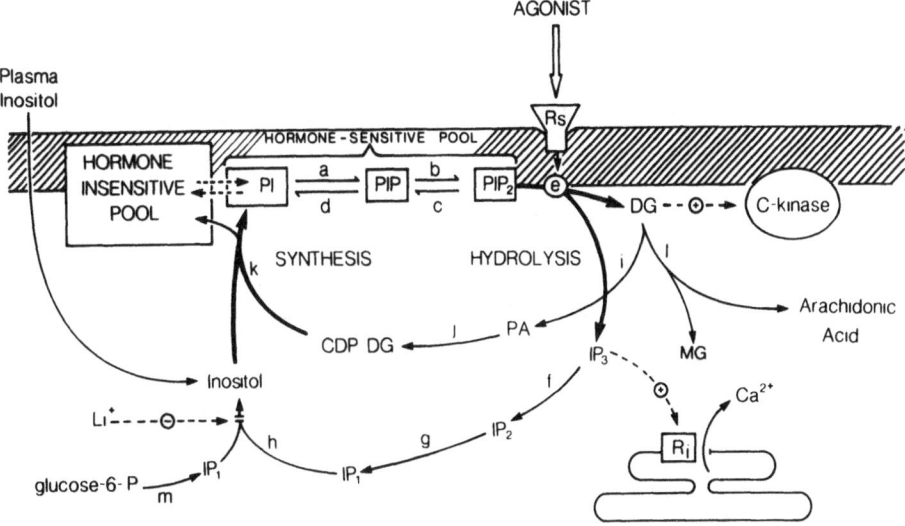

Figure 1. The proposed role of inositol lipids in signal transduction. Agonists acting on surface receptors (R_s) stimulate the hydrolysis of PIP_2, which exists in equilibrium with PIP and PI in a small hormone-sensitive pool that exchanges slowly with remaining cellular inositol lipids (hormone-insensitive pool). Diacylglycerol (DG) activates C-kinase, whereas IP_3 acts on an internal receptor (R_i) to stimulate the release of stored calcium. The size of the hormone-sensitive pool depends on the balance that exists between the rate of hydrolysis of PIP_2 and the rate of PI synthesis. a, PI kinase; b, PIP kinase; c, PIP_2 phosphomonoesterase; d, PIP phosphomonoesterase; e, phospholipase C; f, IP_3 phosphomonoesterase; g, IP_2 phosphomonoesterase; h, IP_1 *myo*-inositol 1-phosphatase; i, DG kinase; j, CTP–phosphatidic acid cytidyltransferase; k, PI synthase; l, DG lipase; m, *myo*-inositol 1-phosphate synthase.

The identification and measurement of inositol phosphates has provided an opportunity to study receptor events at very early time periods (Berridge, 1983; Berridge *et al.*, 1984a). When the salivary gland of the insect was stimulated with 5-HT, there was an immediate and rapid increase in the accumulation of IP_3 and IP_2 (Fig. 2). The increase in these two inositol phosphates is more than fast enough to account for the calcium-dependent physiological response in these glands (Berridge *et al.*, 1984a). Over these early time periods, there were no changes in the level of inositol monophosphate (IP_1) (Fig. 2), which is the expected product should PI be the substrate for the receptor mechanism. Almost identical changes in the three inositol phosphates have been described in cloned GH_3 cells following stimulation with TRH (Martin, 1983; Rebecchi and Gershengorn, 1983). All these observations confirm that the polyphos-

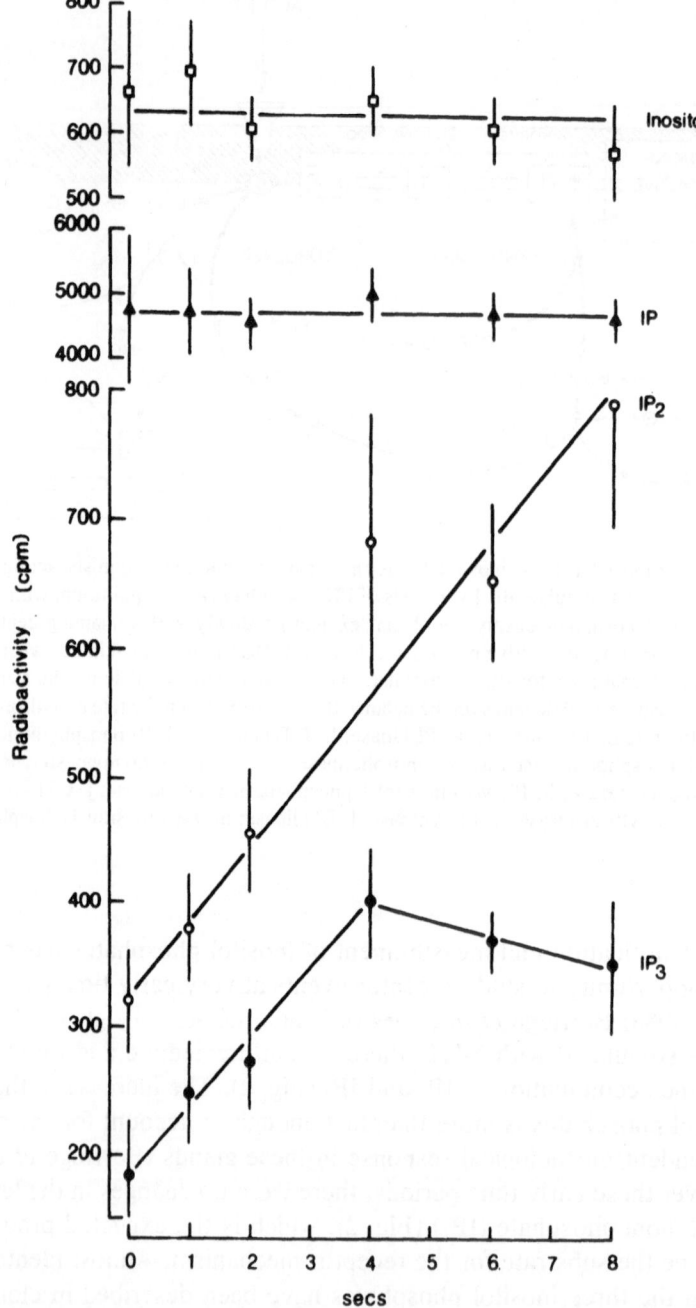

Figure 2. Effect of 5-HT (10 μM) on the accumulation of inositol phosphates and free inositol in an insect salivary gland. (From Berridge *et al.*, 1984a.)

phoinositides, particularly PIP_2, are the immediate substrates used by the receptor mechanism to generate IP_3.

In keeping with these rapid changes in the level of the inositol phosphates, there is an equally rapid increase in formation of DG, which is the other product formed by hydrolyzing PIP_2. Very rapid increases in the level of DG have been recorded in blood platelets (Bell *et al.*, 1979; Rittenhouse-Simmons, 1979; Prescott and Majerus, 1983), thyroid gland (Igarashi and Kondo, 1980), 3T3 cells (Habenicht *et al.*, 1981), pituitary cells (Martin, 1983; Rebecchi *et al.*, 1983), and liver (Thomas et al., 1983). A characteristic feature of this response is that the level of DG reaches a peak in less than a minute then falls back towards the basal level. Either the rate of formation of DG declines because the supply of PIP_2 becomes limiting or there is an increase in the enzymic mechanisms that remove DG.

3. Removal of Diacylglycerol and Inositol Trisphosphate

If DG and IP_3 are to function as second messengers, it is essential that mechanisms exist within the cell to rapidly remove these internal signals once the external agonist is removed. Studies on the insect salivary gland have revealed that the elevated level of IP_3 declines very rapidly towards its resting level when 5-HT is withdrawn (Berridge, 1983). A similar decline in the level of IP_3 has been observed in the parotid following recovery from cholinergic stimulation (Downes and Wusteman, 1983; Aub and Putney, 1984). An inositol trisphosphate phosphomonoesterase that converts IP_3 to IP_2 (Fig. 1,f) has been described in human erythrocyte membranes (Downes *et al.*, 1982). A similar enzyme has been described in the parotid and in the insect salivary gland (Berridge *et al.*, 1983). The latter also contains an inositol bisphosphate phosphomonoesterase (Fig. 1,g) that converts IP_3 to IP_2. There thus appears to be an inositol phosphate cycle (Fig. 1) whereby the IP_3 produced by the receptor mechanism is dephosphorylated in a stepwise manner to liberate the free inositol used to resynthesize PI. The final step in this cascade is the dephosphorylation of IP_1 to free inositol by the enzyme inositol 1-phosphatase. The latter is markedly inhibited by Li^+ (Hallcher and Sherman, 1980), which means that this ion will slow down the resynthesis of PI and could thus interfere with those receptor mechanisms that use these lipids, as is described in Section 7. The critical enzyme with regard to terminating the intracellular action of IP_3 is the inositol trisphosphatase. As yet, the only known inhibitor of this enzyme is 2,3-bisphosphoglycerate (Downes *et al.*, 1982), although there are some suggestions that the enzyme might be inhibited by high concentrations of lithium (Thomas *et al.*, 1984).

There are two major pathways for removing DG. It can be converted to phosphatidic acid by a DG kinase as part of the mechanism for the resynthesis of PI (Fig. 1,i), or it can be hydrolyzed by a DG lipase with the liberation of arachidonic acid (Fig. 1,l). These removal mechanisms are extremely effective because, as mentioned earlier, the level of DG never rises very high and, after an initial increase immediately following stimulation, returns close to that found at rest.

4. Mode of Action of Diacylglycerol and Inositol Trisphosphate

An important function of receptors that use these inositol lipids is to generate a calcium signal (Michell, 1975). The primary calcium signal in many cell systems is obtained by a release of internal calcium, but the link between surface receptors and the intracellular calcium stores was missing. Recently, Berridge (1983) has suggested that the IP_3 released when the receptor hydrolyzes PIP_2 may function as a second messenger to mobilize intracellular calcium (Fig. 1). The fact that IP_3 formation occurs fast enough to account for the onset of the calcium signal in the blowfly salivary gland (Berridge *et al.*, 1984a) is consistent with its proposed role as a second messenger.

In order to obtain evidence for the idea that IP_3 mobilizes intracellular calcium, various techniques have been used to permeabilize the plasma membrane so that this putative second messenger could gain access to the internal stores of calcium. Isolated rat pancreatic acinar cells become very permeable when incubated in a low-calcium medium (Streb and Schulz, 1983). When such leaky cells are incubated in a low-calcium medium in the presence of ATP, they begin to sequester calcium such that the extracellular calcium concentration (as monitored with a calcium electrode) falls well below 1×10^{-6} M (Fig. 3). Once a steady state is achieved, the addition of IP_3 (2.5 μM) results in a large release of calcium to the bathing medium (Fig. 3) (Streb *et al.*, 1983). Half maximal release was achieved at approximately 0.5 μM and was apparently specific for inositol 1,4,5-trisphosphate because there was no release on addition of inositol 1,4-bisphosphate, inositol 1-monophosphate, inositol 1,2-cyclic phosphate, or free inositol (Streb *et al.*, 1983). Very similar results have been obtained in liver cells that have been permeabilized using saponin (Burgess *et al.*, 1984; Joseph *et al.*, 1984). Leaky Swiss 3T3 cells also respond to IP_3 with a large release of calcium (Berridge *et al.*, 1984b).

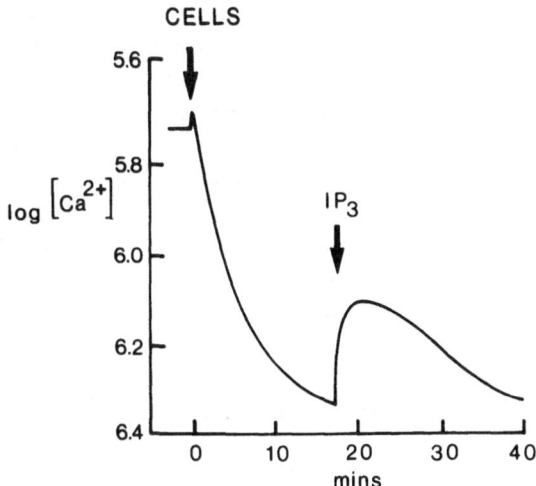

Figure 3. Release of intracellular calcium by IP$_3$ (2.5 μM) in permeabilized rat pancreatic cells (Redrawn from Streb *et al.*, 1983).

One important action of IP$_3$, therefore, appears to be the mobilization of intracellular calcium.

The other major product of PIP$_2$ hydrolysis is DG, which remains within the plane of the membrane to function as a second messenger to activate protein kinase C (Takai *et al.*, 1979; Kishimoto *et al.*, 1980; Kuo *et al.*, 1980; Nishizuka, 1983). The latter has a wide distribution in animal tissues and is particularly prevalent in brain. The activation of C-kinase by DG is a complex process that requires calcium and phosphatidylserine as cofactors. Of particular interest is the observation that this enzyme can be activated by tumor-promoting phorbol esters such as 12-O-tetra-decanoylphorbol-13-acetate (TPA) (Castagna *et al.*, 1982). During this activation of C-kinase by TPA, there is a rapid decrease in soluble C-kinase with a corresponding increase in the amount of enzyme bound to the plasma membrane (Kraft and Anderson, 1983). Despite this close association with the membrane, the enzyme is capable of phosphorylating both particulate and soluble proteins. The search is now on to identify these proteins and to establish how they contribute to the final cellular response.

The initial work on C-kinase was done on blood platelets, where thrombin was known to increase the phosphorylation of a 40k and a 20k protein (Lyons and Atherton, 1979). The former appears to be a specific substrate for C-kinase (Kaibuchi *et al.*, 1982, 1983), whereas the 20k protein, which appears to be the light chain of myosin, can be phosphorylated by both C-kinase and the Ca^{2+}/calmodulin-sensitive myosin light chain kinase (Naka *et al.*, 1983). In liver, hormones such as vasopressin and angiotensin II stimulate the hydrolysis of PIP_2 to activate the two signal pathways shown in Fig. 1. The two second messengers seem to act through separate proteins in that calcium stimulates the phosphorylation of seven proteins, whereas DG is responsible for phosphorylating three separate proteins (Garrison *et al.*, 1984). The fact that these two second messengers (Ca^{2+} and DG) can phosphorylate separate proteins might explain how they can sometimes act synergistically with each other, as is described in the next section.

5. Functional Interactions between Diacylglycerol and Calcium

The most important characteristic of these receptors that utilize inositol lipids is that they generate two functional second messengers that can interact with each other to control many different cellular processes. Although the full range of these interactions remains to be mapped out, it is already evident that the existence of two separate signal pathways greatly enhances the flexibility of the control mechanism. The two control pathways may act cooperatively with each other, or the interaction may be more subtle, leading to some remarkable synergistic responses.

The first example of a synergistic interaction between these two signals emerged from studies on blood platelets. It was found that low concentrations of calcium ionophore or 1-oleoyl-2-acetylglycerol (a DG derivative that activates C-kinase) that alone caused little secretion were able to produce a full secretory response when added in combination with each other (Kaibuchi *et al.*, 1982, 1983). Another example of such synergism is the activation of lymphocytes by a combination of the ionophore A23187 and the phorbol ester TPA (Mastro and Smith, 1983; Kaibuchi *et al.*, 1984). Such synergism between DG and calcium could also explain the ability of TPA to potentiate the secretory effect of gliclazide on insulin-secreting islet cells (Malaisse *et al.*, 1983).

Studies on blood platelets with the fluorescent dye quin-2, which monitors the intracellular level of calcium, have explored this synergism further and have suggested that some agonists might be capable of activating one signal pathway independently of the other (Rink *et al.*, 1982, 1983). When blood platelets were stimulated with a low dose of ionomycin in the absence of external calcium, there was a small increase in intracellular calcium but no aggregation or secretion. However, on addition of a low dose of thrombin, both aggregation and secretion occurred in the absence of any further change in the level of calcium (Rink *et al.*, 1982). It appeared as if thrombin was capable of activating these cellular events through a calcium-independent pathway, which was subsequently identified as the C-kinase pathway (Rink *et al.*, 1983). Activation of this C-kinase pathway with either TPA or 1-oleoyl-2-acetylglycerol was able to stimulate secretion without any change in the resting level of calcium. Sha'afi *et al.* (1983) have made similar observations in neutrophils, where phorbol esters were found to stimulate secretion without any change in the level of intracellular calcium. It is important to point out, however, that in the case of blood platelets, the onset of secretion in response to TPA was somewhat delayed and had a slower time course, which was improved by raising the resting level of calcium (Rink *et al.*, 1983), suggesting that a normal response usually depends on the activation of both signal pathways. However, these studies on blood platelets do raise the intriguing possibility that although many agonists act through both calcium and DG, certain agonists might be capable of switching on one pathway independently of the other. For example, collagen appears to be capable of stimulating secretion in platelets without any apparent change in the intracellular level of calcium (Rink *et al.*, 1983).

Just how these two signal pathways interact with each other to produce synergistic responses remains to be established and will depend on the identification of the various proteins that are phosphorylated by either calcium or DG. Whitfield *et al.* (1973) suggested that phorbol esters may sensitize the effector system, which in this case is DNA synthesis, to the stimulatory action of calcium. Such a mechanism may exist in adrenal cells, where DG is thought to act by altering the affinity of exocytosis for calcium (Knight and Baker, 1983). A direct demonstration of such an increase in sensitivity to calcium has been described using permeabilized blood platelets (Knight and Scrutton, 1983). In control cells, half-maximal secretion required 1.9 μM Ca^{2+}, but in the presence of thrombin this was reduced to 0.4 μM. This heightened sensitivity of the secretory mechanism to calcium was also apparent following addition of either 1-oleoyl-2-acetylglycerol or TPA (Knight and Scrutton, 1983).

6. Oncogenes and Phosphoinositide Metabolism

The action of certain growth factors in stimulating cellular proliferation seems to depend on the hydrolysis of the inositol lipids. Stimulation of Swiss 3T3 cells with PDGF results in the hydrolysis of PIP$_2$ to form DG (Habenicht *et al.*, 1981) and IP$_3$ (Berridge *et al.*, 1984b). Berridge *et al.* (1984b) have proposed that these two second messengers may cooperate with each other to regulate some of the ionic events responsible for initiating cell growth. Growth factors not only raise the level of calcium but also act to switch on a neutral Na^+-H^+ exhange carrier (Rozengurt, 1983). It has been proposed that IP_3 mobilizes intracellular calcium, whereas DG may act to stimulate the Na^+-H^+ antiport (Berridge *et al.*, 1984b).

The fact that phosphoinositide metabolism might be responsible for mediating the action of growth factors such as PDGF raises the fascinating possibility that some of the oncogenes might function through this receptor mechanism. We already know that the *sis* gene codes for a protein that is almost identical to PDGF (Doolitte *et al.*, 1983; Waterfield *et al.*, 1983), whereas the *erb*-B gene produces a protein that is very similar to a part of the receptor for EGF (Downward *et al.*, 1984). It is not too far-fetched, therefore, to propose that some of the other oncogene products might be the enzymes (Fig. 1,a–m) involved in the inositol lipid events responsible for mediating the action of agents such as PDGF.

7. Phosphoinositide Levels and Receptor Sensitivity

The primary event of this receptor mechanism that utilizes the inositol lipids is the hydrolysis of PIP_2 to give IP_3 and DG (Fig. 1). The sensitivity of the receptor mechanism will thus depend on the availability of PIP_2, which could be altered by a number of different factors. It is worth considering some of these factors because there are indications that changes in receptor sensitivity that occur on denervation may result from alterations in the inositol lipids (Abdel-Latif *et al.*, 1979; Takenawa *et al.*, 1983; Downes *et al.*, 1983). On the basis of the model outlined in Fig. 1, it is possible to predict several ways in which the availability of PIP_2 might be altered. One way of altering the level of PIP_2 is to adjust the equilibrium that exists among these inositol lipids. The levels of the three inositol lipids within the plasma membrane will be set by the balance that exists between the kinases (Fig. 1,a and b) that phosphorylate PI to PIP_2 and the phosphomonoesterases (Fig. 1,c and d) that convert these two polyphosphoinositides back to PI. Subtle alterations in the activities of these opposing reactions will shift the position of the equilibrium, resulting in more or less PIP_2. Any change that shifts the equilibrium to the right, resulting in an increase in the size of the PIP_2 pool, may lead to supersensitivity, whereas any decrease in the available pool may result in desensitization. There is an intriguing possibility that some hormones might act to alter receptor sensitivity through such changes in the size of the pool of PIP_2. For example, ACTH may modulate the metabolism of polyphosphoinositides in brain by acting on a protein kinase that alters the activity of the enzyme that converts PIP to PIP_2 (Jolles *et al.*, 1980, 1981).

Another way of altering the level of PIP_2 is by changing the overall level of the inositol lipids within the membrane. The first point to appreciate is that not all the PI present in the cell is freely available to contribute to the receptor mechanism. There appears to be a relatively small hormone-sensitive pool of phosphoinositide, perhaps that confined to the plasma membrane, that equilibrates rather slowly with the remaining hormone-insensitive pool (Fig. 1). Such small hormone-sensitive pools of inositol lipids have been described in the blowfly salivary gland (Fain and Berridge, 1979), in a rat mammary tumor cell line (WRK-1) (Monaco, 1982), and in blood platelets (Billah and Lapetina, 1982b). The size of this pool will depend on the balance that exists between the rate at which PIP_2 is hydrolyzed by the receptor mechanism and the rate at which PI is resynthesized (Fig. 1). During intense stimulation, therefore, PIP_2 will be rapidly degraded and is replaced by drawing on PI. However, since this hormone-sensitive pool is rather small, the supply of PI is limited and will rapidly begin to decline unless the pool is constantly replenished by the resynthesis of PI. There are a number of conditions, both physiological and pathophysiological, that can interfere with this resynthesis of PI that is essential to maintain the levels of PIP_2 necessary to generate the second messengers IP_3 and DG.

The first indication that the size of this hormone-sensitive pool might be important in determining receptor sensitivity emerged from studies on the blowfly salivary gland (Berridge and Fain, 1979; Fain and Berridge, 1979). During stimulation with 5-hydroxytryptamine (5-HT), not only is there a rapid breakdown of the inositol lipids, but there is a simultaneous inhibition of PI synthesis through calcium. If the glands are stimulated with a high dose of 5-HT, the hormone-sensitive pool begins to decline, and this could explain the inactivation of calcium signaling that occurs during periods of prolonged stimulation (Berridge and Fain, 1979). There was little resensitization of the receptor mechanism unless the glands were provided with free inositol in order to replenish the hormone-sensitive pool (Fain and Berridge, 1979). These experiments on the fly gland suggested that receptor sensitivity might depend on the level of lipids within a small hormone-sensitive pool. A critical factor for keeping this hormone-sensitive pool fully charged is the rate of synthesis of PI.

In order for PI synthesis to occur, cells require relatively high levels of inositol, which was once thought of as a vitamin in that it is essential for cell growth and survival. Just why this simple molecule should play such a vital role in cell function has been somewhat of a mystery, but it is now possible to speculate that its importance depends on its participation in signal transduction. In nerve endoneurium, a small reduction in the level of inositol caused a large reduction in the utilization of energy as measured by changes in oxygen consumption, which returned to normal when the level of inositol was restored (Simmons *et al.*, 1982). It was argued that inositol is required to synthesize a small pool of PI that is turning over rapidly and is responsible for controlling a range of energy-requiring processes. This small pool of PI can probably be equated with the hormone-sensitive phosphoinositide pool described earlier (Fig. 1). Further evidence for this idea comes from studies on regenerating liver, where the function of α_1-adrenergic receptors is severely impaired 3 days after hepatectomy but can be restored simply by adding free inositol (Huerta-Bahena and Garcia-Sainz, 1983). A high level of intracellular inositol seems to be essential in order for cells to maintain the sensitivity of those receptors that employ inositol lipids for signal transduction.

The *myo*-inositol used for the synthesis of PI is derived from three main sources (Fig. 1). It is formed as an end product of the inositol phosphate cycle, resulting from the hydrolysis of the phosphoinositides. Any loss of *myo*-inositol can be made good either by *de novo* synthesis from glucose-6-phosphate or by use of dietary inositol that is carried to cells in the plasma. The uptake of *myo*-inositol from plasma is available to all cells except those in the brain or testis, which are closeted behind blood barriers that are relatively impermeable to *myo*-inositol. When [³H]-inositol is injected into plasma, very little label enters the CSF (Margolis *et al.*, 1971). Even nerves in the periphery appear to be cut off from inositol in the plasma because the perineurial membrane functions as a barrier (Gillon and Hawthorne, 1983). Tissues such as the brain that are protected by blood barriers must be capable of making most of their *myo*-inositol, and this synthetic capacity must be considerable because the fluid spaces behind these barriers have very much higher levels of *myo*-inositol than is present in plasma (Table I). These high levels of *myo*-inositol may have a crucial role to play in the normal functioning of both the brain and testis. Since these tissues are very reliant on *de novo* synthesis to maintain their supply of *myo*-inositol, they are much more susceptible to defects in inositol metabolism than other tissues, which are protected by obtaining dietary *myo*-inositol arriving in the blood.

Several biochemical lesions that affect cellular *myo*-inositol levels can severely alter neural function. Galactosemia, which results from a deficiency of the enzyme galactose-1-phosphate uridyltransferase, causes galactose and galactitol to accumulate while the levels of free and lipid-bound *myo*-inositol decline in brain and nerve tissue (Warefield and Segal, 1978). This decline in phosphoinositide metabolism may be responsible for the decrease in nerve conduction and brain development that results

Table I. Concentration of *myo*-Inositol in Different Fluid Compartments

Fluid compartment		*myo*-Inositol concentration (mM)	Reference
Dog	Plasma	0.025	Margolis *et al.*, 1981
Hamster	Plasma	0.085	Hinton *et al.*, 1980
Rat	Plasma	0.062	Allison, 1978
Dog	Cerebrospinal fluid	0.790	Margolis *et al.*, 1981
Rat	Seminiferous tubule	1.83	Hinton *et al.*, 1980
Hamster	Ductus deferens	83.0	Hinton *et al.*, 1980

from such galactose toxicity. Alterations in phosphoinositide metabolism may contribute to the onset of mental retardation found in infants that have suffered from galactosemia.

This link between low inositol levels and neurotoxicity is also apparent in diabetic neuropathy (Greene *et al.*, 1975; Natarajan *et al.*, 1981; Bell *et al.*, 1982; Mayhew *et al.*, 1983). Chronic diabetes in rats can be induced by injecting the drug streptozotocin, which causes abnormal peripheral nerve function such as a marked reduction in nerve conduction velocities and various morphological changes such as a loss of myelinated axons. As in galactosemia, the onset of experimental diabetes is associated with a decline in the neuronal levels of *myo*-inositol (Greene *et al.*, 1975). There is also a fall in the level of PI and a change in the metabolism of the polyphosphoinositides (Palmano *et al.*, 1977; Natarajan *et al.*, 1981; Bell *et al.*, 1982). Of particular interest is the finding that in diabetic rats there was a significant decline in the PIP kinase (Fig. 1,b), which converts PIP to PIP$_2$, as well as in PI synthase (Fig. 1,k) (Whiting *et al.*, 1979). Both defects would serve to lower the level of PIP$_2$. The impairment of peripheral nerve function in these streptozotocin-treated rats can be prevented by supplementing their diet with *myo*-inositol, which restores the intracellular level to normal (Greene *et al.*, 1975). This reversal of diabetic neuropathy by dietary *myo*-inositol occurs despite the fact that the rats continue to be hyperglycemic. Normal neural function seems to depend on the maintenance of high levels of *myo*-inositol in order to sustain the synthesis of the phosphoinositides required for signal transduction.

The therapeutic action of lithium in controlling manic-depressive illness may depend on changes in the hormone-sensitive pool of inositol lipids brought about through a reduction in the synthesis of PI. Allison and Stewart (1971) provided the first link between lithium and the phosphoinositides by showing that there was a considerable reduction in the level of inositol in the brain of lithium-treated rats. This decline in the level of free inositol resulted from a severe inhibition by lithium of the enzyme inositol 1-phosphatase (Hallcher and Sherman, 1980). This reduction in enzyme activity results not only in a lowering of the level of inositol but in an accumulation of inositol 1-phosphate (Allison et al., 1976; Sherman et al., 1981; Berridge et al., 1982). Lithium inhibits the production of inositol originating not only from the inositol phosphate cycle but also from de novo synthesis (Fig. 1). The only other supply of inositol is by uptake from plasma (Fig. 1), which is not freely available to cells in the brain because of the blood–brain barrier. This severe disruption of inositol metabolism in the brain by lithium will lead to a decline in the level of the phosphoinositides and might represent its therapeutic action in the control of manic–depressive illness (Hallcher and Sherman, 1980; Shermann et al., 1981; Berridge et al., 1982). The suggestion is that lithium acts by lowering the level of myo-inositol, which results in a decline in the synthesis of the PI necessary to maintain the hormone-sensitive pool of phosphoinositide.

When lithium is first administered, it will have little effect since nerve cells start off with a normal pool of phosphoinositides that can be used for signalling. However, as this pool is used up and is not replaced because lithium blocks the production of myo-inositol, it will begin to decline, resulting in a decrease in the effectiveness of the calcium-mobilizing receptors discussed earlier. The fact that the inhibitory action of lithium will take time to develop is entirely consistent with clinical observations, because the therapeutic action of lithium is not immediate but develops over several days. This hypothesis suggests, therefore, that lithium controls manic–depressive illness by a partial inactivation of those receptors that use phosphoinositides as part of their transducing mechanism. The hypothesis also predicts that manic–depressive illness may arise as a result of a supersensitivity or hyperactivity of such calcium-mobilizing receptors. There are indications in the literature that the cyclic nature of manic–depressive disorders may result from fluctuations in receptor sensitivity (Bunney et al., 1977). A more specific suggestion is that depression arises as a result of the existence of hypersensitive postsynaptic 5-HT receptors (Aprison et al., 1978). If these 5-HT receptors function through phosphoinositides, then lithium could well reduce their sensitivity by lowering the level of the hormone-sensitive pool of lipid necessary for the operation of this ubiquitous receptor mechanism.

An important outcome of this lithium hypothesis is that manic–depressive illness may be caused by the abnormal activation of calcium-mobilizing receptors. The existence of such receptors in the brain has already been established, and it is likely that many other transmitters in the CNS may share this ubiquitous transducing mechanism (Downes,

1982). Indeed, the identification of such receptors can be facilitated by using lithium as a means of amplifying the responses obtained from receptors that use the phosphoinositides (Berridge *et al.*, 1982; Brown *et al.*, 1984). By inhibiting the hydrolysis of IP$_1$, lithium enables this inositol phosphate to accumulate to levels that are easy to detect. These studies on lithium not only provide a plausible explanation for its therapeutic action in controlling manic–depressive illness but have also introduced a novel and sensitive technique for monitoring the activity of calcium-mobilizing receptors.

References

Abdel-Latif, A. A., Akhtar, R. A., and Hawthorne, J. N., 1977, Acetylcholine increases the breakdown of triphosphoinositide of rabbit iris muscle prelabelled with [^{32}P] phosphate, *Biochem. J.* **162:**61–73.

Abdel-Latif, A. A., Green, K., and Smith, J. P., 1979, Sympathetic denervation and the trisphosphoinositide effect in the iris smooth muscle: A biochemical method for the determination of alpha-adrenergic receptor denervation supersensitivity, *J. Neurochem.* **32:**225–228.

Agranoff, B. W., Murthy, P., and Seguin, E. B., 1983, Thrombin-induced phosphodiesteratic cleavage of phosphatidylinositol bisphosphate in human platelets. *J. Biol. Chem.* **258:**2076–2078.

Akhtar, R. A., and Abdel-Latif, A. A., 1980, Requirement for calcium ions in acetylcholine-stimulated phosphodiesteratic cleavage of phosphatidyl-*myo*-inositol 4,5-bisphosphate in rabbit iris smooth muscle, *Biochem. J.* **192:**783–791.

Allison, J. H., 1978, Lithium and brain *myo*-inositol metabolism, in: *Cyclitols and Phosphoinositides* (W. W. Wells, and F. Eisenberg, eds.), Academic Press, New York, pp. 507–519.

Allison, J. H., and Stewart, M. A., 1971, Reduced brain inositol in lithium-treated rats, *Nature (New Biol.)* **233:**267–268.

Allison J. H., Blisner, M. E., Holland, W. H., Hipps, P. P., and Sherman W. R., 1976, Increased brain *myo*-inositol 1-phosphate in lithium-treated rats, *Biochem. Biophys. Res. Commun.* **71:**664–670.

Aprison, M. H., Takahashi, R., and Tachiki, K., 1978, Hypersensitive serotonergic receptors involved in clinical depression—a theory, in: *Neuropharmacology and Behaviour* (B. Haber and M. H. Aprison, eds.), Plenum Press, New York, pp. 23–53.

Aub, D. L., and Putney, J. W., 1984, Metabolism of inositol phosphates in parotid cells: Implications for the pathways of the phosphoinositide effect and for the possible messenger role of inositol trisphosphate, *Life Sci.* **34:**1347–1355.

Bell, R. L., Kennerly, D. A., Stanford, N., and Majerus, P. W., 1979, Diglyceride lipase: A pathway for arachidonate release from human platelets, *Proc. Natl. Acad. Sci. U.S.A.* **76:**3238–3241.

Bell, M. E., Peterson, R. G., and Eichberg, J., 1982, Metabolism of phospholipids in peripheral nerve from rats with chronic streptozotocin-induced diabetes: Increased turnover of phosphatidylinositol-4,5-bisphosphate, *J. Neurochem.* **39:**192–200.

Berridge, M. J., 1983, Rapid accumulation of inositol trisphosphate reveals that agonists hydrolyse polyphosphoinositides instead of phosphatidylinositol, *Biochem. J.* **212:**849–858.

Berridge, M. J., and Fain, J. N., 1979, Inhibition of phosphatidylinositol synthesis and the inactivation of calcium entry after prolonged exposure of the blowfly salivary gland to 5-hydroxytryptamine, *Biochem. J.* **178:**59–69.

Berridge, M. J., Downes, C. P., and Hanley, M. R., 1982, Lithium amplifies agonist-dependent phosphatidylinositol responses in brain and salivary gland, *Biochem. J.* **206:**587–595.

Berridge, M. J., Dawson, R. M. C., Downes, C. P., Heslop, J. P., and Irvine, R. F., 1983, Changes in the levels of inositol phosphates after agonist-dependent hydrolysis of membrane phosphoinositides, *Biochem. J.* **212:**473–482.

Berridge, M. J., Buchan, P. B., and Heslop, J. P., 1984a, Relationship of polyphosphoinositide metabolism to the hormonal activation of the insect salivary gland to 5-hydroxytryptamine, *Mol. Cell. Endocrinol.* **36:**37–42.

Berridge, M. J., Heslop, J. P., Irvine, R. F., and Brown, K. D., 1984b, Inositol trisphosphate formation and calcium mobilization in Swiss 3T3 cells in response to platelet-derived growth factor *Biochem. J.* **222:**195–206.

Billah, M. M., and Lapetina, E. G., 1982a, Rapid decrease of phosphatidylinositol 4,5-bisphosphate in thrombin-stimulated platelets, *J. Biol. Chem.* **257:**12705–12708.

Billah, M. M., and Lapetina, E. G., 1982b, Evidence for multiple metabolic pools of phosphatidylinositol in stimulated platelets, *J. Biol. Chem,* **257:**11856–11859.

Brown, E., Kendall, D. A., and Nahorski, S. R., 1984, Inositol phospholipid hydrolysis in rat cerebral cortical slices: I. Receptor characterization, *J. Neurochem.* **42:**1379–1387.

Bunney, W. E., Jr., Post, R. M., Andersen, A. E., and Kopanda, T., 1977, A neuronal receptor sensitivity mechanism in affective illness. *Commun. Psycopharmacol.* **1:**393–405.

Burgess, G. M., Godfrey, P. P., McKinney, J. S., Berridge, M. J., Irvine, R. F., and Putney, J. W., 1984, The second messenger linking receptor activation to internal Ca release in liver, *Nature,* **309:**63–66.

Castagna, M., Takai, Y., Kaibuchi, K., Sano, K., Kikkawa, U., and Nishizuka, Y., 1982, Direct activation of calcium-activated phospholipid-dependent protein kinase by tumour-promoting phorbol esters. *J. Biol. Chem.* **257:**7847–7851.

Creba, J. A., Downes, C. P., Hawkins, P. T., Brewster, G., Michell, R. H., and Kirk, C. J., 1983, Rapid breakdown of phosphatidylinositol 4-phosphate and phosphatidylinositol 4,5-bisphosphate in rat hepatocytes stimulated by vasopressin and other Ca^{2+}-mobilizing hormones, *Biochem. J.* **212:**733–747.

Doolittle, R. F., Hunkapiller, M. W., Hood, L. E., Devare, S. G., Robbins, K. S., Aaronson, S. A., and Antoniades, H. N., 1983, Simian sarcoma virus *onc* gene, v-*sis*, is derived from the gene (or genes) encoding a platelet-derived growth factor, *Science* **221:**275–277.

Downes, C. P., 1983, Inositol phospholipids and neurotransmitter–receptor signalling mechanisms, *Trends Neurosci.* **6:**313–316.

Downes, C. P., Mussat, M. C., and Michell, R. H., 1982, The inositol trisphosphate phosphomonoesterase of the human erythrocyte membrane, *Biochem. J.* **203:**169–177.

Downes, C. P., Dibner, M. D., and Hanley, M. R., 1983, Sympathetic denervation impairs agonist-stimulated phosphatidylinositol metabolism in rat parotid glands, *Biochem. J.* **214:**865–870.

Downes, C. P., and Wusteman, M. M., 1983, Breakdown of polyphosphoinositides and not phosphatidylinositol accounts for muscarinic agonist-stimulated inositol phospholipid metabolism in rat parotid glands, *Biochem. J.* **216:**633–640.

Downward, J., Yarden, Y., Mayes, E., Scrace, G., Totty, N., Stockwell, P., Ullrich, A., Schlessinger, J., and Waterfield, M. D., 1984, Close similarity of epidermal growth factor receptor and v-*erb*-B oncogene protein sequences, *Nature* 307:521–527.

Durell, J., Sodd, M. A., and Friedel, R. O., 1968, Acetylcholine stimulation of the phosphodiesterase cleavage of guinea pig brain phosphoinositides, *Life Sci.* 7:363–368.

Fain, J. N., and Berridge, M. J., 1979, Relationship between phosphatidylinositol synthesis and recovery of 5-hydroxytryptamine-responsive Ca^{2+} flux in blowfly salivary gland, *Biochem. J.* 180:655–661.

Garrison, J. C., Johnsen, D. E., and Campanile, C. P., 1984, Evidence for the role of phosphorylase kinase, protein kinase C, and other Ca^{2+}-sensitive protein kinases in the response of hepatocytes to angiotensin II and vasopressin, *J. Biol. Chem.* 259:3283–3292.

Gillon, K. R. W., and Hawthorne, J. N., 1983, Transport of *myo*-inositol into endoneurial preparations of sciatic nerve from normal and streptozotocin-diabetic rats, *Biochem. J.* 210:775–781.

Greene, D. A., De Jesus, P. V. Jr., and Winegrad, A. I., 1975, Effects of insulin and dietary myoinositol on impaired peripheral motor nerve conduction velocity in acute streptozotocin diabetes, *J. Clin. Invest.* 55:1326–1336.

Habenicht, A. J. R., Glomset, J. A., King, W. C., Nist, C., Mitchell, C. D., and Ross, R., 1981, Early changes in phosphatidylinositol and arachidonic acid metabolism in quiescent Swiss 3T3 cells stimulated to divide by platelet-derived growth factor, *J. Biol. Chem.* 256:12329–12335.

Hallcher, L. M., and Sherman, W. R., 1980, The effects of lithium ion and other agents on the activity of *myo*-inositol-1-phosphatase from bovine brain, *J. Biol. Chem.* 255:10896–10901.

Hinton, B. T., White, R. W., and Setchell, B. P., 1980, Concentrations of *myo*-inositol in the luminal fluid of the mammalian testis and epididymis, *J. Reprod. Fertil.* 58:395–399.

Huerta-Bahena, J., and Garcia-Sainz, J. A., 1983, Inositol administration restores the sensitivity of liver cells formed during liver regeneration to alpha₁-adrenergic amines, vasopressin and angiotensin II, *Biochim. Biophys. Acta* 763:125–128.

Igarashi, Y., and Kondo, Y., 1980, Acute effect of thryotropin on phosphatidylinositol degradation and transient accumulation of diacylglycerol in isolated thyroid follicles, *Biochem. Biophys. Res. Commun.* 97:759–765.

Jolles, J., Zwiers, H., van Dongen, C. J., Schotman, P., Wirtz, K. W. A., and Gispen, W. H., 1980, Modulation of brain polyphosphoinositide metabolism by ACTH-sensitive protein phosphorylation, *Nature* 286:623–625.

Jolles, J., Zwiers, H., Dekker, A., Wirtz, K. W. A., and Gispen, W. H., 1981, Corticotropin-(1–24)-tetracosapeptide affects protein phosphorylation and polyphosphoinositide metabolism in rat brain, *Biochem. J.* 194:283–291.

Joseph, S. K., Thomas, A. P., Williams, R. J., Irvine, R. F., and Williamson, J. R., 1984, *Myo*-inositol 1,4,5-trisphosphate: A second messenger for the hormonal mobilization of intracellular Ca^{2+} in liver, *J. Biol. Chem.* 259:3077–3081.

Kaibuchi, K., Sano, K., Hoshijima, M., Takai, Y., and Nishizuka, Y., 1982, Phosphatidylinositol turnover in platelet activation: Calcium mobilization and protein phosphorylation, *Cell Calcium* 3:323–335.

Kaibuchi, K., Takai, Y., Sawamura, M., Hoshijima, M., Fujikura, T., and Nishizuka, Y., 1983, Synergistic functions of protein phosphorylation and calcium mobilization in platelet activation, *J. Biol. Chem.* 258:6701–6704.

Kaibuchi, K., Sawamura, M., Kalakami, Y., Kikkawa, U., Takai, Y., and Nishizuka, Y.,

1985, Calcium and inositol phospholipid degradation in signal transduction, in: *Inositol and Phosphoinositides* (J. E. Bleasdale, J. Eichberg, and E. Hauser, ed.), Humana Press, Clifton, New Jersey (in press).

Kishimoto, A., Takai, Y., Mori, T., Kikkawa, U., and Nishizuka, Y., 1980, Activation of calcium and phospholipid-dependent protein kinase by diacylglycerol, its possible relation to phosphatidylinositol turnover, *J. Biol. Chem.* **255**:2273–2276.

Knight, D. E., and Baker, P. F., 1983, The phorbol ester TPA increases the affinity of exocytosis for calcium in "leaky" adrenal medullary cells, *FEBS Lett.* **160**:98–100.

Knight, D. E., and Scrutton, M. C., 1983, Direct evidence that cyclic nucleotides and 1,2-diacylglycerol modulate the Ca^{2+} sensitivity of secretion in platelets, *J. Physiol. (Lond.)* 49P.

Kraft, A. S., and Anderson, W. B., 1983, Phorbol esters increase the amount of Ca^{2+}, phospholipid-dependent protein kinase associated with plasma membrane, *Nature* **301**:621–623.

Kuo, J. F., Andersson, R. G. G., Wise, B. C., Mackerlova, L., Salomonsson, I., Brackett, N. L., Katoh, N., Shoji, M., and Wrenn, R. W., 1980, Calcium-dependent protein kinase: Widespread occurrence in various tissues and phyla of the animal kingdom and comparison of effects of phospholipids, calmodulin, and trifluoperazine. *Proc. Natl. Acad. Sci. U.S.A.* **77**:7039–7043.

Litosch, I., Lee, H. S., and Fain, J. N., 1984, Phosphoinositide breakdwon in blowfly salivary gland, *Am. J. Physiol.* **246**:c141–c147.

Lyons, R. M., and Atherton, R. M., 1979, Characterization of a platelet protein phosphorylated during the thrombin-induced release reaction, *Biochemistry* **18**:544–552.

Malaisse, W. J., Lebrun, P., Herchuelz, A., Sener, A., and Malaisse-Lagae, F., 1983, Synergistic effect of a tumor-promoting phorbol ester and a hypoglycemic sulfonylurea upon insulin release, *Endocrinology* **113**:1870–1877.

Martin, T. F. J., 1983, Thyrotropin-releasing hormone rapidly activates the phosphodiester hydrolysis of polyphosphoinositides in GH_3 pituitary cells, *J. Biol. Chem.* **258**:14816–14822.

Margolis, R. U., Press, R., Altszuler, N., and Stewart, M. A., 1981, Inositol production by the brain in normal and alloxan-diabetic dogs, *Brain Res.* **28**:535–539.

Mastro, A. M., and Smith, M. C., 1983, Calcium-dependent activation of lymphocytes by ionophore, A23187 and a phorbol ester tumor promoter, *J. Cell Physiol.* **116**:51–56.

Mayhew, J. A., Gillon, K. R. W., and Hawthorne, J. N., 1983, Free and lipid inositol, sorbitol and sugars in sciatic nerve obtained post-mortem from diabetic patients and control subjects, *Diabetologia* **24**:13–15.

Michell, R. H., 1975, Inositol phospholipids and cell surface receptor function, *Biochem. Biophys. Acta* **415**:81–147.

Michell, R. H., Kirk, C. J., Jones, L. M., Downes, C. P., and Creba, J. A., 1981. The stimulation of inositol lipid metabolism that accompanies calcium mobilization in stimulated cells: Defined characteristics and unanswered questions, *Phil. Trans. R. Soc. Lond. [Biol.]* **296**:123–137.

Monaco, M. E., 1982, The phosphatidylinositol cycle in WRK-1 cells, *J. Biol. Chem.* **257**:2137–2139.

Naka, M., Nishikawa, M., Adelstein, R. S., and Hidaka, H., 1983, Phorbol ester-induced activation of human platelets is associated with protein kinase C phosphorylation of myosin light chains, *Nature* **306**:490–492.

Natarajan, V., Dyck, P. J., and Schmid, H. H. O., 1981, Alterations of inositol lipid metabolism of rat sciatic nerve in streptozotocin-induced diabetes, *J. Neurochem.* **36**:413–419.

Nishizuka, Y., 1983, Phospholipid degradation and signal translation for protein phosphorylation, *Trends Biochem. Sci.* **8**:13–16.

Palmano, K. P., Whiting, P. H., and Hawthorne, J. N., 1977, Free and lipid *myo*-inositol in tissues from rats with acute and less severe streptozotocin-induced diabetes, *Biochem. J.* **167**:229–235.

Prescott, S. M., and Majerus, P. W., 1983, Characterization of 1,2-diacylglycerol hydrolysis in human platelets, *J. Biol. Chem.* **258**:764–769.

Putney, J. W., Jr., Burgess, G. M., Halenda, S. P., McKinney, J. S., and Rubin, R. P., 1983, Effects of secretagogues on [^{32}P] phosphatidylinositol 4,5-bisphosphate metabolism in the exocrine pancreas, *Biochem. J.* **212**:483–488.

Rebecchi, M. J., and Gerschengorn, M. C., 1983, Thyrotropin-releasing hormone stimulates rapid hydrolysis of phosphatidylinositol 4,5-bisphosphate by a phosphodiesterase in rat mammatropic pituitary cells: Evidence for an early calcium-independent action, *Biochem. J.* **216**:299–308.

Rebecchi, M. J., Kolesnick, R. N., and Gershengorn, M. C., 1983, Thyrotropin-releasing hormone stimulates rapid loss of phosphatidylinositol and its conversion to 1,2-diacylglycerol and phosphatidic acid in rat mammotropic cells, *J. Biol. Chem.* **258**:227–234.

Rhodes, D., Prpić, V., Exton, J. H., and Blackmore, P. F., 1983, Stimulation of phosphatidylinositol 4,5-bisphosphate hydrolysis in hepatocytes by vasopressin, *J. Biol. Chem.* **258**:2770–2773.

Rink, T. J., Smith S. W., and Tsien, R. Y., 1982, Cytoplasmic free Ca^{2+} in human platelets: Ca^{2+} thresholds and Ca-independent activation for shape-change and secretion, *FEBS Lett.* **148**:21–26.

Rink, T. J., Sanchez, A., and Hallam, T. J., 1983, Diacylglycerol and phorbol ester stimulate secretion without raising cytoplasmic free calcium in human platelets, *Nature* **305**:317–319.

Rittenhouse-Simmons, S., 1979, Production of diglyceride from phosphatidylinositol in activated human platelets. *J. Clin. Invest.* **63**:580–587.

Rozengurt, E., 1983, Growth factors, cell proliferation and cancer: An overview, *Mol. Biol. Med.* **1**:169–181.

Sha'afi, R. I., White, J. R., Molski, T. F. P., Shefcyk, J., Volpi, M., Naccache, P. H., and Feinstein, M. B., 1983, Phorbol 12-myristate 13-acetate activates rabbit neutrophils without an apparent rise in the level of intracellular free calcium, *Biochem. Biophys. Res. Commun.* **114**:638–645.

Sherman, W. R., Leavitt, A. L., Honchar, M. P., Hallcher, L. M., and Phillips, B. E., 1981, Evidence that lithium alters phosphoinositide metabolism: Chronic administration elevates primarily d-*myo*-inositol-1-phosphate in cerebral cortex of the rat. *J. Neurochem.* **36**:1947–1951.

Simmons, D. A., Winegrad, A. I., and Martin, D. B., 1982, Significance of tissue *myo*-inositol concentration in metabolic regulation in nerve, *Science* **217**:848–851.

Streb, H., and Schulz, I., 1983, Regulation of cytosolic free Ca^{2+} concentration in acinar cells of rat pancreas, *Am. J. Physiol.* **245**:G347–G357.

Streb, H., Irvine, R. F., Berridge, M. J., and Schulz, I., 1983, Release of Ca^{2+} from a nonmitochondrial intracellular store in pancreatic acinar cells by inositiol 1,4,5-trisphosphate, *Nature* **306**:67–69.

Takai, Y., Kishimoto, A., Iwasa, Y., Kawahara, Y., Mori, T., and Nishizuka, Y., 1979, Calcium-dependent activation of a multifunctional protein kinase by membrane phospholipid, *J. Biol. Chem.* **254**:3692–3695.

Takenawa, T., Masaki, T., and Goto, K., 1983, Increase in norepinephrine-induced formation of phosphatidic acid in rat vas deferens after denervation, *J. Biochem.* **93:**303–306.

Thomas, A. P., Marks, J. S., Coll, K. E., and Williamson, J. R., 1983, Quantitation and early kinetics of inositol lipid changes induced by vasopressin in isolated and cultured hepatocytes, *J. Biol. Chem.* **258:**5716–5725.

Thomas, A. P., Alexander, J., and Williamson, J. R., 1984, Relationship between inositol polyphosphate production and the increase of cytosolic free Ca^{2+} induced by vasopressin in isolated hepatocytes, *J. Biol. Chem.* **259:**5574–5584.

Warfield, A. S., and Segal, S., 1978, *Myo*-inositol and phosphatidylinositol metabolism in synaptosomes from galactose-fed rats, *Proc. Natl. Acad. Sci. U.S.A.* **75:**4568–4572.

Waterfield, M. D., Scrace, G. T., Whittle, N., Stroobant, P., Johnsson, A., Wasteson, Å., Westermark, B., Heldin, C.-H., Huang, J. S., and Deuel, T. F., 1983, Platelet-derived growth factor is structurally related to the putative transforming protein p28[sis] of simian sarcoma virus, *Nature* **304:**35–39.

Whitfield, J. F., MacManus, J. P., and Gillan, D. J., 1973, Calcium-dependent stimulation by a phorbol ester (PMA) of thymic lymphoblast DNA synthesis and proliferation, *J. Cell. Physiol.* **82:**151–156.

Whiting, P. H., Palmano, K. P., and Hawthorne, J. N., 1979, Enzymes of *myo*-inositol and inositol lipid metabolism in rats with streptozotocin-induced diabetes, *Biochem. J.* **179:**549–553.

Ionic Signal Transduction by Growth Factors

W. H. MOOLENAAR and S. W. de LAAT

1. Introduction

The cell surface plays a crucial role in the regulation of cell metabolism in general and in the regulation of cell proliferation in particular. Receptor proteins in the plasma membrane receive and transduce growth-stimulatory signals that trigger a cascade of biochemical and physiological changes in the cell. Ultimately, this chain of events culminates in replicative DNA synthesis and cell division. Among the most extensively studied mitogens are the polypeptide hormones epidermal growth factor (EGF) and platelet-derived growth factor (PDGF). Although much has been learned about the interaction of EGF and PDGF with their specific receptors on the cell surface, relatively little is known about the ensuing intracellular signals transmitting the mitogenic response from the plasma membrane to the cell nucleus. In the search for putative mitogenic signals, much attention has been focused on the earliest detectable cellular reactions following the addition of growth factors to appropriate target cells. Immediate consequences of growth factor–receptor interaction include tyrosine-specific protein phosphorylations (Ushiro and Cohen, 1980; Ek et al., 1982; Cooper et al., 1982), breakdown of inositol phospholipids (Sawyer and Cohen, 1981; Habenicht et al., 1981), and increased transport of Na^+, K^+, and Ca^{2+} across the plasma membrane (Rozengurt and Heppel, 1975; Moolenaar et al., 1981a, 1982a,b; Sawyer and Cohen, 1981).

The purpose of this chapter is to focus on the ionic aspects of growth factor action. We do not intend to cover all these aspects but limit our-

W. H. MOOLENAAR and S. W. de LAAT • Hubrecht Laboratory, 3584 CT Utrecht, The Netherlands.

selves to the potential signaling role of Na^+-H^+ exchange, cytoplasmic pH (pH_i), and cytoplasmic free Ca^{2+} (Ca^{2+}_i) in the action of EGF, PDGF, and serum. In addition, we briefly discuss the electrical membrane events that accompany serum stimulation of quiescent cells.

2. Monovalent Ions in Growth Factor Action

2.1. Rapid Electrical Events

The first direct evidence that mitogenic stimulation of resting mammalian cells may be accompanied by dramatic changes in the ionic permeability ("conductance") of the plasma membrane was presented by Hülser and Frank (1971), who monitored changes in transmembrane potential during serum stimulation of quiescent rat cells. The electrical membrane potential, as measured with intracellular microelectrodes, is a direct reflection of (1) the asymmetric distribution of Na^+ and K^+ across the plasma membrane and (2) the ratio of Na^+ to K^+ permeability. As a consequence, stimulus-induced changes in Na^+ or K^+ permeability manifest themselves as alterations in membrane potential and membrane resistance. Using conventional electrophysiological techniques, Hülser and Frank showed that growth-promoting serum protein(s) immediately reduce(s) the membrane potential of rat fibroblasts from about -50 mV to -17 mV, but the underlying ionic mechanisms were not further examined.

In our electrophysiological studies on mitogen-induced ion movements in mouse neuroblastoma cells and human fibroblasts (Moolenaar et al., 1979, 1981a, 1982a,b), we extended the earlier observations to show that serum stimulation of these cells elicits a rapid but transient membrane depolarization, and it was established that this effect is attributable to a rather unselective increase in membrane conductance for various ions including Na^+, K^+, and perhaps Ca^{2+} (Fig. 1). Whether this sudden ionic leakiness of the plasma membrane is caused by a mitogen-induced phosphorylation reaction or another pathway is involved remains to be elucidated. It is interesting to note that an acute membrane depolarization is also measured when quiescent fibroblasts are stimulated to grow by "wound formation" in a confluent monolayer (Cone and Tongier, 1973).

Although an intriguing phenomenon in itself, a similar depolarization is not evoked by mitogenic doses of EGF, insulin, or nerve growth factor in various mammalian cells (Moolenaar et al., 1982a,b; Boonstra et al., 1983), while the possible electrophysiological effects of PDGF remain to be investigated. Whereas EGF does elicit a small membrane depolar-

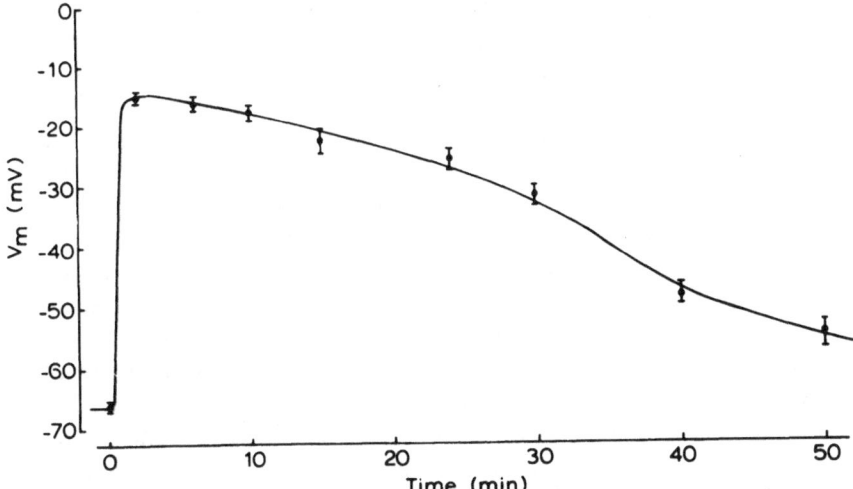

Figure 1. Transient membrane depolarization during serum stimulation of quiescent human fibroblasts as measured by intracellular microelectrodes. Dialyzed fetal calf serum (10%) was added at time zero. Data points represent averages of five to ten impalements in different cells. For full details see Moolenaar *et al.* (1982a).

ization of unknown ionic origin in an epithelial cell line (Rothenberg *et al.*, 1982), the available evidence nevertheless suggests that mitogen-induced changes in ionic conductance and membrane potential are not a general phenomenon accompanying growth stimulation of previously quiescent cells. It thus seems plausible to assume that rapid alterations in electrical membrane properties are not essential for the generation of a mitogenic signal by external stimuli.

2.2. The Na^+-K^+ Pump and Na^+-H^+ Exchange

A common early event following mitogenic activation of various cell types is the acceleration of the Na^+-K^+ pump (Na^+-K^+ ATPase) in the plasma membrane, as evidenced by an increase in the initial rate of ouabain-sensitive K^+ uptake (e.g., Rozengurt and Heppel, 1975; Kaplan, 1978; Moolenaar *et al.*, 1981a, 1982a; Frantz *et al.*, 1981). Enhanced Na^+-K^+ pumping usually leads to a 10–20% increase in intracellular K^+ content, but it is generally agreed that relatively small changes in intracellular K^+ (and Na^+) concentrations are not likely to serve as mitogenic signals.

It is now realized that the observed acceleration of the Na^+-K^+ pump is a secondary event, the direct consequence of a prior increase in

Na$^+$ influx (Smith and Rozengurt, 1978; Mendoza *et al.*, 1980; Moolenaar *et al.*, 1981a, 1982a,b; Pouysségur *et al.*, 1982; Boonstra *et al.*, 1983). What is the nature of this mitogen-induced Na$^+$ influx? An important clue was obtained from our electrophysiological and tracer flux studies on growth-stimulated cells (Moolenaar *et al.*, 1981, 1982a,b; Boonstra *et al.*, 1983). Comparison of the electrophysiological data with Na$^+$ uptake measurements led to the conclusion that mitogen-induced Na$^+$ influx is mediated, at least in large part, by an electrically silent transport system that is distinct from the conductive Na$^+$ permeability. This Na$^+$ transporter was identified as an amiloride-sensitive Na$^+$–H$^+$ exchanger (Moolenaar *et al.*, 1981a,b, 1982; Pouysségur *et al.*, 1982), a H$^+$-extruding system that normally contributes to the regulation of pH$_i$ by virtue of its sensitivity to cytoplasmic H$^+$ (Roos and Boron, 1981; Aronson *et al.*, 1982; Moolenaar *et al.*, 1983, 1984a). Recent advances in pH$_i$ monitoring techniques, particularly the use of intracellularly trapped fluorescent indicators (Thomas *et al.*, 1979; Rink *et al.*, 1982), have allowed us to obtain a fairly good understanding of the role of Na$^+$–H$^+$ exchange in pH$_i$ regulation in general and in the action of growth factors in particular (Moolenaar *et al.*, 1983, 1984a). It appears that activation of the otherwise quiescent Na$^+$–H$^+$ exchanger by growth stimulants leads to a significant rise in pH$_i$ of up to 0.2 pH unit. Thus, the Na$^+$–H$^+$ exchanger is a strong candidate for participating in transmembrane signal transduction by growth factors.

2.3. Changes in Cytoplasmic pH

It may be obvious that insight into the potential signaling role of Na$^+$–H$^+$ exchange and pH$_i$ in growth factor action requires a detailed understanding of the ionic basis of pH$_i$ homeostasis in general. Therefore, we first summarize some of the most relevant characteristics of the pH$_i$-regulating mechanisms in human fibroblasts as determined by the use of the intracellularly trapped pH$_i$-sensitive dye bis(carboxyethyl)carboxyfluorescein (BCECF) (Rink *et al.*, 1982; Moolenaar *et al.*, 1983, 1984a).

2.3.1. Ionic Basis of pH$_i$ Homeostasis

The mechanism underlying pH$_i$ regulation in various cells are usually examined by analyzing the recovery of pH$_i$ towards its resting value after a sudden cytoplasmic acidification (Roos and Boron, 1981). In human fibroblasts, pH$_i$ recovery from an acute acid load, as induced by an NH$_4$$^+$ prepulse, follows an exponential time course and is entirely mediated by

Na^+-H^+ exchange as evidenced by the following findings: (1) pH_i recovery is accompanied by Na^+ influx and net H^+ exit, roughly with a 1:1 stoichiometry; (2) pH_i recovery and concomitant Na^+ and H^+ fluxes are reversibly blocked by amiloride (half-maximal effect at ~0.1 mM); (3) the rate of pH_i recovery depends critically on the external Na^+ concentration (half-maximal rate at ~30 mM Na^+) but not on such other external anions as Cl^- and HCO_3^-; and (4) Li^+ can substitute for Na^+ in pH_i recovery and net H^+ extrusion, but other cations cannot (Moolenaar *et al.*, 1983, 1984a).

The exponentiality of pH_i recovery from an acid load is an important feature. It implies that the rate of the Na^+-H^+ exchanger, being proportional to $d(pH_i)/dt$, is linearly dependent on pH_i provided that the cytoplasmic buffering power is relatively independent of pH_i (Roos and Boron, 1981; Boron, 1983; Moolenaar *et al.*, 1984a). The value of the resting pH_i then reflects the balance between metabolic acidification and H^+ influx on the one hand and H^+ exit via the Na^+-H^+ exchanger on the other. It should be emphasized, however, that the resting pH_i reflects an apparent "shut off" of the Na^+-H^+ exchanger rather than thermodynamic equilibrium. Under normal ionic conditions there is still sufficient energy in the Na^+ and H^+ gradients to raise pH_i nearly 1 pH unit more alkaline.

Aronson *et al.* (1982) have presented evidence that the pH_i sensitivity of the Na^+-H^+ exchanger reflects allosteric activation by internal H^+ at a site that is distinct from the internal transport site. In contrast, external H^+ seems to compete with Na^+ and amiloride for binding at the same external transport site. The external H^+ sensitivity of the exchanger manifests itself in the dependence of the resting pH_i on sudden changes in external pH (pH_o). As a rule, pH_i responds to changes in pH_o by approximately 50% of the externally applied pH shift (Moolenaar *et al.*, 1984a).

2.3.2. pH_i in Growth Factor Action

Figure 2 shows typical examples of alkaline pH_i shifts after addition of EGF or PDGF to human fibroblasts loaded with the pH_i-sensitive dye BCECF. The pH_i shift is detectable within 30 sec and is usually complete by 10–15 min. It is interesting to note that insulin markedly potentiates the effects of EGF and PDGF without affecting pH_i when present alone (Moolenaar *et al.*, 1984a; Fig. 2). The following lines of evidence allow the conclusion that the observed rise in pH_i is mediated by the Na^+-H^+ exchanger in the plasma membrane: (1) the pH_i shift and concomitant electroneutral Na^+ influx are both inhibited by amiloride; (2) the rise in

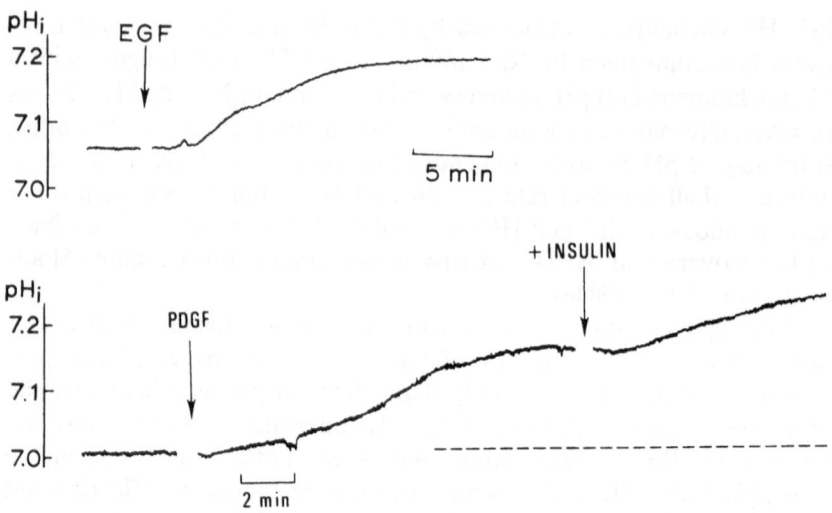

Figure 2. Time course of cytoplasmic alkalinization induced by 50 ng/ml EGF and 40 ng/ml PDGF, respectively. Quiescent fibroblasts were loaded with the fluorescent pH indicator BCECF. Potentiating effect of insulin (5 μg/ml) is shown in PDGF recording. For full details see Moolenaar *et al.* (1983).

pH_i is converted into a fall in pH_i when the direction of the transmembrane Na^+ gradient is reversed; and (3) the pH_i response can be mimicked, at least qualitatively, by the N^+-H^+ ionophore monensin (Moolenaar *et al.*, 1983, 1984a; Cassel *et al.*, 1983). Furthermore, we have verified the theoretical prediction that the amiloride-sensitive Na^+ influx accompanying the pH_i shift has a transient character, its peak being reached after ~5 min of mitogen exposure, when $d(pH_i)/dt$ is maximal (unpublished experiments). In the continuous presence of growth factors, the elevated pH_i persists for at least 1–2 hr, whereas pH_i only slowly returns to its resting level after mitogen washout. It thus seems that whereas activation of Na^+-H^+ exchange occurs rapidly, deactivation of the exchanger is a relatively slow process. Figure 3 summarizes the sequence of monovalent ion movements following receptor occupancy.

Two major questions that now arise are: (1) by which mechanism(s) do growth factors activate Na^+-H^+ exchange, and (2) what is the significance of an early rise in pH_i for the initiation of a proliferative response? A clue to the first question was obtained by kinetic analysis of pH_i recovery from a sudden cytoplasmic acidification in growth-stimulated versus quiescent cells (Moolenaar *et al.*, 1983). As illustrated in Fig. 4, the overall rate of pH_i recovery from an NH_4^+ acid load remains unaltered after growth stimulation, but the value at which pH_i stabilizes is

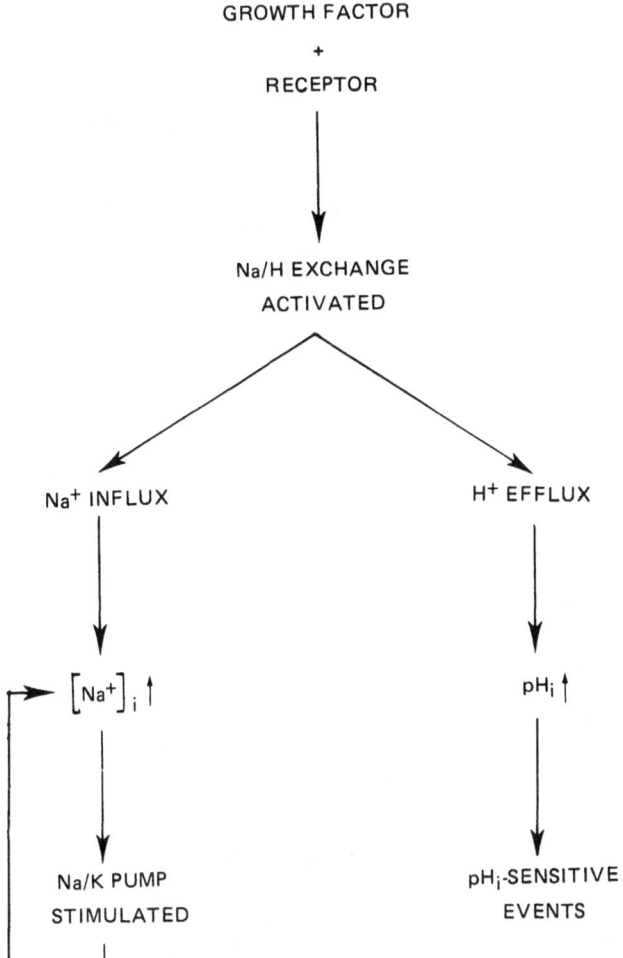

Figure 3. Scheme of ionic events following growth factor–receptor interaction. pH_i-sensitive events are discussed in the text.

~0.2 pH unit more alkaline. This is expressed more quantitatively in Fig. 4B, where Na^+-H^+ exchange activity is measured as $d(pH_i)/dt$. It is seen that growth stimulation simply induces a shift of the function $d(pH_i)/dt$ versus pH_i to the right. In other words, growth factors enhance the cytoplasmic H^+ sensitivity of the Na^+-H^+ exchanger. This intrinsic modification of the exchanger presumably concerns the regulatory H^+ site at the inner side of the plasma membrane (cf. Aronson *et al.*, 1982), resulting in an increased apparent affinity for intracellular H^+ (Moolenaar *et al.*, 1983).

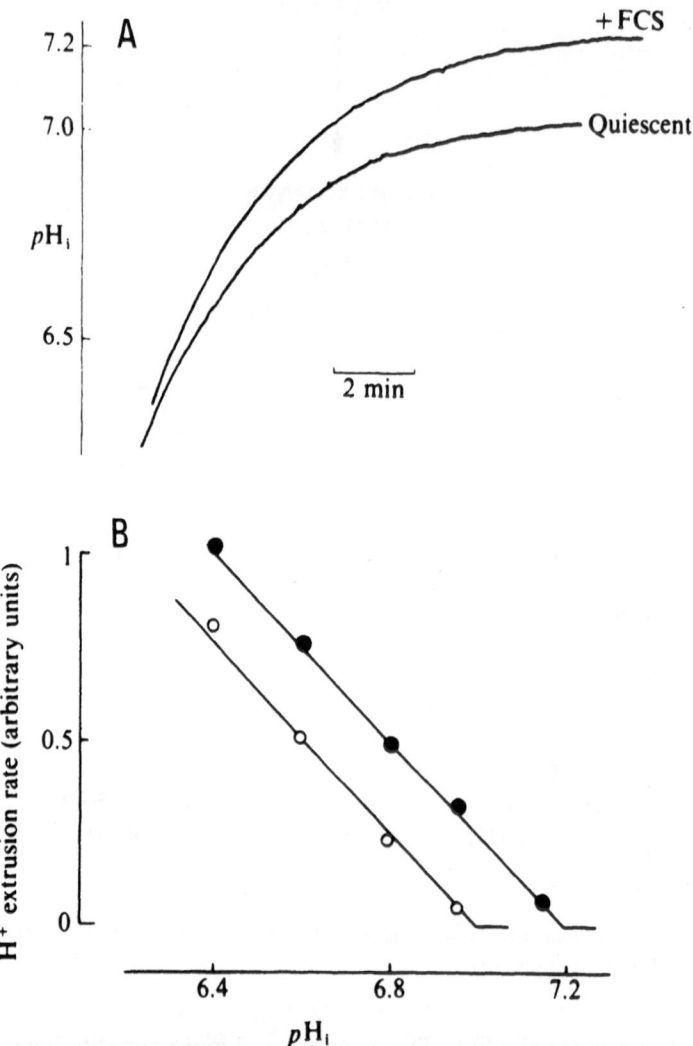

Figure 4. A: Superimposed recordings of exponential pH_i recovery from an NH_4^+ acid load in quiescent and serum-stimulated fibroblasts, respectively. After the first pH_i recovery, 10% fetal calf serum (FCS) was added, and 10 min later a second NH_4^+ pulse was given, and pH_i recovery was monitored. B: Relationship between rate of H^+ extrusion, i.e., Na^+–H^+ exchange activity, and pH_i of the experiment in A. Rate of H^+ extrusion was measured as $d(pH_i)/dt$, estimated as the rise in pH_i over 10-sec intervals. (Reproduced from Moolenaar *et al.*, 1983, with permission of Macmillan Journals Ltd., London.)

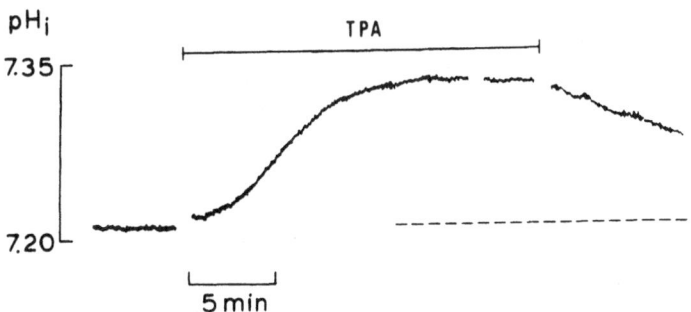

Figure 5. Effect of tumor promoter TPA (200 ng/ml) on pH_i in serum-deprived mouse neu-roblastoma cells (clone N1E-115) loaded with the fluorescent pH_i indicator BCECF. For experimental details see Moolenaar *et al.* (1983).

2.3.3. *Evidence for a Link to Inositol Lipid Breakdown*

How does receptor occupancy lead to activation of Na^+–H^+ exchange? An important key comes from our recent experiments using the tumor promoter 12-O-tetradecanoylphorbol-13-acetate (TPA), which binds to and directly activates a novel phospholipid-dependent protein kinase, kinase C (Castagna *et al.*, 1982). Under normal conditions, this enzyme is activated by endogenous diacylglycerol derived from the prior breakdown of inositol phospholipids (reviewed by Nishizuka, 1983). Our studies indicate that TPA and synthetic diacylglycerol mimics growth factors in raising pH_i in various cell types (W. H. Moolenaar *et al.*, 1984c). A typical example is shown in Fig. 5, where TPA is seen to raise pH_i by ~0.15 unit within 10 min in serum-deprived mouse neuroblastoma cells. Similar responses to TPA have been observed in human fibroblasts and HeLa cells (not shown). These TPA effects are inhibited by amiloride and by removal of external Na^+. Furthermore, we have verified that a non-tumor-promoting TPA analogue that has no affinity for protein kinase C (Castagna *et al.*, 1982) fails to affect pH_i.

The above findings strongly suggest that protein kinase C is responsible for modifying the pH_i sensitivity of the Na^+–H^+ exchanger, resulting in a rise in pH_i. The simplest explanation is that protein kinase C directly phosphorylates the exchanger, but experimental support for this hypothesis is lacking as yet. These results also raise the interesting possibility that an alkaline shift in pH_i is not uniquely induced by growth factors but may be a common cellular response to those surface stimulants that provoke an immediate breakdown of inositol phospholipids.

Table I. Dependence of the Relative Rate of Glycolosis, Protein Synthesis, and DNA Synthesis under Various Conditions Known to Raise pH_i[a]

Conditions	Lactate production (30 min)	Protein synthesis (4 hr)	DNA synthesis (20 hr)
pH_0 7.4	1	1	1
+ FCS	1.5	2.0	15
+ Monensin	1.8	1.2	(toxic)
pH_0 7.9	1.3	1.4	1.7

[a] Aerobic lactate production (0–30 min) was determined fluorimetrically using the Sigma lactate assay kit. Protein synthesis was estimated from the amount of incorporated [^{3}H]-leucine into cellular protein (0–4 hr). DNA synthesis was determined by [^{3}H]-thymidine incorporation over a 20-hr period. Fetal calf serum (FCS) was added at 10%, and monensin at 1 µM.

2.4. Metabolic Effects of a Rise in pH_i

An early rise in pH_i seems indeed to be a fairly common response of metabolically dormant cells to appropriate stimuli. It has been described for the fertilization of sea urchin eggs (Johnson et al., 1976), the action of insulin on frog muscle (Moore, 1979), the activation of platelets by thrombin (Horne et al., 1981), and the chemotactic response of human neutrophils (Molski et al., 1980). What may be the physiological significance of an early alkaline shift in pH_i? From studies on various cell types, it is known that an increase in pH_i may participate in the regulation of such diverse cellular processes as protein synthesis (Grainger et al., 1979), cytoskeletal reorganization (Begg and Rebhun, 1979; Regula et al., 1981), cell-to-cell communication (Spray et al., 1981), oxygen consumption, and cell motility (Christen et al., 1982; Lee et al., 1983). Moore and colleagues have presented convincing evidence that the stimulation of glycolosis by insulin in frog muscle is mediated by a rise in pH_i (Moore et al., 1979; Fidelman et al., 1982). Indeed, the rate-limiting enzyme of the glycolytic pathway, phosphofructokinase (PFK), is extremely pH sensitive (Trivedi and Danforth, 1966). It seems a plausible assumption that regulation of kinase activity by pH_i is not limited to the case of PFK. For example, Pouysségur et al. (1982) have hypothesized that the amiloride-sensitive phosphorylation of ribosomal protein S6 is mediated by a rise in pH_i. This claim has been questioned, however, on account of nonspecific effects of amiloride on the activity of protein kinases (Holland et al., 1983).

We have tested the possibility that the rate of aerobic glycolosis, protein synthesis and DNA synthesis may be modulated by a permanent elevation in pH_i. As shown in Table I, artificially raising pH_i by either alkaline external pH or monensin stimulates lactate production in quiescent fibroblasts. Also, the incorporation of $[^3H]$-leucine into protein and of $[^3H]$-thymidine into DNA is significantly increased by alkaline pH. These results, although preliminary, support the view that an alkaline shift in pH_i may contribute to the metabolic activation of resting cells. However, it is obvious that although pH_i may have an apparent regulatory role, a pH_i shift in itself is not a sufficient mitogenic signal, and mitogens utilize additional signals for the initiation of a full proliferate response.

3. Calcium Mobilization by Growth Factors

It is generally accepted that cytoplasmic free Ca^{2+} is a key regulator of numerous cellular processes. A rise in $[Ca^{2+}]_i$ has been implicated as a second messenger in the action of various surface stimulants, but relatively little is known about a possible role of Ca^{2+}_i in growth factor action. Indirect evidence suggests that growth factors may induce rapid alterations in cellular Ca^{2+} pools, as inferred from $^{45}Ca^{2+}$ tracer flux measurements (Tupper *et al.*, 1978; Sawyer and Cohen, 1981; Owen and Villereal, 1983), but changes in $[Ca^{2+}]_i$ have not previously been measurable.

Using the intracellularly trapped fluorescent Ca^{2+} indicator quin-2, Tsien *et al.* (1982) have recently described an early approximately twofold rise in $[Ca^{2+}]_i$ in lectin-stimulated lymphocytes. Using the same technique, we have monitored the effects of PDGF, EGF, and serum on $[Ca^{2+}]_i$ in quiescent cultures of diploid human fibroblasts. Figure 6 illustrates that these mitogens induce an immediate up to threefold rise in $[Ca^{2+}]_i$. This Ca^{2+} signal is initiated without a detectable lag period; it reaches a maximum by 20–40 sec and then slowly declines to a new steady level ~30% above the normal resting level of 100–150 nM (Moolenaar *et al.*, 1984b). The growth-factor-induced Ca^{2+} response is not prevented by removal of external Ca^{2+} (Fig. 6), indicating that transmembrane Ca^{2+} influx does not participate in the initial rise in $[Ca^{2+}]_i$. We therefore conclude that mitogens act by triggering the release of Ca^{2+} from intracellular stores (e.g., mitochondria, endoplasmatic reticulum, plasma membrane). This Ca^{2+} signal is the fastest response to growth factors thus far described. For example, the PDGF-evoked peak in $[Ca^{2+}]_i$ precedes the maximum of the PDGF-stimulated tyrosine kinase activity in intact cells by several minutes (cf. Cooper *et al.*, 1982). These findings strongly suggest that Ca^{2+} is a primary messenger in the action of EGF and PDGF. Unlike the mitogen-induced pH_i shift, the rise in $[Ca^{2+}]_i$ cannot be mimicked by TPA (Moolenaar *et al.*, 1984b), consistent with the fact that TPA acts at a point distal from Ca^{2+} mobilization. This result

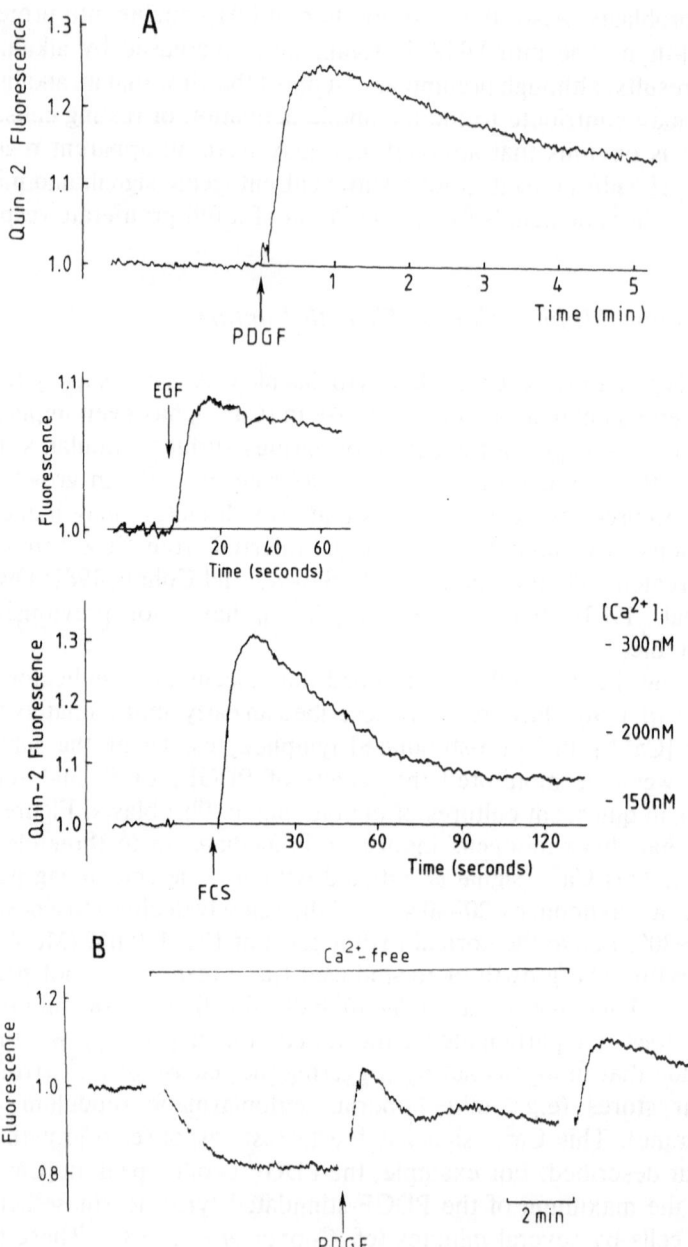

Figure 6. A: Response of quiescent human fibroblasts loaded with the fluorescent Ca^{2+} indicator quin-2 to PDGF (40 ng/ml), EGF (50 ng/ml), and FCS (10%), respectively. Peak transients represent an up to threefold increase in $[Ca^{2+}]_i$ as indicated in the FCS recording. B: Response to PDGF of quin-2-loaded fibroblasts in nominally Ca^{2+}-free medium containing 0.5 mM EGTA. For further details see Moolenaar *et al.* (1984b).

Table II. Effects of Various Stimulants on Cytoplasmic pH and Free Ca^{2+} in Human Fibroblasts[a]

Additions	ΔpH_i (relative to control)	Peak $[Ca^{2+}]_i$	
		nM	Relative to control
None	0	140	(1)
FCS (10%)	0.22	360	(2.6)
EGF (50 ng/ml)	0.10	190	(1.4)
PDGF (40 ng/ml)	0.15	300	(2.1)
Insulin (5 μg/ml)	0	140	(1)
EGF + insulin	0.15		
PDGF + insulin	0.20		
TPA (200 ng/ml)	0.14	140	(1)

[a] Quiescent cultures were loaded with BCECF or quin-2 and assayed for shifts in pH_i and $[Ca^{2+}]_i$ as described in the text. Numbers are averages of 5–12 different experiments (S.E. less than 9%).

clearly indicates that a rise in $[Ca^{2+}]_i$ is not required for the activation of Na^+-H^+ exchange by growth factors.

How do PDGF and EGF trigger the release of intracellular Ca^{2+}? A striking feature of Ca^{2+}-mobilizing hormones is that they provoke an immediate breakdown of inositol phospholipids in their target cells (Michell, 1975; Berridge, 1981), and, indeed, growth factors are no exception (Sawyer and Cohen, 1981; Habenicht *et al.*, 1981). In particular, the very rapid formation of inositol trisphosphate seems to constitute a key signal for the release of Ca^{2+} from nonmitochondrial stores (Streb *et al.*, 1983) and thus may well underlie the Ca^{2+} signals described above. Further studies are required to test this possibility and to establish whether an immediate rise in $[Ca^{2+}]_i$ plays a functional role in the initiation of a proliferative response.

4. Concluding Remarks

Various molecular events have been identified over the past few years that may participate in postreceptor signal transduction by polypeptide growth factors. These include tyrosine-specific protein phosphorylation, inositol lipid breakdown, and the two major ionic events discussed above, namely, elevation of $[Ca^{2+}]_i$ and activation of Na^+-H^+ exchange with a resultant rise in pH_i. Table II summarizes the effects of various stimulants on pH_i and $[Ca^{2+}]_i$ in quiescent human fibroblasts. Accumulating evidence suggests that both ionic events are a direct consequence of the receptor-linked breakdown of inositol lipids, as schematically illustrated

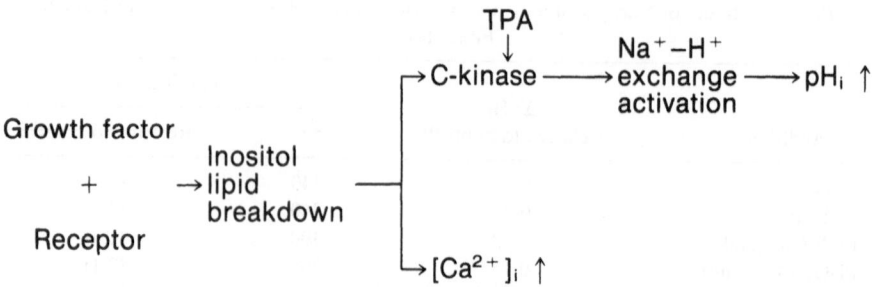

Figure 7. Proposed scheme of the generation of intracellular ionic signals by growth factors such as EGF and PDGF and by the tumor promoter TPA. See text for details.

in Fig. 7. Although uncertainty still exists about the relevance of each of these steps in the eventual initiation of DNA synthesis, it seems likely that the various events evoked by lipid breakdown act in concert with tyrosine-specific protein phosphorylations to stimulate cell proliferation. Selective pharmacological inhibitors without nonspecific side effects would be of great help in elucidating the relevance of each of the signal pathways, but such agents are lacking at present. It is hoped that the use of various monoclonal antibodies against mitogen receptors, Na^+-H^+ exchanger, etc. will greatly facilitate work on this problem in the near future.

Another major area for future research concerns the question of whether the ionic events induced by growth factors have their correlates in the action of transforming oncogene products. Some of these transforming proteins are known to resemble permanently activated mitogen receptors in that they are tyrosine-specific kinases located at the inner side of the plasma membrane. In particular, the use of cells infected with temperature-sensitive transforming mutants of avian sarcoma virus would offer the unique opportunity to examine the overlap between "normal" growth factor action and malignant growth induced by altered oncogene expression.

ACKNOWLEDGMENTS. This work was supported by the Netherlands Cancer foundation (Koningin Wilhelmina Fonds). We are grateful to Dr. R. Y. Tsien for gifts of BCECF and to Dr. C.-H. Heldin for supplying PDGF. We thank Lilian Joosen and Leon Tertoolen for technical assistance.

References

Aronson, P. S., Nee, J., and Suhm, M., 1982, Modifier role of internal H^+ in activating the Na^+-H^+ exchanger in renal microvillus membrane vesicles, *Nature* **299**:161–163.

Begg, D. A., and Rebhun, L. A., 1979, pH regulates the polymerization of actin in the sea urchin egg cortex, *J. Cell Biol.* **83:**241–248.

Berridge, M. J., 1981, Phosphatidylinositol hydrolysis: A multifunctional transducing mechanism, *Mol. Cell. Endocrinol.* **24:**115–140.

Boonstra, J., Moolenaar, W. H., Harrison, P., Moed, P., van der Saag, P. T., and de Laat, S. W., 1983, Ionic responses and growth stimulation induced by NGF and EGF in rat PC12 cells, *J. Cell Biol.* **97:**92–98.

Boron, W. F., 1983, Transport of H^+ and of ionic weak acids and bases, *J. Membr. Biol.* **72:**1–11.

Cassel, D., Rothenberg, P., Zhuang, Y., Deuel, T. F., and Glaser, L., 1983, PDGF stimulates Na^+/H^+ exchange and induces cytoplasmic alkalizination in NR6 cells, *Proc. Natl. Acad. Sci. U.S.A.* **80:**6224–6228.

Castagna, M., Takai, Y., Kaibuchi, K., Sano, K., Kikkawa, U., and Nishizuka, Y., 1982, Direct activation of calcium-activated, phospholipid-dependent protein kinase by tumor-promoting phorbol esters, *J. Biol. Chem.* **257:**7847–7851.

Christen, R., Schackmann, R. W., and Shapiro, B. M., 1982, Elevation of the intracellular pH activates respiration and motility of sperm of the sea urchin, *Strongylocentrotus purpuratus, J. Biol. Chem.* **257:**14881–14890.

Cone, C. D., and Tongier, M., 1973, Contact inhibition of division: Involvement of the electrical membrane potential, *J. Cell. Physiol.* **82:**373–386.

Cooper, J. A., Bowen-Pope, D. F., Raines, E., Ross, R., and Hunter, T., 1982, Similar effects of PDGF and EGF on the phosphorylation of tyrosine in cellular proteins, *Cell* **31:**263–273.

Ek, B., Westermark, B., Wasteson, Å., and Heldin, C.-H., 1982, Stimulation of tyrosine-specific phosphorylation by platelet-derived growth factor, *Nature* **295:**419–420.

Fidelman, M. L., Seeholzer, S. H., Walsh, K. B., and Moore, R. D., 1982, Intracellular pH mediates action of insulin on glycolosis in frog skeletal muscle, *Am. J. Physiol.* **242:**C87–C93.

Frantz, C. N., Nathan, D. G., and Scher, C. D., 1981, Intracellular univalent cations and the regulation of the 3T3 cell cycle, *J. Cell Biol.* **88:**51–56.

Grainger, J. L., Winkler, M. M., Shen, S. S., and Steinhardt, R. A., 1979, Intracellular pH controls protein synthesis rate in the sea urchin and early embryo, *Dev. Biol.* **68:**396.

Habenicht, A., Glomset, J., King, W., Nist, C., Mitchell, C., and Ross, R., 1981, Early changes in phosphatidylinositol and arachidonic acid metabolism in quiescent Swiss 3T3 cells stimulated to divide by PDGF, *J. Biol. Chem.* **256:**12329–12335.

Holland, R., Woodgett, J. R., and Hardie, D. G., 1983, Evidence that amiloride antagonises insulin-stimulated protein phosphorylation by inhibiting protein kinase activity, *FEBS Lett.* **154:**269.

Horne, W. C., Norman, N. E., Schwartz, D. B., and Simons, E. R., 1981, Changes in cytoplasmic pH and in membrane potential in thrombin-stimulated human platelets, *Eur. J. Biochem.* **120:**295–302.

Hülser, D. F., and Frank, W., 1971, Electrophysiological measurements at the cell surface membrane, *Z. Naturforsch.* **26B:**1046–1048.

Johnson, J. D., Epel, D., and Paul, M., 1976, Intracellular pH and activation of sea urchin eggs after fertilization, *Nature* **262:**661–664.

Kaplan, J. G., 1978, Membrane cation transport and the control of proliferation of mammalian cells, *Annu. Rev. Physiol.* **40:**19–46.

Lee, H. C., Johnson, C., and Epel, D., 1983, Changes in internal pH associated with initiation of motility and acrosome reaction of sea urchin sperm, *Dev. Biol.* **95:**31.

Mendoza, S. A., Wigglesworth, N. M., Pohjanpelto, P., and Rozengurt, E., 1980, Na entry and Na–K pump activity in murine, hamster and human cells, *J. Cell. Physiol.* **103:**17–27.

Michell, R. H., 1975, Inositol phospholipids and cell surface receptor function, *Biochim. Biophys. Acta* **415:**81–147.

Molski, T., Naccache, P., Volpi, M., Wolpert, L., and Sha'afi, R., 1980, Specific modulation of the intracellular pH of rabbit neutrophils by chemotactic factors, *Biochem. Biophys. Res. Commun.* **94:**508–514.

Moolenaar, W. H., de Laat, S. W., and van der Saag, P. T., 1979, Serum triggers a sequence of rapid ionic conductance changes in quiescent neuroblastoma cells, *Nature* **279:**721–723.

Moolenaar, W. H., Mummery, C. L., van der Saag, P. T., and de Laat, S. W., 1981a, Rapid ionic events and the initiation of growth in serum-stimulated neuroblastoma cells, *Cell* **23:**789–798.

Moolenaar, W. H., Boonstra, J., van der Saag, P. T., and de Laat, S. W., 1981b, Sodium/proton exchange in mouse neuroblastoma cells, *J. Biol. Chem.* **256:**12883–12887.

Moolenaar, W. H., Yarden, Y., de Laat, S. W., and Schlessinger, J., 1982a, EGF induces electrically silent Na^+ influx in human fibroblasts, *J. Biol. Chem.* **257:**8502–8506.

Moolenaar, W. H., Mummery, C. L., van der Saag, P. T., and de Laat, S. W., 1982b, in: *Membranes in Tumour Growth* (T. Galeotti, A. Cittadini, G. Neri, and S. Papa, eds.), Elsevier, Amsterdam, pp. 413–418.

Moolenaar, W. H., Tsien, R. Y., van der Saag, P. T., and de Laat, S. W., 1983, Na^+/H^+ exchange and cytoplasmic pH in the action of growth factors, *Nature* **304:**645–648.

Moolenaar, W. H., Tertoolen, L. G. J., and de Laat, S. W., 1984a, The regulation of cytoplasmic pH in human fibroblasts, *J. Biol. Chem.* **259:**7563–7569.

Moolenaar, W. H., Tertoolen, L. G. J., and de Laat, S. W., 1984b, Growth factors immediately raise cytoplasmic free Ca^{2+} in human fibroblasts, *J. Biol. Chem.* **259:**8066–8069.

Moolenaar, W. H., Tertoolen, L. G. J., and de Laat, S. W., 1984c, Phorbol ester and diacylglycerol mimick growth factors in raising cytoplasmic pH, *Nature* **312:**371–374.

Moore, R. D., 1979, Elevation of intracellular pH by insulin in frog skeletal muscle, *Biochem. Biophys. Res. Commun.* **91:**900–904.

Moore, R. D., Fidelman, M. L., and Seeholzer, S. H., 1979, Correlation between insulin action upon glycolysis and change in intracellular pH, *Biochem. Biophys. Res. Commun.* **91:**905–910.

Nishizuka, Y., 1983, Phospholipid degradation and signal translation for protein phosphorylation, *Trends Biochem. Sci.* **8:**13–16.

Owen, N., and Villereal, M., 1983, Efflux of $^{45}Ca^{2+}$ from human fibroblasts in response to serum or growth factors, *J. Cell. Physiol.* **117:**23–29.

Pouysségur, J., Chambard, J. C., Franchi, A., Paris, S., and van Obberghen-Schilling, E., 1982, Growth factor activation of an amiloride-sensitive Na^+/H^+ exchange system in quiescent fibroblasts: Coupling to ribosomal protein S6 phosphorylation, *Proc. Natl. Acad. Sci. U.S.A.* **79:**3935–3939.

Regula, C. S., Pfeiffer, J. R., and Berlin, R. D., 1981, Microtubule assembly and disassembly at alkaline pH, *J. Cell Biol.* **89:**45–53.

Rink, T. J., Tsien, R. Y., and Pozzan, T., 1982, Cytoplasmic pH and free Mg^{2+} in lymphocytes, *J. Cell Biol.* **95:**189–196.

Roos, A., and Boron, W., 1981, Intracellular pH, *Physiol. Rev.* **61:**296–411.

Rothenberg, P., Reuss, L., and Glaser, L., 1982, Serum and EGF transiently depolarize quiescent BSC-1 epithelial cells, *Proc. Natl. Acad. Sci. U.S.A.* **79:**7783–7787.

Rozengurt, E., and Heppel, L. A., 1975, Serum rapidly stimulates ouabain-sensitive $^{86}Rb^+$ influx in quiescent 3T3 cells, *Proc. Natl. Acad. Sci. U.S.A.* **72:**4492–4495.

Sawyer, S. T., and Cohen, S., 1981, Enhancement of calcium uptake and phosphatidyli-nositol turnover by EGF in A-431 cells, *Biochemistry* **20:**6280–6282.

Smith, J. B., and Rozengurt, E., 1978, Serum stimulates the Na^+,K^+ pump in quiescent fibroblasts by increasing Na^+ entry, *Proc. Natl. Acad. Sci. U.S.A.* **75:**5560–5564.

Spray, D. C., Harris, A. L., and Bennet, M. V. L., 1981, Gap junctional conductance is a simple and sensitive function of intracellular pH, *Science* **211:**712–715.

Streb, H., Irvine, R. F., Berridge, M. J., and Schulz, I., 1983, Release of Ca^{2+} from a nonmitochondrial intracellular store in pancreatic acinar cells by inositol-1,4,5-tris-phosphate, *Nature* **306:**67–69.

Trivedi, B., and Danforth, W. H., 1966, Effect of pH on the kinetics of frog muscle phos-phofructokinase, *J. Biol. Chem.* **241:**4110–4114.

Tsien, R. Y., Pozzan, T., and Rink, T. J., 1982, T-cell mitogens cause early changes in cytoplasmic free Ca^{2+} and membrane potential in lymphocytes, *Nature* **295:**68–71.

Tupper, J. T., Del Rosso, M., Hazelton, B., and Zorgniotti, F., 1978, Serum-stimulated changes in calcium transport and distribution in mouse 3T3 cells and their modification by db-cAMP, *J. Cell. Physiol.* **95:**71–84.

Ushiro, H., and Cohen, S., 1980, Identification of phosphotyrosine as a product of EGF-activated protein kinase in A-431 cell membranes, *J. Biol. Chem.* **255:**8363–8365.

10

Guanine-Nucleotide-Binding Regulatory Proteins

Membrane-Bound Information Transducers

ALFRED G. GILMAN, MURRAY D. SMIGEL,
GARY M. BOKOCH, and JANET D. ROBISHAW

1. Introduction

Studies of the hormone-sensitive adenylate cyclase complex and of mechanisms of transduction of visual information have led to the discovery of a family of guanine-nucleotide-binding regulatory proteins. This family is now known to include at least four highly homologous members. We can, at this time, formulate general definitions and hypotheses about the family and its functions.

2. The G-Protein Family

Receptors that regulate adenylate cyclase activity do so by interacting with a pair of membrane-bound guanine-nucleotide-binding regulatory proteins (Ross and Gilman, 1980; Smigel *et al.*, 1984). One of these, G_s, is responsible for stimulation of the activity of the catalyst of adenylate cyclase. The other, G_i, is responsible for inhibition. A stylized view of the components of the adenylate cyclase system is shown in Fig. 1. Although the evidence for the precise orientations shown in Fig. 1 is far

ALFRED G. GILMAN, MURRAY D. SMIGEL, GARY M. BOKOCH, and JANET D. ROBIS-
HAW • Department of Pharmacology, Southwestern Graduate School, University of
Texas Health Science Center at Dallas, Dallas, Texas 75235.

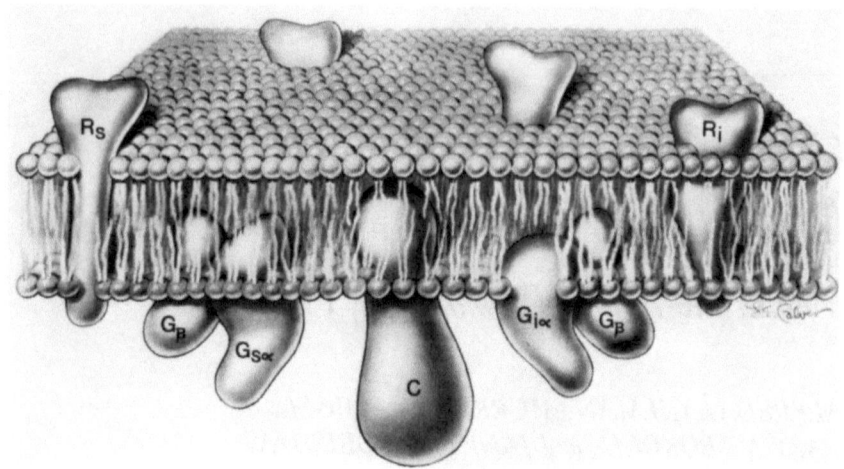

Figure 1. Model of the components of the hormone-sensitive adenylate cyclase complex. (From Gilman, 1984, with permission.) See text for explanation.

from complete, it seems very likely that the receptors are transmembrane proteins with both extracellular and cytoplasmic domains. Less well established is the view that the G proteins and the catalyst lack extracellular domains. The relative stoichiometry of the components is also not shown in Fig. 1. Roughly, this appears to be 1 receptor : 10 G_s : 50 G_i. The catalyst has not yet been purified, and there is no way to label it specifically for purposes of identification and quantitation. However, it is doubtful that it is abundant. It is also possible that the entity shown as the catalyst represents a complex of polypeptides.

The retinal rod outer segment contains an analogous system (Stryer *et al.*, 1981). The receptor for photons, rhodopsin, is known to stimulate the activity of a cGMP-specific phosphodiesterase. The intermediary in this pathway is transducin, a guanine-nucleotide-binding regulatory protein related to G_s and G_i.

The three G proteins mentioned above share a number of properties. All are membrane bound. All have an oligomeric structure and consist of one α, one β, and one γ subunit. Ligand-induced dissociation of the subunits of these oligomeric proteins plays an important role in their regulatory functions.

The α subunits of the three G proteins are clearly different, although they appear to be very homologous. Molecular weights range from 39,000 to 52,000. Each α subunit has a single site for high-affinity binding of

Table I. Components of the Hormone-Sensitive Adenylate Cyclase System and the Analogy with Transducin

Component	M_r	Ligand binding sites	Comment
Stimulatory receptors (R_s)	Various	β-Adrenergic agents, adrenocorticotropin, gonadotropins, many others	—
Inhibitory receptors (R_i)	Various	α-Adrenergic agents, muscarinic agents, opioids, others	—
Stimulatory G protein (G_s)	α: 45k[a] β: 35k γ: 8k[b]	GTP; AlF_4^-	α: Activator of C; site of ADP-ribosylation by cholera toxin β: Deactivates $G_{s\alpha}$
Inhibitory G protein (G_i)	α: 41k β: 35k γ: 8k[b]	GTP; AlF_4^- (?)	α: Site of ADP-ribosylation by IAP β: Deactivates $G_{s\alpha}$
Brain G protein (G_o)	α: 39k β: 35k γ: 8k[b]	GTP; AlF_4^- (?)	α: Site of ADP-ribosylation by IAP; functions unknown
Catalytic unit (C)	?	Forskolin[c] Adenosine[c]	Converts ATP $\rightarrow$ cAMP
Transducin	α: 39k β: 35k γ: 8k	GTP; AlF_4^- (?)	α: Activates a cGMP phosphodiesterase; independently ADP-ribosylated by cholera toxin and IAP β: Deactivates α

[a] G_s from certain sources contains a 52,000-da α subunit in addition to the 45,000-da polypeptide. It is likely that they are products of the same gene.
[b] An 8000-da peptide is usually found in preparations of G_i, G_s, and G_o. It is probable that it is a third (γ) subunit. A similar subunit exists in transducin. Its function is unknown.
[c] Since the catalyst has not been purified, it is not certain that these ligands bind to this polypeptide.

guanine nucleotides. Each is capable of hydrolyzing GTP to GDP—a reaction that is responsible for "deactivation" of the G protein. Each serves as a substrate for ADP-ribosylation catalyzed by a specific bacterial toxin. G_s is ADP-ribosylated by cholera toxin; G_i is ADP-ribosylated by islet-activating protein (one of the toxins of *Bordetella pertussis*); transducin has two independent sites for ADP-ribosylation—one (an arginine residue) for cholera toxin, the other for IAP.

The α subunit of G_s also interacts with F^-, a ubiquitous activator of adenylate cyclases from higher organisms. The actual ligand that interacts with $G_{s\alpha}$ appears to be a complex of F^- and Al^{3+}, most likely AlF_4^- (Sternweis and Gilman, 1982). The significance of this odd phenomenon is unclear. The existence of a similar site on G_i and transducin is suspected but has not been explored carefully.

The β subunits of the G proteins appear to have a very high degree of homology based on amino acid compositions, analysis of peptides produced by proteolysis, and a modest amount of sequence information (Manning and Gilman, 1983). Molecular weights are approximately 35,000. At least one important function of the β subunit, the ability to bind to $G_{s\alpha}$, is a property that is shared by the β subunits from G_s, G_i, and transducin.

Each G protein also appears to contain a third, γ, subunit. This small polypeptide (M_r 8000) binds tightly to β; its function is unknown (Bokoch et al., 1984; Hildebrandt et al., 1984).

A fourth member of the family has apparently been discovered recently (Sternweis and Robishaw, 1984). This protein, termed G_o for the moment, is present in relatively high concentrations in a crude particulate fraction from brain. It contains a 39,000-da α subunit that binds guanine nucleotides and that can be ADP-ribosylated by IAP; the β and γ subunits are apparently similar to those described above. Although the function of this protein is not yet known, it can interact with muscarinic cholinergic receptors from brain in reconstituted lipid vesicles and alter the affinity of the receptors for agonists in a guanine-nucleotide-dependent manner (Florio and Sternweis, 1984). The abundance of this protein in brain should facilitate characterization of the family.

The properties of the G proteins are summarized in Table I.

3. Regulation of Adenylate Cyclase Activity by G_s and G_i

The resolved catalyst of adenylate cyclase displays little or no activity in the presence of its physiological substrate, MgATP. Modest activity is observed when G_s and GTP are added. Near-maximal activities are observed with a mixture of G_s and the catalyst in the presence of nonhydrolyzable analogues of guanine nucleotides [e.g., GTPγS or Gpp(NH)p] or F^- plus Al^{3+}.

Incubation of G_s with GTPγS or F^- causes "activation" of the regulatory protein. This activated state reverses only slowly, and it is readily detectable by observation of the stimulatory effect of activated G_s on the catalyst in the absence of a significant concentration of free GTPγS or F^-. Activation of G_s by GTPγS proceeds in parallel with high-affinity

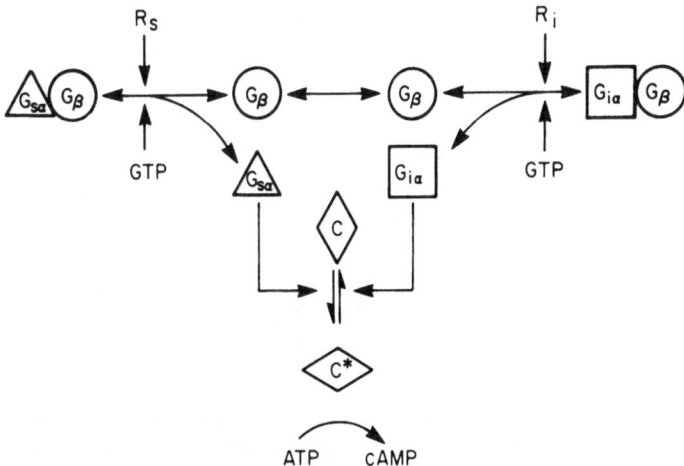

Figure 2. Mechanism of regulation of adenylate cyclase activity by G_s and G_i. (From Gilman, 1984, with permission.) See text for explanation.

binding of the nucleotide to the protein. However, the reaction does not proceed in a simple bimolecular fashion; furthermore, it is slow and is accompanied by a major change in the hydrodynamic properties of the protein. A great deal of evidence indicates that the activation reaction is the result of ligand-promoted dissociation of the subunits of G_s:

$$G_{s\alpha} \cdot \beta \cdot \gamma + GTP\gamma S \rightleftharpoons GTP\gamma S \cdot G_{s\alpha} \cdot \beta \cdot \gamma \xrightarrow{Mg^{2+}} GTP\gamma S \cdot G_{s\alpha} + \beta \cdot \gamma$$

This reaction can be accelerated greatly by Mg^{2+} or, more interestingly, by β-adrenergic agonists in a reconstituted system containing purified β-adrenergic receptors, pure G_s, and appropriate lipids (Pedersen and Ross, 1982; Brandt *et al.*, 1984).

When the dissociated subunits of G_s were resolved from each other and tested for their ability to activate the catalyst, it was found that $GTP\gamma S \cdot G_{s\alpha}$ was the active species (Northup *et al.*, 1983a). The $\beta \cdot \gamma$ subunit complex inhibits activation of G_s and stimulates deactivation of $G_{s\alpha}$ by promoting formation of the G_s oligomer (Fig. 2) (Northup *et al.*, 1983b).

The homologous structures of G_i and G_s suggested that there might be similarities in their mechanisms of action. Indeed, it was found that guanine nucleotides also promoted dissociation of the subunits of G_i (Bokoch *et al.*, 1983) and that such dissociation greatly increased the inhibitory activity of this G protein when reconstituted into platelet or S49 cell plasma membranes (Katada *et al.*, 1984a). Further analogy with G_s suggested that the appropriately liganded form of $G_{i\alpha}$ would be the inhibitor of the catalyst of adenylate cyclase. In fact, $GTP\gamma S \cdot G_{i\alpha}$ has been shown to inhibit adenylate cyclase activity in platelet and wild-type S49 lymphoma cell membranes. However, it is a relatively impotent inhibitor of adenylate cyclase (Katada et al., 1984a,b,c). Of considerable interest is the fact that the inhibitory effect of the resolved β subunit of G_i accounted for most of the activity of the dissociated G_i oligomer. This was shown to result from the interaction of the β subunit of G_i with $G_{s\alpha}$; these data indicated that the β subunits of G_s and G_i were functionally interchangeable (Fig. 2) (Katada et al., 1984a,b,c). Thus, the predominant inhibitory effect of G_i appears to result from the β subunit:

$$\beta + G_{s\alpha} \rightleftharpoons G_{s\alpha} \cdot \beta$$

There appears to be considerably more G_i than G_s in rabbit liver membranes, lending credence to the feasibility of this mechanism.

Additional evidence indicates that the irreversible inhibition of membrane-bound adenylate cyclase activity that results from exposure to guanine nucleotide analogues is caused by the activity of the β subunit. When platelet or wild-type S49 cell membranes are incubated with low concentrations of $GTP\gamma S$ (for short times at low concentrations of Mg^{2+}), a state of persistent inhibition is induced. If the resolved β subunit is reconstituted with these membranes, there is no further inhibition of enzymatic activity. If unliganded $G_{i\alpha}$ is added, the inhibition is overcome completely. The most likely explanation for this is that $G_{i\alpha}$ associates with β and, thus, that the inhibited state of the membranes was caused by the presence of free β (Katada et al., 1984c).

4. Speculations

We know that those receptors that stimulate adenylate cyclase activity do so via G_s and that $G_{s\alpha}$ appears to be the activator of the catalyst. Is this the only function of G_s? Insight comes from the data of Maguire and co-workers (Maguire, 1984). Maguire and Erdos (1980) have described an effect of β-adrenergic agonists and prostaglandin E_1 to inhibit the influx of Mg^{2+} into S49 cells. Although the physiological significance of this phenomenon is unclear, it does serve as a useful marker of an additional effect that is mediated by the receptors that stimulate adenylate cyclase activity in S49 cells. The S49 cell system offers a great advantage, since various mutants in the cAMP system are available. Thus, it was found that mutants that are devoid of cAMP-dependent protein kinase activity function normally (i.e., isoproterenol or PGE_1 inhibits Mg^{2+} influx) and that cAMP or its analogues have no effect on Mg^{2+} influx.

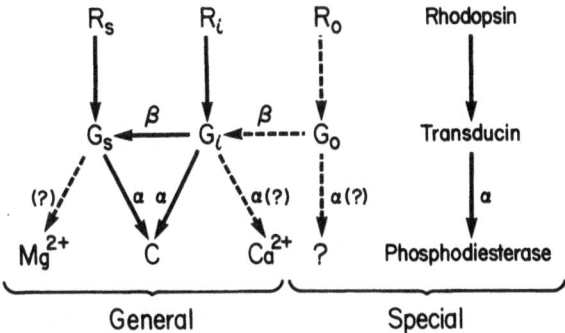

Figure 3. Hypothetical role of G proteins as integrators and coordinators of various functions of the membrane. Regulation of adenylate cyclase activity by G_s and G_i occurs in most cells. Hypothetically (dashed lines), G_s and G_i can also modulate the functions of specific ion channels in the membrane. Transducin is the specialized G protein in the retinal rod outer segment. G_o may represent a specialized G protein modulating as yet unknown functions but also capable of inhibiting adenylate cyclase activity by means of release of its β subunit.

However, mutants that have an altered $G_{s\alpha}$ fail to show the response. These include the *cyc⁻* mutant, in which the activity of $G_{s\alpha}$ is undetectable, and the *UNC* mutant, which has an altered $G_{s\alpha}$ that cannot interact with appropriate receptors. The logical hypothesis is that the response requires receptors and G_s but that the catalyst and cAMP are not involved. This possibility is indicated in Fig. 3.

Interesting hypotheses can also be formulated about G_i. In particular, the α subunit of G_i appears to act only as a rather impotent inhibitor of the catalyst of adenylate cyclase, and its primary role insofar as this enzyme is concerned may be to sequester β (i.e., it acts as the antiinhibitor). Perhaps $G_{i\alpha}$ has more important functions in the membrane. Hints come from experiments utilizing IAP, since this toxin ADP-ribosylates G_i with resultant blockade of its function. Data from both our laboratory (Bokoch and Gilman, 1984) and that of Michio Ui (Okajima and Ui, 1984) indicate that IAP interferes with the effects of the chemotactic peptide formyl-methionyl-leucyl-phenylalanine on polymorphonuclear leukocytes. Responses to the peptide normally include release of arachidonic acid, release of hydrolytic enzymes, and generation of superoxide. These effects are dependent on extracellular Ca^{2+}, and the pathway can also be activated by exposure of cells to the Ca^{2+} ionophore A23187. After exposure of cells to IAP, responses to the peptide are blocked almost completely, but those to the ionophore remain intact. Data from others indicate that the binding of the peptide to its receptor is modulated by guanine nucleotides, strongly suggesting an interaction between the agonist–receptor complex and a G protein (Koo *et al.*, 1983). Labeling of membranes from polymorphonuclear leukocytes with [^{32}P]-NAD and IAP reveals only a 41,000-da polypeptide that comigrates with $G_{i\alpha}$. Although more direct experiments must be done, the data obtained to date are consistent with the hypothesis that G_i mediates these effects of formyl-methionyl-leucyl-phenylalanine, that they are dependent on Ca^{2+} influx, and that they are independent of alterations of cyclic nucleotide metabolism. Perhaps $G_{i\alpha}$ can interact with and control the function of polypeptides that are involved in the regulation of Ca^{2+} flux (Fig. 3). This hypothesis is strengthened by the work of Gomperts (1983), who introduced guanine nucleotide analogues into permeabilized mast cells. This treatment caused the cells to undergo exocytotic secretion in response to the addition of extracellular Ca^{2+}.

The subunit dissociation mechanism for the G proteins, coupled with the fact that their β subunits appear to be functionally indistinguishable (at least insofar as interactions with $G_{s\alpha}$ are concerned) suggest an interesting view of the large number of agents that inhibit adenylate cyclase activity. It is difficult to believe that this is the primary mechanism of action of all of these neurotransmitters and hormones. Although inhibition of adenylate cyclase activity may be an important facet of the mechanism of some of these regulatory ligands, it is attractive to speculate that they exert at least two effects simultaneously. The α subunit of G_i, G_o, or other as yet undiscovered G proteins could be the primary effector of other functions of the membrane, particularly regulation of ion permeability, whereas the β subunit could inhibit the cAMP system in response to demands from other counterregulatory pathways. This hypothesis puts the G proteins in the position of acting as both integrators and coordinators. Thus, a given G protein may serve not only as a common focal point for a number of receptors that all influence the same reaction but also as a branch point in the membrane for coordinated regulation of other functions.

The common characteristics of the G proteins suggest obvious strategies for detection of additional members of the family. These searches will be aided greatly by antibodies, which are now becoming available. Data obtained to date indicate that the products of the *ras* oncogenes may at least be cousins. The *ras* gene products are membrane-bound proteins with a high-affinity binding site for guanine nucleotides. Although their molecular weights are about half of those of the α subunits described above, the fact that transducin has two independent sites for ADP-ribosylation hints at some type of duplication of a smaller precursor. Sequence information is limited. However, a portion of the sequence of the ADP-ribosylated octapeptide found in transducin is represented in *ras* (Manning *et al.*, 1984).

Transducin: Lys/Arg-Glu-Asn-Leu-Lys-Asn (ADP-ribose)-
v-K-*ras*:   Arg-Glu-Gln-Leu-Lys-Arg

Preliminary data obtained in collaboration with J. Hurley and M. Simon indicate extensive homology among $G_{o\alpha}$, the α subunit of transducin, and *ras* at the amino termini. This information will be useful and exciting if the known properties and functions of the G proteins suggest strategies for deducing the function of *ras*.

ACKNOWLEDGMENTS. Work from the authors' laboratory was supported by United States Public Health Service grant NS18153 and by grant BC240F from the American Cancer Society.

References

Bokoch, G. M., and Gilman, A. G., 1984, Inhibition of β receptor-mediated release of arachidonic acid by pertussis toxin. *Cell* **39**:301–308.

Bokoch, G. M., Katada, T., Northup, J. K., Hewlett, E. L., and Gilman, A. G., 1983, Identification of the predominant substrate for ADP-ribosylation by islet activating protein, *J. Biol. Chem.* **258**:2072–2075.

Bokoch, G. M., Katada, T., Northup, J. K., Ui, M., and Gilman, A. G., 1984, Purification and properties of the inhibitory guanine nucleotide-binding regulatory component of adenylate cyclase, *J. Biol. Chem.* **259**:3560–3567.

Brandt, D. R., Asano, T., Pedersen, S. E., and Ross, E. M., 1983, Reconstitution of catecholamine-stimulated GTPase, *Biochemistry* **22**:4357–4362.

Florio, V., and Sternweis, P. C., 1984, Reconstitution of resolved muscarinic cholinergic receptors with purified GTP-binding proteins, *J. Biol. Chem.* **260**:3477–3483.

Gilman, A. G., 1984, Minireview. G proteins and dual control of adenylate cyclase, *Cell* **36**:577–579.

Gomperts, B. D., 1983, Involvement of guanine nucleotide-binding protein in the gating of Ca^{2+} by receptors, *Nature* **306**:64–66.

Hildebrandt, J. D., Codina, J., Risinger, R., and Birnbaumer, L, 1984, Identification of a γ subunit associated with the adenylyl cyclase regulatory proteins N_s and N_i, *J. Biol. Chem.* **259**:2039–2042.

Katada, T., Bokoch, G. M., Northup, J. K., Ui, M., and Gilman, A. G., 1984a, The inhibitory guanine nucleotide-binding regulatory component of adenylate cyclase. Properties and function of the purified protein, *J. Biol. Chem.* **259**:3568–3577.

Katada, T., Bokoch, G. M., Smigel, M. D., Ui, M., and Gilman, A. G., 1984b, The inhibitory guanine nucleotide-binding regulatory component of adenylate cyclase. Subunit dissociation and the inhibition of adenylate cyclase in S49 lymphoma *cyc* and wild type membranes, *J. Biol. Chem.* **259**:3586–3595.

Katada, T., Northup, J. K., Bokoch, G. M., Ui, M., and Gilman, A. G., 1984c, The inhibitory guanine nucleotide-binding regulatory component of adenylate cyclase. Subunit dissociation and guanine nucleotide-dependent hormonal inhibition, *J. Biol. Chem.* **259**:3578–3585.

Koo, C., Lefkowitz, R. J., and Synderman, R., 1983, Guanine nucleotides modulate the binding affinity of the oligopeptide chemoattractant receptor on human polymorphonuclear leukocytes, *J. Clin. Invest.* **72**:748–753.

Maguire, M. E., 1984, Hormone-sensitive magnesium transport and magnesium regulation of adenylate cyclase, *Trends Pharmacol. Sci.* **5**:73–78.

Maguire, M. E., and Erdos, J. J., 1980, Inhibition of magnesium uptake by β-adrenergic agonists and prostanglandin E_1 is not mediated by cyclic AMP, *J. Biol. Chem.* **255**:1030–1035.

Manning, D. R., and Gilman, A. G., 1983, The regulatory components of adenylate cyclase and transducin: A family of structurally homologous guanine nucleotide binding proteins, *J. Biol. Chem.* **258**:7059–7063.

Manning, D. R., Fraser, B. A., Kahn, R. A., and Gilman, A. G., 1984, ADP-ribosylation of transducin by islet activating protein. Identification of asparagine as the site of ADP-ribosylation, *J. Biol. Chem.* **259**:3560–3567.

Northup, J. K., Smigel, M. D., Sternweis, P. C., and Gilman, A. G., 1983a, The subunits of the stimulatory regulatory component of adenylate cyclase. Resolution of the activated 45,000-dalton (α) subunit, *J. Biol. Chem.* **258**:11369–11376.

Northup, J. K., Sternweis, P. C., and Gilman, A. G., 1983b, The subunits of the stimulatory regulatory component of adenylate cyclase. Resolution, activity, and properties of the 35,000-dalton (β) subunit, *J. Biol. Chem.* **258**:11361–11368.

Okajima, F., and Ui, M., 1984, ADP-ribosylation of the specific membrane protein by islet-activating protein, pertussis toxin, associated with inhibition of a chemotactic peptide-induced arachidonate release in neutrophils. A possible role of the toxin substrate in Ca^{2+}-mobilizing biosignaling, *J. Biol. Chem.* **259**:3863–3871.

Pedersen, S., and Ross, E. M., 1982, Functional reconstitution of β-adrenergic receptors and the stimulatory GTP-binding protein of adenylate cyclase. *Proc. Natl. Acad. Sci. U.S.A.* **79**:7228–7232.

Ross, E. M., and Gilman, A. G., 1980, Biochemical properties of the regulatory component of adenylate cyclase, *Annu. Rev. Biochem.* **49**:533–564.

Smigel, M. D., Ross, E. M., and Gilman, A. G., 1984, Role of the β-adrenergic receptor in the regulation of adenylate cyclase, in: *Cell Membranes: Methods and Reviews* (E. L. Elson, W. A. Frazier, and L. Glaser, eds.) Plenum, New York, pp. 247–294.

Sternweis, P. C., and Gilman, A. G., 1982, Aluminum: A requirement for activation of the regulatory component of adenylate cyclase by fluoride, *Proc. Natl. Acad. Sci. U.S.A.* **79**:4888–4891.

Sternweis, P. C., and Robishaw, J. D., 1984, Isolation of two major proteins with high affinity for guanine nucleotides from membranes of bovine brain. *J. Biol. Chem.* **259**:13806–13813.

Stryer, L., Hurley, J. B., and Fung, B. K.-K., 1981, First stage of amplification in the cyclic-nucleotide cascade of vision, *Curr. Top. Membr. Transp.* **15**:93–108.

Role of Cyclic-AMP-Dependent Protein Kinase in the Regulation of Cellular Processes

EDWIN G. KREBS, DONALD K. BLUMENTHAL,
ARTHUR M. EDELMAN, and C. NICHOLAS HALES

1. Criteria for Evaluating the Role of Protein Phosphorylation in Cyclic-AMP-Mediated Processes

Approximately 15 years ago it was shown that the stimulatory effect of cAMP on the phosphorylase kinase activation reaction *in vitro* was mediated by a distinct cAMP-dependent protein kinase, which exhibited broader specificity than that of a phosphorylase kinase kinase in that it could also catalyze the phosphorylation of casein and protamine (Walsh *et al.*, 1968). Huijing and Larner (1966) had also concluded that the target for cAMP-mediated regulation of glycogen synthase was probably the synthase kinase itself, and a few years later it was determined that an identical cAMP-dependent protein kinase is involved in the phosphorylase kinase and glycogen synthase phosphorylation reactions (Schlender *et al.*, 1969; Soderling *et al.*, 1970). Noting a widespread distribution of the newly discovered cAMP-dependent protein kinase in various cell types and tissues, Kuo and Greengard (1969) postulated that all of the actions of cAMP are mediated by this enzyme. This hypothesis is still considered valid for most, if not all, eucaryotic cell functions controlled by cAMP, but it does not hold for procaryotes, in which the cAMP receptor protein, CRP

EDWIN G. KREBS, DONALD K. BLUMENTHAL, ARTHUR M. EDELMAN, and C. NICHOLAS HALES • Howard Hughes Medical Institute Laboratory, Department of Pharmacology, University of Washington, Seattle, Washington 98195.   *Present address of C.N.H.:* Department of Clinical Biochemistry, University of Cambridge, Cambridge, England.

(CAP), binds to DNA and regulates the catabolite-repressible operons (Pastan and Adhya, 1976).

Because cAMP is involved in so many different cellular functions, knowledge that its actions are mediated by protein phosphorylation reactions served as a strong stimulus for research aimed at the elucidation of specific systems in which this process might be involved. During the past decade an ever increasing number of papers have appeared describing a great variety of proteins believed to serve as targets for the cAMP-dependent protein kinase. In addition, any number of cellular processes known to be regulated by cAMP, and hence assumed to involve the cAMP-dependent protein kinase, have been reported.

Early in the period of explosive growth of work on protein phosphorylation, one of us (Krebs, 1973) formulated a set of criteria that should be satisfied before it could be concluded that a given function of cAMP is mediated by the cAMP-dependent protein kinase. It was hoped that such criteria might prove useful to investigators in the same sense as those cited for justifying claims that a given hormone produces its effect as a result of the stimulation of adenylate cyclase (Robison *et al.*, 1971). The criteria advanced for relating protein phosphorylation to cAMP functions were as follows:

1. The cell type involved should be shown to contain a cAMP-dependent protein kinase.
2. A phosphorylatable protein substrate bearing a functional relationship to the cAMP-mediated process should be identified.
3. It should be shown that phosphorylation of this substrate alters its function.
4. It should be demonstrated that the protein substrate is modified *in vivo* in response to cAMP.
5. A phosphoprotein phosphatase that can reverse the phosphorylation process should be shown to exist.

These criteria were later reevaluated by Nimmo and Cohen (1977), who pointed out that the first criterion had become irrelevant inasmuch as cAMP-dependent protein kinases clearly had been shown to be distributed in all types of eucaryotic cells. In addition, the fifth criterion was also considered superfluous, although these authors included reversibility and phosphatase action as parts of other criteria. The set of criteria to be met before an effect mediated by cAMP can be said to occur through phosphorylation of a protein as described by Nimmo and Cohen are given in Table I.

The criteria of Table I were further extended and modified to make them applicable to protein phosphorylation–dephosphorylations in gen-

Table I. Criteria to Be Met before an Effect Mediated by Cyclic AMP Is Attributed to the Phosphorylation of a Given Protein

---

1. A protein substrate for cyclic AMP-dependent protein kinase should exist which bears a functional relationship to the process mediated by cyclic AMP. The rate of phosphorylation of that protein, in its native state, should be adequate to account for the speed at which the process occurs *in vivo* in response to cyclic AMP.
2. The function of the protein should be shown to undergo a reversible alteration *in vitro* by phosphorylation and dephosphorylation, catalyzed by cyclic AMP-dependent protein kinase and a protein phosphatase.
3. A reversible change in the function of the protein should occur *in vivo* in response to cyclic AMP.
4. Phosphorylation of the protein should occur *in vivo* in response to a hormone at the same site(s) phosphorylated by cyclic AMP-dependent protein kinase *in vitro*.

---

eral (Krebs and Beavo, 1979). Manning *et al.* (1980) have reviewed the particular problems faced by investigators attempting to meet criteria 3 and 4.

In reference to application of the criteria listed in Table I, it should be noted that investigators do not necessarily start out with knowledge that a particular physiological event involves cAMP and then embark on studies carried out more or less in the order prescribed by the four criteria. They may have other entry points. For example, a study may be prompted by the observation that a given protein can be phosphorylated by the cAMP-dependent protein kinase. If this occurs in an intact cell system, the investigator may assume that it has a physiological meaning and set out to determine what that might be. Under circumstances in which the protein phosphorylation reaction is observed *in vitro*, if the V_{max} is high and the K_m for the protein substrate is low, the investigator may also conclude that further work is warranted. Under these circumstances the tables are turned, and the investigator is seeking a physiological function to fit a phosphorylation event. In still other instances, evidence obtained by investigators studying cAMP-mediated protein phosphorylation can be of help in determining that cAMP itself is involved in a given physiological process, e.g., see Section 3.3 below.

Are the criteria proposed in Table I still valid? Do they need to be modified or extended to take into account new findings bearing on cAMP-mediated processes? Have new techniques emerged that make it desirable to formulate additional and/or more stringent criteria? How well are the criteria being met with respect to studies on specific systems? These and related questions constitute the major thrust of the present chapter. Although we focus our attention on protein phosphorylation reactions regulated by cAMP, any principles that can be established in reference to this particular second messenger should in most instances be applicable to other messenger-controlled protein phosphorylations.

2. Additional Approaches

2.1. Measurement of the Activity of Cyclic-AMP-Dependent Protein Kinases in Vivo

A regulatory effect of cAMP mediated by protein phosphorylation implies activation of a cAMP-dependent protein kinase as an intermediate reaction in the process. It is possible, at least in theory, to determine whether such an activation has taken place in an intact tissue and, furthermore, to determine whether it has taken place within a time scale that is consistent with an intermediate role in the overall process. Since activation of cAMP-dependent protein kinases is believed to involve dissociation of the catalytic and regulatory subunits of the holoenzyme, direct measurement of the degree of dissociation chromatographically provides one method of studying activation. A second and more convenient method is to measure what is termed the "activity ratio" of cAMP-dependent protein kinases. The activity ratio is defined as the activity of the cAMP-dependent protein kinase measured in the absence of cAMP divided by that measured in the presence of cAMP (added in an amount sufficient to achieve maximal activation). If a process is mediated by cAMP-dependent protein phosphorylation, then the activity ratio should increase. Therefore, strictly speaking, one of the criteria to be satisfied in ascribing a regulatory effect to cAMP-dependent protein phosphorylation is the demonstration of activation of one or more cAMP-dependent protein kinases. The difficulties that have been encountered in providing an entirely rigorous demonstration of activation are considerable.

Methods for the estimation of the degree of activation of cAMP-dependent protein kinases were first provided by Corbin et al. (1973). These workers discussed several potential sources of error in obtaining valid measurements, including artifactual dissociation or association of the subunits of the enzymes during homogenization or assay. Both the physical separation method and the measurement of activity ratios were used to study adipose tissue cAMP-dependent protein kinase. The importance of using 0.5 M NaCl to prevent reassociation of the enzyme was stressd. It was also pointed out that in some tissues high salt concentrations could cause dissociation of the holoenzyme.

Possible activation of cAMP-dependent protein kinase by the production or release of cAMP not accessible to the kinase in vivo has been investigated by the addition of charcoal during homogenization (Keely *et al.*, 1975). In experiments with rat heart, it was shown that the activating effect of exogenously added cAMP could be prevented by this procedure but that the presence of charcoal did not change the basal activity ratio. As a further check on the validity of measurements of cAMP-dependent protein kinase activity ratios, Palmer *et al.* (1980) recommended the addition of exogenous cAMP-dependent kinase as an internal standard to monitor whether the extraction and assay conditions prevented artifactual association or dissociation of the holoenzyme. The methods and pitfalls of measuring cAMP-dependent protein kinase activity ratios in intact tissues have recently been reviewd by Corbin (1983).

Depending on the species of animal and tissue investigated, protein kinase measurements may include the activity of cAMP-independent kinases. The extent to which these enzymes interfere with the measurements will depend on the protein substrate used and is best measured as activity remaining in the presence of the specific heat-stable polypeptide inhibitor of the catalytic subunit of cAMP-dependent protein kinase (Corbin *et al.*, 1973). Ideally, all measurements of the activity ratio should be calculated after the contribution of the cAMP-independent enzymes has been subtracted, although this correction has often been ignored (Nimmo and Cohen, 1977). It is also possible that up to 20% of the catalytic activity of the cAMP-dependent kinase may be inhibited by a heat-stable inhibitor *in vivo* (Beavo *et al.*, 1974), and the presence and stability of such a complex will be of importance to the measurement of the activity ratio. No systematic attempt has been made to investigate this aspect of the problem and its contribution to measurements of the activity ratio.

The affinity of the regulatory subunit for the catalytic subunit of cAMP-dependent protein kinase is dependent on the amount of cAMP bound. The two cAMP binding sites on each regulatory subunit are both apparently involved in activation, and the binding characteristics of the sites show cooperativity (Robinson-Steiner and Corbin, 1983; Ogreid *et al.*, 1983). However, it has also been known for several years that certain substrates, in particular, histone (Miyamoto *et al.*, 1971) and protamine (Tao, 1972), may activate the cAMP-dependent protein kinase in the absence of the cAMP, presumably by displacing the regulatory subunit from the holoenzyme. The apparent K_a for cAMP when casein or glycogen synthase is used as a substrate is 0.2 μM, whereas with histone it is 0.06μM or lower depending on the assay conditions (Beavo *et al.*, 1974). The state of protein substrates with regard to the binding of their allosteric effectors may also change the rate of phosphorylation by the catalytic subunit (El-Maghrabi *et al.*, 1983). A measured activity ratio therefore may only strictly relate to the substrate used in the assay. Furthermore, the *in vivo* pattern of cAMP-dependent protein phosphorylation could be determined not only by the actual concentration of cAMP produced but also by the ability of substrate proteins to displace the regulatory subunit at any particular concentration of cAMP.

The evaluation of *in vivo* phosphorylation is still further complicated by the subcellular distribution of cAMP-dependent protein kinases. It has been known for many years that the protein kinase exists in particulate and soluble forms (Corbin *et al.*, 1977). At least part of the particulate form of the enzyme is associated with the plasma membrane in heart (Jones *et al.*, 1980). It is not at present certain whether still more complex patterns of distribution within the cell occur. For example, binding of cAMP-dependent protein kinase to a microtubule-associated protein has been described (Miller *et al.*, 1982).

Cyclic AMP is produced by the plasma membrane enzyme adenylate cyclase. At low levels of stimulation, preferential activation of plasma-membrane-bound kinase may occur, and this may not be detected in an assay of whole homogenate activity ratios. In this manner, it has been suggested, compartmentalization of hormonal control could be achieved (Corbin *et al.*, 1977). Experimental evidence was recently presented that separate pools of protein kinase may be activated by different agents acting in the same tissue to elevate cAMP content by similar amounts (Hayes *et al.*, 1980; Buxton and Brunton, 1983). Rat hearts perfused with isoproterenol exhibited an increased cAMP content and activation of phosphorylase, whereas hearts perfused with prostaglandin E_1 showed comparable increases of cAMP but no increase in phosphorylase activity. Isoproterenol increased the soluble protein kinase activity ratio, increased the amount of cAMP in the particulate fraction, and decreased the percentage of particulate protein kinase. In contrast, prostaglandin E_1 did not increase particulate cAMP or decrease particulate protein kinase activity. The differences in the effects of these two agents are clearly not a consequence of the amount of cAMP generated but may reflect the presence of "multiple receptor–adenylate cyclase populations which are capable of generating cyclic AMP into specific intracellular spaces" (Hayes *et al.*, 1980).

Attempts to investigate the role of specific protein phosphorylation in the mediation of the effects of cAMP must therefore take account of the potential complexities considered above. It is clear that subcellular compartmentalization of cAMP and cAMP-dependent protein kinases exists. Even the measurement of whole-tissue activity ratios may be considered "semiquantitative" (Corbin, 1983). In view of all these uncertainties, it would perhaps be unrealistic at the present time to insist that the demonstration of a change in a cAMP-dependent protein kinase activity ratio is an absolutely essential criterion to be satisfied in defining the role of protein phosphorylation in the action of cAMP.

2.2. Introduction of Kinase Subunits into Cells

A recent addition to the methodological arsenal for examining the role of cAMP in cellular processes is the use of the direct introduction of intact protein molecules into cells. Three such approaches, which have been applied successfully and are illustrated below, are the direct microinjection of proteins into intact cells, the use of permeabilized ("skinned") cells, and the fusion of cells with protein-loaded vehicles such as red blood cell ghosts. This is by no means intended to provide a list of all conceivable approaches or to describe in detail and critique the technical aspects of these procedures but rather to illustrate cases in which the criteria in Table I may need to be modified when the involvement of cAMP in a physiological response can be shown by predominantly *in vivo* approaches. The phrase "introduction of kinase subunits . . ." is here broadly defined to include all those protein components relating to the action of cAMP, including, for example, catalytic (C) and regulatory (R) subunits of cAMP-dependent protein kinase, heat-stable protein kinase inhibitor (PKI), cyclic nucleotide phosphodiesterases, and potentially other molecules such as antibodies raised to any of the components of the system.

2.2.1. Microinjection into Intact Cells

The large, easily identifiable neurons of invertebrates provide preparations in which the intracellular injection of protein components can be combined with pharmacological manipulation and stimulation and intracellular recording of the neurons in a defined circuit. Kandel and co-workers (for a review see Kandel and Schwartz, 1982) have proposed a molecular model for presynaptic facilitation in the marine mollusc *Aplysia* that involves the cAMP-dependent phosphorylation of a specific K^+ channel or associated protein in presynaptic terminals of sensory neurons. This hypothesis was initially based on findings that dibutyryl cAMP or intracellularly injected cAMP mimicked the effect of stimulation of facilitatory neurons in producing increased transmitter release from sensory neurons (measured by recording from cells postsynaptic to the sensory neurons) (Brunelli *et al.*, 1976). Futhermore, endogenous cAMP content was increased by stimulation (Cedar *et al.*, 1972). To test the model more directly, Castellucci *et al.* (1980) pressure injected into the sensory neurons purified C-subunit and observed, as predicted by the model, decreases in the K^+ current, an increased Ca^{2+} influx, and increased neurotransmitter release. As a further test, PKI was injected into the sensory neurons and found to antagonize the effects produced by stimulation of the facilitatory neurons (Castellucci et al., 1982). Effects of intracellular injection of C-subunit on specific K^+ currents in other invertebrate systems such as bag cell neurons of Aplysia and photoreceptor cells of Hermissenda have also been described (Kaczmarek et al., 1980; Alkon et al., 1983). In the case of neuron R15 of Aplysia, serotonin-induced changes in K^+ conductance were reported to be selectively blocked by PKI (Adams and Levitan, 1982).

A different type of system in which microinjection into intact cells has been utilized is the study of progesterone regulation of amphibian oocyte maturation (for a review see Maller, 1983). The large size (1.4 mm diameter) and simple visual assay for maturation (pigmentation changes) make the oocyte well suited for study by intracellular injection. Microinjection of purified C and R subunits of cAMP-dependent protein kinase, cyclic nucleotide phosphodiesterase, and PKI has provided strong support for a model in which decreases in cAMP levels caused by progesterone initiate a sequence of events leading to oocyte maturation (Maller and Krebs, 1977; Huchon et al., 1981; Foerder et al., 1982).

2.2.2. Introduction of Protein Components into Permeabilized ("Skinned") Cells

Kerrick and collaborators have described procedures for permeabilizing muscle fibers and mouse lymphoma cells to allow the introduction of protein components by diffusion from the bathing medium (Kerrick et al., 1981; Kerrick and Bourguignon, 1984). Exposure of the cells to "skinning" solutions such as a high-EGTA buffer causes the plasma membrane to become "leaky" by an as yet unknown mechanism. This allows precise manipulation of the ionic and protein composition surrounding the remaining nondiffusable intracellular structures. With these techniques, the introduction of exogenous C-subunit was shown to inhibit the development of tension in skinned smooth muscle fibers (Bridenbaugh et al., 1981; Kerrick et al., 1981) and capping in the permeable lymphoma cells (Kerrick and Bourguignon, 1984), effects apparently mediated through phosphorylation of myosin light-chain kinase (MLCK). Although the physiological role of this phosphorylation is currently controversial (see Section 3.2), permeabilization of cells represents a technique by which the role of cAMP-dependent protein kinase may be evaluated in cells difficult to microinject, albeit with the important qualification that the permeabilization process may remove from the cell additional endogenous regulatory components.

2.2.3. Fusion of Cells with Protein-Loaded Vehicles

Physiological responses involving cAMP that require the monitoring of thousands of cells to be detected obviously cannot be studied by means of microinjection. A promising technique for this purpose has been developed by Schlegel and Rechsteiner (1978). This involves fusing cells with red blood cell ghosts ("vehicles") loaded with the protein to be introduced. In this procedure, nonnucleated red cells are loaded by hypotonic lysis in the presence of the protein to be introduced followed by dialysis against isotonic phosphate-buffered saline. The protein is then transferred by fusing the loaded ghosts with recipient cells using polyethylene glycol or sendai virus as fusogen. Using these techniques, Boney and co-workers (1983) demonstrated an induction of tyrosine aminotransferase in cultured rat hepatoma cells on the introduction of C-subunit. This effect lent support to the hypothesis that the induction in these cells in response to dibutyryl cAMP involves the cAMP-dependent protein kinase and not a distinct cAMP binding protein as in *E. coli*.

2.2.4. General Comments

Although powerful, these techniques are not without hazard. For example, the introduction of C-subunit without data concerning cAMP levels, responses to dibutyryl cAMP, or introduction of PKI may be misleading because of nonspecific effects caused by the ability of the enzyme to phosphorylate sites that *in vivo* may be phosphorylated by a different kinase or by active contaminants in the C-subunit preparation or by the ATPase activity of C-subunit. The last two artifacts become more likely in situations in which the amount of C-subunit introduced, because of difficulties in calculation of intracellular concentrations, is in great excess over the amount normally present.

Nevertheless, these techniques have been shown by the investigators cited above to be capable of elucidating the role of cAMP in cellular processes by virtually entirely *in vivo* experiments. Thus, the requirements of criteria 2 and 4 of Table I that *in vitro* phosphorylation of the protein substrate be examined appear unnecessary in these cases. They may in fact be misleading when, for example, in the case of an ion channel, phosphorylation *in vitro* in the absence of the lipid bilayer in which the channel is normally embedded may have different characteristics from the phosphorylation *in vivo*. In these situations, therefore, studies concerned with *in vivo* phosphorylation of an identified protein substrate become all the more critical to establish unequivocally the mechanism by which cAMP mediates the effect of a given first messenger.

2.3. Stoichiometry of Protein Phosphorylations

Are considerations of stoichiometry important in studies involving protein phosphorylation reactions? The obvious answer is yes, but this should not be construed as implying that it is necessarily essential to demonstrate that a putative cAMP-dependent protein kinase substrate is phosphorylated stoichiometrically *in vitro* and *in vivo* by the kinase. Clearly, if stoichiometric phosphorylation cannot be achieved *in vitro*, some explanation is in order. *In vivo*, however, it is probably rare that protein phosphorylation reactions ever proceed to completion. Relevant to measurement of the extent of phosphorylation under either condition is the question of whether the actual phosphorylation state of the protein substrate being investigated is masked by the presence of endogenous unlabeled phosphate. Thus, when possible, it is often desirable to determine phosphate content by direct analysis.

Can any generalizations be made as to how extensive the phosphorylation of a given protein substrate should be *in vivo* or in intact cells in order to be considered physiologically significant? The important consideration here is that the extent of phosphorylation and any accompanying functional change in the protein should be in keeping with the physiological change caused by cAMP. In some instances, the change in protein phosphorylation might be directly proportional to the extent of alteration of the physiological process. However, in many instances, the physiological change caused by cAMP may be several steps removed from the initial phosphorylation event, as in cascade systems involving more than one protein kinase, and under such circumstances a very small extent of phosphorylation could produce a major end result (Stadtman *et al.*, 1976). The best studied example of this type is the phosphorylation and activation of phosphorylase that results from elevation of tissue cAMP levels. Here, relatively small cAMP-dependent protein-kinase-mediated changes in the extent of phosphorylation and activation of phosphorylase kinase can lead to much larger changes in the conversion of phosphorylase *b* to phosphorylase *a* (Posner *et al.*, 1965; Drummond, *et al.*, 1969; Stull and Mayer, 1971; Yeaman and Cohen, 1975).

Another situation in which a very low level of protein phosphorylation might nonetheless be of great significance would be under conditions in which compartmentation occurs. If a minor portion of a given protein is undergoing phosphorylation because of its special localization in the cell but, when analyzed, is mixed with the entire cellular pool of that protein, then the extent of phosphorylation might appear to be very small.

Finally, if the phosphorylation of a protein were to serve the purpose of programming that protein for participation in some irreversible event, e.g., destruction by proteolysis, then a very low level of phosphorylation could be highly significant. One cAMP-dependent protein kinase substrate that is known to be more readily degraded in its phosphorylated form than in its dephosphorylated form is pyruvate kinase (reviewed in Engstrom, 1980). In this instance, however, regulation of degradation would hardly be thought of as a *raison d'être* for the enzymes serving as substrate, since pyruvate kinase is one of the best characterized examples of an enzyme whose activity is regulated by cAMP-mediated phosphorylation (Engstrom, 1980). In a very recent report, Siekierka *et al.* (1984) describe another example in which nonstoichiometric phosphorylation can nonetheless be highly significant. The GDP exchange factor (GEF), which has a catalytic roll in the formation of the ternary complex involved in initiation of protein synthesis, can be "trapped" as the phosphorylated form of eIF-2, i.e., as eIF-2(αp). Since there is more eIF-2 than GEF, it suffices that only a small portion (20–40%) of eIF-2 be phosphorylated to block protein synthesis.

3. Selected Examples of Cyclic-AMP-Mediated Protein Phosphorylation

To what extent have the various criteria for physiological relevance been met in the study of actual cAMP-dependent processes believed to be regulated by protein phosphorylation? It is far beyond the scope of this chapter to examine all of the processes that have at one time or another been mentioned in this context, but it does appear useful to examine several specific systems to see "how well they score" at this time.

3.1. Skeletal Muscle Phosphorylase Kinase

Skeletal muscle phosphorylase kinase and glycogen synthase were the first substrates for the cAMP-dependent protein kinase that were recognized. As such, they have received much attention over the years. Phosphorylase kinase, which is discussed here, has served as a prototype for workers interested in examining cAMP-mediated phosphorylations, and as recently as 1977 it was cited as the only clearly recognized example of an enzyme that had been shown to be phosphorylated by the cAMP-dependent protein kinase *in vivo* as well as *in vitro* (Nimmo and Cohen, 1977).

Skeletal muscle phosphorylase kinase is a large molecule with a molecular weight of 1.3×10^6 (Hayakawa *et al.*, 1973a). It is made up of four types of subunits and has a quaternery structure generally considered to be, $\alpha_4\beta_4\gamma_4\delta_4$ (Cohen, 1973; Hayakawa *et al.*, 1973b; Cohen *et al.*, 1978), although this laboratory in its original estimation of the number of individual subunits had obtained data indicative of more than four γ subunits (Hayakawa *et al.*, 1973b). The γ subunit is recognized as being the catalytic subunit based on the work of Skuster *et al.* (1980) as well as the fact that its amino acid sequence is homologous to that of the catalytic subunit of the cAMP-dependent protein kinase (Reimann *et al.*, 1984).

It was shown very early that phosphorylase kinase exists in nonactivated and activated forms and that the nonactivated form of the enzyme can be converted to the activated form in a protein phosphorylation reaction that is stimulated by cAMP (reviewed by Krebs *et al.*, 1966). The activated form of the kinase is distinguished from the nonactivated form by virtue of the fact that it has a higher pH 6.8 to pH 8.2 activity ratio. From the onset it was appreciated that two types of phosphorylation, autophosphorylation and phosphorylation catalyzed by a separate enzyme, were involved in the *in vitro* activation process (Krebs *et al.*, 1966).

In 1968 it was determined that the latter enzyme was a cAMP-dependent protein kinase capable of phosphorylating not only phosphorylase kinase but other protein substrates as well, hence the name "cAMP-dependent protein kinase" rather than "phosphorylase kinase kinase" (Walsh *et al.*, 1968). When the cAMP-stimulated phosphorylation and activation of phosphorylase kinase was studied in more detail using the purified catalytic subunit of the cAMP-dependent kinase, it was found that this enzyme catalyzed a stoichiometric phosphorylation of the α and β subunits (Cohen, 1973; Hayakawa *et al.*, 1973b). Phosphorylation of the β subunit correlated more closely with activation of the kinase than did that of the α subunit in keeping with the concept of a specific "activation site" in the enzyme (Riley *et al.*, 1968). It is still not entirely clear, however, whether cAMP-mediated phosphorylation and activation of phosphorylase kinase occurs solely through phosphorylation of a single site in the β subunit or whether it also involves phosphorylation of the α subunit (Hayakawa *et al.*, 1973a; Singh *et al.*, 1982).

Cohen *et al.* (1975) developed methods for examining the specific sites in the α and β subunits that are phosphorylated by the cAMP-dependent protein kinase. In addition to activation brought about through autophosphorylation and phosphorylation catalyzed by the cAMP-dependent protein kinase, it is known that cGMP-dependent protein kinase (Lincoln and Corbin, 1977; Khoo *et al.*, 1977), protein kinase C (Nishizuka *et al.*, 1978), and glycogen synthase (casein) kinase 1 (Singh *et al.*, 1982) also catalyze the phosphorylation and activation of phosphorylase kinase. Glycogen synthase (casein) kinase 1 appears to phosphorylate the specific "activation site" of the subunit (Singh *et al.*, 1982), and recently it has been shown that this site is also phosphorylated in the autophosphorylation reaction (King *et al.*, 1983). Autophosphorylation and the accompanying activation of phosphorylase kinase are known, however, to involve many sites other than the activation site on the β subunit (Hayakawa, 1973b; Carlson and Graves, 1976; Wang *et al.*, 1976; Singh and Wang, 1977; Singh *et al.*, 1982; King *et al.*, 1983; Hallenbeck and Walsh, 1983).

Abundant evidence is available that *in vitro* skeletal muscle phosphorylase kinase exhibits all of the properties that would be expected of an enzyme whose activity can be regulated by the cAMP-dependent protein kinase; i.e., criteria 1 and 2 of Table I have been met (for reviews see Gross and Mayer, 1974; Krebs and Preiss, 1976; Nimmo and Cohen, 1977; Carlson *et al.*, 1979). What is known about the reversible activation and site-specific phosphorylation of phosphorylase kinase that occurs *in vivo* in response to activation of the cAMP-dependent protein kinase by cAMP? How well have criteria 3 and 4 been met?

In 1964 experiments carried out in anesthetized rats and pithed frogs showed that epinephrine administration increased skeletal muscle cAMP levels and caused phosphorylase kinase activation as indicated by an increase in the pH 6.8/8.2 activity ratio of the enzyme (Posner *et al.*, 1964), an observation that was confirmed by Drummond *et al.* (1969). In the first of these reports, slight kinase activation was also seen with electrical stimulation of muscle contraction, but this was not confirmed in the later study. Neither set of investigators found any effect of electrical stimulation on cAMP levels. Mayer and Krebs (1970) carried out an additional study in which they attempted to show net phosphate incorporation into phosphorylase kinase accompanying its *in vivo* activation in response to epinephrine, but these experiments were inconclusive. Based on the totality of the above studies, however, it became generally accepted that epinephrine-induced conversion of phosphorylase *b* to phosphorylase *a* in muscle occurs by cAMP-dependent protein-kinase-mediated phosphorylation of phosphorylase kinase but that the conversion of phosphorylase *b* to phosphorylase *a* seen with muscle contraction does not involve phosphorylase kinase phosphorylation but rather results from its stimulation by calcium (no futher discussion of this aspect of phosphorylase kinase regulation is presented here).

In 1971 Stull and Mayer carried out a very thorough study of muscle phosphorylase kinase activation and the conversion of phosphorylase *b* to *a* in rabbit gracilis muscle in response to isoproterenol administered *in vivo*. They found that at very low levels of isoproterenol, conversion of phosphorylase *b* to *a* could be seen without any change in cAMP levels or phosphorylase kinase activation. At higher levels of isoproterenol, elevation of cAMP levels and phosphorylase kinase activation did occur, but these changes lagged behind phosphorylase activation. They concluded that conversion of phosphorylase *b* to phosphorylase *a* is more sensitive to stimulation by the β-adrenergic agents than cAMP formation or phosphorylase kinase activation in terms of both dose and time after injection. These studies suggested that an alternative mechanism other than phosphorylation of the kinase is operative at low doses of isoproterenol. The time courses of Stull and Mayer were not long enough to determine whether phosphorylase kinase activation was reversible *in vivo*.

Yeaman and Cohen (1975) administered epinephrine to intact rabbits and then isolated phosphorylase kinase to determine the phosphate content of specific peptides representing the cAMP-dependent protein kinase phosphorylation sites in the α and β subunits. Their results showed actual phosphate incorporation into these sites, which for the β subunit site was commensurate with the extent of kinase activation caused by epinephrine. Apart from the fact that 6–7 min elapsed between the time that the rabbits were injected with epinephrine and the point at which the muscle tissue came in contact with EDTA and fluoride, which block *in vitro* interconversions, the experiments of these investigators appeared to offer conclusive proof of *in vivo* cAMP-dependent protein-kinase-mediated phosphorylation of phosphorylase kinase. It could still be argued, however, that completely foolproof evidence that the cAMP-dependent protein kinase was involved as the catalyst for the phosphorylations that they observed must still be obtained. This follows since it is now known that at least the β subunit site can be phosphorylated by casein kinase I or in the autophosphorylation reaction (see above). Thus far, nobody has carried out experiments actually showing that phosphorylase kinase activated *in vivo* returns to the nonactivated state *in vivo*, although Yeaman and Cohen (1975) did demonstrate that enzyme activated *in vivo* could be dephosphorylated and inactivated *in vitro*.

Despite the fact that the phosphorylation and activation of skeletal muscle phosphorylase kinase have been studied in many different laboratories for more than 20 years, it is clear that there is still more to be learned about this system. This applies particularly to the activation process *in vivo* or in the intact cell. It can probably be stated with certainty that activation of the enzyme does occur in intact muscle in response to phosphorylation by the cAMP-dependent protein kinase, but further quantitation of the process would be desirable. With the finding that the cAMP-dependent protein kinase is not the only enzyme that can catalyze phosphorylation of the "activation site" on the β subunit, the possible role of enzymes other than the cAMP-dependent kinases participating in the activation process still needs to be kept in mind. The role of autophosphorylation bears further investigation, as does the possibility that second messengers other than cAMP, e.g., Ca^{2+}, might have a function in β-adrenergic stimulation of muscle in order to account for the observations of Stull and Mayer (1971) with low doses of isoproterenol.

3.2. Cyclic-AMP-Dependent Phosphorylation of Smooth Muscle Myosin Light-Chain Kinase

Relaxation of many types of smooth muscle can be induced by β-adrenergic agonists and other agents that increase intracellular concentrations of cAMP (reviewed by Andersson *et al.*, 1975; Diamond, 1978). Several mechanisms have been proposed that can explain the actions of cAMP in a plausible manner; however, one hypothesis in particular has recently received widespread attention. The primary target of the cAMP-dependent protein kinase in this mechanism is the enzyme myosin light-chain kinase (MLCK), which catalyzes the phosphorylation of a specific serine residue of one of the "light-chain" subunits of myosin. The activity of MLCK is dependent on micromolar concentrations of Ca^{2+} and the calcium-binding protein calmodulin (Dabrowska, *et al.*, 1978; Adelstein and Klee, 1981).

A large body of evidence now indicates that phosphorylation of myosin by MLCK is obligatory for contraction in smooth muscle (reviewed by Stull, 1980; Adelstein *et al.*, 1981). Early experiments by Adelstein *et al.* (1978) demonstrated that purified turkey gizzard smooth muscle MLCK was rapidly and stoichiometrically phosphorylated *in vitro* by the catalytic subunit of the cAMP-dependent protein kinase. This phosphorylation was associated with a two-fold decrease in specific activity of MLCK when isolated gizzard myosin light chains were used as substrate. Subsequently, Silver and DiSalvo (1979) demonstrated that addition of cAMP and cAMP-dependent protein kinase to preparations of bovine aortic native actomyosin (which contain endogenous calmodulin and MLCK) depressed the level of phosphorylation of myosin light chains. The inhibition of light-chain phosphorylation was associated with phosphorylation of a protein of 100,000 molecular weight (presumably MLCK) and a decrease in actomyosin Mb^{2+} ATPase activity (an *in vitro* index of actin–myosin interaction). Studies using chemically skinned smooth muscle fiber preparations, which retain many enzymatic activities as well as the ability to contract in a Ca^{2+}-dependent manner, demonstrated that relaxation could be induced by the addition of cAMP (Mrwa *et al.*, 1979) or purified catalytic subunit of cAMP-dependent protein kinase (Kerrick and Hoar, 1981; Rüegg and Paul, 1982). The effect of cAMP or cAMP-dependent protein kinase was reversed by its removal from the medium (Mrwa *et al.*, 1979; Rüegg and Paul, 1982) or by the addition of calmodulin to the medium (Kerrick and Hoar, 1981).

The mechanism by which phosphorylation inhibited the activity of smooth muscle MLCK was shown by Conti and Adelstein (1981) to involve a markedly decreased affinity of MLCK for its activator, calmodulin. Phosphorylation of gizzard MLCK by cAMP-dependent protein kinase in the absence of calmodulin results in 1.4–2.0 moles phosphate/mole MLCK and a ten-to 20-fold increase in the concentration of calmodulin required for half-maximal activation of MLCK. Little or no change in maximal velocity is observed. Phosphorylation of MLCK in the presence of calmodulin results in the incorporation of 0.4–1.0 mole phosphate/mole MLCK and no change in calmodulin dependence. Thus, only one of the phosphorylation sites appears to regulate enzyme activity. The effect of phosphorylation on the activity of MLCK can be reversed by dephosphorylation using a protein phosphatase purified from smooth muscle (Pato and Adelstein, 1980; Adelstein *et al.*, 1981). Phosphorylation of purified MLCK from bovine trachea (Miller *et al.*, 1983) and bovine aorta (Vallet *et al.*, 1981) has also been demonstrated to have comparable changes in calmodulin activation properties.

A plausible scheme consistent with these data was proposed by Conti and Adelstein (1981) to explain the mechanism by which cAMP might regulate smooth muscle contraction. In resting muscle, Ca^{2+} concentrations are low ($<10^{-6}$ M), and calmodulin is mostly dissociated from MLCK. The MLCK is therefore largely inactive under these conditions. An increase in cAMP concentrations under resting conditions would result in phosphorylation of MLCK at both sites and would decrease the affinity of MLCK for calmodulin. Subsequent contraction resulting from an increase in Ca^{2+} concentration would be inhibited, since MLCK activation is required for tension generation and MLCK would be more difficult to activate in its phosphorylated state. In order to explain how an increase in cAMP concentration might relax a contracted smooth muscle, one must assume that a fraction of the MLCK is not associated with calmodulin so that it might be phosphorylated at the "regulatory" site. Furthermore, dissociation of calmodulin from unphosphorylated MLCK must occur frequently enough so that a significant fraction of MLCK can eventually become phosphorylated (and thus "inactivated") within the time frame of the relaxation phenomenon

To date, there are only limited data from intact-muscle studies in support of the Conti and Adelstein mechanism. Indeed, there appear to be more data that refute the hypothesis. Studies by Silver and Stull (1982) with bovine tracheal smooth muscle strips examined the effect of the β-adrenergic agonist isoproterenol on carbachol-mediated isometric tension generation, myosin light-chain phosphorylation, and phosphorylase *a* formation. Pretreatment with isoproterenol inhibited the rate and extent of both tension development and light-chain phosphorylation during stimulation with carbachol. These findings are consistent with the Conti and Adelstein hypothesis but are also consistent with mechanisms involving cAMP-dependent effects on cytoplasmic calcium concentrations.

Studies by Murphy and co-workers (Gerthoffer and Murphy, 1983; Gerthoffer *et al.*, 1984) using infected swine carotid artery preparations have shown that relaxation consists of at least two components: a rapid phase, which corresponds to myosin dephosphorylation, and a slower, Ca^{2+}-sensitive phase, which operates through an unidentified regulatory mechanism. Under appropriate conditions, tension in this preparation can be maintained even though myosin phosphorylation has returned to basal (resting) values (0.08 ± 0.02 mole phosphate/mole light chain). The relaxant effects of several vasodilators that are thought to act via cAMP (adenosine, 3-isobutyl-1-methylxanthine, forskolin) were examined under these conditions. These agents induced relaxation without futher reducing the levels of myosin phosphorylation, indicating that phosphorylation of MLCK is not the primary mechanism of relaxation by cAMP under certain experimental conditions.

In order to test criterion 3, Stull and co-workers (Miller *et al.*, 1983) have developed an activity ratio assay to measure changes in calmodulin activation properties of MLCK that might occur with *in vivo* phosphorylation by cAMP-dependent protein kinase. To determine this activity ratio, MLCK activity in dilute muscle extracts is assayed under two conditions: maximally activated (high $[Ca^{2+}]$ and high [calmodulin]) and submaximally activated (low $[Ca^{2+}]$ and high [calmodulin]). Changes in this activity ratio reflect a change in Ca^{2+} sensitivity and therefore a change in calmodulin sensitivity of MLCK. The procedure is analogous to activity ratio measurements that have been used to determine phosphorylation states of phosphorylase, phosphorylase kinase, glycogen synthetase, and hormone-sensitive lipase.

In bovine tracheal smooth muscle strips, no significant change in activity ratio was observed when the strips were exposed to a concentration of isoproterenol sufficient to relax the muscle and increase phosphorylase *a* formation. A 16-fold higher concentration of isoproterenol did affect the activity ratio, with a change corresponding to a twofold change in calmodulin affinity. More convincing evidence that isoproterenol does not mediate relaxation via a change in MLCK activity was obtained in an experiment from the same study in which a bovine tracheal smooth muscle strip was exposed to carbachol for 2 hr (Miller *et al.*, 1983). Following exposure to carbachol isometric tension in the muscle strip reached a maximum value within 3 min and was sustained for the 2-hr period. Phosphorylation of myosin light chain increased rapidly from a resting value of 0.1 mole phosphate/mole light chain to 0.8 mole phosphate/mole light chain in less than 2 min. However, the phosphate content of the myosin light chain gradually decreased and eventually returned to the resting value after 2 hr. Administration of isoproterenol at this time resulted in marked and immediate relaxation of tension with no further change in myosin light chain phosphate content. Thus, β-adrenergic relaxation of intact smooth muscle did not appear necessarily to involve changes in phosphorylation of either myosin light chains or MLCK.

In summary, an attractive hypothesis to explain the mechanism of β-adrenergic relaxation of smooth muscle was proposed and supported by a large body of *in vitro* experimental data. Indeed, criteria 1 and 2 of Table I were thoroughly investigated and shown to obtain for the case of cAMP-dependent phosphorylation of myosin light-chain kinase. Preliminary results (de Lanerolle *et al.*, 1983) from intact tracheal smooth muscle indicate possible changes in MLCK phosphorylation, suggesting that criterion 4 might also be satisfied. However, substantial evidence has been presented to indicate that criterion 3 is not satisfied under certain conditions. More extensive investigations involving intact muscle preparations will be necessary to clarify the relative importance of phosphorylation of MLCK as a mechanism for cAMP-mediated regulation of smooth muscle contraction.

3.3. Activation of Tyrosine Hydroxylase by Neuronal Depolarization

Tyrosine hydroxylase (TH) catalyzes the conversion of L-tyrosine to DOPA in the presence of O_2 and a reduced pterin cofactor (Nagatsu *et al.*, 1964), a reaction generally regarded as the rate-limiting step in catecholamine biosynthesis (Levitt *et al.*, 1965). Regultion of TH activity is, therefore, a mechanism by which catecholaminergic neurons could maintain intracellular transmitter levels in the face of increased impulse flow. This has been abundantly demonstrated by studies showing that in sympathetic nerves (Alousi and Weiner, 1966; Gordon *et al.*, 1966) and central noradrenergic (Roth *et al.*, 1975) and dopaminergic neurons (Murrin *et al.*, 1976), impulse flow is correlated with an acceleration of catecholamine synthesis. In addition, in *in vitro* preparations such as brain slices (Harris and Roth, 1971), isolated nerve endings (synaptosomes) (Patrick and Barchas, 1974), and isolated retinal (Iuvone and Marshburn, 1982) and pheochromocytoma cell suspensions (Vaccaro *et al.*, 1980), K^+- or veratridine-induced depolarization leads to increased catecholamine biosynthesis. The stimulation-induced acceleration of catecholamine biosynthesis is, in all cases examined, mediated by an effect at the tyrosine hydroxylase step (Gordon *et al.*, 1966; Sedvall and Kopin, 1967; Chieuh and Moore, 1974).

A variety of observations, discussed below, prompted the hypothesis that an alteration in the catalytic capacity of TH itself, perhaps as a result of cAMP-dependent protein phosphorylation, underlies the depolarization-induced acceleration of catecholamine biosynthesis (Roth *et al.*, 1975; Weiner *et al.*, 1978). This discussion is restricted to an examination of this hypothesis. Other mechanisms by which TH may be regulated are covered in a recent review (Masserano and Weiner, 1983). As is discussed, the issue as it historically developed was not so much one of ascribing a known role for cAMP to a specific protein phosphorylation but rather whether the stimulation-induced rise in TH activity is mediated by cAMP and, if so, whether by cAMP-dependent phosphorylation of TH.

Early experiments by Berkowitz and co-workers (1970) and Waldeck (1971) found increased brain catecholamine synthesis after administration of cyclic-nucleotide phosphodiesterase inhibitors. In 1973, Goldstein and co-workers demonstrated that dibutyryl cAMP was capable of accelerating the synthesis of dopamine in slices of rat striatum. Subsequently, the dibutyryl and 8-bromo derivatives of cAMP were shown to accelerate catecholamine synthesis in a wide variety of tissues including brain cortical slices (Taneda *et al.*, 1974), synaptosomes from a variety of brain areas (Harris *et al.*, 1974; Ebstein *et al.*, 1974; Patrick and Barchas, 1976), isolated superior cervical ganglia (Keen and McLean, 1974), and isolated vasa deferentia (Taneda *et al.*, 1974). A number of these reports localized the effect to the tyrosine hydroxylase step. Keen and McLean (1974) further demonstrated that this was an activation rather than an induction of TH. Agents (cholera toxin, forskolin) that increase endogenous cAMP levels in cells were also found to activate TH (Chalfie *et al.*, 1979; El-Mestikawy *et al.*, 1983). The activation process is reversible *in situ*. Removal of the cAMP derivative from the bathing medium leads to a return to base line of either catecholamine synthesis measured in the intact tissue (Meligeni *et al.*, 1982) or tyrosine hydroxylase activity in soluble extracts prepared from the tissue (Simon and Roth, 1979).

Harris and co-workers (1974) were the first to demonstrate that cAMP could activate TH in soluble extracts, an effect dependent on the presence of MgATP and cAMP-dependent protein kinase (Lovenberg et al., 1975; Morgenroth et al., 1975). Letendre et al. (1977) reported that TH incorporated $^{32}P_i$ in organ cultures of rat tissues, but the relationship of this to cAMP-dependent phosphorylation was not explored.

Tyrosine hydroxylase has now been highly purified from a variety of sources and shown to be activated and phosphorylated by the cAMP-dependent protein kinase (Edelman et al., 1978, 1981; Joh et al., 1978; Yamauchi and Fujisawa, 1979a; Vulliet et al., 1980; Markey et al., 1980). Activated TH is similar in its kinetic characteristics whether the activation is by incubation of intact tissue with membrane-permeable cAMP derivatives followed by in vitro assay or by exposure of crude or purified enzyme cAMP-dependent phosphorylating conditions. Although some laboratories have reported changes in V_{max} (Joh et al., 1978) or K_m for substrate (Harris et al., 1974), the consensus is that activation is expressed as a reduced K_m for pterin cofactor and an increased K_i for catecholamines (Lloyd and Kaufman, 1975; Lovenberg et al., 1975; Goldstein et al., 1976; Edelman et al., 1978; Markey et al., 1980; Vulliet et al., 1980). It has been argued, however, that under more physiological assay conditions (pH ~7) the activation is manifested primarily as a change in V_{max} (Pollock et al., 1981; Lazar et al., 1982). The stoichiometry of cAMP-dependent phosphorylation is 0.7–0.9 mole ^{32}P incorporated/mole subunit (Vulliet et al., 1980; Edelman et al., 1981), although in some preparations lower values were apparently found (Joh et al., 1978; Markey et al., 1980). The extent of activation parallels the extent of phosphorylation initially (Vulliet et al., 1980); however, because of increased thermolability of the phosphorylated enzyme (Lazar et al., 1981; Vrana et al., 1981), activity eventually decreases as phosphorylation proceeds (Vulliet et al., 1980). The activation by phosphorylation is also reversed by dephosphorylation, and the deactivated and dephosphorylated enzyme can be subsequently rephosphorylated and reactivated (Yamauchi and Fujisawa, 1979b).

A considerable body of evidence can be marshaled to directly relate the depolarization- and cAMP-induced increases in catecholamine synthesis. Weiner and co-workers (1978) and Masserano and Weiner (1979) presented evidence that tyrosine hydroxylase normally exists in tissues in activated and nonactivated states and that the effect of nerve stimulation is to convert TH molecules to the activated state, a state kinetically similar to that observed with TH isolated from the same tissue after incubation of the intact tissue with 8-methylthio cAMP or incubation of soluble extracts in the presence of cAMP, ATP, and Mg^{2+}. Earlier, Roth and collaborators demonstrated that stimulation of central nonadrenergic (Roth et al., 1975) or dopaminergic (Murrin et al., 1976) cell bodies resulted in TH activity in extracts prepared from the neurons of the terminal fields that was kinetically altered in a similar fashion to the enzyme cAMP-treated extracts from nonstimulated neurons. Similarly, TH in retina, activated by photic stimulation *in vivo* or exposure to cAMP-dependent protein phosphorylation *in vitro*, showed similar alterations in its kinetic constants (Iuvone *et al.*, 1982). The activation of TH by nerve stimulation in all of these studies was expressed as K_m (and/or K_i) rather than V_{max} changes; however, since, as mentioned above, assay conditions (e.g., pH) may determine the way the activation is expressed, it is perhaps more important to note that impulse flow and cyclic nucleotides produced similar changes in a given assay system, keeping in mind the difficulty of extrapolating values for the kinetic constants to *in vivo* conditions.

Other studies that may be cited in support of the involvement of cAMP in depolarization-induced activation of TH examined the extent to which TH can be activated *in situ*. If nerve stimulation and cAMP activate through the same mechanism, it should be possible to show that the two types of treatment are nonadditive when applied in concert to the same preparation (provided at least one of the treatments results in full activation). Nonadditivity of TH activation by neuronal stimulation and cAMP has been reported in central dopaminergic neurons (Murrin *et al.*, 1976), adrenal gland (Masserano and Weiner, 1979), and retina (Iuvone *et al.*, 1982). The involvement of cAMP is also suggested by findings that in some catecholaminergic tissues depolarization is reported to increase cAMP levels (brain slices) (Kakiuchi *et al.*, 1969) or to activate cAMP-dependent protein kinase (adrenal medulla) (Masserano and Weiner, 1979). In addition, Weiner *et al.* (1978) reported that the activation of tyrosine hydroxylase by electrical stimulation of the hypogastric nerve–vas deferens preparation was potentiated by stimulation of the nerve in the presence of the cyclic-nucleotide phosphodiesterase inhibitor isobutylmethylxanthine. Finally, perhaps the most critically important data in support of this hypothesis come from *in vivo* labeling experiments. Haycock *et al.* (1982b) reported that in isolated adrenal chromaffin cells preloaded with $^{32}P_i$, administration of acetylcholine (the neurotransmitter that activates catecholamine synthesis in the gland *in vivo*) leads to an increased incorporation of $^{32}P_i$ into immunoprecipitated tyrosine hydroxylase. Similarly, the nicotinic agonist dimethylphenylpiperazinium causes an increased incorporation of ^{32}P into tyrosine hydroxylase in these cells (Frye and Holz, 1983). Meligeni and co-workers (1981) found that adrenal TH prepared from acutely stressed animals incorporated less ^{32}P than TH from nonstressed control on *in vitro* incubation under cAMP-dependent phosphorylating conditions. The inference, therefore, was that nervous stimulation of the gland caused by the stress procedure led to an increased *in vivo* phosphorylation of TH, making it less susceptible to subsequent *in vitro* phosphorylation.

It seems, therefore, that a strong case could be made for an obligatory role of cAMP-dependent phosphorylation in the physiological activation of TH. Criteria 1–3 of Table I are largely satisfied, and much additional supporting evidence can be supplied. Although the *in vivo* rate of TH phosphorylation has not been calculated and directly compared to the time course of activation evoked by nerve stimulation (criterion 1), Masserano and Weiner (1979) have calculated that the extent of activation of cAMP-dependent protein kinase in adrenal gland after neurogenic stimulation may be adequate to account for the activation of TH seen after this stimulation.

In spite of this, some evidence has accumulated casting doubt on the cAMP hypothesis or at least suggesting that the regulation may be more complex than previously appreciated. Some preparations (synaptosomes, pheocromocytoma cells) can respond to depolarization with increases in tyrosine hydroxylase activity with no changes in cAMP levels or rates of production of $[^3H]$-cAMP from $[^3H]$-adenine (De Belleroche *et al.*, 1974; Patrick and Barchas, 1974; Chalfie *et al.*, 1979). One could, however, argue that increases may be transient and small because of compartmentalization of components, and in many instances protein kinase activity ratios, which may be more sensitive, have not been performed.

Although experiments cited above reported nonadditive effects of neuronal stimulation and cAMP on TH activation, additive effects of the two treatments have also been reported in striatal synaptosomes (Goldstein *et al.*, 1976) or slices (Simon and Roth, 1979; El-Mestikawy *et al.*, 1983) and superior cervical ganglia (Horwitz and Perlman, 1984). Finding additivity is perhaps more difficult to interpret since, because of intracellular compartmentation, two treatments may be apparently additive simply because one (e.g., dibutyryl cAMP) may saturate a transport mechanism and therefore activate a small percentage of the total number of activatable enzyme molecules. Nevertheless, a careful study by El-Mestikawy *et al.* (1983) found that tyrosine hydroxylase, which by a number of criteria was fully phosphorylated by the cAMP-dependent protein kinase *in situ*, could still be activated *in situ* by K^+ depolarization. Since the depolarization-induced rise in TH activity is Ca^{2+} dependent (Harris and Roth, 1971; Patrick and Barchas, 1974), it has been suggested that Ca^{2+}, rather than cAMP, is the second messenger mediating this process (Morgenroth *et al.*, 1974). Tyrosine hydroxylase can be phosphorylated and activated *in vitro* by the Ca^{2+}/lipid-activated kinase (kinase C) (Raese *et al.*, 1979) and the Ca^{2+}/calmodulin-dependent kinase (Yamauchi and Fujisawa, 1981). El-Mestikawy *et al.* (1983) observed that TH in striatal homogenates could be activated by ATP, Mg^{2+}, and Ca^{2+} but that TH in striatal slices activated *in situ* by depolarization could not be further activated by ATP, Mg^{2+}, and Ca^{2+} in a homogenate prepared from the depolarized tissue. The enzyme was, however, activatable by cAMP-dependent phosphorylating conditions following depolarization. Yet a fourth candidate for regulation is a cAMP- and Ca^{2+}-independent kinase capable of activating TH in striatum (Andrews *et al.*, 1983).

The activity of TH can be influenced by a wide variety of experimental treatments (reviewed by Masserano and Weiner, 1983), and it may be that the mode of TH regulation varies with differing conditions. Length of time of stimulation, for example, is a factor not always varied in the studies cited above. Iuvone and collaborators (1979) demonstrated that photic stimulation for brief periods of time decreased the apparent K_m of retinal TH for cofactor without changing the apparent V_{max}, whereas longer stimulation returned the K_m to base-line values and led to an increased V_{max}. The latter was not caused by an increase in the total number of immunoprecipitable TH molecules and may therefore represent a time-dependent shift to a different regulatory mechanism. Mode of regulation may vary with tissue and subcellular localization as well. Impulse flow is thought to increase cAMP production postsynaptically rather than presynaptically (Nathanson, 1977). This presumably explains why, in the studies cited above, depolarization could increase cAMP in brain slices (which contain postsynaptic elements) but not in synaptosomes (pinched-off nerve endings). In areas of the cell such as dendrites and cell bodies, TH may therefore be more likely to be regulated by cAMP, whereas at the nerve endings Ca^{2+}-dependent processes might operate. In this regard, Lewander *et al.* (1977) found that following oxotremorine treatment, TH was activated in CNS noradrenergic cell bodies but not in axons.

Definitive understanding of the mechanism of TH activation may ultimately be provided by studies of the *in vivo* sites of phosphorylation after nerve stimulation (criterion 4 of Table I). Haycock and co-workers (1982a) recently reported that in adrenal chromaffin cells prelabeled with $^{32}P_i$, two distinct peptides prepared from immunoprecipitated TH incorporated $^{32}P_i$ after stimulation by acetylcholine, but only one of these two peptides was labeled after treatment of the cells with dibutyryl cAMP. It remains to be determined if the cAMP-stimulated phosphorylation site is different from that stimulated by acetylcholine on the common fragment. If this site is truly unique, it would rule out at least in this system the involvement of cAMP. If not (as is suggested by the data of Meligeni *et al.*, 1981, with the adrenal system), other approaches (such as microinjection of the heat-stable inhibitor of cAMP-dependent protein kinase) may be required to solve this difficult problem. Nevertheless, it appears that the study of *in vivo* sites of phosphorylation of TH will prove an important step in the resolution of this problem and illustrate the benefit of applying criterion 4 in cases in which multiple kinases have access *in vivo* to the physiologically regulated step.

3.4. Cyclic-AMP-Dependent Protein Phosphorylation and Nuclear Events

A diverse range of functions involving the nucleus have been investigated with regard to possible regulatory effects of cAMP. These include the control of cell growth and cell division (Boynton and Whitfield, 1983), oocyte maturation (Maller, 1983), regulation of histone function(s) (Langan *et al.*, 1981), RNA synthesis (Lee and Jungmann, 1981), and specific enzyme induction (Rosenfield and Barrieux, 1979). In contrast to many of the other processes thought to involve cAMP-dependent protein phosphorylation, up to the present time the effort of workers in these fields has been mainly concentrated on providing unequivocal proof that cAMP itself is indeed involved. Relatively little attention has been devoted to defining specific protein phosphorylation. This priority of research to date is readily understood, since there is only a very limited understanding of the role of nuclear proteins in the regulation of cell growth and division, differentiation, and the induction of repression of specific protein synthesis in eucaryotes. This situation is likely to change quite dramatically in the next few years.

There is perhaps more reason to question the role of cAMP-dependent protein phosphorylation in relation to the control of intranuclear events that in many other systems. The much-studied actions of cAMP in control of DNA expression in bacteria (recently reviewed by Ullmann and Danchin, 1983) do not appear to involve protein phosphorylation but rather binding to a protein termed catabolite activator protein (CAP). The binding of the cAMP–CAP complex to specific regions of DNA then determines the rate of transcription of specific operons. It is reasonable to wonder whether a similar mechanism might operate in the nucleus of eucaryotic cells. In this case, it would suggest that the complex of cAMP with the regulatory subunit or a related protein would be more important in subsequent regulatory events than the kinase effects of the released catalytic subunit. Protein sequencing has shown that the cAMP-binding domains of the regulatory subunit (R_{II}) of cAMP-dependent protein kinase and of CAP are homologous (Weber *et al.*, 1982). However, the carboxyl-terminal domain of CAP, which shows homologies to the DNA-binding domains of other regulatory proteins, has no significant homology with R_{II}.

It has been known for many years that certain nuclear proteins are excellent *in vitro* substrates for the cAMP-dependent protein kinase. A number of studies have addressed the basic question of whether the holoenzyme or subunits of cAMP-dependent protein kinase may exist in or be translocated to the nucleus. Early work on the problem was reviewed by Jungmann *et al.* (1975). It was concluded that the free catalytic subunit can translocate to the nucleus. Using an immunocytochemical approach, Steiner *et al.* (1978) found cAMP in the nuclear membrane of rat liver cells. The catalytic subunit of cAMP-dependent protein kinase was also located in the nucleus, including the nucleolus. The regulatory subunit showed a clumped distribution in the nucleus. During liver regeneration, the tissue content of cAMP increased. In parallel with these changes, increased nuclear fluorescence by antibodies to the catalytic and type II regulatory subunits was demonstrated. More recently, Harper *et al.* (1981) reported that treatment of rats with ACTH led to changes in the nuclear staining of adrenal cells that were incubated with antisera raised aginst the catalytic and regulatory subunits of the cAMP-dependent protein kinase. The greatest increase in staining of the catalytic and type II regulatory subunits was seen 11 hr after ACTH administration. It is evident, therefore, that not only may the subunits of cAMP-dependent protein kinase be found in the nucleus but also their content may change during regulatory processes.

The possible role of cAMP in the regulation of cell proliferation has recently been reviewed (Boynton and Whitfield, 1983). These authors point out that whereas initial studies of this subject were preoccupied with effects of cAMP to inhibit cell proliferation, it is now reasonable to conclude that cAMP may be positively involved in some aspect or aspects of events leading to DNA synthesis in a wide variety of cells. It is possible that the nucleotide is also involved in subsequent events leading to mitosis and cell division. One of the difficulties leading to present uncertainties as to cAMP involvement has been the need to define precisely how and when the intracellular concentrations of cAMP change in relation to the various phases of the cell cycle. The search for proteins whose phosphorylation might change in a cAMP-dependent process is hampered by lack of understanding of the basic mechanisms involved.

Cyclic AMP may also be implicated in the overall control of nuclear RNA synthesis. It has been known for many years that certain hormones whose actions are accompanied by elevation of intracellular cAMP, e.g., FSH, increase nuclear RNA synthesis. Possible sites of such an effect could be control of the state of the DNA template or control of the actions of proteins mediating RNA synthesis. Studies of the phosphorylation of histone and nonhistone proteins in the nucleus and of RNA polymerase relate to these possibilities. They are discussed in detail in the references quoted at the beginning of this section and are not considered further here.

A separate consideration in relation to effects of cAMP on RNA synthesis is whether the specific enzyme inductions produced by hormones known to elevate cell cAMP content could be mediated by cAMP-dependent protein phosphorylation. This is the process most closely related to the effect of cAMP on enzyme induction in bacteria and therefore perhaps the one in which most careful attention should be paid to the possibility of control by a mechanism not involving phosphorylation. However, recent studies have provided fairly convincing evidence that a phosphorylation reaction may be involved at least as part of the overall process.

Tyrosine aminotransferase (E.C. 2.6.1.5) activity in the liver is subject to regulation by a variety of hormones, some of which, e.g., glucagon, act via changes in the cell cAMP content. Continuing RNA synthesis appears to be essential for cAMP-induced increases in tyrosine amino-transferase activity (Noguchi *et al.*, 1982), although the precise site(s) of the action(s) of cAMP remain unclear (Lewis *et al.*, 1982). Hepatoma cells exposed to 8-bromo cAMP exhibited an increase in tyrosine aminotransferase messenger RNA as judged by *in vitro* translation (Culpepper and Liu, 1983). However, the use of cDNA probes is required to investigate the precise basis of this change of messenger RNA activity. Whatever the site of action of cAMP in this process, there is evidence that it acts through phosphorylation rather than a reaction involving a cAMP binding protein. Boney *et al.* (1983) introduced purified cAMP-dependent protein kinase catalytic subunit into cultured hepatoma cells by fusing the cells with human red blood cell ghosts loaded with the subunit. A two-to threefold increase of tyrosine aminotransferase activity was induced. This effect could be blocked by the specific heat-stable protein kinase inhibitor and by purified regulatory subunit.

Thus, the evidence as far as the cAMP-induced changes in tyrosine aminotransferase are concerned favors a protein phosphorylation effect, although it is still not certain that this leads to changes in specific messenger RNA synthesis. On the other hand, in the case of phosphoenol-pyruvate carboxykinase, another enzyme whose activity appears to be cAMP regulated, the evidence in favor of an increase in specific nuclear RNA production is much more clear cut (Chrapkiewics *et al.*, 1982). Dibutyryl cAMP acting on cultured hepatoma cells was shown by hybridization to increase both cytoplasmic and nuclear phosphoenolpyruvate carboxykinase RNA sequences, the latter preceding the former. It is not known whether similar effects can be achieved by the administration of the catalytic subunit of the cAMP-dependent protein kinase.

In view of the remaining uncertainties as to the site of action of cAMP in increasing tyrosine aminotransferase activity and the lack of information as to whether the catalytic subunit is involved in the effect on phosphoenolpyruvate carboxykinase messenger RNA production, it is not possible at present to be confident that protein phosphorylation is exclusively involved in specific enzyme induction by cAMP.

4. Conclusions

Based on the sampling of specific examples selected for review in this chapter, it appears that in most instances investigators are more successful in obtaining information at the biochemical level regarding the potential regulatory role of a given phosphorylation event than at the cellular or organismal level. In some cases serious attempts have not been made to collect relevant data using intact-cell preparations; however, in many cases, sensitive and quantitative techniques are not available for testing criteria 3 and 4 of Table I. Thus, the literature is full of potentially significant protein phosphorylations that have not been or cannot be fully explored.

Future developments, particularly in the field of molecular biology, promise to greatly increase the number of techniques available for study of the role of cAMP functions at all levels. For the present, attempts should be made to take advantage of recent methodological developments, such as are described in Section 2.2, to increase our understanding of the role of cAMP-mediated phenomena. As more methods become available, it should be possible to add additional criteria to those in Table I. These additional techniques and criteria will be important for improving our understanding of well-characterized cAMP-mediated phenomena, distinguishing between alternative phenomena that may occur in the intact cell, and discovering new phenomena regulated by cAMP.

References

Adams, W. B., and Levitan, I. B., 1982, Intracellular injection of protein kinase inhibitor blocks the serotonin-induced increase in K^+ conductance in *Aplysia* neuron R15, *Proc. Natl. Acad. Sci. U.S.A.* **79:**3877–3880.

Adelstein, R. S., and Klee, C. B., 1981, Purification and characterization of smooth muscle myosin light chain kinase, *J. Biol. Chem.* **256:**7501–7509.

Adelstein, R. S., Conti, M. A., Hathaway, D. R., and Klee, C. B., 1978, Phosphorylation of smooth muscle myosin light chain kinase by the catalytic subunit of adenosine 3′:5′-monophosphate-dependent protein kinase, *J. Biol. Chem.* **253:**8347–8350.

Adelstein, R. S., Pato, M. D., and Conti, M. A., 1981, The role of phosphorylation in regulating contractile proteins, *Adv. Cyclic Nucleotide Res.* **14:**361–373.

Alkon, D. L., Acosta-Urquidi, J., Olds, J., Kuzma, G., and Neary, J. T., 1983, Protein kinase injection reduces voltage-dependent potassium currents, *Science* **219:**303–306.

Alousi, A., and Weiner, N., 1966, The regulation of norepinephrine synthesis in sympathetic nerves: Effect of nerve stimulation, cocaine and catecholamine-releasing agents, *Proc. Natl. Acad. Sci. U.S.A.* **56:**1491–1496.

Andersson, R., Nilsson, K., Wikberg, J., Johansson, S., Mohme-Lundholm, E., and Lundholm, L., 1975, Cyclic nucleotides and the contraction of smooth muscle, *Adv. Cyclic Nucleotide Res.* **5:**491–518.

Andrews, D. W., Langan, T. A., and Weiner, N., 1983, Evidence for the involvement of a cyclic AMP-independent protein kinase in the activation of soluble tyrosine hydroxylase from rat striatum, *Proc. Natl. Acad. Sci. U.S.A.* **80:**2097–2101.

Beavo, J. A., Bechtel, P. J., and Krebs, E. G., 1974, Activation of protein kinase by physiological concentrations of cyclic AMP, *Proc. Natl. Acad. Sci. U.S.A.* **71:**3580–3583.

Berkowitz, B. A., Tarver, J. H., and Spector, S., 1970, Release of norepinephrine in the central nervous system by theophylline and caffeine, *Eur. J. Pharmacol.* **10:**64–71.

Boney, C., Fink, D. Schlichter, D., Carr, K., and Wicks, W. D., 1983, Direct evidence that the protein kinase catalytic subunit mediates the effects of cAMP on tyrosine aminotransferase synthesis, *J. Biol. Chem.* **258:**4911–4918.

Boynton, A. L., and Whitfield, J. F., 1983, The role of cyclic AMP in cell proliferation: A critical assessment of the evidence, *Adv. Cyclic Nucleotide Res.* **15:**193–294.

Bridenbaugh, R. L., Hoar, P. E., and Kerrick, W. G. L., 1981, Phosphorylation of myosin light chain kinase by C-subunit and its effects on smooth muscle, *Biophys. J.* **33:**235a.

Brunelli, M., Castellucci, V., and Kandel, E. R., 1976, Synaptic facilitation and behavioral sensitization in *Aplysia:* Possible role of serotonin and cyclic AMP, *Science* **194:**1178–1181.

Buxton, I. L. O., and Brunton, L. L., 1983, Compartments of cyclic AMP and protein kinase in mammalian cardiomyocytes, *J. Biol. Chem.* **258:**10233–10239.

Carlson, G. M., and Graves, D. J., 1976, Stimulation of phosphorylase kinase autophosphorylation by peptide analogs of phosphorylase, *J. Biol. Chem.* **251:**7480–7486.

Carlson, G. M., Bechtel, P. J., and Graves, D. J., 1979, Chemical and regulatory properties of phosphorylase kinase and cyclic AMP-dependent protein kinase, *Adv. Enzymol.* **50:**41–115.

Castellucci, V. F., Kandel, E. R., Schwartz, J. H., Wilson, F. D., Nairn, A. C., and Greengard, P., 1980, Intracellular injection of the catalytic subunit of cyclic AMP-dependent protein kinase simulates facilitation of transmitter release underlying behavioral sensitization in *Aplysia,*, *Proc. Natl. Acad. Sci. USA.* **77:**7492–7496.

Castellucci, V. F., Nairn, A., Greengard, P., Schwartz, J. H., and Kandel, E. R., 1982, Inhibitor of adenosine 3′:5′-monophosphate-dependent protein kinase blocks presynaptic facilitation in *Aplysia, J. Neurosci.* **2:**1673–1681.

Cedar, H., Kandel, E. R., and Schwartz, J. H., 1972, Cyclic adenosine monophosphate in the nervous system of *Aplysia californica* I. Increased synthesis in response to synaptic stimulation, *J. Gen. Physiol.* **60:**558–569.

Chalfie, M., Settipani, L., and Perlman, R. L., 1979, The role of adenosine 3′:5′-monophosphate in the regulation of tyrosine 3-monooxygenase activity, *Mol. Pharmacol.* 15:263–270.

Chieuh, C. C., and Moore, K. E., 1974, *In vivo* release of endogenously synthesized catecholamines from the rat brain evoked by electrical stimulation and by *d*-amphetamine, *J. Neurochem.* 23:159–168.

Chrapkiewics, N. B., Beale, E. G., and Granner, D. K., 1982, Induction of the messenger ribonucleic acid coding for phosphoenolpyruvate carboxykinase in H4-II-E cells, *J. Biol. Chem.* 257:14428–14432.

Cohen, P., 1973, The subunit structure of rabbit skeletal muscle phosphorylase kinase and the molecular basis of its activation reactions, *Eur. J. Biochem.* 34:1–14.

Cohen, P., Watson, D. C., and Dixon, G. H., 1975, The hormonal control of activity of skeletal muscle phosphorylase kinase, *Eur. J. Biochem.* 51:79–92.

Cohen, P., Burchell, A., Foulkes, J. G., Cohen, P. T. W., Vanaman, T. C., and Nairn, A. C., 1978, Identification of the Ca^{2+}-dependent modulator protein as the fourth subunit of rabbit skeletal muscle phosphorylase kinase, *FEBS Lett.* 92:287–292.

Conti, M. A., and Adelstein, R. S., 1981, The relationship between calmodulin binding and phosphorylation of smooth muscle myosin kinase by the catalytic subunit of 3′:5′ cAMP-dependent protein kinase, *J. Biol. Chem.* 256:3178–3181.

Corbin, J. D., 1983, Determination of the cAMP-dependent protein kinase activity ratio in intact tissues, *Methods Enzymol.* 99:227–232.

Corbin, J. D., Soderling, T. R., and Park, C. R., 1973, Regulation of adenosine 3′,5′-monophosphate-dependent protein kinase. 1. Preliminary characterization of the adipose tissue enzyme in crude extracts, *J. Biol. Chem.* 248:1813–1821.

Corbin, J. D., Sugden, P. H., Lincoln, T. M., and Keely, S. L., 1977, Compartmentalization of adenosine 3′:5′-monophosphate and adenosine 3′:5′-monophosphate-dependent protein kinase in heart tissue, *J. Biol. Chem.* 252:3854–3861.

Culpepper, J. A., and Liu, A. Y.-C., 1983, Pretranslational control of tyrosine aminotransferase synthesis by 8-bromo-cyclic AMP in H-4 rat hepatoma cells, *J. Biol. Chem.*, 258:13812–13819.

Dabrowska, R., Sherry, J. M. F., Aromatorio, D. K., and Hartshorne, D. J., 1978, Modulator protein as a component of the myosin light chain kinase from chicken gizzard, *Biochemistry* 17:253–258.

De Belleroche, J. S., Das, I., and Bradford, H. F., 1974, Absence of an effect of histamine, noradrenaline and depolarizing agents on the levels of adenosine 3′-5′-monophosphate in nerve endings isolated from cerebral cortex, *Biochem. Pharmacol.* 23:835–843.

de Lanerolle, P., Nishikawa, M., and Adelstein, R. S., 1983, Protein phosphorylation and cAMP-levels in forskolin-treated tracheal smooth muscle, *Biophys. J.* 41:152a.

Diamond, J., 1978, Role of cyclic nucleotides in control of smooth muscle contraction, *Adv. Cyclic Nucleotide Res.* 9:327–339.

Drummond, G. I., Harwood, J. P., and Powell, C. A., 1969, Studies on the activation of phosphorylase in skeletal muscle by contraction and by epinephrine, *J. Biol. Chem.* 214:4235–4240.

Ebstein, B., Roberge, C., Tabachnick, J., and Goldstein, M., 1974, The effect of dopamine and apomorphine on db-cAMP-induced stimulation of synaptosomal tyrosine hydroxylase, *J. Pharm. Pharmacol.* 26:975–977.

Edelman, A. M., Raese, J. D., Lazar, M. A., and Barchas, J. D., 1978, *In vitro* phosphorylation of a purified preparation of bovine corpus striatal tyrosine hydroxylase, *Commun. Psychopharmacol.* 2:461–465.

Edelman, A. M., Raese, J. D., Lazar, M. A., and Barchas, J. D., 1981, Tyrosine hydroxylase: Studies on the phosphorylation of a purified preparation of the brain enzyme by the cyclic AMP-dependent protein kinase, *J. Pharmacol. Exp. Ther.* **216**:647–653.

El-Maghrabi, M. R., Claus, T. H., and Pilkis, S. J., 1983, Substrate-directed regulation of cAMP-dependent phosphorylation, *Methods Enzymol.* **99**:581–591.

El-Mestikawy, S., Glowinski, J., and Hamon, M., 1983, Tyrosine hydroxylase activation in depolarized dopaminergic terminals—involvement of Ca^{2+}-dependent phosphorylation, *Nature* **302**:830–833.

Engstrom, L., 1980, Regulation of liver pyruvate kinase by phosphorylation–dephosphorylation, in: *Molecular Aspects of Cellular Regulations*, Vol. 1 (P. Cohen, ed.), Elsevier/North-Holland Biomedical Press, Amsterdam, pp. 11–31.

Foerder, C. A., Martins, T. J., Beavo, J. A., Krebs, E. G., 1982, Induction of *Xenopus laevis* oocyte maturation by cyclic GMP-stimulated phosphodiesterase, *J. Cell. Biol.* **95**:304a.

Frye, R. A., and Holz, R. W., 1983, Phospholipase A_2 inhibitors block catecholamine secretion and calcium uptake in cultured bovine adrenal medullary cells, *Mol. Pharmacol.* **23**:547–550.

Gerthoffer, W. T., and Murphy, R. A., 1983, Ca^{2+}, myosin phosphorylation, and relaxation of arterial smooth muscle, *Am. J. Physiol.* **245**:C271–C277.

Gerthoffer, W. T., Trevethick, M. A., and Murphy, R. A., 1984, Myosin phosphorylation and cyclic adenosine 3′,5′-monophosphate in relaxation of arterial smooth muscle by vasodilators, *Circ. Res.* **54**:83–89.

Goldstein, M., Anagnoste, B., and Shirron, C., 1973, The effect of trivastal, haloperidol and dibutyryl cyclic AMP on [^{14}C]dopamine synthesis in rat striatum, *J. Pharm. Pharmacol.* **25**:348–351.

Goldstein, M., Bronaugh, R. L., Ebstein, B., and Roberge, C., 1976, Stimulation of tyrosine hydroxylase activity by cyclic AMP in synaptosomes and in soluble striatal enzyme preparations, *Brain Res.* **109**:563–574.

Gordon, R., Reid, J. V. O., Sjoerdsma, A., and Udenfriend, S., 1966, Increased synthesis in the rat heart on electrical stimulation of the stellate ganglia, *Mol. Pharmacol.* **2**:610–613.

Gross, S. R., and Mayer, S. E., 1974, Regulation of phosphorylase *b* to *a* conversion in muscle, *Life Sci.* **14**:401–414.

Hallenbeck, P. C., and Walsh, D. A., 1983, Autophosphorylation of phosphorylase kinase: Divalent metal cation and nucleotide dependency, *J. Biol. Chem.* **258**:13493–13501.

Harper, J. F., Wallace, R. W., Cheung, W. Y., and Steiner, A. L., 1981, ACTH-stimulated changes in the immunocytochemical localization of cyclic nucleotides, protein kinases and calmodulin, *Adv. Cyclic Nucleotide Res.* **14**:581–591.

Harris, J. E., and Roth, R. H., 1971, Potassium-induced acceleration of catecholamine biosynthesis in brain slices. 1. A study on the mechanism of action, *Mol. Pharmacol.* **7**:593–604.

Harris, J. E., Morgenroth, V. H. III, Roth, R. H., and Baldessarini, R. J., 1974, Regulation of catecholamine synthesis in the rat brain *in vitro* by cyclic AMP, *Nature* **252**:156–158.

Hayakawa, T., Perkins, J. P., and Krebs, E. G., 1973a, Studies on the subunit structure of rabbit skeletal muscle phosphorylase kinase, *Biochemistry* **12**:574–580.

Hayakawa, T., Perkins, J. P., Walsh, D. A., and Krebs, E. G., 1973b, Physicochemical properties of rabbit skeletal muscle phosphorylase kinase, *Biochemistry* **12**:567–573.

Haycock, J. W., Bennett, W. F., George, R. J., and Waymire, J. C., 1982a, Multiple site phosphorylation of tyrosine hydroxylase: Differential regulation *in situ* by 8-bromo-cAMP and acetylcholine, *J. Biol. Chem.* **257**:13699–13703.

Haycock, J. W., Meligeni, J. A., Bennett, W. F., and Waymire, J. C., 1982b, Phosphorylation and activation of tyrosine hydroxylase mediate the acetylcholine-induced increase in catecholamine biosynthesis in adrenal chromaffin cells, *J. Biol. Chem.* **257**:12641–12648.

Hayes, J. S., Brunton, L. L., and Mayer, S. E., 1980, Selective activation of particulate cAMP-dependent protein kinase by isoproterenol and prostaglandin E_1, *J. Biol. Chem.* **255**:5113–5119.

Horwitz, J., and Perlman, R. L., 1984, Stimulation of DOPA synthesis in the superior cervical ganglion by veratridine, *J. Neurochem.* **42**:384–389.

Huchon, D., Ozon, R., Fischer, E. H., and DeMaille, J. G., 1981, The pure inhibitor of cAMP-dependent protein kinase initiates *Xenopus leavis* meiotic maturation: A 4-step scheme for meiotic maturation, *Mol. Cell. Endocrinol.* **22**:211–222.

Huijing, F., and Larner, J., 1966, On the effect of adenosine 3′,5′cyclophosphate on the kinase of UDGP α-1,4-glucan α-4-glucosyl transferase, *Biochem. Biophys. Res. Commun.* **23**:259–263.

Iuvone, P. M., and Marshburn, P. B., 1982, Regulation of tyrosine hydroxylase activity in retinal cell suspensions: Effects of potassium and 8-bromo-cyclic AMP, *Life Sci.* **30**:85–91.

Iuvone, P. M., Joh, T. H., and Neff, N. H., 1979, Regulation of retinal tyrosine hydroxylase: Long term exposure to light increased the apparent V_{max} without a concomitant increase of immunotitratable enzyme molecules, *Brain. Res.* **178**:191–195.

Iuvone, P. M., Rauch, A. L., Marshburn, P. B., Glass, D. B., and Neff, N. H., 1982, Activation of retinal tyrosine hydroxylase *in vitro* by cyclic AMP-dependent protein kinase: Characterization and comparison to activation *in vivo* by photic stimulation, *J. Neurochem.* **39**:1632–1640.

Joh, T. H., Park, D. H., and Reis, D. J., 1978, Direct phosphorylation of brain tyrosine hydroxylase by cyclic AMP-dependent protein kinase: Mechanism of enzyme activation, *Proc. Natl. Acad. Sci. U.S.A.* **75**:4744–4748.

Jones, L. R., Maddock, S. W., and Besch, H. R., 1980, Unmasking effect of alamethicin on the (Na^+,K^+)-ATPase, β-adrenergic receptor-coupled adenylate cyclase, and cAMP-dependent protein kinase activities of cardiac sarcolemmal vesicles, *J. Biol. Chem.* **255**:9971–9980.

Jungmann, R. A., Lee, S. G., and DeAngelo, A. B., 1975, Translocation of cytoplasmic protein kinase and cyclic adenosine monophosphate-binding protein to intracellular acceptor sites, *Adv. Cyclic Nucleotide Res.* **5**:281–306.

Kaczmarek, L., Jennings, K., Strumwasser, F., Nairn, A., Walter, U., Wilson, F., and Greengard, P., 1980, Microinjection of catalytic subunit of cyclic AMP-dependent protein kinase enhances calcium action potentials of bag cell neurons in cell culture, *Proc. Natl. Acad. Sci. U.S.A.* **77**:7487–7491.

Kakiuchi, S., Rall, T. W., and McIlwain, H., 1969, The effect of electrical stimulation upon the accumulation of adenosine 3′:5′-phosphate in isolated cerebral tissue, *J. Neurochem.* **16**:485–491.

Kandel, E. R., and Schwartz, J. H., 1982, Molecular biology of learning: Modulation of transmitter release, *Science* **218**:433–443.

Keely, S. L., Corbin, J. D., and Park, C. R., 1975, Regulation of adenosine 3′:5′-monophosphate-dependent protein kinase. Regulation of the heart enzyme by epinephrine, glucagon, insulin and 1-methyl-3-isobutylxanthine, *J. Biol. Chem.* **250**:4832–4840.

Keen, P., and McLean, W. G., Effect of dibutyryl-cyclic AMP and dexamethasone on noradrenaline synthesis in isolated superior cervical ganglia, *J. Neurochem.* **22**:5–10.

Kerrick, W. G. L., and Bourguignon, L. Y. W., 1984, Regulation of receptor capping in mouse lymphoma T cells by CA^{2+}-activated myosin light chain kinase, *Proc. Natl. Acad. Sci. U.S.A.* **81**:165–169.

Kerrick, W. G. L., and Hoar, P. E., 1981, Inhibition of smooth muscle tension by cyclic AMP-dependent protein kinase, *Nature* **292**:253–255.

Kerrick, W. G. L., Hoar, P. E., Cassidy, P. S., and Bridenbaugh, R. L., 1981, Skinned muscle fibers: Functional significance of phosphorylation and calcium activated tension, in: *Protein Phosphorylation: Cold Spring Harbor Conferences on Cell Proliferation,* Vol. 8 (O. M. Rosen and E. G. Krebs, Eds.), Cold Spring Harbor Laboratory, Cold Spring Harbor, New York, pp. 887–900.

Khoo, J. C., Sperry, P. J., Gill, G. N., and Steinberg, D., 1977, Activation of hormone-sensitive lipase and phosphorylase kinase by purified cyclic GMP-dependent protein kinase, *Proc. Natl. Acad. Sci. U.S.A.* **74**:4843–4847.

King, M. M., Fitzgerald, T. J., and Carlson, G. M., 1983, Characterization of initial autophosphorylation events in rabbit skeletal muscle phosphorylase kinase, *J. Biol. Chem.* **258**:9925–9930.

Krebs, E. G., 1973, The mechanism of hormonal regulation by cyclic AMP, *Excerpta Medica Int. Cong. Ser.* **273**:17–29.

Krebs, E. G., and Beavo, J. A., 1979, Phosphorylation–dephosphorylation of enzymes, *Annu. Rev. Biochem.* **48**:923–959.

Krebs, E. G., and Preiss, J., 1976, Regulatory mechanisms in glycogen metabolism, in: *Biochemistry of Carbohydrates,* Biochemistry Ser. 1, Vol. 5 (W. J. Whelan, ed.), Butterworths–University Park Press, Baltimore, pp. 337–389.

Krebs, E. G., DeLange, R. J., Kemp, R. G., and Riley, W. D., 1966, Activation of skeletal muscle phosphorylase, *Pharmacol. Rev.* **18**:163–171.

Kuo, J. F., and Greengard, P., 1969, An adenosine 3′,5′-monophosphate-dependent protein kinase from *Escherichia coli, J. Biol. Chem.* **244**:3417–3419.

Langan, T. A., Zeilig, C., and Leichtling, B., 1981, Characterization of multiple-site phosphorylation of H1 histone in proliferating cells, in: *Protein Phosphorylation: Cold Spring Harbor Conferences on Cell Proliferation,* Vol. 8 (O. M. Rosen and E. G. Krebs, eds.), Cold Spring Harbor Laboratory, Cold Spring Harbor, New York, pp. 1039–1052.

Lazar, M. A., Truscott, R. J. W., Raese, J. D., and Barchas, J. D., 1981, Thermal denaturation of native striatal tyrosine hydroxylase: Increased thermolability of the phosphorylated form of the enzyme, *J. Neurochem.* **36**:677–682.

Lazar, M. A., Lockfeld, A. J., Truscott, R. J. W., and Barchas, J. D., 1982, Tyrosine hydroxylase from bovine striatum: Catalytic properties of the phosphorylated and non-phosphorylated forms of the purified enzyme, *J. Neurochem.* **39**:409–442.

Lee, S., and Jungmann, R. A., 1981, Isoproterenol-induced selective phosphorylation *in vivo* of the 214,000 dalton subunit of rat C6 glioma cell RNA polymerase II, *Biochem. Biophys. Res. Commun.* **102**:538–544.

Letendre, C. H., MacDonnell, P. C., and Guroff, G., 1977, The biosynthesis of phosphorylated tyrosine hydroxylase by organ cultures of rat adrenal medulla and superior cervical ganglia, *Biochem. Biophys. Res. Commun.* **74**:891–897.

Levitt, M., Spector, S., Sjoerdsma, A., and Udenfriend, S., 1965, Elucidation of the rate-limiting step in norepinephrine biosynthesis in the perfused guinea-pig heart, *J. Pharmacol. Exp. Ther.* **148**:1–8.

Lewander, T., Joh, T. H., and Reis, D. J., 1977, Tyrosine hydroxylase: Delayed activation in central noradrenergic neurons and induction in adrenal medulla elicited by stimulation of central cholinergic receptors, *J. Pharmacol. Exp. Ther.* **200**:523–534.

Lewis, E. J., Calie, P., and Wicks, W. D., 1982, Differences in rates of tyrosine aminotransferase deinduction with cyclic AMP and glucocorticoids, *Proc. Natl. Acad. Sci. U.S.A.* **79**:5778–5782.

Lincoln, T. M., and Corbin, J. D., 1977, Adenosine 3':5'-cyclic monophosphate- and guanosine 3':5'-cyclic monophosphate-dependent protein kinases: Possible homologous proteins, *Proc. Natl. Acad. Sci. U.S.A.* **74**:3239–3243.

Lloyd, T., and Kaufman, S., 1975, Evidence for the lack of direct phosphorylation of bovine caudate tyrosine hydroxylase following activation by exposure to enzymatic phosphorylating conditions, *Biochem. Biophys. Res. Commun.* **66**:907–913.

Lovenberg, W., Bruckwick, E. A., and Hanbauer, I., 1975, ATP, cyclic AMP and magnesium increase the affinity of rat striatal tyrosine hydroxylase for its cofactor, *Proc. Natl. Acad. Sci. U.S.A.* **72**:2955–2958.

Maller, J. L., 1983, Interaction of steroids with the cyclic nucleotide system in amphibian oocytes, *Adv. Cyclic Nucleotide Res.* **15**:295–336.

Maller, J. L., and Krebs, E. G., 1977, Progesterone-stimulated meiotic cell division in *Xenopus* oocytes: Induction by regulatory subunit and inhibition by catalytic subunit of adenosine 3':5'-monophosphate-dependent protein kinase, *J. Biol. Chem.* **252**:1712–1718.

Manning, D. R., DiSalvo, J., and Stull, J. T., 1980, Protein phosphorylation: Quantitative analysis *in vivo* and in intact cell systems, *Mol. Cell. Endocrinol.* **19**:1–19.

Markey, K. A., Kondo, S., Shenkman, L., and Goldstein, M., 1980, Purification and characterization of tyrosine hydroxylase from a clonal pheochromocytoma cell line, *Mol. Pharmacol.* **17**:79–85.

Masserano, J., and Weiner, N., 1979, Similarities between the *in vivo* activation of adrenal tyrosine hydroxylase and the *in vitro* activation of the enzyme by an adenosine 3',5'-monophosphate dependent protein phosphorylating system, in: *Catecholamines: Basic and Clinical Frontiers*, Vol. 1 (E. Usdin, I. J. Kopin, and J. D. Barchas, eds.), Pergamon Press, New York, pp. 100–102.

Masserano, J. M., and Wiener, N., 1983, Tyrosine hydroxylase regulation in the central nervous system, *Mol. Cell. Biochem.* **53/54**:129–152.

Mayer, S. E., and Krebs, E. G., 1970, Studies on the phosphorylation and activation of skeletal muscle phosphorylase and phosphorylase kinase *in vivo*, *J. Biol. Chem.* **245**:3153–3160.

Meligeni, J., Tank, A. W., Stephens, J. K., Dreyer, E., and Weiner, N., 1981, *In vivo* phosphorylation of rat adrenal tyrosine hydroxylase during acute decapitation stress, in: *Protein Phosporylation: Cold Spring Harbor Conferences on Cell Proliferation*, Vol. 8 (O. M. Rosenthal and E. G. Krebs, ed.), Cold Spring Harbor Laboratory, Cold Spring Harbor, New York, pp. 1377–1389.

Meligeni, J. A., Haycock, J. W., Bennett, W. F., and Waymire, J. C., 1982, Phosphorylation and activation of tyrosine hydroxylase mediate the cAMP-induced increase in catecholamine biosynthesis in adrenal chromaffin cells, *J. Biol. Chem.* **257**:12632–12640.

Miller, J. R., Silver, P. J., and Stull, J. T., 1983, The role of myosin light chain kinase phosphorylation in *beta*-adrenergic relaxation of tracheal smooth muscle, *Mol. Pharmacol.* **24**:235–242.

Miller, P., Walter, U., Theurkauf, W. E., Vallee, R. B., and DeCamili, P., 1982, Frozen tissue sections as an experimental system to reveal specific binding sites for the regulatory subunit of type II cAMP-dependent protein kinase in neurons, *Proc. Natl. Acad. Sci. U.S.A.* **79:**5562–5566.

Miyamoto, E., Petzold, G. L., Harris, J. S., and Greengard, P., 1971, Dissociation and concomitant activation of adenosine 3′,5′-monophosphate-dependent protein kinase by histone, *Biochem. Biophys. Res. Commun.* **44:**305–312.

Morgenroth, V. H. III, Boadle-Biber, M., and Roth, R. H., 1974, Tyrosine hydroxylase: Activation by nerve stimulation, *Proc. Natl. Acad. Sci. U.S.A.* **71:**4283–4287.

Morgenroth, V. H. III, Hegstrand, L. R., Roth, R. H., and Greengard, P., 1975, Evidence for involvement of protein kinase in the activation by adenosine 3′:5′-monophosphate of brain tyrosine 3-monooxygenase, *J. Biol. Chem.* **250:**1946–1948.

Mrwa, U., Troschka, M., and Ruegg, J. C., 1979, Cyclic AMP-dependent inhibition of smooth muscle actomyosin, *FEBS Lett.* **107:**371–374.

Murrin, L. C., Morgenroth, V. H. III, and Roth, R. H., 1976, Dopaminergic neurons: Effects of electrical stimulation on tyrosine hydroxylase, *Mol. Pharmacol.* **12:**1070–1081.

Nagatsu, T., Levitt, M., and Udenfriend, S., 1964, Tyrosine hydroxylase, the initial step in norepinephrine biosynthesis, *J. Biol. Chem.* **239:**2910–2917.

Nathanson, J. A., 1977, Cyclic nucleotides and nervous system function, *Physiol. Rev.* **57:**157–256.

Nimmo, H. G., and Cohen, P., 1977, Hormonal control of protein phosphorylation, *Adv. Cyclic Nucleotide Res.* **8:**145–255.

Nishizuka, Y., Takai, Y., Kishimoto, A., Hashimoto, E., Inoue, M., Yamamoto, M., Criss, W. E., and Kuroda, Y., 1978, A role of calcium in the activation of a new protein kinase system, *Adv. Cyclic Nucleotide Res.* **9:**209–220.

Noguchi, T., Disterhaft, M., and Granner, D., 1982, Evidence for a dual effect of dibutyryl cyclic AMP on the synthesis of tyrosine amino transferase in rat liver, *J. Biol. Chem.* **257:**2386–2390.

Ogreid, D., Doskeland, S. O., and Miller, J. P., 1983, Evidence that cyclic nucleotides activating rabbit muscle protein kinase I interact with both types of cAMP binding sites associated with the enzyme, *J. Biol. Chem.* **258:**1041–1049.

Palmer, W. K., McPherson, J. M., and Walsh, D. A., 1980, Critical controls in the evaluation of cAMP-dependent protein kinase activity ratios as indices of hormonal action, *J. Biol. Chem.* **255:**2663–2666.

Pastan, I., and Adhya, S., 1976, Cyclic adenosine 3′-5′-monophosphate in *Escherichia coli*, *Bacteriol. Rev.* **40:**527–551.

Pato, M. D., and Adelstein, R. S., 1980, Dephosphorylation of the 20,000 dalton light chain of myosin by two different phosphatases from smooth muscle, *J. Biol. Chem.* **255:**6535–6538.

Patrick, R. L., and Barchas, J. D., 1974, Stimulation of synaptosomal dopamine synthesis by veratridine, *Nature* **250:**737–739.

Patrick, R. L., and Barchas, J. D., 1976, Dopamine synthesis in rat brain synaptosomes. II. Dibutyryl cyclic adenosine 3′:5′-monophosphoric acid and 6-methyltetrahydropterine-induced synthesis increases without an increase in endogenous dopamine release, *J. Pharmacol. Exp. Ther.* **197:**97–104.

Pollock, R. J., Kapatos, G., and Kaufman, S., 1981, Effect of cyclic AMP-dependent protein phosphorylating conditions on the pH-dependent activity of tyrosine hydroxylase from beef and rat striata, *J. Neurochem.* **37:**855–860.

Posner, J. B., Stern, R., and Krebs, E. G., 1965, Effects of electrical stimulation and epinephrine on muscle phosphorylase, phosphorylase *b* kinase, and adenosine-3′,5′-phosphate, *J. Biol. Chem.* **240:**982–985.

Raese, J. D., Edelman, A. M., Makk, G., Bruckwick, E. A., Lovenberg, W., and Barchas, J. D., 1979, Brain striatal tyrosine hydroxylase: Activation of the enzyme by cyclic AMP-independent phosphorylation, *Commun. Psychopharmacol.* 3:295–301.

Reimann, E. M., Titani, K., Ericsson, L. H., Wade, R. D., Fischer, E. H., and Walsh, K. A., 1984, Homology of the γ subunit of phosphorylase *b* kinase with cAMP-dependent protein kinase, *Biochemistry* 23:4185–4192.

Riley, W. D., DeLange, R. J., Bratvold, G. E., and Krebs, E. G., 1968, Reversal of phosphorylase kinase activation, *J. Biol. Chem.* 243:2209–2215.

Robinson-Steiner, A. M., and Corbin, J. D., 1983, Probable involvement of both intrachain cAMP binding sites in activation of protein kinase, *J. Biol. Chem.* 258:1032–1040.

Robison, G. A., Butcher, R. W., and Sutherland, E. W., 1971, *Cyclic AMP*, Academic Press, New York, London.

Rosenfield, M. G., and Barrieux, A., 1979, Regulation of protein synthesis by polypeptide hormones and cyclic AMP, *Adv. Cyclic Nucleotide Res.* 10:205–264.

Roth, R. H., Morgenroth, V. H. III, and Salzman, P. M., 1975, Tyrosine hydroxylase: Allosteric activation induced by stimulation of central noradrenergic neurons, *Naunyn-Schmiedebergs Arch. Pharmacol.* 289:327–343.

Ruegg, J. C., and Paul, R. J., 1982, Vascular smooth muscle: Calmodulin and cyclic AMP-dependent protein kinase alter calcium sensitivity in porcine carotid skinned fibers, *Circ. Res.* 50:394–399.

Schlegel, R. A., and Rechsteiner, M. D., 1978, Red cell-mediated microinjection of marcomolecules into mammalian cells, in: *Methods in Cell Biology*, Vol. 20 (D. M. Prescott, ed.), Academic Press, New York, pp. 341–354.

Schlender, K. K., Wei, S. H., and Villar-Palasi, C., 1969, UDP-glucose:glycogen α-4-glucosyltransferase I kinase activity of purified muscle protein kinase: Cyclic nucleotide specificity, *Biochim. Biophys. Acta* 191:272–278.

Sedvall, G. C., and Kopin, I. J., 1967, Acceleration of norepinephrine synthesis in the rat submaxillary gland *in vivo* during sympathetic nerve stimulation, *Life Sci.* 6:45–51.

Siekierka, J., Manne, V., and Ochoa, S., 1984, Mechanism of translational control by partial phosphorylation of the α subunit of eukaryotic initiation factor 2, *Proc. Natl. Acad. Sci. U.S.A.* 81:352–356.

Silver, P. J., and DiSalvo, J. D., 1979, Adenosine 3′:5′-monophosphate-mediated inhibition of myosin light chain phosphorylation in bovine aortic actomyosin, *J. Biol. Chem.* 254:9951–9954.

Silver, P. J., and Stull, J. T., 1982, Regulation of myosin light chain and phosphorylase phosphorylation in tracheal smooth muscle, *J. Biol. Chem.* 257:6145–6150.

Simon, J. R., and Roth, R. H., 1979, Striatal tyrosine hydroxylase: Comparison of the activation produced by depolarization and dibutyryl-cAMP, *Mol. Pharmacol.* 16:224–233.

Singh, T. J., and Wang, J. H., 1977, Effect of Mg^{2+} concentration on the cAMP-dependent protein kinase-catalyzed activation of rabbit skeletal muscle phosphorylase kinase, *J. Biol. Chem.* 252:625–632.

Singh, T. J., Akatsuka, A., and Huang, K. P., 1982, Phosphorylation and activation of rabbit skeletal muscle phosphorylase kinase by a cyclic nucleotide- and Ca^{2+}-independent protein kinase, *J. Biol. Chem.* 257:13379–13384.

Skuster, J. R., Chan, K. F. J., and Graves, D. J., 1980, Isolation and properties of the catalytically active γ subunit of phosphorylase *b* kinase, *J. Biol. Chem.* 255:2203–2210.

Soderling, T. R., Hickenbottom, J. P., Reimann, E. M., Hunkeler, F. L., Walsh, D. A., and Krebs, E. G., 1970, Inactivation of glycogen synthetase and activation of phosphorylase kinase by muscle adenosine 3',5'-monophosphate-dependent protein kinase, *J. Biol. Chem.* **245**:6317–6328.

Stadtman, E. R., Chock, P. B., and Adler, S. P., 1976, Metabolic regulation of coupled covalent modification cascade systems, in: *Metabolic Interconversion of Enzymes, 1975* (S. Shaltiel, ed.), Springer-Verlag, Berlin, Heidelberg, New York, pp. 142–149.

Steiner, A. L., Koide, Y., Earp, H. S., Bechtel, P. J., and Beavo, J. A., 1978, Comparmentalization of cyclic nucleotides and cyclic AMP-dependent protein kinases in rat livers: Immunocytochemical demonstration, *Adv. Cyclic Nucleotide Res.* **9**:691–705.

Stull, J. T., 1980, Phosphorylation of contractile proteins in relation to muscle function, *Adv. Cyclic Nucleotide Res.* **13**:39–93.

Stull, J. T., and Mayer, S. E., 1971, Regulation of phosphorylase activation in skeletal muscle *in vivo*, *J. Biol. Chem.* **246**:5716–5723.

Taneda, M., Izumi, F., and Oka, M., 1974, Effect of dibutyryl adenosine 3',5'-monophosphate on catecholamine synthesis in rat brain cortical slices and isolated vasa deferentia, *Jpn. J. Pharmacol.* **24**:934–936.

Tao, M., 1972, Dissociation of rabbit red blood cell cyclic AMP-dependent protein kinase by protamine, *Biochem. Biophys. Res. Commun.* **46**:56–61.

Ullmann, A., and Danchin, A., 1983, Role of cyclic AMP in bacteria, *Adv. Cyclic Nucleotide Res.* **15**:1–53.

Vaccaro, K. K., Liang, B. T., Perelle, B. A., and Perlman, R. L., 1980, Tyrosine 3-monoxygenase regulates catecholamine synthesis in pheochromocytoma cells, *J. Biol. Chem.* **255**:6539–6541.

Vallet, B., Molla, A., and Demaille, J. G., 1981, Cyclic adenosine 3':5'-monophosphate-dependent regulation of purified bovine aortic calcium/calmodulin-dependent myosin light chain kinase, *Biochim. Biophys. Acta* **674**:256–264.

Vrana, K. E., Allhiser, C. L., and Roskoski, R., 1981, Tyrosine hydroxylase activation and inactivation by protein phosphorylating conditions, *J. Neurochem.* **36**:92–100.

Vulliet, P. R., Langan, T. A., and Weiner, N., 1980, Tyrosine hydroxylase: A substrate of cyclic AMP-dependent protein kinase, *Proc. Natl. Acad. Sci. U.S.A.* **77**:92–96.

Waldeck, B., 1971, Some effects of caffeine and aminophylline on the turnover of catecholamines in the brain, *J. Pharm. Pharmacol.* **23**:824–830.

Walsh, D. A., Perkins, J. P., and Krebs, E. G., 1968, An adenosine 3'-5'-monophosphate-dependent protein kinase from rabbit skeletal muscle, *J. Biol. Chem.* **243**:3763–3765.

Wang, J. H., Stull, J. T., Huang, T. S., and Krebs, E. G., 1976, A study on the autoactivation of rabbit muscle phosphorylase kinase, *J. Biol. Chem.* **251**:4521–4527.

Weber, I. T., Takio, K., Titani, K., and Steitz, T. A., 1982, The cAMP-binding domains of the regulatory subunit of cAMP-dependent protein kinase and the catabolite gene activator protein are homologous, *Proc. Natl. Acad. Sci. U.S.A.* **79**:7679–7683.

Weiner, N., Lee, F.-L., Dreyer, E., and Barnes, E., 1978, The activation of tyrosine hydroxylase in noradrenergic neurons during acute nerve stimulation, *Life Sci.* **22**:1197–1216.

Yamauchi, T., and Fujisawa, H., 1979a, *In vitro* phosphorylation of bovine adrenal tyrosine hydroxylase by adenosine 3':5'-monophosphate-dependent protein kinase, *J. Biol. Chem.* **254**:503–507.

Yamauchi, T., and Fujisawa, H., 1979b, Regulation of bovine adrenal tyrosine 3-monooxygenase phosphorylation–dephosphorylation reaction catalyzed by adenosine 3':5'-monophosphate dependent protein kinase and phosphoprotein phosphatase, *J. Biol. Chem.* **254**:6408–6413.

Yamauchi, T., and Fujisawa, H., 1981, Tyrosine 3-monooxygenase is phosphorylated by Ca^{2+}-calmodulin-dependent protein kinase, followed by activation by activator protein, *Biochem. Biophys. Res. Commun.* **100**:807–813.

Yeaman, S. J., and Cohen, P., 1975, The hormonal control of activity of skeletal muscle phosphorylase kinase: Phosphorylation of the enzyme at two sites *in vivo* in response to adrenalin, *Eur. J. Biochem.* **51**:93–104.

The Homogeneity and Discreteness of Membrane Domains

ZANVIL A. COHN and RALPH M. STEINMAN

1. Introduction

This volume has given us an excellent opportunity to examine the flow and interactions of cellular membranes. It is clear that scientists with diverse interests are now involved in "membranology," and their special expertise adds new dimensions to the field. Endocrinologists, immunologists, physical chemists, and cell biologists are all reviewing their systems in a common forum.

As in much of science, advances in this field have a saltatory quality. New observations are often dependent on a novel technique or a highly specialized target cell. This knowledge will occasionally have general applications.

In this chapter we focus on some of our own pet systems in which we are able to examine both the influx and efflux of plasma-membrane-derived organelles. This occurs in a "professionally" disposed endocytic cell in which these events serve important roles in the physiology of most multicellular organisms. In particular, using labeling techniques and monoclonal antibodies, we have examined some of the major and minor membrane polypeptides in the murine macrophage (Steinman *et al.*, 1983).

2. Membrane Traffic and Cytoplasmic Compartmentalization

It is now clear that exogenous solutes, acid hydrolases, and secretory products are rigidly compartmentalized in membrane-bound vesicles. In

ZANVIL A. COHN and RALPH M. STEINMAN • Laboratory of Cellular Physiology and Immunology, The Rockefeller University, New York, New York 10021.

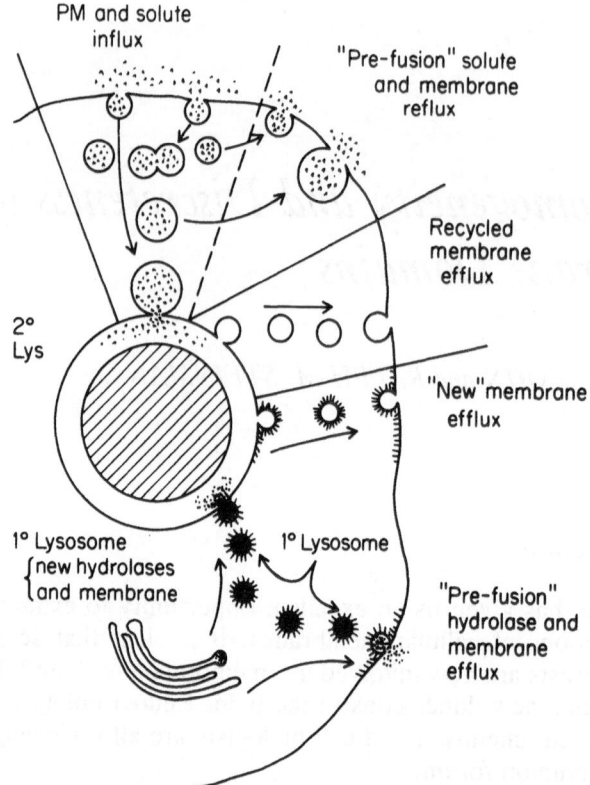

Figure 1. Bidirectional flow of membranes originating either from the Golgi–RER complex or the plasma membrane. Mixing of vesicle contents and membrane polypeptides occurs by fusions that link discrete compartments. (From Cohn and Steinman, 1982.)

the case of endocytosis, environmental molecules and particulates are captured within segments of the plasma membrane as illustrated in Fig. 1. Many of these surface-derived vesicles are initially formed with a cytoplasmic decoration of clathrin, which we discuss in Section 8. However, in a cell such as the macrophage, pinocytosis is a constitutive event with a carefully regulated rate. Under *in vitro* conditions, this cell is taking up a volume of fluid equal to 25% of its total volume each hour. Perhaps even more surprising is the fact that it is interiorizing an equivalent of its entire surface membrane ($\sim 850 \ \mu m^2$) each 33 min. The influx of both fluid and membrane occurs constantly over many days in culture with little change in the total surface area and volume of the whole cell (Steinman *et al.*, 1976).

3. Endocytosis and the Vacuolar Apparatus

We are aware of two pathways for macrophage vesicle contents. First, during the first minutes after generation, contents may be discharged back to the environment after fusion with the plasma membrane in a reflux or short circuit (Besterman *et al.*, 1981). The extent of this pathway could correspond to one-third to two-thirds of pinocytosed contents. The other pathway involves the flow of vesicles into the peri-Golgi zone and fusion with primary and secondary lysosomes. Shortly after its generation and prior to apparent lysosomal fusion, the contents of the vesicle are continually acidified so that the secondary lysosome maintains a pH of 4.6–4.7. Also, prior to lysosomal fusion, vesicle-to-vesicle fusion can occur, resulting in a pinosome with a transiently greater volume.

The two pathways followed by endocytosed contents may represent separate reflux and delivery pathways, or some reflux may be an obligatory component of membrane recycling during delivery. Evidence for a direct short-circuit pathway has been obtained in *Entamoeba histolytica*, where most pinocytosed contents never reach an acidic, perinuclear, or degradative compartment and seem to be regurgitated directly (Aley *et al.*, 1984).

Both the interiorized fluid and surface membrane rapidly enter a lysosomal compartment, which, under steady-state conditions, is maintained at a constant volume and surface area. The secondary lysosomal compartment occupies 2.5% of the volume of the entire cell and contains approximately 18% of the total surface area. It is apparent, therefore, that there must be an extensive and constant efflux of water and electrolytes as well as some mechanism to replenish the plasma membrane.

The plasma membrane, endocytic vesicles, and lysosomes are in constant interaction and form a unit known as the vacuolar apparatus. Each discrete compartment is linked by membrane fusions that in turn mix both their luminal contents and membrane constituents. Under steady-state conditions, the volumes and membrane surface areas of the three components are rather constant. However, with the interiorization of large particles or nondigestible solutes, the secondary lysosome compartment may enlarge with a concomitant decrease in the cell surface area. This may lead to cell rounding. With more prolonged incubation, new membrane will be accreted, and the cell shape changes are reversed (Werb and Cohn, 1972).

4. Membrane Recycling within the Vacuolar Apparatus

These observations strongly suggested that a compensatory flow of membrane to the cell surface was required to maintain total surface area.

For this purpose, we designed labeling procedures that allowed us to specifically iodinate tyrosyl residues of polypeptides on the luminal face of phagolysosomes (Hubbard and Cohn, 1975; Muller et al., 1980a,b). This technique, performed at low temperature to minimize vesicle movement, results in the labeling of approximately two dozen glycoproteins that are discernible on unidimensional (Muller et al., 1980a,b; 1983) or two-dimensional gels. Isotopic grains can then be discerned by electron micrographic autoradiography. This is done initially after low-temperature labeling and then after subsequent incubation at 37°C in pulse-chase-type experiments. The labeling procedure depends on the introduction of polystyrene beads containing a covalently bound coat of lactoperoxidase (Muller et al., 1980a,b). These are rapidly ingested and transferred to an acidified, acid-phosphatase-positive lysosomal compartment. In the presence of hydrogen peroxide and ^{125}I, selective labeling takes place. In addition, the efficiency of labeling can be enhanced tenfold by treating the cells with chloroquine, thereby raising the intralysosomal pH to about 6.0, a value close to the pH optimum of the lactoperoxidase (Muller et al., 1983).

Immediately after labeling, 90% of the grains are over the lysosomes. As the temperature is raised to 37°C, there is a rapid distribution of grains to the cell surface. Within minutes, grains are randomly distributed over the plasma membrane with a concomitant reduction of grains over the lysosomes. This, then, is evidence for a centrifugal flow of membrane components back to the cell surface.

The nature of the transported label was next examined. For this purpose, intralysosomally labeled macrophages were incubated at 37°C until the surface and cytoplasmic labeled compartments reached equilibrium (20 min). Segments of plasma membrane were then captured by promoting the ingestion of very-low-density butadiene beads, which can be separated from LPO-latex beads on density gradients. After homogenization, the butadiene beads were purified, and the membrane polypeptides were solubilized in SDS and displayed on gels. This showed that all the polypeptides labeled in the lysosomal compartment were now present in the plasma membrane surrounding the butadiene beads. This constituted proof that the recycling from the lysosomal compartment encompassed a representative group of tyrosine-containing glycoproteins. From this type of experiment and others that examined the polypeptide composition of pinocytic vesicles (Mellman et al., 1980), it is our conclusion that all membranes of the vacuolar apparatus have considerable homology, are in constant association via fusions, and are flowing in a bidirectional fashion.

Additional experiments (Muller et al., 1983) have compared the membrane polypeptides of secondary lysosomes and plasma membrane by means of monoclonal antibodies. These have indicated that the majority of antigens are present within the intracytoplasmic compartment in amounts similar to their distribution in the plasma membrane. Obviously, these techniques examine only the surface of the plasma membrane and the analogous luminal surface of pinocytic vesicles and secondary lysosomes. Homology of the cytoplasmic face of these structures has not been examined.

5. The Synthetic Compartment

Detailed information concerning the input of acid hydrolases and membranous components into the vacuolar apparatus is scanty at best. Acid hydrolases have been detected in segments of the Golgi saccules and vesicles or GERL, and the apparent fusion of these vesicles with endocytic vacuoles has been suggested. However, these are static observations, and the kinetics and rate-limiting factors are not understood. The lower portion of Fig. 1 suggests the role of the much quoted and seldom observed primary lysosome—as a link from the synthetic pathway. In addition, it coats the vesicles with a fuzzy coat. What it does focus on is the fact that vesicular traffic from the RER-Golgi must carry membrane as well as luminal contents such as acid hydrolases. It is therefore possible that this represents a pathway for the quantal input of vacuolar membrane polypeptides. These in turn, by means of fusions and recycling, would be distributed throughout the secondary lysosomes, endocytic vesicles, and plasma membrane. Both vesicle fusion and peptide diffusion in the plane of the membrane would rapidly homogenize their contents.

6. Surveillance and Reconstitution

Although we propose a rather basic homogeneity of vacuolar membrane polypeptide composition, there are indications of selective enrichment and down-regulation of individual components. As seen in the upper portion of Fig. 2, ligands that are recognized by membrane receptors may be clustered prior to uptake, thereby enriching the content of both receptor and ligand within the confines of the particular vacuole. This is particularly true of multivalent ligand–receptor interactions, as in the Fc-mediated uptake of antibody-coated particles by macrophages: $\geq 50\%$ of the cell surface Fc receptors are occupied prior to uptake. Then both

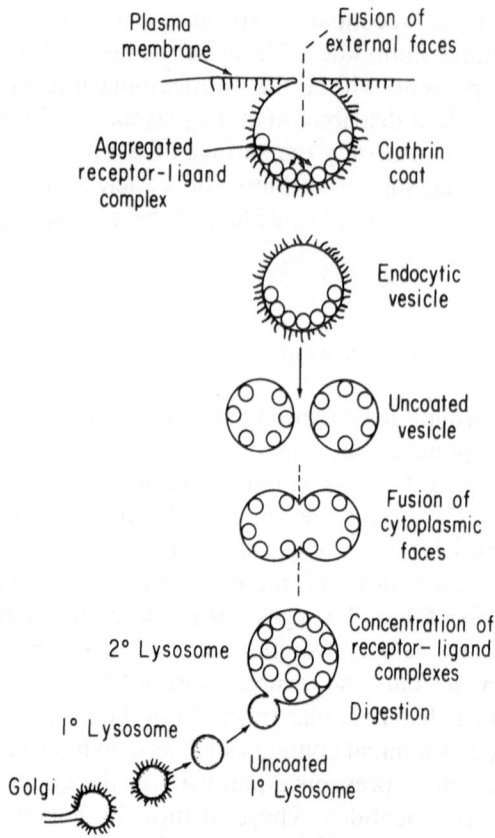

Figure 2. The fusions of coated and uncoated vesicles in the vacuolar apparatus. Coated vesicles arising from the plasma membrane and Golgi saccules are generated by the apposition of their carbohydrate-rich or surface faces. Uncoated vesicle fusion occurs commonly between endocytic vesicles and lysosomes.

receptor and ligand are internalized, delivered to lysosomes, and rapidly degraded, leading to a selective loss of ≥50% of surface Fc receptors (Mellman *et al.*, 1983). Therefore, new receptor synthesis and insertion would be required for continuing ligand surveillance. In contrast, other plasma membrane receptors may be internalized but then continually recycle back to the plasma membrane (Mellman *et al.*, 1980). The macrophage Fc receptor, in the absence of ligand, seems to be internalized and recycled constitutively. In spite of the interest in receptor-mediated uptake, there are still few direct studies of receptor distribution in the presence and absence of ligand. The use of Fab fragments of high-affinity monoclonal antireceptor antibodies would seem ideal in this regard.

The resynthesis of receptors after "down-regulation" is a useful method of following membrane dynamics. Many questions exist. For example, if membrane components are inserted in quanta, do new receptors appear in conjunction with and at the same rate as other membrane polypeptides, or are receptors selectively produced, transported, and inserted? One would expect that in the former case, the rate of receptor appearance would approximate the normal turnover rate of both receptor and other membrane polypeptides.

7. The Generation of Vesicles: Membrane Faces

The fusion of vesicles serves to link the discontinuous compartments of the vacuolar and Golgi apparatus, allowing bidirectional flow. Such membrane-to-membrane fusion is also necessary for the generation of vesicles from the plasma membrane, the ends of Golgi saccules, and presumably from the secondary lysosomes during the recycling process. Although these events appear similar in certain respects, they are in fact quite dissimilar when one considers the sidedness of the structures involved. For example, the external or surface (s) face of the plasma membrane comes together and fuses to form the pinocytic vesicle. On the other hand, fusion of endocytic vesicles with one another and with secondary lysosomes involves the apposition of the cytoplasmic (c) faces of the vesicles. Surface-to-surface (s–s) fusion brings together those highly decorated, carbohydrate-rich segments of the transmembrane proteins. This is illustrated schematically in Fig. 3A. In contrast, c–c fusion would see a carbohydrate-poor domain.

We speculate that the Golgi-derived primary lysosome of Fig. 2 would have a configuration similar to the plasma membrane. One could imagine that intrinsic membrane polypeptides have already been inserted through the bilayer at the level of the ribosome-studded rough endoplasmic reticulum. In the area of the Golgi, enzymes in the lumen would decorate these proteins appropriately, leading to the complete glycoproteins destined for the vacuolar apparatus and plasma membrane. To generate a vesicle from Golgi saccules would, in the sense of membrane sidedness, be similar to the formation of a pinocytic vesicle at the cell surface: the apposition of carbohydrate-rich domains.

Another example would be the "recycling vesicle" generated from the secondary lysosome and destined for the cell surface. Since the luminal face of the lysosome is derived from the cell surface (s–s of Fig. 3B), glycoprotein-rich areas would come together at the time of fusion.

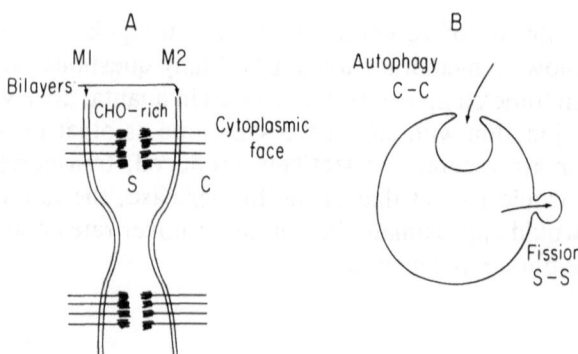

Figure 3. Schematic aspects of membrane fusion. A: The apposition of two carbohydrate-rich membrane domains. In this speculative view, transmembrane glycoproteins are aggregated, allowing closer contact of the lipid bilayers. B: Examples of vesicle generation from the secondary lysosome. In the upper pole, membrane invagination leads to the formation of an autophagic vacuole through the contacts of cytoplasmic faces (c–c). In contrast, the formation of a recycling vesicle would require close contact of the luminal or s–s faces.

In contrast, the formation of an autophagic vacuole (Fig. 3B) would lead to the apposition of cytoplasmic faces (c–c).

8. Membrane Coats and the Generation of Vesicles

There is evidence in the macrophage that large segments on the cytoplasmic surface of the plasma membrane are lined with clathrin (Aggeler *et al.*, 1982). Many pinocytic vesicles, as they are forming, demonstrate "coats," and even large phagosomes have clathrin on portions of the forming vacuole. In each instance, clathrin is associated with the forming vesicle (Fig. 2, top) and, soon after vesicle generation, is lost from this structure. Another area in which "coats" are prominent is on the budding vesicles of the Golgi saccules (Fig. 2, bottom). Again, soon after their generation, coats are lost from the small vesicles of the peri-Golgi zone. Whether this coat is identical to that of the forming endocytic vesicle is not clear.

We have, then, an association of transient vesicle coats with the process of vesicle generation and apposition of surface (s–s) membranes. Perhaps for the efficient interaction and fusion of membrane bilayers, as illustrated in Fig. 3A, a mechanism is necessary for the clustering of membrane glycoproteins and the exposure of the bilayer. Clathrin may be involved in this process, and this implies an energy-consuming reaction in which the clathrin basketry is assembled and leads to the aggregation of glycoproteins via their transmembrane domains. In contrast, the fusion of the cytoplasmic, carbohydrate-poor face of the membrane occurs without the apparent presence of a fuzzy coat or clathrin.

9. Other Pathways

We have omitted from consideration many questions that relate to the synthesis and traffic of membrane polypeptides and vesicle contents. Starting from the rough-surfaced endoplasmic reticulum, polypeptides are inserted into the membrane of the ER and begin a centrifugal journey leading to more distal sites in the cell. Obviously, mechanisms are required for sorting membrane components and contents for extracellular secretion or maintenance within the vacuolar apparatus. How the Golgi is compartmentalized, the concentrating mechanisms within it, and the eventual sorting of products all remain largely unknown. Other chapters in this volume address some of these important questions.

References

Aggeler, J., and Werb, Z., 1982, Initial events during phagocytosis by macrophages viewed from outside and inside the cell membrane–particle interactions and clathrin, *J. Cell Biol.* **94**:613–623.

Aley, S. B., Cohn, Z. A., and Scott, W. A., 1984, Endocytosis in *Entamoeba histolytica*. Evidence for a unique non-acidified compartment, *J. Exp. Med.* **160**:724–737.

Besterman, J. M., Airhart, J. A., Woodworth, R. C., and Low, R. B., 1981, Exocytosis of pinocytosed fluid in cultured cells: Kinetic evidence for rapid turnover and compartmentation, *J. Cell Biol.* **91**:716–727.

Cohn, Z. A., and Steinman, R. M., 1982, Phagocytosis and fluid phase pinocytosis, *Ciba Found. Symp.* **92**:15–28.

Hubbard, A. L., and Cohn, Z. A., 1975, Externally disposed plasma membrane proteins. II. Metabolic fate of iodinated polypeptides of mouse L cells, *J. Cell Biol.* **64**:461–479.

Mellman, I. S., Steinman, R. M., Unkeless, J. C., and Cohn, Z. A., 1980, Selective iodination and polypeptide composition of pinocytic vesicles, *J. Cell Biol.* **86**:712–722.

Mellman, I. S., Plutner, H., Steinman, R. M., Unkeless, J. C., and Cohn, Z. A., 1983, Internalization and degradation of macrophage Fc receptors during receptor-mediated phagocytosis, *J. Cell Biol.* **93**:887–895.

Muller, W. A., Steinman, R. M., and Cohn, Z. A., 1980a, The membrane proteins of the vacuolar system. I. Analysis by a novel method of intralysosomal iodination, *J. Cell Biol.* **86**:292–303.

Muller, W. A., Steinman, R. M., and Cohn, Z. A., 1980b, The membrane proteins of the vacuolar system. II. Bidirectional flow between secondary lysosomes and plasma membrane, *J. Cell Biol.* **86**:304–314.

Muller, W. A., Steinman, R. M., and Cohn, Z. A., 1983, The membrane proteins of the vacuolar system. III. Further studies on the composition and recycling of endocytic vacuole membrane in cultured macrophages, *J. Cell Biol.* **96:**29–36.

Steinman, R. M., Brodie, S. E., and Cohn, Z. A., 1976, Membrane flow during pinocytosis. A stereologic analysis, *J. Cell Biol.* **68:**665–687.

Steinman, R. M., Mellman, I. S., Muller, W. A., and Cohn, Z. A., 1983, Endocytosis and the recycling of plasma membrane, *J. Cell Biol.* **96:**1–27.

Werb, Z., and Cohn, Z. A., 1972, Plasma membrane synthesis in the macrophage following phagocytosis of polystyrene latex particles, *J. Biol. Chem.* **247:**2439–2446.

Internalization and Processing of Peptide Hormone Receptors

PEDRO CUATRECASAS

1. Receptor-Mediated Endocytosis

There are two major mechanisms of a general nature by which macro-molecules can gain entrance into intact, living cells (Pastan and Willing-ham, 1981a). One of these is generally referred to as fluid-phase endo-cytosis, which is a nonconcentrative, nonadsorptive mechanism that internalizes anything that happens to be in the surrounding medium of the cell. There are two basic organelles and pathways generally recognized for this process. In one of these, very small invaginations of the cell membrane at the surface pinch off and internalize as very small endocytic vesicles. These vesicles go directly to lysosomes, generally reaching them within minutes, where they fuse and dislodge their contents into these lysosomes. The other general process involves the fusing of large ruffles of the cell membrane, thus trapping into very large vesicles quite substantial quantities of the exterior medium. These large vacuoles or vesicles again travel rapidly to the area of lysosomal structures, and on impact these appear to shatter into many small vesicles, which subsequently fuse with the lysosomes.

The other, totally different mechanism by which external macrom-olecules can traverse cell membranes is called receptor-mediated endocytosis, and, unlike the first, this process is a selective, concentrative, or adsorptive one that involves considerable selectivity of the external macromolecules to be internalized (Cuatrecasas, 1982; Herzog, 1981; Cuatrecasas and Roth, 1983; King and Cuatrecasas, 1981; Pastan and Wil-

PEDRO CUATRECASAS • Wellcome Research Laboratories, Burroughs Wellcome Company, Research Triangle Park, North Carolina 27709.

lingham, 1981a,b; Brown *et al.*, 1983). The general mechanism of receptor-mediated endocytosis involves the participation of coated pits, so called because of their characteristic coating or lining with a protein called clathrin. With the major exception of low-density lipoprotein, receptors for external ligands and proteins appear to be distributed diffusely and evenly on the cell surface. On binding of a ligand, the receptor–ligand complex is localized selectively to a coated pit, whereupon it is subsequently found quite rapidly within endosomes or receptosomes. There is some controversy as to whether these endosomes result from pinching off of the coated pits, forming coated vesicles, which subsequently become smooth-walled endosomes by shedding their clathrin coat, which then returns to the cell surface, or alternatively by invagination from the coated pit of membrane-containing vesicles without any clathrin protein.

In many instances, the ligand is delivered to lysosomes, where it is degraded by proteolysis. In most cases, the receptor for the ligand appears to be returned to the cell surface, where it is reutilized. In some cases, such as with low-density lipoprotein and asialoglycoprotein, the entrance and exit of the receptor itself is so fast that there is no decrease in the total number of receptors on the cell surface on binding even large quantities of the ligand; thus, the recycling occurs extremely rapidly and is not rate limiting.

Generally, when cells are exposed at 37°C to a receptor-bound protein such as epidermal growth factor, insulin, low-density lipoprotein, and others, the hormone–receptor complex can be localized to a coated pit within the first minute or two. It is not necessary to invoke a selective process for directing the occupied receptor into coated pits, since the diffusion constant of most receptors measured on cell membranes is fast enough to predict that even when the receptors are empty they will enter and leave coated pits at least two or three times each minute. What is needed, therefore, is that one evoke a process by which the receptor, once occupied with ligand, does not leave the coated pit once it arrives there so long as it is in the form of the liganded complex. In this way, a coated pit could rapidly accumulate 20 to 30 receptor–ligand complexes, which is about the number generally felt to be present in a coated pit.

At 37°C, the ligand–receptor complexes are observed in smooth-lined endosomes within 5 min. Subsequent to this, the endosomes appear to migrate to the Golgi region, where they subsequently become associated with either microtubular processes at or near the coated, clathrin-containing regions of the Golgi (Willingham *et al.*, 1984). Recent data suggest that sorting out of endosomal contents first occurs soon after endocytosis, just before arrival at the Golgi region but after dissociation of the ligand–receptor complex. The Golgi area is where there may be segregation of vesicles into those that will be returned to the cell surface or go to another region of the cell and those that are destined to be localized in lysosomes. The former are found in the microtubular region, and the latter appear to associate with the coated regions of the Golgi.

In many cases, the function or purpose of receptor-mediated endocytosis is quite clear and elegant. For example, with low-density lipoprotein, this is a mechanism by which cholesterol can be carried in the blood in a water-soluble form inside of a large protein complex and delivered to lysosomes, where the protein is degraded and the cholesterol is delivered to the cytosol to affect the biosynthetic pathway for cholesterol; thus, it is possible to fine-tune the regulation of cholesterol biosynthesis by this type of feedback process. Furthermore, in this case, clinical diseases of a genetic nature are known that express defects in this receptor-processing mechanism and simultaneously reflect abnormal and pathological consequences. For example, in one type of disorder, there is a marked reduction in the total number of receptors for low-density lipoprotein, whereas in another the receptors are present but do not localize to coated pits regardless of the presence of ligand, and in both cases there is associated serum hypercholesterolemia and premature atherosclerosis with consequent coronary and other arteriovascular disease. Similarly, for systems such as asialoglycoprotein, in which serum proteins are targeted for degradation because of a particular, specific change in their carbohydrate composition, the molecules are selectively delivered to lysosomes for this to occur. There are a number of other transport proteins for which the purpose is very clear and the mechanism well adapted; this is the case, for example, for yolk proteins such as phosvitin and lipovetellin (Cuatrecasas and Roth, 1983) and for vitamin-binding proteins such as transcobalamin and transferrin.

It should be noted that the precise mechanism of endocytosis is probably adjusted to the specific needs of the particular system, and probably very few processes would be completely identical in all details. For example, in the endocytosis of maternal immunoglobulin from the intestine or the transport of immunoglobulin A in hepatic cells, the purpose is not to deliver the protein to a particular cell organelle but rather to deliver it from one surface of the cell across the interior to another surface of the cell (i.e., transcellular transport) (Schiff *et al.*, 1984). In the case of the intestine, the objective is for the immunoglobulin to go from the luminal membrane surface to the basolateral surface of the cell and into the bloodstream, whereas in the liver cell, it must traverse from the vascular portion of the cell across the sinusoidal membrane for delivery of immunoglobulin A to the bile. Also unique for these systems is the fact that the receptor, on reaching the surface of its ultimate destination, appears not to be reutilized by recycling back to the original site but, rather, is released into the external medium by proteolysis, presumably so it can be reutilized at that particular region.

2. Endocytosis of Hormones

Unlike the examples cited above, in the case of mitogenic or transforming proteins or hormones that cause or induce differentiation, as well as some other peptide hormones that act through adenylate cyclase or yet unknown mechanisms, the particular role of and function resulting from receptor-mediated endocytosis are not clear (Cuatrecasas, 1982, 1983a; King and Cuatrecasas, 1981). In some cases, such as epidermal growth factor, the receptors in cultured cells do not appear to recycle once endocytosed, and the complex is instead delivered to lysosomes where both the ligand and the receptor are degraded (Cuatrecasas, 1982; King and Cuatrecasas, 1981). Consequently, there is a significant decrease in the number of surface receptors. This generally occurs and is maximal within 1 hr, and the receptors can fall to a level of approximately 20% of the original complement. The process is generally referred to as "downregulation," and the obvious "purpose," that of decreasing the number of receptors to decrease the sensitivity to subsequent hormone exposure, has not been proved despite its attractive nature. Such an explanation is also not very convincing when the data indicate that only a very small proportion of the total number of surface receptors need normally be occupied to induce a maximal biological effect.

In the case of insulin, the receptors may be endocytosed and degraded or returned intact to the surface depending on the conditions and the cell type examined (Jacobs and Cuatrecasas, 1983). In most other cases, the receptors appear to be primarily recycled or to be degraded at a very slow rate that does not exceed the rate of new receptor synthesis. In the case in which the receptors are degraded, the new steady state is dependent on new protein and receptor synthesis and reinsertion into the cell surface.

There is a distinct possibility that in the case of peptide hormones the endocytosis is related to the transfer of information through the delivery of structural components to intracellular organelles. For example, in the case of insulin, it is possible that this could be a mechanism for delivering the insulin molecule to nuclear receptors or for delivering receptors themselves to regions where they might affect certain specific functions. In the case of EGF and perhaps other hormones, it may be the delivery of a peptide fragment derived from proteolytic digestion of a receptor that serves as an intracellular signal for control of gene expression or protein synthesis (Cuatrecasas, 1983b; Fox and Das, 1979).

3. Acidic Nature of Endosomal Compartments: Consequences

One of the characteristic features of the endosome or receptosome, as well as all distal endocytic compartments including the lysosome, is their acidic nature. In addition to permitting the selective function of lysosomal enzymes, the acidic nature of the lysosome also plays other important roles such as dissociating ligand–receptor complexes. For example, in the case of insulin or epidermal growth factor, the binding is highly pH sensitive so that the complex dissociates at a pH of 6. In fact, this sharp pH dependence of binding is generally the rule rather than the exception with most ligand–receptor complexes. By this mechanism, it is possible to subsequently segregate the direction of transfer of the membrane components (receptor) from the internal protein (hormone or protein) to be degraded. A notable exception to this is transferrin, which utilizes a unique mechanism to achieve its biological function, the delivery of iron to intracellular sites (Karin and Mintz, 1981; Brown *et al.*, 1983; Dautry-Varsat *et al.*, 1983; Klausner *et al.*, 1983; May *et al.*, 1984). In this case, the iron–transferrin complex binds to surface receptors and is internalized into acidic endosomes. As a consequence of this fall in pH, the iron dissociates, but the apotransferrin–receptor complex remains intact, whereupon it is rapidly returned to the cell surface. At the cell surface, the apotransferrin dissociates rapidly since its affinity for the receptor at neutral pH is very low. In contrast, the affinity of the apo- or non-iron-containing transferrin for the receptor is very high at relatively acidic pH values.

In certain cases, the acidic endosomal system is opportunistically usurped to achieve nonphysiological but biochemically effective objectives. For example, it appears that the transport of the A subunit of diphtheria toxin into the cell interior depends on the acidic environment of the endosomal system (Cuatrecasas and Roth, 1983; Sandvig and Olsnes, 1981). The intact diphtheria toxin, which itself has no inherent biological or enzymatic activity, appears first to bind to a surface receptor and subsequently to be internalized, where in an acidic environment the protein undergoes proteolytic cleavage into the A and B chains; the B chain is that region responsible for recognizing and binding to the receptor, whereas the A chain is the enzymatic or catalytically active moiety. Once inside a vesicle of low pH, it appears that the B subunit, which has strong hydrophobic properties, spontaneously penetrates through the membrane, forming a pore through which the hydrophilic A chain traverses as if it were an ionophore selectively made to transfer the peptide from one side of the membrane to the other. Once inside the cytosol, the A chain can inactivate the elongation factor P_2 through ADP-ribosylation. The key condition is the low pH, which drives transport of the peptide from one side of the membrane to the other.

Predictably, it is possible to block the action of diphtheria toxin by chemicals such as methylamine, chloroquine, and others that neutralize or effectively prevent acidification of the endocytic compartment. Similarly, the entrance into endosomal systems can be circumvented artificially if the nicked or cleaved toxin is added to the cell so that is binds to surface receptors and there is a subsequent exposure of the entire cell to pH values in the region of 6 for very short time periods. By this procedure, the A chain is actively transported across the surface membrane, where it gains access to the cytoplasm.

Certain coated viruses such as influenza and VSV also appear to require an acidic membrane compartment for penetration of cell membranes (Cuatrecasas and Roth, 1983). In these cases, the viruses apparently bind to coated pits, from which they are internalized into endosomal compartments, where in the presence of an acidic environment there is a spontaneous fusion of the viral membrane with the cellular membrane, thus emptying the contents of the virion into the cell cytoplasm. Compounds that neutralize the acidic nature of lysosomes and other intracellular membrane components can block the action and duplication of such viruses. Amantadine, a known antiinfluenza drug, has this effect, and it can be shown to inhibit the degradation and update of radioactively labelled VSV (Schlegal *et al.*, 1982).

4. Endocytosis and Hormone Receptor Activation

It is very possible that there are some analogous physiological processes that utilize the acidic endosomal environment to provide some chemical signal associated with receptor activation. For example, it has recently been reported that epidermal growth factor bound to cell surface receptors, after a period of activation, can, after exposure to external acidic conditions, lead to a mitogenic response 20 hr later in the absence of exogenously added hormone (King and Cuatrecasas, 1982). Under usual conditions, the peptide hormone must be present continually for at least 8 to 10 hr to obtain the mitogenic or DNA-stimulating response. Consistent with this, these biological effects of epidermal growth factor can be blocked with compounds that neutralize endosomal systems such as methylamine and dansylcadaverine but not by compounds such as leupeptin that simply inhibit lysosomal proteases. It appears, however, that even in the presence of methylamine, hormone added to the cell can bypass the endosomal system and thus be active if the intact cell is artifically exposed for short periods of time to acidic buffers.

These data suggest the possibility, originally proposed by Fox and his colleagues a few years ago (Fox and Das, 1979), that a proteolytic fragment of the EGF receptor may be released subsequent to receptor activation and that this peptide in the cytosol conveys biological messages (Cuatrecasas, 1983b). In this peptide signal hypothesis, one might speculate that the hormone–receptor complex leads to proteolytic nicking or cleavage without dissociation of the fragment of the receptor. When the complex is then internalized into an acidic compartment, however, there would be dissociation of the complex and release of the peptide inside the cells. It is logical to suggest that the nicking occurs on the cytoplasmic or cellular part of the receptor, which is known to be transmembrane (Cuatrecasas, 1983b). It is also this region that is known to be phosphorylated and to contain kinase activity, making this region even more attractive as a reservoir for providing a unique chemical message in the form of a peptide. The nature of the signal for such proteolysis is of course not known, although one might speculate that receptor aggregation or dimerization might be required. It is interesting, and provocative, that the receptor for epidermal growth factor is generally found in two forms, of about 150,000 and 170,000 molecular weight, with the 150,000 being a degradation product of the 170,000. It is also known that a region of the native receptor is susceptible to proteolytic cleavage to release an approximately 20,000-mol.-wt. component.

5. Other Mechanisms in Receptor Activation

The actions of a hormone such as epidermal growth factor are extremely diverse and include some long-term consequences such as mitogenesis or stimulation of cell growth or, in other systems, stimulation of differentiation. Some very short-term, immediate effects can also be detected, such as stimulation of breakdown of phosphatidylinositol and increased transport of glucose, amino acids, and other nutrients. The general pleiotypic effects include the general stimulation of protein and RNA synthesis as well as the induction of some specific enzymes such as ornithine decarboxylase. With such diverse biological effects, it is not likely that everything can be explained on the basis of endocytosis. The speculations described above are particularly attractive for explaining some of the long-term effects, which require the continual exposure of cells to the hormone. There are other known biochemical activities associated with this and other hormone receptors, as well as changes induced by ligands and their binding, that must be taken into account and considered as mechanisms to explain the generation of at least certain of these biological activities. These include the receptor-associated tyrosine protein kinase and the calcium- and phospholipid-dependent protein kinase C activities, which are described next, as well as the unique role of receptor aggregation and dimerization, which are discussed in Section 8.

6. Receptor Tyrosine Kinase Activity

Recently, receptor phosphorylation has received much attention as a possible mechanism for transmembrane signaling (Kolata, 1982; Sefton et al., 1980; Rosen et al., 1983). Several mitogenic hormones including epidermal growth factor (Cohen et al., 1980), platelet-derived growth factor (Denton et al., 1981; Cooper et al., 1982), insulin (Zick et al., 1983a,b), and somatomedin C (Kull et al., 1983) appear to induce a tyrosine-specific protein kinase that is stimulated by hormone binding and is capable of autophosphorylation of these hormone receptors. In the case of isolated or purified insulin and epidermal growth factor receptors, the kinase activity can also be expressed using other substrates, and the possibility arises that this tyrosine-specific phosphorylation capacity in vivo be expressed on other important and selective proteins. This mechanism has some intriguing resemblance to the activation of certain tyrosine-specific kinases by products of oncogenes such as that of the sarc gene. Although with isolated or purified receptors, the phosphorylation appears to be fairly well restricted to tyrosine residues, phosphorylation stimulated by hormones in intact cells appears to occur on serine and threonine as well as tyrosine. The mechanism and significance of this nontyrosine phosphorylation are not known.

7. Protein Kinase C, Phorbol Esters, Polyinositides, and Receptors

The receptors for epidermal growth factor in insulin also appear to be specifically phosphorylated in intact cells by protein kinase C (Niedel et al., 1983), an enzyme of recent increased interest because of its apparent activation by phorbol esters. This enzyme is a calcium- and phospholipid-requiring enzyme that is activated by diacylglycerol or by phorbol esters and phosphorylates serine and threonine residues of proteins (Kawahara *et al.*, 1980; Takai *et al.*, 1981). There is evidence that this may be the receptor or target for the tumor-promoting phorbol esters (Castagna *et al.*, 1982; Niedel *et al.*, 1983; Sharkey *et al.*, 1984). In platelets, the protein kinase C activity has been related to the specific and major phosphorylation of a 40,000k protein whose function is not known, as well as (transiently) a region of the 20,000-mol.-wt. myosin light chain (Lapetina and Siegel, 1983; Lapetina and Siess, 1983; Kawahara *et al.*, 1980; Takai *et al.*, 1981; Kaibuchi *et al.*, 1983; Siess *et al.*, 1983). The enzyme is activated as a consequence of stimulation and activation of various receptors such as by thrombin, arachidonic acid, and platelet-activating factor. Presumably, these actions are a consequence of the production of diacylglycerol.

Recently, as mentioned elsewhere in this volume, several stimuli such as thrombin, serotonin, and thyroid-stimulating hormone appear to activate extremely rapidly (within seconds) a phosphodiesterase or phospholipase C that is specific for phosphatidylinositol 4,5-P_2. Activation of this enzyme leads to the production of diacylglycerol, which can stimulate protein kinase C, and which is subsequently converted quite rapidly to phosphatidic acid. Phosphatidic acid, in turn, may have some activities of its own such as stimulating phospholipases of the A_2 variety, and it is of course also converted, perhaps quite quickly, back to phosphatidylinositol by conjugation with CTP.

The other product of phosphodiesterase hydrolysis, inositol triphosphate (IP_3), appears to be an extremely potent and selective agent for the release of calcium from the endoplasmic reticulum. Thus, IP_3 is a good candidate for a second messenger in situations in which the cellular stimulus is one that leads to the mobilization of calcium and to activation of calcium-dependent processes.

Recently, the findings of Rovera *et al.* (1982) that phorbol esters can rapidly decrease the number of receptors for transferrin on the surface of HL60 promyelocytic leukemia cells have been confirmed (May *et al.*, 1984). This effect is caused by a decreased number of receptors rather than by an effect on affinity and is explained by an apparent translocation of the receptors to an intracellular compartment (May *et al.*, 1984). This sequestration of the receptor is accompanied by stimulation of receptor phosphorylation by approximately ten-to 20-fold (by phorbol esters). The phosphorylation occurs on threonine and serine residues, and it occurs rapidly and simultaneously with the loss of receptors. This generally is more than 50% complete within 15 min. The continued presence of the phorbol ester, importantly, induces the cells to differentiate into monocytic, macrophage-type cells, and this transition is complete in approximately 12 hrs; by this time, the transferrin receptor has virtually disappeared from the cell surface. Receptor phosphorylation does not occur with the ligand transferrin, nor does the ligand inhibit the phorbol-ester-induced phorphorylation. Thus, the possible role of phosphorylation appears to be unrelated to the usual cycling mechanism of the transferrin-receptor complex, which is initiated by ligand binding.

It is intriguing to speculate that the effects of phorbol esters on inducing cellular differentiation might result from the initial changes in the status of phosphorylation of the transferrin receptor or perhaps other surface proteins. It is possible that a specific region of a receptor, when phosphorylated, could control or exert an influence on the processes related to endocytosis and/or recycling back to the plasma membrane following such internalization. Thus, a likely specific role for the phorbol-ester-enhanced phosphorylation of the transferrin receptor may be as a signal for receptor internalization and intracellular sequestration in a manner that regulates the proportion of the receptor that is expressed at the cell surface.

Importantly, the receptors for insulin, somatomedin C, and EGF also appear to be phosphorylated on serine and threonine residues by phorbol esters (Jacobs *et al.*, 1983). In the case of insulin and EGF, this effect of phorbol esters is accompanied by a very major decrease in the affinity of the receptor for the hormone (Grunberger and Gorden, 1982; Shoyab *et al.*, 1979; Thomopoulas *et al.*, 1982); this is in contrast with the effect described above on transferrin receptors. The phosphorylation of the insulin and somatomedin C receptors appears to be on the β subunit and is not caused by phorbol esters that are devoid of tumor-promoting activity (Jacobs *et al.*, 1983). The physiological consequences of such protein kinase C-dependent phosphorylation of these receptors are not known but might be related to the general processes described above for transferrin; i.e., the relative proportion of receptor present on the cell surface versus that which is sequestered inside the cell may be altered under influences that regulate this kind of phorphorylating activity.

8. Receptor Aggregation and Dimerization

There are considerable data suggesting that receptor microaggregation, dimerization, or some related type of cross linking may be important in the process by which activation occurs. In the case of epidermal growth factor, a synthetic chemical derivative was produced on treatment with cyanogen bromide that showed retention of receptor binding but was virtually without biological activity on DNA synthesis (Shechter *et al.*, 1979b). If the cyanogen bromide EGF is added to cells and subsequently bivalent (but not monovalent Fab fragments) antibodies are added, then it is possible to reconstitute the biological activity of epidermal growth factor. With fluorescently labeled derivatives of EGF treated with cyanogen bromide, it is possible to show that instead of retaining diffuse distribution of labeling (even after 2 hr at 37°C) after cross linking with anti-EGF antibodies, there was rapid formation of patches and endocytosis. It is likely that the primary defect, at least that observed by microscopy, is one that prevents internalization or endocytosis of the hormone–receptor complex.

Cells have been prepared in tissue culture (after many passages) that demonstrate a normal receptor number for EGF but a diminished biological response. In such cells, the addition of anti-EGF antibody can restore biological activity and increase by a remarkably large factor the sensitivity of the cells to the hormone (Shechter *et al.*, 1979a). Under conditions in which the half-maximal response is obtained with 1 ng/ml of EGF, in the presence of antibody the curve can be shifted by two to three orders of magnitude to lower concentrations. In addition, in the case of EGF as well as with insulin, monoclonal antibodies directed against the receptor itself have been shown to possess full biological activity when in a divalent state. Although Fab fragments do not show biological activity, when these are cross linked with anti-Fab antibodies, the biological activity is observed again.

Interestingly, these monoclonal antireceptor antibodies also cause receptor phosphorylation. These experiments of course also indicate that it is possible to observe the full complement of biological activity in the absence of any hormone at all, indicating that the basic information is present totally within the structure of the receptor rather than the ligand. It is likely that the hormone molecules simply act to activate the receptor and to allow expression by the receptor of its informational potential. In addition to the cross linking described above, similar insulinlike effects have been observed with plant lectins such as concanavalin A and wheat germ agglutinin, which are capable of cross linking cell surface glycoproteins.

The results described above do not necessarily mean that hormones such as insulin act to actively recruit diffuse, nonaggregated receptors into an active state of the aggregated variety (Cuatrecasas, 1983a, 1985). In a given cell there may be an equilibrium between single and aggregated or activated groups of receptors that are ready to be triggered on binding of the hormone. In such a situation, the hormone, when bound to isolated or single receptors, would either bind with low affinity or would not cause activation; activation would occur only when the hormone perturbs a preaggregated state of the receptor. In the case of external cross-linking reagents such as antibodies and plant lectins, the situation would be somewhat different in that diffuse or single, nonaggregated receptors would be actively recruited into aggregates or patches, which would be simultaneously perturbed and thus activated by the external cross-linking reagent. In this case, the latter would be serving two purposes, creating a larger reservoir of preaggregated and potentially active receptors and then interacting with them and activating them in a fashion similar to that which occurs with the hormone.

It is important to appreciate that simple dimerization of the receptor is not the absolute activating trigger; activation does not occur unless the receptors are aggregated in just the exact manner and simultaneously properly perturbed. For example, the anti-acetylcholine-receptor antibodies found in certain patients with myasthenia gravis can, when bound to receptors, inhibit the activation by acetylcholine, although they do not prevent the binding of the ligand to the receptor active site. When the antireceptor antibody is digested with papain to retain a dimeric antireceptor cross-linking antibody, the inhibition is still observed, but the inhibition is lost once Fab fragments of a monovalent type are made. These monomeric Fab fragments can be shown to inhibit the action of acetylcholine if they are secondarily cross linked with anti-Fab antibodies.

The findings described above also have some potential and provocative implications regarding the nature of receptor agonists and antagonists (Cuatrecasas, 1983a, 1985). We have speculated for some time (Shechter *et al.*, 1979b) that the primary distinguishing feature is related to the ability of an agonist to induce receptor dimerization or cross linking. An antagonist would presumably bind to the receptor but would not induce or allow the proper cross linking to occur. In this connection, it has been possible to observe conversion of an antagonist to an agonist on divalent coupling or by functional conversion in the case of opiate agonists as well as GNRH.

9. Conclusions

In summary, some of the salient features of this presentation are described diagrammatically in Fig. 1. Occupation of a receptor with a ligand would, in the case of peptide hormone receptors, induce microaggregation or dimerization, perhaps as a result of ligand-induced phosphorylation of tyrosine residues. The properly dimerized hormone receptor, while still on the cell surface, would exert certain biological effects

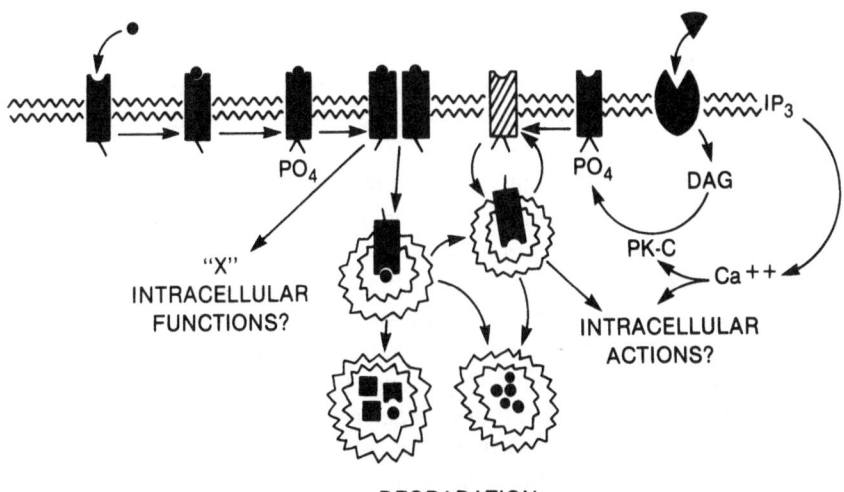

Figure 1. Hormone-dependent receptor phosphorylation and endocytosis. Following hormone binding, the receptor is phosphorylated and forms dimers and/or microclusters in coated pits, leading to internalization. The internalized hormone dissociates from the receptor in the acidic endosome, and the hormone is delivered in endosomes to the lysosome (via the Golgi complex) for degradation and extrusion to the medium. The endosome containing the receptor can be recycled to the cell surface or delivered to lysosomes for degradation. Proteolytic cleavage (nicking) of the receptor might result in the generation of a peptide fragment, perhaps a phosphorylated one facing the cytoplasmic surface, which is released in the cytosol and mediates intracellular actions (signal hypothesis). Other ligands can lead to the production of 1,2-diacylglycerol (DAG), presumably by activating a specific phosphodiesterase that acts on phosphatidylinositol 4′,5′-diphosphate. The DAG can turn on protein kinase C (PK-C), a phospholipid-dependent enzyme that also requires Ca^{2+}. The other product of the phosphodiesterase hydrolysis, 2-PO_4-inositol-4,5-triphosphate (IP_3), can cause the release of Ca^{2+} from intracellular stores and thus also activate PK-C. The receptor also recycles in the absence of ligand, presumably without being altered. It is possible that the receptor phosphorylated by PK-C enters more rapidly (or exits more slowly) so that there is a balance toward more receptor being inside rather than on the surface of the cell. This can lead to "loss" of receptors (down-regulation) or can give the overall impression of lower hormone affinity if the recycling process is fast enough, especially if the acidic endosomes lead to dissociation of hormone from the receptor. In addition, properly activated receptors may generate small-molecular-weight chemical signals ("X"), which mediate intracellular responses.

by yet unknown mechanisms possibly related to the ability to generate a chemical signal in the fashion of a low-molecular-weight substance. The cell surface microaggregated receptor would be internalized via coated pits into endosomes, appearing there within minutes. Subsequently, the hormone dissociates from the receptor as a consequence of the acidic content of the endosome. Depending on the conditions, the receptor and the hormone can be delivered to lysosomes for degradation, or, alternatively, a segregation and sorting-out process occurs that leads to delivery of the dissociated ligand to lysosomes and the receptor into vesicles, which fuse with the cell membrane to rapidly recycle the receptor for reutilization.

It is also possible that intracellular processing, including proteolysis, could lead to the formation of receptor-derived peptides that may have independent biological effects in the cell cytoplasm. The hormone receptor cycles in and out of the cell even in the absence of ligand. This process can be influenced by phosphorylation on serine or threonine residues by protein kinase C, a phospholipid- and calcium-dependent enzyme. Such phosphorylation would presumably change the equilibrium or steady-state distribution of receptor such that a larger proportion of the population of receptors would be in intracellular compartments compared with those expressed on the cell surface; possibly such phosphorylated receptors would not be reinserted in the cell membrane at all.

The signal of activation by protein kinase C would presumably be derived from diacylglycerol. Diacylglycerol is produced on hydrolysis of phosphotidylinositol 4,5-diphosphate by a specific phosphodiesterase. Such a phosphodiesterase activity, whose cellular localization is not yet known, would presumably occur on activation of other receptors by other ligands such as perhaps by bradykinin, collagen, serotonin, thrombin, and platelet-derived growth factor. Consequent to the production of diacylglycerol is the formation and release of inositol triphosphate (inositol 2,4,5P). This molecule would act to mobilize calcium from intracellular storage sites and would be involved in processes that require calcium, including the activation of protein kinase C. Thus, such signals would produce two of the modulators of this enzyme, diacylglycerol and calcium. The activation of protein kinase C and its regulation appear also to involve a translocation from one organelle to the other; possibly, the transfer of a soluble enzyme to a cell-membrane-bound enzyme accompanies activation and facilitates action on the proper substrates.

References

Brown, M. S., Anderson, R. G. W., and Goldstein, J. L., 1983, Recycling receptors: The round-trip itinerary of migrant membrane proteins, *Cell* **32**:663–667.

Castagna, M., Takai, Y., Kaibuchi, K., Sano, K., Kikkawa, U., and Nishizuka, Y., 1982, Direct activation of calcium-activated, phospholipid-dependent protein kinase by tumor-promoting phorbol esters, *J. Biol. Chem.* **257**:7847–7851.

Cohen, S., Carpenter, G., and King, L., 1980, Epidermal growth factor–receptor–protein kinase interactions, *J. Biol. Chem.* **255**:4834–4842.

Cooper, J.A., Bowen-Pope, D. F., Raines, E., Ross, R., and Hunter, T., 1982, Similar effects of platelet-derived growth factor on the phosphorylation of tyrosine in cellular protein, *Cell* **31**:263–273.

Cuatrecasas, P., 1982, Epidermal growth factor: Uptake and fate, *Ciba Found. Symp.* **92**:96–108.

Cuatrecasas, P., 1983a, Emerging concepts in the mechanism of action of membrane receptors, in: *Affinity Chromatography and Biological Recognition* (I. M. Chaiken, M. Wilchek, and I. Parikh, eds.), Academic Press, Orlando, pp. 29–42.

Cuatrecasas, P., 1983b, Developing concepts in receptor research, *Drug Intell. Clin Pharm.* **17**:357–366.

Cuatrecasas, P., 1985, Emerging methods and concepts in the action of membrane receptors, in: *Proceedings of the 1st International Conference New Methods in Drug Research*, Vol 1 (A. Makriyannis, ed.) J. R. Prous Science, Barcelona, Spain, pp. 103–111.

Cuatrecasas, P., and Roth, T., (eds.), 1983, *Receptor-Mediated Endocytosis, Receptors and Recognition Series*, Chapman and Hall, London.

Dautry-Varsat, A., Ciechanover, A., and Lodish, H., 1983, pH and the recycling of transferrin during receptor-mediated endocytosis, *Proc. Natl. Acad. Sci. U.S.A.* **80**:2258–2262.

Denton, R. M., Brownsey, R. W., and Belsham, G. J., 1981, A partial view of the mechanism of insulin action, *Diabetologia* **21**:347–362.

Fox, C. F., and Das, M., 1979, Internalization and processing of the EGF receptor in the induction of DNA synthesis in cultured fibroblasts: The endocytic activation hypothesis, *J. Supramol. Struct.* **10**:199–214.

Grunberger, G., and Gorden, P., 1982, Affinity alteration of insulin receptor induced by a phorbol ester, *Am. J. Physiol.* **243**:E319–E324.

Herzog, V., 1981, Pathways of endocytosis in secretory cells, *Trends Biochem Sci.* **1981**:319–322.

Jacobs, S., and Cuatrecasas, P., 1983, Insulin receptors, *Annu. Rev. Pharmacol. Toxicol.* **23**:461–479.

Jacobs, S., Sahyoun, N. E., Saltiel, A. R., and Cuatrecasas, P., 1983, Phorbol esters stimulate the phosphorylation of receptors for insulin and somatomedin C, *Proc. Natl. Acad. Sci. U.S.A.* **80**:6211–6213.

Kaibuchi, K., Takai, Y., Sawamura, M., Hoshijima, M., Fujikura, T., and Nishizuka, Y., 1983, Synergistic functions of protein phosphorylation and calcium mobilization in platelet activation, *J. Biol. Chem.* **258**:6701–6704.

Karin, M., and Mintz, B., 1981, Receptor-mediated endocytosis of transferrin in developmentally totiputent mouse teratocarcinoma stem cells, *J. Biol. Chem.* **256**:3245–3252.

Kawahara, Y., Takai, Y., Minakuchi, R., Sano, K., and Nishizuka, Y., 1980, Phospholipid turnover as a possible transmembrane signal for protein phosphorylation during human platelet activation by thrombin, *Biochem. Biophys. Res. Commun.* **97**:309–317.

King, A. C., and Cuatrecasas, P., 1981, Peptide hormone-induced receptor mobility, aggregation, and internalization, *N. Engl. J. Med.* **305**:77–88.

King, A. C., and Cuatrecasas, P., 1982, Exposure of cells to an acidic environment reverses the inhibition by methylamine of the mitogenic response to epidermal growth factor, *Biochem. Biophys. Res. Commun.* **106**:479–485.

Klausner, R. D., Renswoude, J. V., Ashwell, G., Kempf, C. Schechter, A., Dean, A., and Bridges, R., 1983, Receptor-mediated endocytosis of transferrin in K562 cells, *J. Biol. Chem.* **58**:4715–4724.

Kolata, G., 1982, Is tyrosine the key to growth control, *Science* **219**:377–378.

Kull, F. C., Jr., Jacobs, S., Su, Y.-F., Svoboda, M. E., Van Wyk, J. J., and Cuatrecasas, P., 1983, Monoclonal antibodies to receptors for insulin and somatomedin-C, *J. Biol. Chem.* **258**:6561–6566.

Lapetina, E. G., and Siegel, F. L., 1983, Shape change induced in human platelets by platelet-activating factor, *J. Biol. Chem.* **258**:7241–7244.

Lapetina, E. G., and Siess, W., 1983, The role of phospholipase C in platelet responses, *Life Sci.* **33**:1011–1018.

May, W. S., Jacobs, S., and Cuatrecasas, P., 1984, The association of phorbol ester induced hyperphosphorylation and reversible regulation of transferrin membrane receptors in HL60 cells, *Proc. Natl. Acad. Sci. U.S.A.* **81**:2016–2020.

Niedle, J. E., Kuhn, L. J., and Vandenbark, G. R., 1983, Phorbol diester receptor copurifies with protein kinase C, *Proc. Natl. Acad. Sci. U.S.A.* **80**:36–40.

Pastan, I. H., and Willingham, M. C., 1981a, Receptor-mediated endocytosis of hormones in cultured cells, *Annu. Rev. Physiol.* **43**:239–250.

Pastan, I. H., and Willingham, M. C., 1981b, Journey to the center of the cell: Role of the receptosome, *Science* **214**:504–509.

Rosen, O. M., Herrera, R., Olowe, Y., Petruzzelli, L. M., and Cobb, M. H., 1983, Phosphorylation activates the insulin receptor tyrosine protein kinase, *Proc. Natl. Acad. Sci. U.S.A.* **80**:3237–3240.

Rovera, G., Ferreo, D., Pagliardi, G. L., Vartikar, J., Pessano, S., Bottero, L., Abraham, S., and Lebman, D., 1982, Induction of differentiation of human myeloid leukemias by phorbol diesters: Phenotypic changes and mode of action, *Ann. N.Y. Acad. Sci.* **397**:211–220.

Sandvig, K., and Olsnes, S., 1981, Rapid entry of nicked diphtheria toxin into cells at low pH: Characterization of the entry process and effects of low pH on the toxin molecule, *J. Biol. Chem.* **256**:9068–9076.

Schiff, J. M., Fisher, M. M., and Underdown, B. J., 1984, Receptor-mediated biliary transport of immunoglobulin A and asialoglycoprotein: Sorting and missorting of ligands revealed by two radiolabeling methods, *J. Cell Biol.* **98**:79–89.

Schlegal, R., Dickson, R. B., Willingham, M. C., and Pastan, I. H., 1982, Amantadine and dansylcadaverine inhibit vesicular stomatitis virus uptake and receptor-mediated endocytosis of α_2-macroglobulin, *Proc. Natl. Acad. Sci. U.S.A.* **79**:2291–2295.

Sefton, B. M., Hunter, T., Beemon, K., and Eckhart, W., 1980, Evidence that the phosphorylation of tyrosine is essential for cellular transformation by Rous sarcoma virus, *Cell* **20**:807–816.

Sharkey, N. A., Leach, K. L., and Blumberg, P. M., 1984, Competitive inhibition by diacylglycerol of specific phorbol ester binding, *Proc. Natl. Acad. Sci. U.S.A.* **81**:607–610.

Shechter, Y., Chang, K.-J., Jacobs, S., and Cuatrecasas, P., 1979a, Modulation of binding and bioactivity of insulin by anti-insulin antibody: Relation to possible role of receptor self-aggregation in hormone action, *Proc. Natl. Acad. Sci. U.S.A.* **76**:2720–2724.

Shechter, Y., Hernaez, L., Schlessinger, J., and Cuatrecasas, P., 1979b, Local aggregation of hormone–receptor complexes is required for activation by epidermal growth factor, *Nature* **278**:835–838.

Shoyab, M., DeLarco, J. E., and Todaro, G. J., 1979, Biologically active phorbol esters specifically alter affinity of epidermal growth factor membrane receptors, *Nature* **279**:387–391.

Siess, W., Siegel, F. L., and Lapetina, E. G., 1983, Arachidonic acid stimulates the formation of 1,2-diacylglycerol and phosphatidic acid in human platelets, *J. Biol. Chem.* **258**:11236–11242.

Takai, Y., Kaibuchi, K., Matsubara, T., and Nishizuka, Y., 1981, Inhibitory action of guanosine 3',5'-monophosphate on thrombin-induced phosphatidylinositol turnover and protein phosphorylation in human platelets, *Biochem. Biophys. Res. Commun.* **101**:61–67.

Thomopoulos, P., Testa, Y., Gourdin, M., Hervy, C., Titeur, M., and Vainchenker, W., 1982, Inhibition of insulin receptor binding by phorbol esters, *Eur. J. Biochem.* **129**:389–393.

Willingham, M. C., Hanover, J. A., Dickson, R. B., and Pastan, I., 1984, Morphologic characterization of the pathway of transferrin endocytosis and recycling in human KB cells, *Proc. Natl. Acad. Sci. U.S.A.* **81**:175–179.

Zick, Y., Kasuga, M., Kahn, C. R., and Roth, J., 1983a, Characterization of insulin-mediated phosphorylation of the insulin receptor in a cell-free system, *J. Biol. Chem.* **258**:75–80.

Zick, Y., Whittaker, J., and Roth, J., 1983b, Insulin stimulated phosphorylation of its own receptor, *J. Biol. Chem,* **258**:3431–3434.

Sorting and Recycling of Cell Surface Receptors and Endocytosed Ligands

The Asialoglycoprotein and Transferrin Receptors

AARON CIECHANOVER, ALAN L. SCHWARTZ, and HARVEY L. LODISH

1. Summary

With a few exceptions, receptor-mediated endocytosis of specific ligands is mediated through clustering of receptor–ligand complexes in coated pits on the cell surface followed by internalization of the complex into endocytic vesicles. During this process, ligand–receptor dissociation occurs, most probably in a low-pH prelysosomal compartment. In most cases, the ligand is ultimately directed to the lysosomes, wherein it is degraded, while the receptor recycles to the cell surface.

We have studied the kinetics of internalization and recycling of both the asialoglycoprotein receptor and the transferrin receptor in a human hepatoma cell line. By employing both biochemical and morphological/immunocytochemical approaches, we have gained some insight into the complex mechanisms that govern receptor recycling as well as ligand sorting and targeting. We can, in particular, explain why transferrin is

AARON CIECHANOVER and HARVEY L. LODISH • Department of Biology, Massachusetts Institute of Technology, Cambridge, Massachusetts 02139. *ALAN L. SCHWARTZ* • Division of Pediatric Hematology/Oncology, Children's Hospital Medical Center, Dana-Farber Cancer Institute, and Department of Pediatrics, Harvard Medical School, Boston, Massachusetts 02115. *Present address of A.C.:* Unit of Biochemistry, Faculty of Medicine, Technion-Israel Institute of Technology, Haifa 31096, Israel.

exocytosed intact from the cells whereas asialoglycoproteins are degraded in lysosomes. We have also localized the intracellular site at which endocytosed receptor and ligand dissociate.

2. Introduction

Many cells are capable of internalizing macromolecules by receptor-mediated endocytosis. The first step in this multiphase process involves binding of a ligand, such as a hormone, virus, plasma protein, or toxin, to a specific receptor molecule functionally exposed at the cell surface. In most cases, these receptors are distributed diffusely over the cell surface. This has been demonstrated by visualization of fluorescent-labeled ligands such as α_2-macroglobulin, insulin, epidermal growth factor, transferrin, and asialoglycoprotein (Maxfield et al., 1978; Sullivan et al., 1976; Geuze et al., 1982).

Binding of ligand is followed by rapid clustering of the surface ligand–receptor complexes into clathrin-coated pits in the plasma membrane and internalization into coated vesicles (Goldstein et al., 1979). Thereafter, both the ligand and the receptor are found in uncoated vesicles. Many ligands such as asialoglycoproteins, α_2-macroglobulin, low-density lipoprotein (LDL), and insulin are then transported within membrane-limited compartments to lysosomes, where they are rapidly degraded (Geuze et al., 1983; Hartford et al., 1983; Zeitlin and Hubbard, 1982; Schwartz et al., 1982; Brown et al., 1982; Pastan and Willingham, 1981).

By contrast, the receptor almost invariably escapes degradation, recycles to the cell surface, and mediates the internalization of additional ligand molecules. Recycling of receptors was deduced from the early observation that cells continue to internalize receptor-bound ligands at a steady rate for many hours without depleting their surface receptors even when synthesis of new receptor molecules is blocked by protein synthesis inhibitors. More direct evidence for recycling comes from the observation that the process can be inhibited by agents such as weak bases (e.g., chloroquine, NH_4Cl) or carboxylic ionophores (e.g., monensin) that disrupt proton gradients and raise the pH of acidic intracellular compartments. When cells take up ligands in the presence of these agents, the receptors do not return to the cell surface, and the number of receptors on the surface rapidly decreases. Such evidence that surface receptors recycle was obtained for receptors for asialoglycoproteins (Schwartz et al., 1982; Tanabe et al., 1979; Tolleshaug and Berg 1979; Schwartz et al., 1984), mannose-6-phosphate-terminated proteins (Gonzalez-Noreiga et al., 1980), mannose-terminal proteins (Stahl et al., 1980), LDL (Brown *et al.*, 1983), α_2-macroglobulin (van Leuven *et al.*, 1980), insulin (Marshall *et al.*, 1981), and the chemotactic peptide (Zigmond *et al.*, 1982). There are, however, exceptions. For example, the Fc-receptor-mediated phagocytosis of IgG-coated erythrocyte ghosts results in the selective and largely irreversible removal of receptors from the macrophage plasma membrane (Mellman *et al.*, 1983).

An important development in understanding of the mechanism of receptor recycling was the demonstration by Tycko and Maxfield (1982) that following ligand internalization, endosomes rapidly become acidified. These investigators incubated cultured cells with fluorescein-labeled α_2-macroglobulin. Within 15 min of endocytosis, the fluorescein was located within an acidic compartment, as indicated by alteration of its emission spectrum. This interval was too short for the ligand to have reached the lysosomes, a conclusion confirmed by histochemical electron microscopy.

The finding of an acid pH in the endosome suggests a mechanism by which receptor–ligand dissociation may be initiated. Many ligands such as asialoglycoproteins (Ashwell and Morell, 1974), lysosomal enzymes (Gonzalez-Noreiga *et al.*, 1980), LDL (Basu *et al.*, 1978), and insulin (Posner *et al.*, 1978) rapidly dissociate from their respective receptors at pH values below 6. The low pH within the endosome would be expected to cause ligand–receptor dissociation and thereby allow unoccupied receptors to return to the surface.

The assumption that the receptor–ligand dissociation occurs in a low-pH prelysosomal compartment accounts for two important features of receptor recycling. First, it explains how dissociation can occur so rapidly after internalization. The LDL receptor, for example, is estimated to return to the cell surface within 12 min after it enters the cell, a time too short to allow transit to the lysosome (Brown *et al.*, 1982). We found similarly short times (~8 min) for the asialoglycoprotein (Schwartz *et al.*, 1982) and the transferrin receptors (Ciechanover *et al.*, 1983). Second, this finding provides a mechanism whereby a receptor can make many trips into and out of the cell without sustaining proteolytic damage. By dissociating from their ligands prior to fusion with the lysosome, surface receptors are segregated from lysosomal proteases. Acidification of the endosome seems to be an obligate step in receptor-mediated endocytosis and occurs when many different ligands enter the cell: asialoglycoproteins (Tycko *et al.*, 1983), transferrin (van Renswoude *et al.*, 1982), and viruses (Marsh *et al.*, 1983). The process involves an ATP-dependent proton pump (Forgac *et al.*, 1983; Stone *et al.*, 1983) similar (but probably not identical) to that described in lysosomes (Schneider, 1981). Still, many problems remain unsolved in understanding this complex round-trip itinerary of cell surface receptors. What directs ligands only to lysosomes following their dissociation from receptor? What are the mechanisms and signals involved in segregation and recycling of the receptors to the cell surface? Why are certain ligands, such as transferrin, segregated from lysosomal proteases and returned to the cell surface intact?

In order to address some of these problems, we have studied the receptors for asialoglycoproteins and transferrin in a human hepatoma cell line (HepG2) using a combination of biochemical and immunocyto-chemical electron micrographic approaches.

3. The Asialoglycoprotein Receptor

3.1. Recycling of the Asialoglycoprotein Receptor: Biochemical Evidence

We and others have been studying the receptor for galactose-terminal carbohydrates of glycoproteins (asialoglycoproteins) that is localized to the hepatic parenchymal cell (Ashwell and Harford, 1982). Endocytosis of asialoglycoproteins has been studied in considerable detail in whole liver *in vivo*, in perfused liver *in situ*, and in isolated rat hepatocytes (Schwartz *et al.*, 1982, 1984; Ashwell and Morell, 1974; Baenziger and Fiete, 1980). There are as many as 500,000 high-affinity surface receptors in the rat hepatocyte (Zeitlin and Hubbard, 1982; Schwartz *et al.*, 1980). Recent studies have begun to elucidate the characteristics of receptor-mediated endocytosis in this system. Using electron microscopic techniques, Hubbard and colleagues have demonstrated the uptake of galactose-terminal glycoproteins by rat hepatic parenchymal cells and followed their subsequent transfer through a series of endocytic vesicles to lysosomes (Zeitlin and Hubbard, 1982; Hubbard *et al.*, 1979; Wall and Hubbard, 1981). Biochemical studies by Tolleshaug (1977) and others have provided evidence for a receptor-mediated uptake of asialoglyco-proteins by isolated rat hepatocytes.

The human hepatoma cell HepG2 isolated by Knowles (1980) contains abundant asialoglycoprotein receptors (Schwartz *et al.*, 1981). These cells specificially bind [^{125}I]-asialoorosomucoid (ASOR); binding requires the presence of Ca^{2+} and is not substantially affected by the presence of a 100-fold excess (by mass) of orosomucoid or asialoagalactoorosomu-coid. Pretreatment of the cells with neuraminidase renders them incapable of binding [^{125}I]-ASOR. There are 150,000–250,000 high-affinity ASOR-binding sites per cell surface (Schwartz *et al.*, 1982). These data are consistent with the characteristics of the asialoglycoprotein receptor in rat

Table I. Specificity of Release of Surface-Bound ASOR from
Hepatoma Cells[a]

Addition (concentration; time)	[^{125}I]-ASOR bound (%)
None	100
EDTA (5 mM; 3 min)	12
N-Acetylgalactosamine	
(100 mM; 10 min)	11
(50 mM; 10 min)	24
Galactose (100 mM; 10 min)	62
N-Acetylglucosamine (100 mM; 10 min)	100
ASOR (200 μg/ml; 300 min)	95

[a] Dishes were washed and incubated with [^{125}I]-ASOR (2 μg/ml) in the standard manner (2 hr, 4°C; see Schwartz *et al.*, 1981b). After washing, the indicated additions were made for the indicated times at 4°C. Results are expressed as a percentage of the total [^{125}I]-ASOR bound compared with the control (None). Adapted from Schwartz *et al.* (1982).

hepatocytes. In addition, once bound to its receptor, [^{125}I]-ASOR could be readily displaced by either a brief treatment at 4°C with EDTA or N-acetylgalactosamine but only minimally by galactose. N-Acetylglucosamine or ASOR was without effect (Table I). The sensitivity of surface bound [^{125}I]-ASOR to displacement by EDTA or N-acetylgalactosamine provides a sensitive and convenient assay for surface-bound ligand; internalized ligand is resistant to such treatments.

At 37°C, there is a linear increase in the amount of cell-associated [^{125}I]-ASOR during the first 2 hr. A constant level of cell-associated ligand is reached by 2 hr. There is little ^{125}I label in degradation products in the medium before 1 hr, and the linear increase in [^{125}I]-degradation products begins by the second hour. As measured by the sum of cell-associated and degraded ^{125}I radioactivity, the overall rate of cellular uptake of ASOR is constant at 15,000 molecules per cell per minute for at least 6 hr.

Because the rate of ligand flux is dependent on the total cell complement of functional receptors that participate in this process, we have determined the cell receptor distribution by destroying cell surface receptors with protease. Single-cell suspensions of HepG2 prepared by treatment of monolayer cultures with an EDTA solution at 4°C bind at 4°C the same amount of [^{125}I]-ASOR as do cells assessed under standard conditions in monolayer culture. The rate of ligand uptake at 37°C is also unimpaired. However, if trypsin is included in the EDTA solution at 4°C, binding of [^{125}I]-ASOR to cells in inhibited by over 90%, indicating that virtually all surface receptors are destroyed by this protease. When trypsin-treated cells are incubated at 37°C in the absence of protein synthesis, uptake of $[^{125}I]$-ASOR is linear with time (data not shown) but is only $20 \pm 2\%$ of that of control cells. Taking into account that only 94% of the surface receptors are actually destroyed by trypsin, the data indicate that in growing HepG2 cells approximately 88% of the functional receptor is on the surface, and 12% is internal.

It should be pointed out, however, that in some systems such as the transferrin receptor in HepG2 cells (Ciechanover et al., 1983b), the mannose receptor in macrophages (Stahl et al., 1980), the mannose-6-phosphate receptor in CHO cells (Willingham et al., 1983), and the receptor for asialoglycoproteins in rat hepatocytes (Schwartz et al., 1980), the bulk of the receptors are on the inside rather than on the cell surface.

Because the uptake and degradation of ligand continue at a steady state of 15,000 molecules per cell per minute independent of new receptor synthesis for at least 6 hr (Schwartz et al., 1982), and because there are 150,000–200,000 binding sites per cell surface (Schwartz et al., 1981b), either there must exist a large pool of previously synthesized receptor within the cell, or receptor reuse must occur to some extent. If no reuse occurs, then the functional receptor pool within the cell must be at least 30- to 60-fold greater than the number of surface receptors. However, in these HepG2 cells, 88% of all functional receptors are on the cell surface. As calculated from the total number of functional receptors per cell (225,000) and the rate of ligand uptake (15,000 molecules per minute at an ASOR concentration of 2μg/ml), each receptor must recycle the ligand, on the average, every 15.0 min ($225,000/15,000$ min^{-1}). These observations and calculated values were all obtained at a ligand concentration of 50 nM. At higher ligand concentrations, the total cycle time decreases, as does the time required for ligand binding, until a point is reached at which binding is no longer rate limiting. The rate of ligand uptake and degradation at 10–20 μg of $[^{125}I]$-ASOR ml^{-1} is double that at 2 μg/ml (Schwartz et al., 1982), and the receptor cycle time is about half, or 7.9 min.

At 2 μg/ml ASOR, binding of ligand to surface receptors requires a mean time of 8.7 min. Internalization of receptor–ligand complexes requires a mean of 2.2 min, whereas a mean of 4.2 min is required for the internalized receptor to dissociate its ligand and return to the cell surface (Fig. 1; Table II; Schwartz et al., 1982). The sum of these times yields 15.1 min for the total cycle time of the asialoglycoprotein receptor (assessed at 2 μg/ml ligand).

Asialoglycoprotein ligands are therefore capable of being taken up and processed through to the lysosomes at a considerable rate (see also Ashwell and Harford, 1982), whereas the receptor is apparently spared

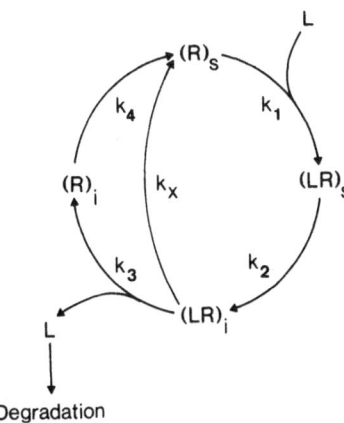

Figure 1. A kinetic model for receptor-mediated endocytosis of asialoglycoprotein receptor. L, ligand $(R)_s$, unoccupied surface receptors; $(LR)_s$, occupied surface receptors; $(LR)_i$, occupied internal receptors; $(R)_i$, unoccupied internal receptors; k_1, rate constant for binding; k_2, rate constant for internalization; k_3, rate constant for dissociation of ligand and receptor within the cell; k_4, rate constant for reappearance of receptor on cell surface; k_x, overall rate constant for dissociation of receptor–ligand complex and return of internal receptor to cell surface ($1/k_x = 1/k_3 + 1/k_4$). (Adapted from Schwartz *et al.*, 1982.)

Table II. Parameters of Endocytosis of Transferrin and Asialoglycoprotein[a]

Parameter	Receptor	
	Transferrin	Asialoglycoprotein
Binding of ligand[b]		
k_1(mole^{-1} min^{-1})	3.02×10^6	2.23×10^6
Mean time (min)[c]	4.3	8.7
Dissociation of ligand		
k_{-1}(min^{-1})	0.09/0.106[d]	<0.00
Mean time (min)	11.1/9.4	—
Internalization of surface receptor		
k_2(min^{-1})	0.20/0.30	0.46
Mean time (min)	5.0/3.3	2.2
Return of receptor to surface		
k_x(min^{-1})	0.14	0.24
Mean time (min)	7.19	4.2
Dissociation of apotransferrin		
k_0(min^{-1})	2.6	—
Mean time (min)	0.38	—
Cycle time (T_c)(min) measured from the rates of iron or asialoorosomucoid uptake	15.8	15.9
Sum of $1/(k_1L) + 1/(k_2) + 1/(k_x) + 1/(k_0)$	16.9	15.1

[a] Adapted from Schwartz *et al.* (1982a) and Ciechanover *et al.* (1983b). The rate constants are defined in Figs. 1 and 11.
[b] Experiments were performed at 50 nM asialoorosomucoid and 77 nM transferrin.
[c] Equals (k_1^{-1})(molar concentration of ligand)$^{-1}$; the mean time required for a surface receptor to bind a ligand at the concentration employed: 50 nM asialoorosomucoid or 77 nM transferrin.
[d] The first measurement was carried out using [^{125}I]-ferrotransferrin and the second using [^{59}Fe]-transferrin. The former values are used in all subsequent calculations.

degradation. Additional biochemical studies have demonstrated that the intracellular half-life of ASGP ligand taken up by receptor-mediated endocytosis is about 15–20 min, whereas that of the receptor is probably greater than 30 hr (Ashwell and Harford, 1982; Schwartz and Rup, 1983).

Importantly, in HepG2 cells, degradation of internalized ligand begins only after 20–30 min, a time much longer than the total cycle time of the receptor. Such studies suggest that receptor is not transferred to lysosomes, a conclusion substantiated by our morphological studies.

If the internalization of a surface receptor for a particular ligand is a specific process, and if all surface receptors are internalized in synchrony, one should be able to show directly a significant and transient depletion of surface receptors until the depleted pool is replaced. Even if the process of receptor internalization occurs continuously in both the presence and absence of bound ligand (Schwartz et al., 1984), it is likely that accelerated internalization or decreased externalization occurs in the presence of ligand, causing a transient depletion of surface receptors. This assumption was based on our previous finding that in the continuous presence of ASOR, there is a substantial redistribution of functional asialoglycoprotein receptors to intracellular sites. The number of functional intracellular receptors is doubled (13 to 28% of total) when HepG2 cells are continuously exposed to 2 µg/ml ASOR (Schwartz et al., 1982).

Indeed, we were able to demonstrate a transient 55% reduction in cell surface asialoglycoprotein binding sites after saturating the cell surface sites with ASOR at 4°C and then warming the cells to 37°C (Fig. 2). Cells were incubated at 4°C for 2 hr in the presence of excess unlabeled ASOR (40 µg/ml; $k_d = 0.4$ µg/ml; Schwartz et al., 1982) and thereafter washed free of unbound ligand. The cells were then warmed to 37°C in the absence of added ligand for various times ranging from 0.5 to 11 min and were then quickly rechilled to 4°C. Surface receptors unoccupied by ligand were quantified by binding [^{125}I]-ASOR to the cells under saturating conditions at 4°C. To measure the total number of receptors present on the cell surface, both those occupied and those unoccupied with ligand, we first stripped the cells of surface bound ASOR by incubation for 3 min at 4°C in ice-cold PBS containing 5mM EDTA. Replicate dishes of cells were then incubated under saturating conditions at 4°C with [^{125}I]-ASOR.

After internalization of ASOR, the total number of surface receptors dropped to 45–55% by 2 min and then returned to its original value within the next 8 min (Fig. 2B). Initially, all of the surface receptors were occupied with ligand. All of the receptors that reappeared on the cell surface after one cycle of endocytosis lacked bound ligand (Fig. 2B). Following a lag of about 1 min, unoccupied receptors reappear on the surface with

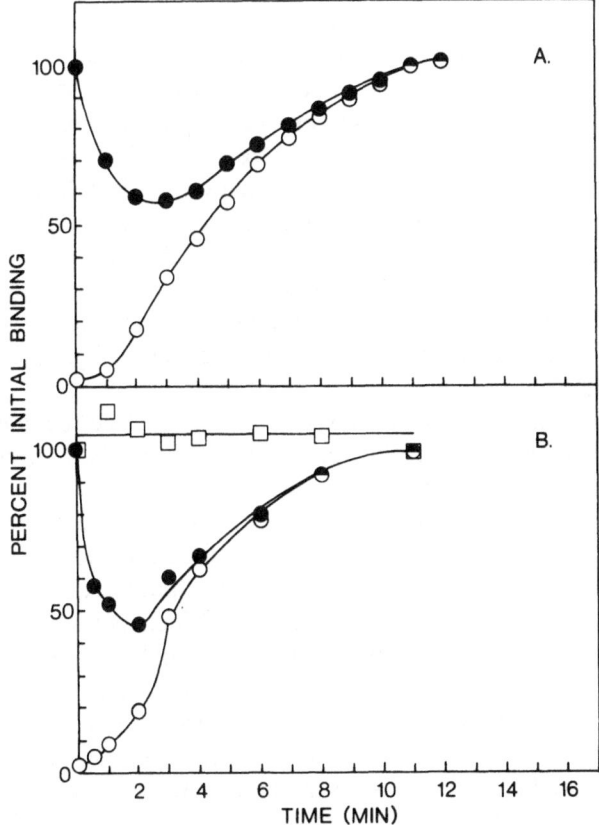

Figure 2. Calculated and observed recycling kinetics of the asialoglycoprotein receptor in HepG2 cells. A: Calculated curve. The differential equation describing our model of receptor internalization and cycling was solved making the following assumptions: At $t = 0$, $(LR)_s$ = 0.87, $(LR)_i$ = 0.13, and R_s = 0; $k_1L = 0$ (i.e., the absence of free ligand precludes the binding of additional ligand to the cell); k_2 = 0.47 min^{-1}; and k_x = 0.23 min^{-1}. The total number of surface receptors (●——●) ($[LR]_s + R_s$) is plotted, normalized to the value of 1.00 at $t = 0$. Since we assume here that $k_1L = 0$, at the end of the experiment all of the receptors will be on the surface (R_s = 1.00), and thus there will be slightly more surface receptors than at $t = 0$. The number of surface receptors free of ligand (R_s) is also plotted (○——○). B: Experimental curve. HepG2 cells were preincubated for 30 min at 37°C in binding medium containing 0.4 mM cycloheximide and then chilled. They were incubated for an additional 2 hr at 4°C with (●——●, ○——○) or without (□——□) 0.5 μM of unlabeled ASOR in the presence of 0.4 mM cycloheximide. The binding medium was removed, and the cells were washed three times in PBS (containing 1.7 mM CaCl₂) at 4°C. The cells were then incubated with 1 ml of prewarmed binding medium (containing 0.4 mM cycloheximide) at 37°C. At the indicated times, the medium was quickly removed, and the cells chilled immediately to 0°C by immersion in ice-cold PBS (containing 1.7 mM CaCl₂). The cells were then treated for 3 min at 4°C in PBS containing 5 mM EDTA (●——●) (to release the surface-bound ligand) or in PBS alone (○——○, □——□). Binding of [^{125}I]-ASOR at 4°C was performed in a medium contained 0.4 mM cycloheximide. (Adapted from Ciechanover *et al.*, 1983a.)

a half-time of about 3.5 min. Since control studies (data not shown) demonstrated no loss of prebound ligand to the medium after 15 min of incubation at either 4°C or 37°C, and since new receptor synthesis was totally abolished by the presence of cycloheximide throughout the experiment, we feel that all of the measured surface ligand-binding sites at the end of the study originate from those that were originally on the surface and recycled or from receptors that were internal at the start of the study.

If ASOR ligand is not prebound to the cell surface at 4°C, there is no alteration in the number of surface ASOR receptors subsequent to warming to 37°C (Fig. 2B). We conclude that both the loss and the reappearance of surface receptors are consequences of ligand binding and internalization.

This experiment directly shows that many surface receptors recycle back to the cell surface. At the start of the 37°C incubation, only 13% of the total population of receptor was internal, and 87% was on the surface (Schwartz et al., 1982), yet 45–55% of the cell surface receptors disappeared and then reappeared. At least 34% of the receptors on the surface at the end of the study must have been those that were originally on the surface and then internalized and recycled (Ciechanover et al., 1983b).

Making use of a simple kinetic model for asialoglycoprotein receptor function (Fig. 1; Schwartz et al., 1982), we can calculate a theoretical curve (Fig. 2A) for the number of surface receptors following ligand internalization. We regard the agreement of the experimental results in Fig. 2B with our prediction (Fig. 2A) as marked confirmation of both our model and the calculated values of k_2 and k_x.

3.2. Recycling of the Asialoglycoprotein Receptor: Immunoelectron Microscopy during Receptor-Mediated Endocytosis

In collaboration with H. J. Geuze at the Center for Electronmicroscopy, Utrecht, The Netherlands, we have used the recently developed double-labeling immunocytochemical electron microscopic technique (Gueze et al., 1981) with antibodies against both asialoglycoprotein ligand and receptor to visualize the compartment in which dissociation of the ligand–receptor complex occurs. Asialofetuin was administered to adult rats by continuous infusion, followed by perfusion fixation with formaldehyde–glutaraldehyde. Cryosectioning and immunolabeling with colloidal gold adsorbed to staphylococcal protein A were essentially as described by Geuze et al. (1981, 1983). Affinity-purified monospecific rabbit antibodies against the purified rat liver asialoglycoprotein receptor (Schwartz et al., 1981a) and against purified asialofetuin were employed.

Both receptor and ligand were found associated with the membrane

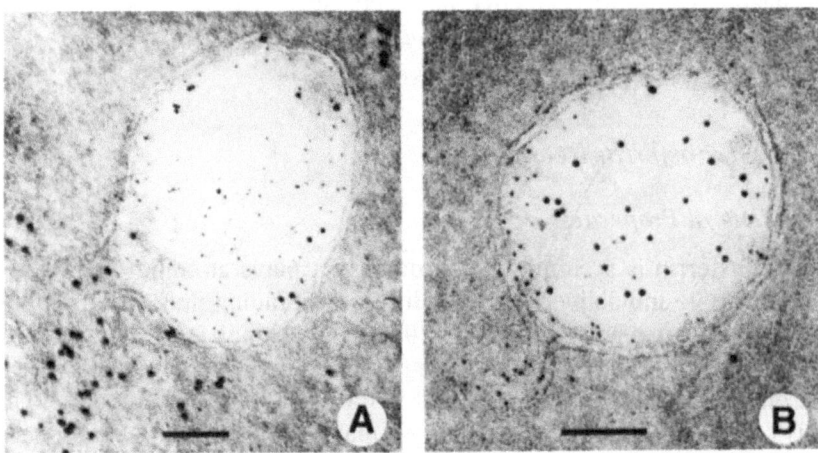

Figure 3. (A) Immunocytochemical electron micrograph of ultrathin cryosections from perfusion-fixed rat liver during continuous infusion of asialofetuin. Ligand was labeled first with antiasialofetuin antibody and then with 5-nm colloidal gold protein A. Thereafter, asialoglycoprotein receptor was immunolabeled with antibody and then with 8-nm colloidal gold protein A. Free ligand can be seen in the lumen of the vesicular portion of this sorting vesicle, which also shows scarce and heterogeneous receptor distribution. Receptor labeling is intense over the connecting tubules. Bar, 0.1 μm. (B) Similar to A except that receptor is labeled with 5-nm gold, whereas ligand is labeled with 8-nm gold. Receptor is located predominantly at the fold where a tubule with heavy receptor labeling is connected. Most of the ligand is present free within the vesicle lumen. Bar, 0.1 μm. (Adapted from Geuze *et al.,* 1983.)

of small clathrin-coated endocytic vesicles close to the cell surface. Little or no free ligand occurred within the lumen of these vesicles.

We also identified other larger vesicles found at some distance from the plasma membrane that contain ligand accumulated within the lumen. The membranes of these latter vesicles contained little receptor, but receptor was concentrated in detached tubular extensions that were largely free of ligand (Fig. 3). No significant receptor labeling was ever found within the vesicle lumen. Interestingly, receptor was not uniformly distributed along the membrane of these larger vesicles but was either dispersed in clusters along the vesicle membrane or appeared as accumulations at the poles, where vesicles and their membranous tubules approximated each other or were continuous. It is in these vesicles that, we believe, the ligand is uncoupled from the receptor; the tubular membranous structures could be intermediates in the recycling of receptor to the cell surface. This double labeling pattern strongly suggests that these curl-tailed vesicles represent the compartment of uncoupling of receptor and ligand. The acronym CURL has been suggested to identify this compartment of dissociation (Geuze *et al.*, 1983).

4. The Transferrin Receptor

4.1. General Properties

Transferrin is a serum glycoprotein that plays an important role in iron transport and delivery to cells. It has two binding sites for ferric ions (reviewed in Aisen and Litowsky, 1980), and it binds to a specific membrane receptor, which appears to be the first step in the complex process of iron uptake (Jandl and Katz, 1963; van Bockxmeer and Morgan, 1977). This receptor glycoprotein is found on many cells and has recently been purified (Enns *et al.*, 1981; Trowbridge and Omary, 1981). It is probably a homodimer of 180,000 molecular weight. The two subunits are linked by disulfide bonds.

All cells require iron as a constituent of respiratory and other heme-containing proteins. Studies of iron uptake by cells and the role of the transferrin receptor have led to some disagreement about the events that occur following binding of transferrin to the cell surface receptor. According to one model, iron enters the cell together with transferrin by receptor-mediated endocytosis. The iron then dissociates from transferrin, probably in a low-pH prelysosomal compartment (Van Renswoude *et al.*, 1982; Dautry-Varsat *et al.*, 1983) and is delivered in a yet unknown way to the iron storage protein ferritin (Klausner *et al.*, 1983a). The apotransferrin recycles to the cell surface and is released intact to the medium to be reutilized as a iron carrier (Ciechanover *et al.*, 1983b; Dautry-Varsat *et al.*, 1983; Klausner *et al.*, 1983a,b).

According to an alternative model, iron is removed from the transferrin at the cell surface, and the iron alone is internalized by an as yet undefined cellular process (Nunez *et al.*, 1983; Cole and Glass, 1983). It is possible that different cells employ different mechanisms of iron uptake.

We became interested in the process of receptor-mediated endocytosis of transferrin because of the many puzzling features of the system. Whereas most ligands endocytosed by a receptor-mediated mechanism are transported to lysosomes and are degraded, intact apotransferrin (after delivery of iron to the cell) is exocytosed intact into the medium (Klausner *et al.*, 1983a; A. Ciechanover and H. F. Lodish, unpublished data). Is apotransferrin dissociated from its receptor within the cell as are other ligands? If so, how does it escape degradation by the lysosome, and how is it secreted into the medium? Or does transferrin remain bound to its receptor in endocytic vesicles? If so, how and when is apotransferrin released from its receptor into the culture medium?

4.2. The Fate of the Transferrin Polypeptide and Iron during a Single Cycle of Endocytosis

Our first experiments focused on the fate of the protein and iron moieties of transferrin during a single cycle of endocytosis in HepG2 cells. In these studies, a saturating amount of $[^{125}I]$- or $[^{59}Fe]$-diferric transferrin is bound at 4°C to the surface of HepG2 cells. Unbound ligand is removed, and the cells are incubated for various times at 37°C. The medium is quickly removed, and the cells are chilled and treated with pronase at 4°C. Only surface-bound ligand is accessible to the proteolytic enzyme, whereas internalized ligand is protected from proteolysis and is recovered with the cell pellet. At least 39% of surface-bound $[^{125}I]$-transferrin is internalized within 5 min and then is exocytosed into the medium (Fig. 4). All of the exocytosed ^{125}I radioactivity is in intact transferrin as shown by sodium dodecylsulfate gel electrophoresis (A. Ciechanover and H. F. Lodish, data not shown). Similar results have been reported by others (Jandl and Katz, 1963; Klausner *et al.*, 1983a; Harding and Stahl, 1983; Ward *et al.*, 1982; Lamb *et al.*, 1983; Karin and Mintz, 1981).

The fate of the iron moiety of transferrin is different. As can be seen from Fig. 5, 63% of the iron of surface-bound $[^{59}Fe]$-transferrin is subject to endocytosis and remains within the cell; 37% of the ^{59}Fe is lost from the cell surface and released directly into the medium as intact $[^{59}Fe]$-transferrin.

4.3. pH and the Recycling of Transferrin and the Transferrin Receptor during Receptor-Mediated Endocytosis

Next, we examined the mechanism(s) involved in the dissociation of Fe and transferrin from the transferrin receptor. Since endocytic vesicles that contain α_2-macroglobulin (Tycko and Maxfield, 1982), transferrin (van Renswoude *et al.*, 1982), and asialoglycoproteins (Tycko *et al.*, 1983) are acidic, we examined the effect of low pH on dissociation of various ligands from their respective receptors. As can be seen in Fig. 6, the stability of the transferrin–receptor complex was not affected by pH, whereas both insulin and asialoorosomucoid are dissociated from their respective receptors at pHs of 5 or less. This result underscores the marked differences between the transferrin–receptor complex and other ligand–receptor complexes.

Because the pH of the endocytic vesicle is approximately 5 (Tycko

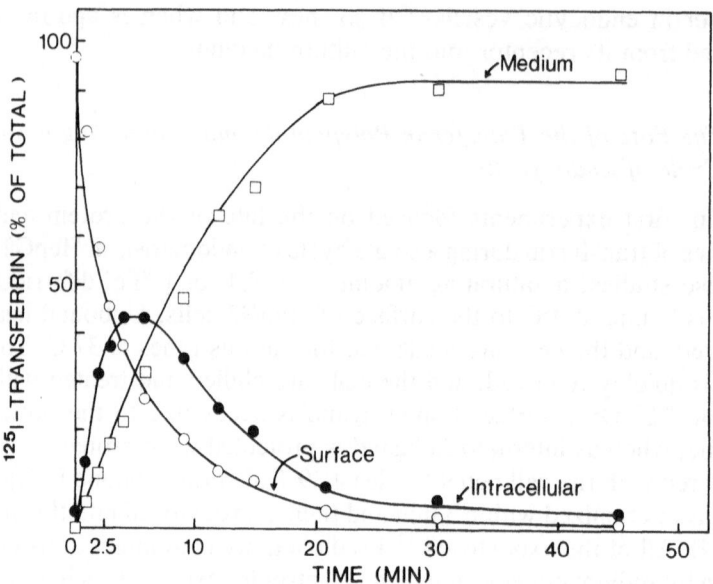

Figure 4. Diacytosis of [^{125}I]-transferrin in HepG2 cells. Transferrin was bound to HepG2 cells at 4°C. After washing off excess unbound ligand, the cells were incubated at 37°C following addition of prewarmed binding medium containing 128 nM unlabeled transferrin. At the indicated times, the medium was quickly removed, and the cells were chilled in ice-cold PBS (containing 1.7 mM CaCl$_2$) and treated with pronase. The radioactivity in the pronase-resistant (internalized) fraction (●——●), pronase-sensitive (cell surface) fraction (○——○), and the medium (□——□) was determined. (Adapted from Ciechanover *et al.*, 1983b.)

and Maxfield, 1982; Tycko *et al.*, 1983; van Renswoude *et al.*, 1982), we examined the possibility that iron dissociates from receptor-bound transferrin at low pH and, more importantly, that the resultant apotransferrin does not dissociate from its receptor under these conditions. As can be seen in Fig. 7, treatment at acid pH in the presence of the iron chelator desferrioxamine released the iron associated with receptor-bound transferrin while the transferrin protein itself remained tightly bound to its receptor.

The above experiment suggested apotransferrin has a high affinity for its receptor at low pH. Concentration-dependent binding of apotransferrin to HepG2 cells at pH 5.4 confirmed this prediction (data not shown). Scatchard analysis of the data gave a similar number of cell surface binding sites as for holotransferrin at pH 5 (not shown) or neutral pH (Ciechanover *et al.*, 1983b; approximately 50,000 sites/cell with K_d of about

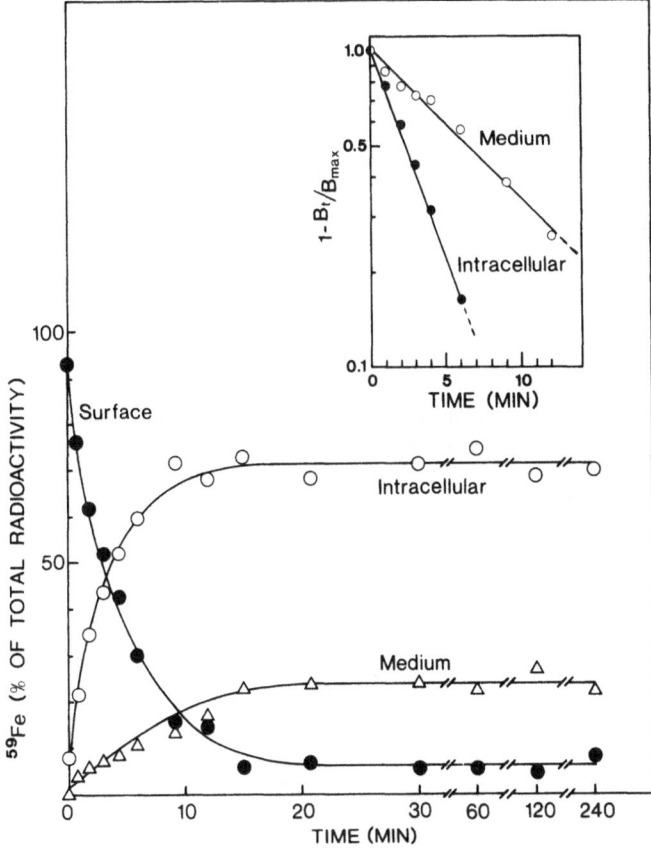

Figure 5. Single-cycle receptor-mediated endocytosis of [^{59}Fe]-transferrin in HepG2 cells. The experimental details are as delineated in the legend to Fig. 4 except that [^{59}Fe]-transferrin was bound to the cells instead of [^{125}I]-transferrin. Inset: The accumulation of ^{59}Fe in the cell and the dissociation of [^{59}Fe]-transferrin into the medium were replotted semilogarithmically as $1-(B_t/B_{max})$ versus time, where B_{max} is the maximum amount in the compartment and B_t is the amount at time t. From the intracellular values, 8% was subtracted to correct for the background of pronase-resistant radioactivity at zero time. (Adapted from Ciechanover *et al.*, 1983b.)

10^{-8} M), suggesting that apotransferrin is tightly bound to the transferrin receptor molecule at low pH.

Importantly, apotransferrin shows very little detectable specific binding to cell surface receptors at pH 7.3. At 30 nM, the saturating concentration for holotransferrin (Ciechanover *et al.*, 1983b), the binding of apotransferrin is at most 5% of that found for holotransferrin. In another experiment, we showed that apotransferrin bound to its receptor at pH

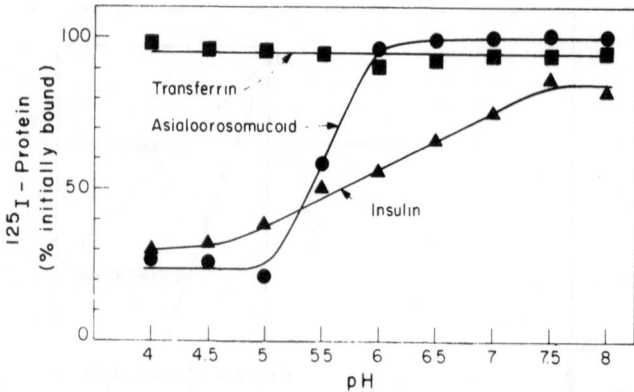

Figure 6. Effect of pH at 4°C on dissociation of transferrin, insulin, and asialoorosomucoid from HepG2 cells. [^{125}I]-Insulin (▲——▲), [^{125}I]-transferrin (■——■), and [^{125}I]-asialoorosomucoid (●——●) were bound to HepG2 cells at 4°C. After washing off unbound ligand, the cells were incubated at 4°C for 5 min at different pHs. (Adapted from Dautry-Varsat *et al.*, 1983.)

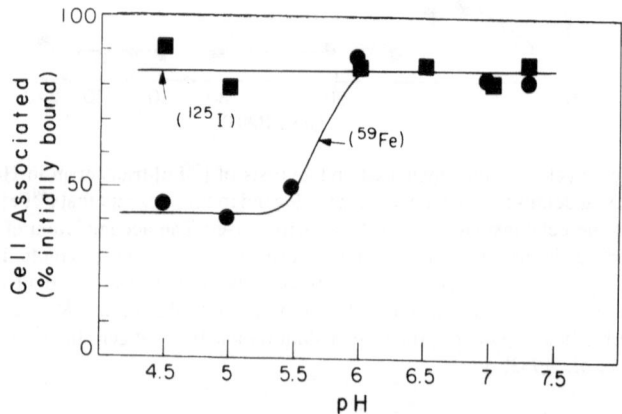

Figure 7. Effect of pH at 37°C on dissociation of transferrin from HepG2 cells. [^{125}I]-Transferrin (■——■) and [^{59}Fe]-transferrin (●——●) were bound at pH 7.2 and 4°C to HepG2 cells in binding medium containing inhibitors of ATP generation. Excess ligand was removed, and incubation buffer was added to each dish. The buffer contained inhibitors of ATP generation to prevent internalization and the iron chelator desferrioxamine. The prewarmed medium was rapidly added while the dishes were transferred to a 37°C water bath. After 2 min at 37°C, the buffer was collected; the cells were counted for radioactivity as well as the buffer. (Adapted from Dautry-Varsat *et al.*, 1983.)

5.4 is rapidly dissociated with a half-time of 16 sec (not shown) when the pH is raised to 7.4. Under the same conditions, diferric transferrin dissociates with a half-time of 7.5 min (not shown).

These results led us to suggest a novel model for the delivery of iron to cells and for cycling of transferrin (Fig. 8; Dautry-Varsat *et al.*, 1983). Iron-loaded transferrin binds to its receptor on the cell surface at neutral pH; under these conditions, binding of apotransferrin is negligible. After binding, diferric transferrin is internalized by receptor-mediated endocytosis. The transferrin–receptor complex moves to an acidic prelysosomal compartment. There, perhaps in the presence of an iron-chelating or reducing component, iron is released from transferrin and is transported in a yet undefined pathway to the iron storage protein ferritin (Klausner *et al.*, 1983a). Apotransferrin remains bound to its receptor, and together they are recycled to the cell surface. On reaching neutral pH (either at the cell surface or just prior to it in an intracellular vesicle), apotransferrin rapidly dissociates from its receptor. The free receptor at the cell surface is available for another cycle of receptor-mediated endocytosis. The released apotransferrin will be transported in the blood to a loading site where two Fe^{3+} ions will be rebound. A similar model has been proposed recently by Klausner and co-workers (1983b).

In order to further probe the role of pH in the dissociation of iron from transferrin and its delivery to the cell, we have utilized several lysosomotropic agents. These are weak bases that increase the intralysosomal (Ohkuma and Poole, 1978) and the endosomal pH (Maxfield, 1982). As was reported recently, these agents perturb the pH-dependent dissociation of asialoorosomucoid from its receptor and its delivery to the lysosome (Harford *et al.*, 1983). One such agent, NH_4Cl, decreases iron uptake into cells (Klausner *et al.*, 1983a; A. Ciechanover and H. F. Lodish, data not shown); however, the fate of iron internalized via transferrin was not determined. These agents do not inhibit binding of transferrin to cell surface receptor, nor do they inhibit internalization of the receptor–ligand complex (Klausner *et al.*, 1983a; data not shown). However, as is evident from Fig. 9, they block the retention of iron in the cell (compare to Fig. 5). All of the endocytosed Fe is secreted into the medium as intact diferric transferrin (not shown). Similar findings were also recently reported by Harding and Stahl (1983).

The principal effect of NH_4Cl is therefore the inhibition of dissociation of iron from the endocytosed transferrin–receptor complex, presumably because it raises the pH of the endocytic vesicle. Iron is dissociated from the transferrin receptor complex only at pH values less than 6. When the pH is perturbed, iron is not dissociated, and intact diferric transferrin is exocytosed. These results suggest that the low pH of the endocytic

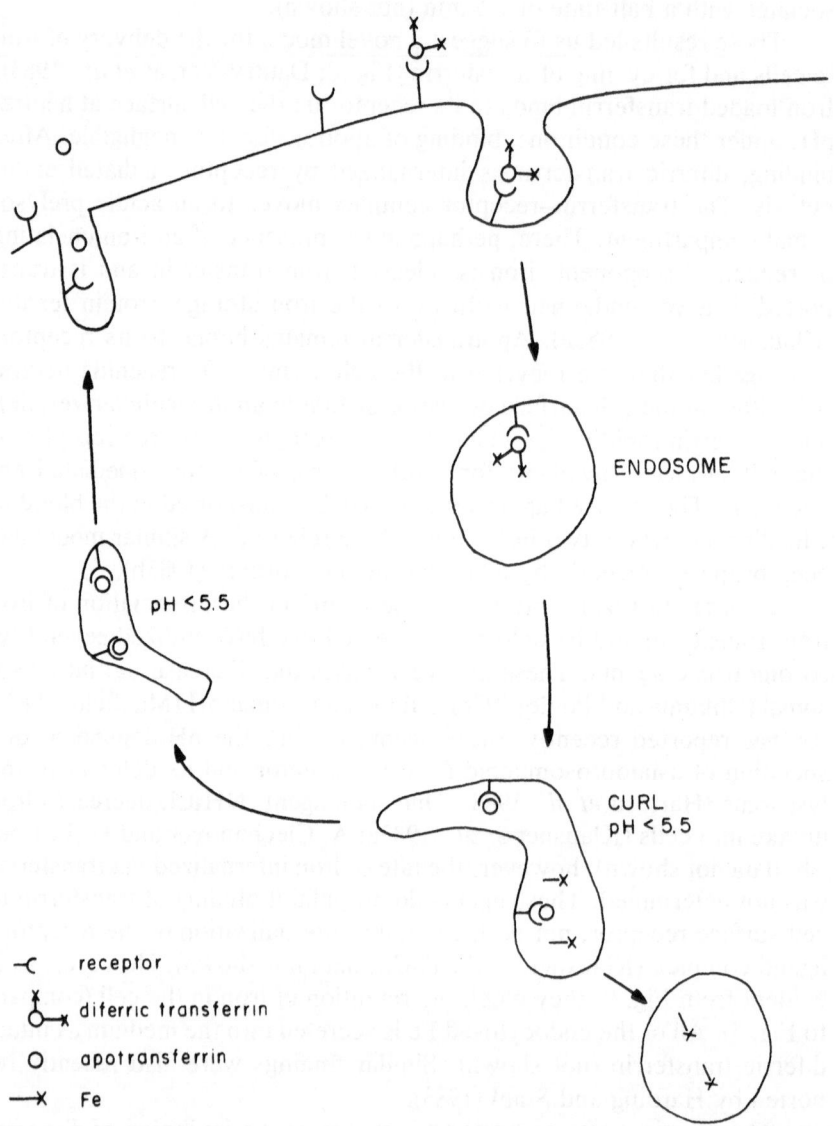

Figure 8. The transferrin cycle. See text for details. (Adapted from Dautry-Varsat *et al.,* 1983.)

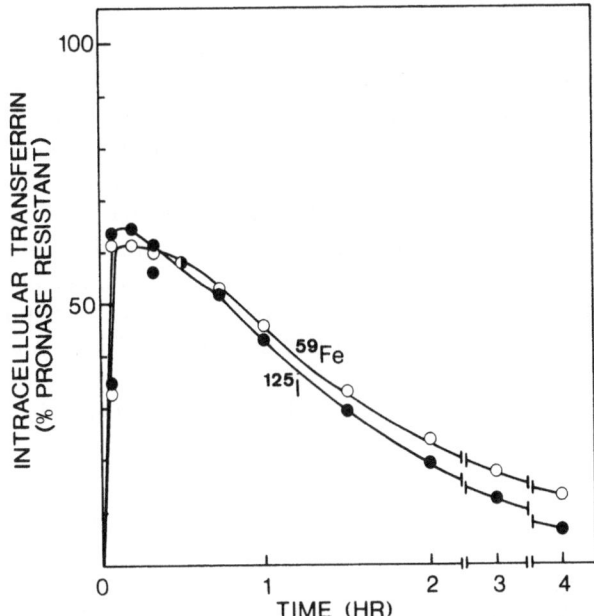

Figure 9. Effect of NH₄Cl on the endocytic cycle of transferrin. Experimental details are the same as described in legends to Figs. 4 and 5 except that 20 mM NH₄Cl was added to all the solutions. Shown are percentages of radioactivity in cell-associated material that is pronase resistant (i.e., internalized). ^{125}I (●——●); ^{59}Ke (○——○). (Adapted from Ciechanover *et al.*, 1983b.)

vesicle is essential for dissociation of iron from the transferrin–receptor complex but is not essential for the recycling of the transferrin polypeptide back to the cell surface. Recently, Klausner *et al.* (1984) have characterized mutants defective in acidification of the endosome. In these cells, they noted similar behavior of transferrin to that seen in NH₄Cl-treated-wild-type cells—diacytosis of diferric transferrin compared to apotransferrin. This finding supports our contention that the low pH is required mainly for removal of Fe from transferrin.

We have recently been able to further probe the intracellular pathway taken by ASOR and transferrin following internalization by making use of density-gradient centrifugation of cell homogenates incubated with double-labeled ligands.

Shortly after internalization (5 min), both ligands are associated with a class of vesicles having a density similar to that of the plasma membranes (1.020 g/ml). Using the density shift technique (in which one ligand is tagged with peroxidase, which is reacted with diaminobenzidine following internalization, thus forming a polymer of high specific density), we have further shown that the two ligands share the same entry vesicles in the initial phase of their intracellular pathway (A. Ciechanover, unpublished data). These data were further confirmed using electron microscopic sections of HepG2 cells labeled with colloidal-gold-labeled ligands (Neutra, unpublished). Subsequently, after the initial entry phase, the two ligands diverged. Whereas ASOR ultimately was transferred to a compartment that cofractionated with lysosomal enzyme (≥ 30 min; density 1.085 g/ml), most of the transferrin was released to the medium from the light, buoyant density fraction; very little was transferred to lysosomes. Similar findings were reported recently by Pastan and co-workers (Dickson et al., 1983; Willingham et al., 1984). These results further corroborate our model of intracellular recycling and sorting of the endocytosed ligands, transferrin, and ASOR.

4.4. Determination of the Cycle Time of Transferrin Receptor

As shown earlier (Figs. 4 and 5), the uptake kinetics of iron in a single endocytic cycle is different from that of the transferrin protein moiety. While iron remains within the cell, the protein diacytoses through the cell and is exocytosed as apotransferrin (Ciechanover et al., 1983b; Klausner et al., 1983b). Thus, the rate of transferrin-mediated iron uptake is a measure of the total rate of transferrin endocytosis. At 4°C, the maximum amount of [^{59}Fe]-transferrin that could be associated with the cell was similar to that obtained with [^{125}I]-ferrotransferrin, as expected, since both bind to the same surface receptors. In contrast, at 37°C, there is a time-dependent uptake of iron that is linear for almost 4 hr (Fig. 10). The rate of uptake is about 19×10^3 iron ions/cell per min. Since each transferrin binds two irons, this represents 9.5×10^3 transferrin molecules that cycle through the cell per minute. Taking into account that there are about 1.5×10^5 functional transferrin binding sites per cell (Ciechanover et al., 1983b; data not shown) and that the experiment is carried out in the presence of cycloheximide to block protein synthesis and therefore synthesis of new receptor molecules, we calculate that the time required for each receptor molecule to traverse an endocytic cycle is $1.5 \times 10^5 \div (9.5 \times 10^3) = 15.8$ min.

The cycle time was also determined from an additional independent series of experiments. In these experiments, we measured separately each step of the endocytic cycle as depicted in Fig. 11 (Ciechanover et al., 1983b; experimental details are not shown). The sum of the mean time for each separate step of the cycle is 16.9 min (see also Table II).

It is of interest to compare the kinetic parameters for endocytosis of

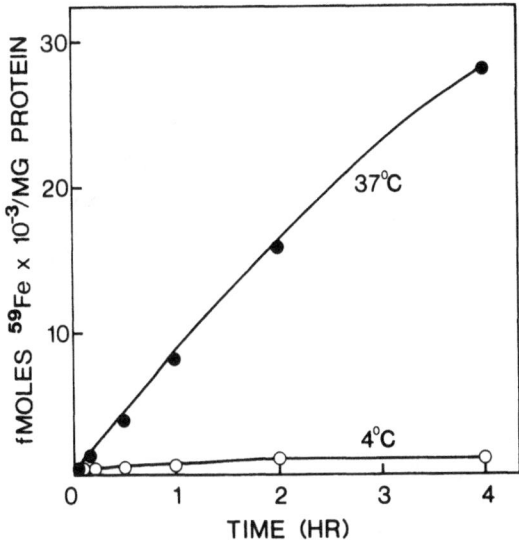

Figure 10. Time course of transferrin-mediated ^{59}Fe uptake into HepG2 cells. [^{59}Fe]-Transferrin was bound to cells for the indicated times. (O———O) 4°C; (●———●) 37°C. (Adapted from Ciechanover *et al.*, 1983.)

transferrin to those for asialoglycoprotein in the same cell line (Table II). As can be seen, the various parameters of the endocytic cycle are similar, although there are some differences. We note that the cycling time of both the transferrin and the asialoglycoprotein receptors is rather short, about 16 min. The time necessary for internalization of the receptor–ligand complex into the low-pH endocytic vesicles and the dissociation of the ligand or the coligand (in the case of transferrin) is even shorter. Dissociation of asialoglycoprotein from its receptor probably occurs in the low-pH endocytic vesicle designated CURL by Geuze *et al.* (1983). However, degradation of many ligands in lysosomes starts only after a lag period of about 30 min (Schwartz *et al.*, 1982; Goldstein and Brown, 1974). This delay is probably caused by slow delivery of the ligand to the lysosome but not by any steps in which the receptor is involved. (A. Ciechanover, unpublished data; see also Dickson *et al.*, 1983).

4.5. Are Cell Surface Receptors Internalized and Recycled Independently?

Most cells contain more than one type of cell surface receptor. Does internalization of one ligand cause internalization of other cell surface

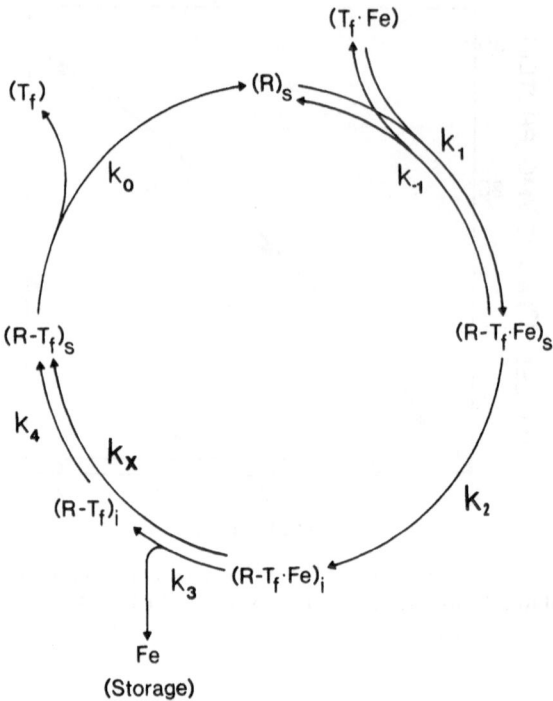

Figure 11. A kinetic model for receptor-mediated endocytosis of transferrin. T_f·Fe, ferrotransferrin; $(R)_s$, unoccupied surface receptor; $(R\text{-}T_f·Fe)_s$, surface ferrotransferrin–receptor complex; $(R\text{-}T_f·Fe)_i$, intracellular ferrotransferrin–receptor complex; $(R\text{-}T_f)_i$, intracellular apotransferrin–receptor complex; $(R\text{-}T_f)_s$, surface apotransferrin–receptor complex; T_f, apotransferrin; k_1, rate constant for binding of ferrotransferrin to surface receptors; k_{-1}, rate constant for dissociation of ferrotransferrin from cell-surface receptors. k_2, rate constant for internalization of surface ferrotransferrin–receptor complex; k_3, rate constant for dissociation of iron from internalized ferrotransferrin–receptor complexes; k_4, rate constant for movement of the receptor–apotransferrin complex to the surface; k_x, overall rate constant for dissociation of iron and movement of the apotransferrin–receptor complex to the cell surface; k_o, rate constant for dissociation of apotransferrin from cell surface receptor. (Adapted from Ciechanover *et al.*, 1983b.)

receptors, or do cell surface receptors internalize and recycle independently of one another? Specifically, are receptors for asialoglycoproteins, transferrin, and insulin internalized and recycled independently?

To address this problem, we first characterized specific insulin receptors on the cell surface of HepG2 cells (Ciechanover *et al.*, 1983; data not shown). We then made use of the experimental protocol described in the legend to Fig. 2B. As can be seen in Fig. 12, binding of ASOR causes a rapid and transient reduction in the number of surface asialoglycoprotein

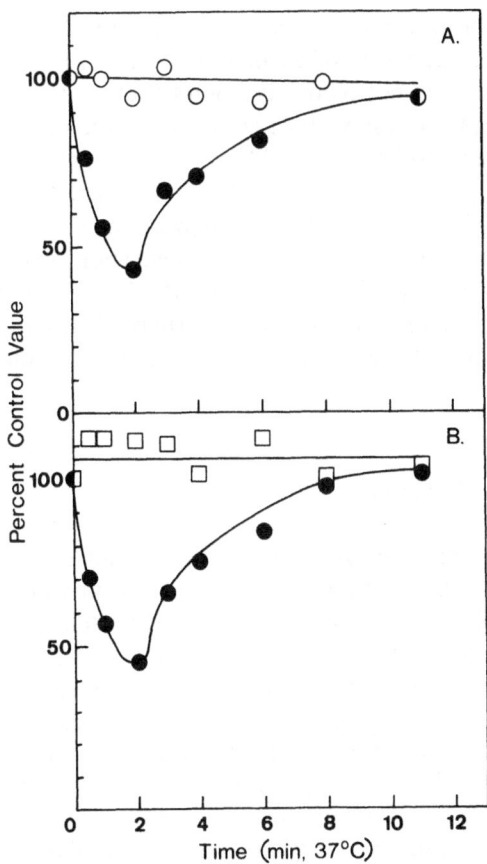

Figure 12. Independent internalization of receptors for asialoglycoprotein, transferrin, and insulin. Cells were treated as described in the legend to Fig. 2B. After "stripping" of the noninternalized ASOR with EDTA, binding of labeled ASOR, transferrin, and insulin was measured as described except that the insulin concentration was 200 nM. (In A, zero-time values for ASOR were 1147 and 209 fmoles/mg protein for total and nonspecific binding, respectively, and 391 and 47 for transferrin. In B, the values were 1096 and 21 for ASOR and 374 and 69 for insulin, respectively.) A: (●——●) ASOR; (○——○) transferrin. B: (●——●) ASOR; (□——□) insulin. All data shown represent averages of quadruplicate determinations and were corrected for nonspecific binding. (Adapted from Ciechanover *et al.*, 1983b.)

receptors. However, there is no alteration in the number of sufrace binding sites for either transferrin (Fig. 12A) or insulin (Fig. 12B). We conclude that binding of ASOR induces a highly specific internalization only of its own receptor.

However, there is an alternative explanation for our results. It is possible that internalization of asialoglycoprotein receptor occurs at the same rate whether or not ligand is bound (Schwartz et al., 1984). The recycling of receptor to the surface could be slowed dramatically if ligand is bound to it, perhaps because a different pathway is used. It should be noted, however, that we were unable to see depletion of cell surface receptors even after very short times of warming (0.5 min) without pre-binding of a ligand. In contrast to our findings, a portion of the LDL (Basu et al., 1981) and the mannose (Tietze et al., 1982) receptors appears to cycle independently of ligand binding. Perhaps the pathway and parameters for recycling of these receptors with or without bound ligand is the same.

5. Concluding Remarks

Our studies, together with many other studies on receptor-mediated endocytosis, have defined an endocytic system in which specific membrane proteins circulate. By moving lipid and protein components within this endocytic system, the cell can regulate some of its more important functions, ingest and degrade macromolecules (such as growth-promoting and structural factors), engage in secretion, and maintain the composition of its organelles and membranes. Furthermore, our studies have defined important modifications of the general pathway of receptor-mediated endocytosis: by differing in their response to a low-pH environment, various receptor–ligand complexes may be sorted and differentially directed within the cell.

However, much is still not known of the complex itinerary of recycling receptors. In particular, the signals that determine the movement of the receptors from one compartment to another are unknown. On the cell surface, these proteins mingle with other proteins that are permanent residents of the plasma membrane. What signals differential segregation such that only the appropriate receptors and other molecules will be directed to the coated pits? What signals internalization of receptor–ligand complexes and their transport to CURL vesicles? How are receptor and ligand segregated from each other following their dissociation? And what signals recycling of the receptor to the cell surface? It is likely that there are signals within the migrant proteins themselves that ticket them for inclusion into transport vesicles to move them to the next location. Since many cell surface receptors probably share the same route, it is not inconceivable that they also share common regulatory signals. Attempts to elucidate these signals will include comparative studies of the primary structure, synthesis, and processing of these proteins together with functional characterization of coated pits and CURL vesicles, the two compartments at which segregation occurs. To this end, we are cloning the gene for the asialoglycoprotein receptor (M. Spiess and H. Lodish, unpublished data) with the aim of genetically dissecting the several key regions of the polypeptide. What now appears simplest to understand are the pH-dependent mechanisms involved in receptor–ligand dissociation, a step that is crucial to receptor recycling.

ACKNOWLEDGMENTS. A.C. is supported by the Melvin Brown Memorial Fellowship through the Israel Cancer Research Fund and by grants from the Leukemia Society of America and the Medical Foundation, Inc. A.L.S. is supported in part by the John and George Hartford Foundation and the National Foundation. This study was supported by grants GM27909-03 and AI08814-14 from the NIH. We thank Miriam Boucher for her skillful help and patience in preparing and typing the manuscript.

References

Aisen, P., and Litowsky, I., 1980, Iron transport and storage proteins. *Annu. Rev. Biochem.* **49**:357–393.

Ashwell, G., and Harford, J., 1982, Carbohydrate-specific receptors of the liver, *Annu. Rev. Biochem.* **51**:531–554.

Ashwell, G., and Morell, A. G., 1974, The role of surface carbohydrates in the hepatic recognition and transport of circulating glycoproteins, *Adv. Enzymol.* **41**:99–128.

Baenziger, J. U., and Fiete, D., 1980, Galactose and N-acetylgalactosamine-specific endocytosis of glycopeptides by isolated rat hepatocytes, *Cell* **22**:611–620.

Basu, S. K., Goldstein, J. L., and Brown, M. S., 1978, Characterization of the low density lipoprotein receptor in membranes prepared from human fibroblasts, *J. Biol. Chem.* **253**:3852–3856.

Basu, S. K., Goldstein, J. L., Anderson, R. G. W., and Brown, M. S., 1981, Monensin interrupts the recycling of low density lipoprotein receptors in human fibroblasts, *Cell* **24**:493–502.

Brown, M. S., Anderson, R. G. W., Basu, S. K., and Goldstein, J. L., 1982, Recycling of cell-suface receptors: Observations from the LDL receptor system, *Cold Spring Harbor Symp. Quant. Biol.* **46**:713–721.

Brown, M. S., Anderson, R. G. W., and Goldstein, J. L., 1983, Recycling receptors: The round-trip itinerary of migrant membrane proteins, *Cell* **32**:663–667.

Ciechanover, A., Schwartz, A. L., and Lodish, H. F., 1983a, The asialoglycoprotein receptor internalizes and recycles independently of the transferrin and insulin receptors, *Cell* **32:**267–275.

Ciechanover, A., Schwartz, A. L., Dautry-Varsat, A., and Lodish, H. F., 1983b, Kinetcis of internalization and recycling of transferrin and the transferrin receptor in HepG2 human hepatoma: Effect of lysosomotropic agents, *J. Biol. Chem.* **258:**9681–9689.

Cole, E. S., and Glass, J., 1983, Transferrin binding and iron uptake in mouse hepatocytes, *Biochim. Biophys. Acta* **762:**102–110.

Dautry-Varsat, A., Ciechanover, A., and Lodish, H. F., 1983, pH and the recycling of transferrin during receptor-mediated endocytosis, *Proc. Natl. Acad. Sci. U.S.A.* **80:**2258–2262.

Dickson, R. B., Hanover, J. A., Willingham, M. C., and Pastan, I., 1983, Prelysosomal divergence of transferrin and epidermal growth factor during receptor-mediated endocytosis, *Biochemistry* **22:**5667–5674.

Enns, C. A., Schindelman, J. E., Tonik, S. E., and Sussman, H. H., 1981, Radioimmunochemical measurement of the transferrin receptor in human trophoblast and reticulocyte membranes with a specific antireceptor antibody, *Proc. Natl. Acad. Sci. U.S.A.* **78:**4222–4225.

Forgac, M., Cantley, L., Wiedenmann,B., Altstiel, L., and Branton, D., 1983, Clathrin-coated vesicles contain an ATP-dependent proton pump, *Proc. Natl. Acad. Sci. U.S.A.* **80:**1300–1303.

Geuze, H. J., Slot, J. W., Van der Lev, P. A., and Scheffer, R. T. C., 1981, Use of colloidal gold particles in double-labeling immunoelectron microscopy of ultrathin frozen tissue sections, *J. Cell Biol.* **89:**653–665.

Geuze, H. J., Slot, J. W., Strous, G. J. A. M., Lodish, H. F., and Schwartz, A. L., 1982, Immunocytochemical localization of the receptor for asialoglycoprotein in rat liver cells, *J. Cell Biol.* **92:**865–870.

Geuze, H. J., Slot, J. W., Strous, G. J. A. M., Lodish, H. F., and Schwartz, A. L., 1983, Intracellular site of asialoglycoprotein receptor–ligand uncoupling: Double-label immunoelectron microscopy during receptor-mediated endocytosis, *Cell* **32:**277–287.

Goldstein, J. L., and Brown, M. S., 1974, Binding and degradation of low density lipoproteins by cultured human fibroblasts, *J. Biol. Chem.* **249:**5153–5162.

Goldstein, J. L., Anderson, R. G. W., and Brown, M. S., 1979, Coated pits, coated vesicles, and receptor mediated endocytosis, *Nature* **279:**679–685.

Gonzalez-Noriega, A., Grubb, J. H., Talkad, V., and Sly, W. S., 1980, Chloroquine inhibits lysosomal enzyme pinocytosis and enhances lysosomal enzyme secretion by impairing receptor recycling, *J. Cell Biol.* **85:**839–852.

Harding, C., and Stahl, P., 1983, Transferrin recycling in reticylocytes: pH and iron are important determinants of ligand binding and processing, *Biochem. Biophys. Res. Commu.* **113:**650–658.

Harford, J., Bridges, K., Ashwell, G., and Klausner, R. D., 1983, Intracellular dissociation of receptor bound asialoglycoproteins in cultured hepatocytes. A pH mediated nonlysosomal event, *J. Biol. Chem.* **258:**3191–3197.

Hubbard, A. L., Wilson, G., Ashwell, G., and Stukenbrok, H., 1979, An electron microscope autoradiographic study of the carbohydrate recognition systems in rat liver, *J. Cell Biol.* **83:**47–64.

Jandl, J. H., and Katz, J. H., 1963, The plasma-to-cell cycle of transferrin, *J. Clin. Invest.* **42:**314–326.

Karin, M., and Mintz, B., 1981, Receptor-mediated endoctyosis of transferrin in developmentally totipotent mouse teratocarcinoma stem cells, *J. Biol. Chem.* **256:**3245–3252.

Klausner, R. D., Ashwell, G., van Renswoude, J., Harford, J. B., and Bridges, K. R., 1983a,

Binding of apotransferrin to K562 cells: Explanation of the transferrin cycle, *Proc. Natl. Acad. Sci. U.S.A.* **80**:2263–2266.

Klausner, R. D., van Renswoude, J., Ashwell, G., Kempf, C., Schechter, A. N., Dean, A., and Bridges, K. R., 1983b, Receptor mediated endocytosis of transferrin in K562 cells, *J. Biol. Chem.* **258**:4715–4724.

Klausner, R. D., van Renswoude, J., Kempf, C., Rao, K., Bateman, J. L., and Robbins, A., 1984, Failure to release iron from transferrin in a Chinese hamster ovary cell mutant pleiotropically defective in endocytosis, *J. Cell Biol.* **98**:1098–1101.

Knowles, b. B., Howe, C. C., and Aden, D. P., 1980, Human hepatocellular carcinoma cell lines secrete the major plasma proteins and hepatitis B surface antigen, *Science* **209**:497–499.

Lamb, J. E., Ray, F., Ward, J. H., Kushner, J. P., and Kaplan, J., 1983, Internalization and subcellular localization of transferrin and transferrin receptors in HeLa cells, *J. Biol. Chem.* **258**:8751–8758.

Marsh, M., Bolzau, E., and Helenius, A., 1983, Penetration of Semiliki Forest virus from acidic prelysosomal compartment, *Cell* **32**:931–940.

Marshall, S., Green, A., and Olefsky, J. M., 1981, Evidence for recycling of insulin receptors in isolated rat adipocytes, *J. Biol. Chem.* **256**:11464–11470.

Maxfield, F. R., 1982, Weak bases and ionophores rapidly and reversibly raise the pH of endocytic vesicles in cultured mouse fibroblasts, *J. Cell Biol.* **95**:676–681.

Maxfield, F. R., Schlessinger, J., Shechter, Y., Pastan, I., and Willingham, M. C., 1978, Collection of insulin, EGF and α_2-macroblogin in the same patches on the surface of cultured fibroblasts and common internalization, *Cell* **14**:805–810.

Mellman, I. S., Plutner, H., Steinman, R. M., Unkeless, J. C., and Cohn, Z. A., 1983, Internalization and degradation of macrophage Fc receptors during receptor mediated endocytosis, *J. Cell Biol.* **96**:887–895.

Nunez, M.-T., Cole, E. S., and Glass, J., 1983, The reticulocyte plasma membrane pathway of iron uptake as determined by the mechanism of α,α'-dipyridyl inhibition, *J. Biol. Chem.* **258**:1146–1151.

Ohkuma, S., and Poole, P., 1978, Fluorescence probe measurement of the intralysosomal pH in living cells and the perturbation of pH by various agents, *Proc. Natl. Acad. Sci. U.S.A.* **75**:3327–3331.

Pastan, I. H., and Willingham, M. C., 1981, Journey to the center of the cell: Role of the Receptosome, *Science* **214**:504–509.

Posner, B. I., Josefberg, Z., and Bergeron, J. J. M., 1978, Intracellular polypeptide hormone receptors, *J. Biol. Chem.* **253**:4067–4073.

Schneider, D. L., 1981, ATP-dependent acidification of intact and disrupted lysosomes, *J. Biol.Chem.* **256**:3858–3864.

Schwartz, A. L., and Rup, D., 1983, Biosynthesis of the human asialoglycoprotein receptor, *J. Biol. Chem.* **258**:11249–11255.

Schwartz, A. L., Rup, D., and Lodish, H. F., 1980, Difficulties in the quantification of asialoglycoprotein receptors on the rat hepatocyte, *J. Biol. Chem.* **255**:9033–9036.

Schwartz, A. L., Marshak-Rothstein, A., Rup, D., and Lodish, H. F., 1981a, Identification and quantification of the rat hepatocyte asialoglycoprotein receptor, *Proc. Natl. Acad. Sci. U.S.A.* **78**:3348–3352.

Schwartz, A. L., Fridovich, S. E., Knowles, B. B., and Lodish, H. F., 1981b, Characterization of the asialoglycoprotein receptor in a continuous hepatoma line, *J. Biol. Chem.* **256**:8878–8881.

Schwartz, A. L., Fridovich, S. E., and Lodish, H. F., 1982a, Kinetics of internalization and recycling of the asialoglycoprotein receptor in a hepatoma cell line, *J. Biol. Chem.* **257**:4230–4237.

Schwartz, A. L., Geuze, H. J., and Lodish, H. F., 1982b, Recycling of the asialoglycoprotein recpetor: Biochemical and immunocytochemical evidence, *Phil. Trans. R. Soc. Lond. [Biol.]* **300**:229–235.

Schwartz, A. L., Bolognesi, A., and Fridovich, S. E., 1984, Recycling of the asialoglyco-protein receptor and the effect of lysosomotropic amines in hepatoma cells, *J. Cell Biol.* **98**:732–738.

Stahl, P., Schlesinger, P. H., Sigardson, E., Rodman, J. S., and Lee, A., 1980, Receptor mediated pinocytosis of mannose glycoconjugates by macrophages: Characterization and evidence for receptor recycling, *Cell* **19**:207–215.

Stone, D. K., Xie, X.-S., and Racker, E., 1983, An ATP-driven proton pump in clathrin-coated vesicles, *J. Biol. Chem.* **258**:4059–4062.

Sullivan, A. L., Grasso, J. A., and Weintraub, L. R., 1976, Micropinocytosis of transferrin by developing red cells: An electron-microscopic study utilizing ferritin-conjugated transferrin and ferritin-conjugated antibodies to transferrin, *Blood* **47**:133–143.

Tanabe, T., Pricer, W. E., and Ashwell, G., 1979, Subcellular membrane topology and turnover of rat hepatic binding protein specific for asialoglycoproteins, *J. Biol. Chem.* **254**:1038–1043.

Tietze, C., Schlesinger, P., and Stahl, P., 1982, Mannose-specific endocytosis receptor of alveolar macrophages: Demonstration of two functionally distinct intracellular pools of receptor and their role in receptor recycling, *J. Cell Biol.* **92**:417–424.

Tolleshaug, H., and Berg, T., 1979, Chloroquine reduces the number of asialoglycoprotein receptors in the hepatocyte plasma membrane, *Biochem. Pharmacol.* **28**:2912–2922.

Tolleshaug, H. T., Berg, T., Nilsson, N., and Noren, K. R., 1977, Uptake and degradation of ^{125}I-labeled asialofetuin by isolated rat hepatocytes, *Biochem. Biophys. Acta* **499**:73–84.

Trowbridge, I. S., and Omary, M. B., 1981, Human cell surface glycoprotein related to cell proliferation is the receptor for transferrin, *Proc. Natl. Acad. Sci. U.S.A.* **78**:3039–3043.

Tycko, B., and Maxfield, F. R., 1982, Rapid acidification of endocytic vesicles containing α_2-macroglobulin, *Cell* **28**:643–651.

Tycko, B., Keith, C. H., and Maxfield, F. R., 1983, Rapid acidification of endocytic vesicles containing asialoglycoprotein in cells of a human hepatoma line, *J. Cell Biol.* **97**:1762–1776.

van Bockxmeer, F. M., and Morgan, E. H., 1977, Identification of transferrin receptors in reticulocytes, *Biochim. Biophys. Acta* **468**:437–450.

van Renswoude, J., Bridges, K. R., Harford, J. B., and Klausner, R. D., 1982, Receptor mediated endocytosis of transferrin and the uptake of Fe in K562 cells: Identification of a nonlysosomal acidic compartment, *Proc. Natl. Acad. Sci. U.S.A.* **79**:6186–6190.

van Leuven, F., Cassiman, J.-J., and Van den Bergh, H., 1980, Primary amines inhibit recycling of α_2M receptors in fibroblasts, *Cell* **20**:37–43.

Wall, D. A., and Hubbard, A. L., 1981, Galactose-specific recognition system of mammalian liver: Receptor distribution on the hepatocyte cell surface, *J. Cell Biol.* **90**:687–696.

Ward, J. H., Kushner, J. P., and Kaplan, J., 1982, Regulation of HeLa cell transferrin receptors, *J. Biol. Chem.* **257**:10317–10323.

Willingham, M. C., Pastan, I. H., and Sahagian, G. G., 1983, Ultrastructural immunocy-tochemical localization of the phosphomannosyl receptor in Chinese hamster ovary (CHO) cells, *J. Histochem. Cytochem.* **31**:1–11.

Willingham, M. C., Hanover, J. A., Dickson, R. B., and Pastan, I., 1984, Morphological characterization of the pathway of transferrin endocytosis and recycling in human KB cells, *Proc. Natl. Acad. Sci. U.S.A.* **81**:175–179.

Zeitlin, P. L., and Hubbard, A. L., 1982, Cell surface distribution and intracellular fate of asialoglycoproteins: A morphological and biochemical study of isolated rat hepatocytes and monolayer cultures, *J. Cell Biol.* **92:**634–647.

Zigmond, S. H., Sullivan, S. J., and Lauffenburger, D. A., 1982, Kinetic analysis of chemotactic peptide receptor modulation, *J. Cell Biol.* **92:**34–43.

The Nicotinic Acetylcholine Receptor

Its Structure, Multiple Binding Sites, and Cation Transport Properties

MICHAEL A. RAFTERY, BIANCA M. CONTI-TRONCONI,
SUSAN M. J. DUNN, REBECCA D. CRAWFORD,
and DAVID MIDDLEMAS

1. Introduction

The acetylcholine receptor (AcChR) from *Torpedo* electroplax has been isolated both in its native membrane-bound state and by affinity chromatography after solubilization (see Conti-Tronconi and Raftery, 1982). Purified membrane fragments can reseal, forming closed, right-side-out vesicles, which can be used for both functional and structural studies. The major physiochemical properties of *T. californica* AcChR protein are summarized in Fig. 1.

2. Subunits of Torpedo AcChR

It is now established that *Torpedo* AcChR is formed from four different proteins, which in *T. californica* have apparent molecular weights of 40,000, 50,000, 60,000, and 65,000 as determined by SDS gel electrophoresis (see Conti-Tronconi and Raftery, 1982) (Fig. 2). This complex subunit structure was first described from this laboratory (Raftery *et al.*,

---
MICHAEL A. RAFTERY, BIANCA M. CONTI-TRONCONI, SUSAN M. J. DUNN, REBECCA D. CRAWFORD, AND DAVID MIDDLEMAS • Division of Chemistry, California Institute of Technology, Pasadena, California 91125.

STRUCTURE

SUBUNIT COMPOSITION ...... 40, 50, 60, 65 x 10³ DALTONS
SUBUNIT STOICHIOMETRY 2:1:1:1
MOLECULAR WEIGHT .... 270 ± 30 x 10³ DALTONS (EXPERIMENTAL)
.... 255 x 10³ DALTONS (CALCULATED)

PHYSICAL PROPERTIES

S VALUE 9S
.... 13.7S AS DIMER
STOKES RADIUS .... 72 Å
pl 4.9
SPECIFIC ACTIVITY .... ONE α-BUTX PER 110 ± 15 x 10³ DALTONS

SEQUENCE HOMOLOGY AT N-TERMINI OF ACETYLCHOLINE RECEPTOR SUBUNITS

```
      2    4    6    8   10  12  14  16  18  20  22  24  26  28  30  32  34  36  38  40  42  44  46  48  50  52  54  56
40K  S E H E T R L V A N L L   E N Y N K V I R P V E H M T H F V D I T V G L Q L I Q L I S V D E V N Q I V E T N V
50K  S V M E D T L L S V L F   E T Y N P K V R P A Q T V G D K V T V R V G L T L T N L L I L N E K I E E M R T N V
60K  E N E E G R L I E K L L   G D Y D K R I I P A K T L D H I I D V T L K L T L T N L I S L N E M E E A L T T N V
65K  V N E E E R L I N D L L I V N K Y N K H V R P V K H N N E V V N I A L S L T L S N L I S L K E T D E T L T S N V
```

CARBOHYDRATE .... ~75 RESIDUES/MOLECULE ⎫
O-SUBSTITUTED SER .... ~22 RESIDUES/MOLECULE ⎪ OF AcChR
O-SUBSTITUTED THR .... ~23 RESIDUES/MOLECULE ⎬
PHOSPHOSERINE .... ~7 RESIDUES/MOLECULE ⎭

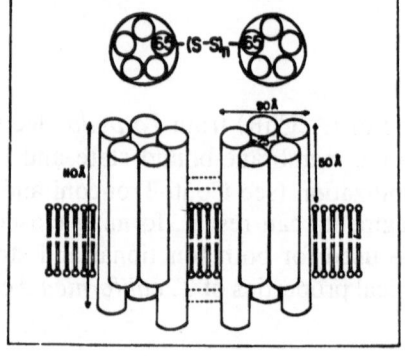

Figure 1. Properties of *Torpedo* AcChR (see Raftery *et al.*, 1980).

Figure 2. Sodium dodecylsulfate gel electrophoresis scans of *T. californica* AcChR (1), *Electrophorus* AcChR (3), and fetal calf AcChR (4): SDS gel scans of purified AcChR-rich membrane fragments before (dotted line) and after (solid line) alkali extraction (Elliott *et al.*, 1980). All the extrinsic membrane proteins are removed by this treatment, and the AcChR subunits are the only polypeptides left. All gels were stained with Coomassie blue. Numbers shown are $M_r \times 10^{-3}$. Negative-staining electron micrographs of these AcChRs are shown in the insets (1,3,4, 250,000×; 2, 125,000×).

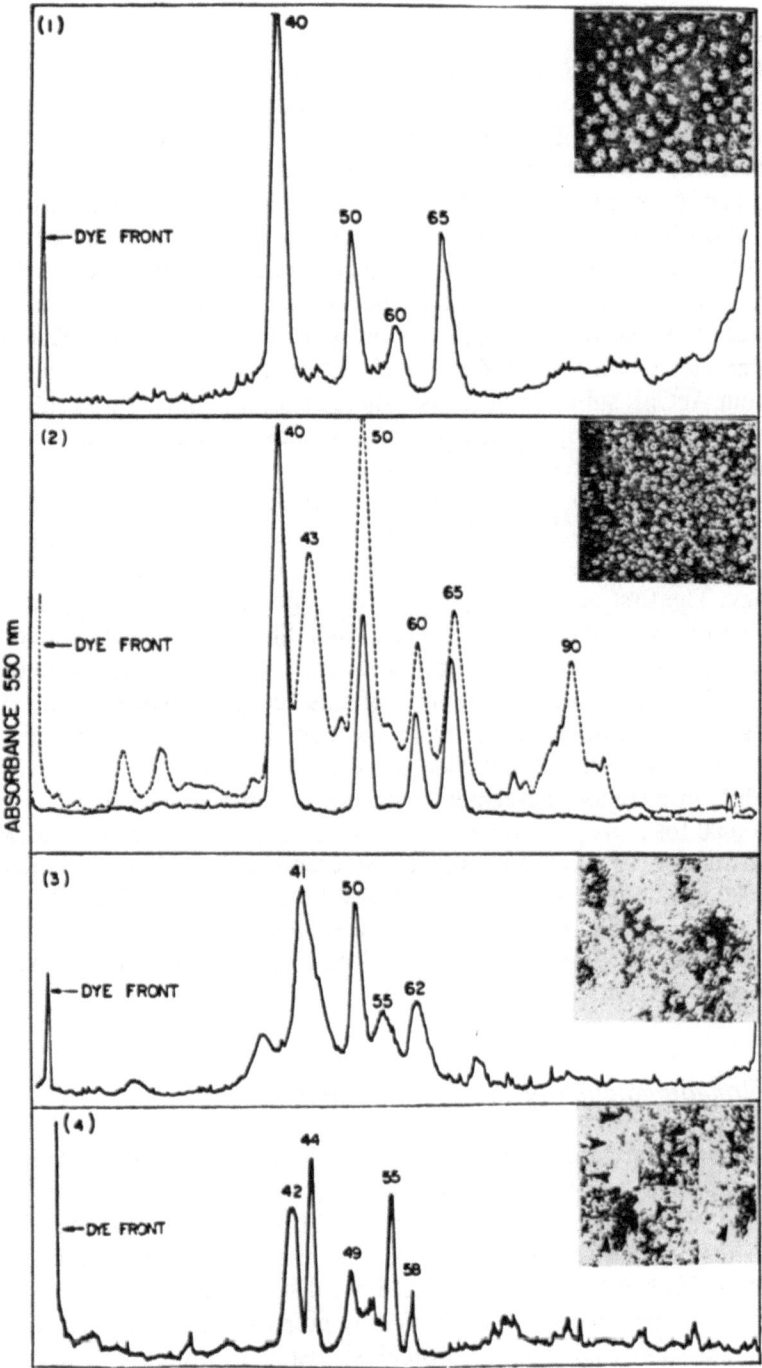

1974) and confirmed shortly thereafter (Weill *et al.*, 1974). This correct model of the AcChR formed by four subunits was challenged for many years by persistent erroneous reports of AcChR preparations composed of only one subunit of M_r 40,000 (Sobel *et al.*, 1977; Heidmann and Changeux, 1978, 1980), suggesting an alternative model of the AcChR as a hexamer of six identical "proteins" (Sobel *et al.*, 1979). This controversy was resolved by the demonstration that the four peptides present in pure AcChR preparations are highly homologous proteins (Raftery *et al.*, 1980) (Fig. 1). By amino-terminal microsequencing it was also possible to demonstrate that the four subunits are present in the AcChR molecule in a stoichiometry of $\alpha_2\beta\gamma\delta$ (Raftery *et al.*, 1980). The extensive homology of the four AcChR subunits suggests a shared ancestry and allows generation within the postsynaptic membrane of a pseudosymmetric protein complex formed by five related proteins.

In the recent past, the complete amino acid sequences of the precursors of all *Torpedo* subunits have been deduced from sequencing of corresponding nucleic acid clones obtained by recombinant DNA technology. The first of these sequences was reported for the α-subunit of *T. marmorata* AcChR (Sumikawa *et al.*, 1982). Shortly thereafter, sequences were published for precursors of the γ-chain of *T. californica* (Claudio *et al.*, 1982) and for all the subunits of this receptor (Noda *et al.*, 1982, 1983a,b). The calculated M_r from the cDNA sequences of all four *T. californica* AcChR subunits leads to an M_r of 268,078 for the intact AcChR, in excellent agreement with our experimental value of 270,000 $\pm$ 30,000 for a preparation consisting of essentially all 9 S AcChR (Martinez-Carrion *et al.*, 1976) and considerably higher than the value of 250,000 usually referred to as the M_r of *T. californica* AcChR (Reynolds and Karlin, 1978). At a later date, the sequence of the precursor of the α-subunit of *T. marmorata* was confirmed by other workers (Devillers-Thiery *et al.*, 1983).

3. Nonequivalence of the Two α-Subunits

Bromoacetylcholine (an agonist) can label one or both α-subunits of *T. californica* AcChR, whereas MBTA (an antagonist) labels only one subunit (Conti-Tronconi and Raftery, 1982). The nonequivalence of these two sites could arise from different microenvironments since each α-chain must be flanked by the other related but distinct subunits. We have purified the α-subunits of *T. californica* by preparative SDS gel electrophoresis and submitted them to proteolysis using V_8 protease. The peptide pattern obtained is identical to that described by Gullick *et al.* (1981).

One feature is that two prominent peptides of approximate M_r 17,000 and 19,000 (V_8 17 and V_8 19) are consistently present in similar amounts. It has been reported (Gullick *et al.*, 1981) that only the V_8 19K peptide can be labeled by MBTA under mild reducing conditions and that only the V_8 17K peptide stains for carbohydrate. We have found that V_8 17 and V_8 19 have the same NH_2-terminal sequence starting at residue 47 of the α-subunit. Since extensive enzymatic hydrolysis does not convert V_8 19 to V_8 17, it is most likely that these peptides differ only in their degree of glycosylation. These results can therefore serve to explain the non-equivalence of high-affinity ligand binding to the α-subunits (Conti-Tronconi and Raftery, 1982), since the region of N-glycosylation is thought to be close to the ligand binding sites on these subunits (Noda *et al.*, 1982).

4. The AcChR is a Transmembrane Protein

A direct demonstration that the AcChR is a transmembrane protein has come from morphological studies of binding of anti-AcChR antibodies to the outside and inside surfaces of *Torpedo* postsynaptic membranes (Strader *et al.*, 1979). In order to determine which of the AcChR subunits is transmembrane, we have studied the effect of proteases acting either inside or outside membrane vesicles containing the AcChR (Strader and Raftery, 1980; Conti-Tronconi *et al.*, 1982a), and all subunits were susceptible to tryptic degradation from either side of the membrane. When trypsin was added outside, all four peptides disappeared with the same time course, whereas when the enzyme acted from within the vesicles, the subunits were degraded at a rate proportional to their molecular weight. This leads to the conclusion that all the AcChR subunits span the membrane and that they protrude outside to about the same extent and inside to an extent approximately proportional to their M_r (scheme of Fig. 1).

5. Exposure of AcChR Subunits to the Lipid Bilayer

All four AcChR subunits in membrane fragments from *T. californica* were labeled using [^{3}H]-adamantanediazirine (Fig. 3). This hydrophobic probe, which on irradiation generates [^{3}H]-adamantylidene, was proposed as an alternative to aryl nitrene probes (Bayley and Knowles, 1978). Carbenes label both saturated and unsaturated membrane lipids more efficiently than nitrenes, suggesting that they may be superior for labeling lipid-exposed regions of intrinsic membrane proteins (Bayley and

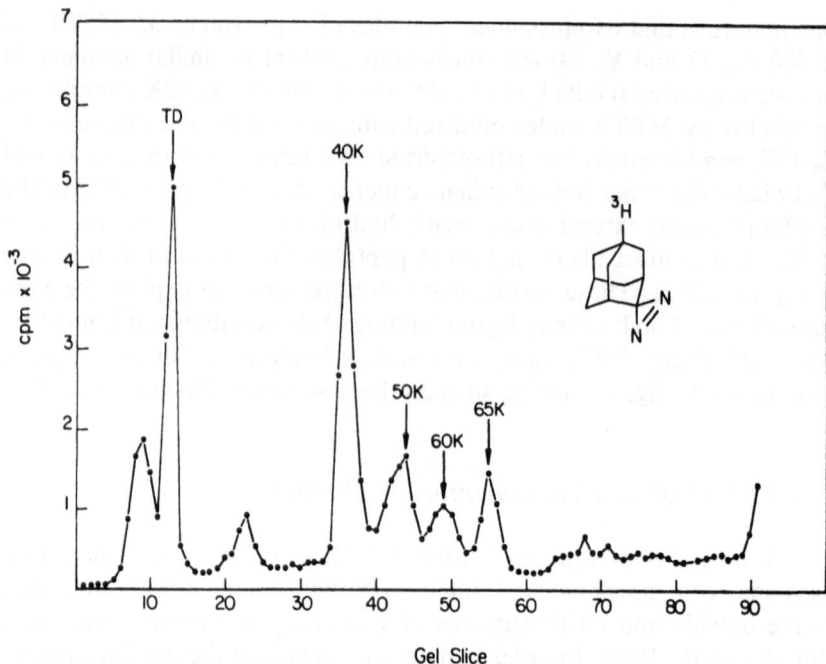

Figure 3. Polyacrylamide gel electrophoresis scan of AcChR-enriched membrane fragments labeled with [³H]-adamantanediazirine (596 mCi/mmole). Membrane fragments were incubated in the dark with 33 μM [³H]-adamantanediazirine and 20 mM reduced glutathione followed by UV irradiation. After electrophoresis, the gel was sliced and counted (TD-tracking dye).

Knowles, 1980). Since all four of the AcChR homologous subunits are in contact with the hydrocarbon core of the lipid bilayer, it is possible that all the subunits interact with the surrounding membrane in a related fashion.

6. The AcChR from Torpedo as a Model for Other Nicotinic AcChRs

It is difficult to study the AcChR from sources other than *Torpedo* because of the much lower AcChR content and the high levels of intrinsic protease activity. Similarities in the pharmacology, antigenicity, morphology, and physical properties as well as observation of complex polypeptide patterns on SDS gel electrophoresis (Conti-Tronconi and Raftery, 1982) reminiscent of the subunit pattern of *Torpedo* AcChR suggest the likelihood of close structural and functional similarities between AcChRs from various sources.

Torpedo (a marine elasmobranch) and *Electrophorus* (a freshwater teleost) are highly diverged species that evolved separately from the primordial vertebrate stock (~400 million years ago), and the presence of electric organs in these two species is a result of convergent evolution. Sufficient AcChR can be isolated from *Electrophorus* for structural analysis of its subunits. In Fig. 2 the SDS-PAGE pattern and the morphology of purified *Electrophorus* AcChR are depicted. This AcChR contains four main polypeptides in the same M_r range as *Torpedo* AcChR, and both types of AcChR have the same rosettelike morphology (Conti-Tronconi and Raftery, 1982). The amino-terminal amino acid sequence was determined for each of the four peptides (Conti-Tronconi *et al.*, 1982) [Fig. 4(1)]. All four subunits have distinct but homologous sequences, and the degree of identity between pairs ranges from 47.5% to 37.5%. Conservative substitutions tend to further increase the degree of similarity between the subunits. In Fig. 4(2), comparison is made between the amino-terminal sequences of *Torpedo* and *Electrophorus* AcChR subunits of comparable M_r, and the extent of sequence identity (up to 62.5%) is indicated. The greatest level of identity was between the corresponding

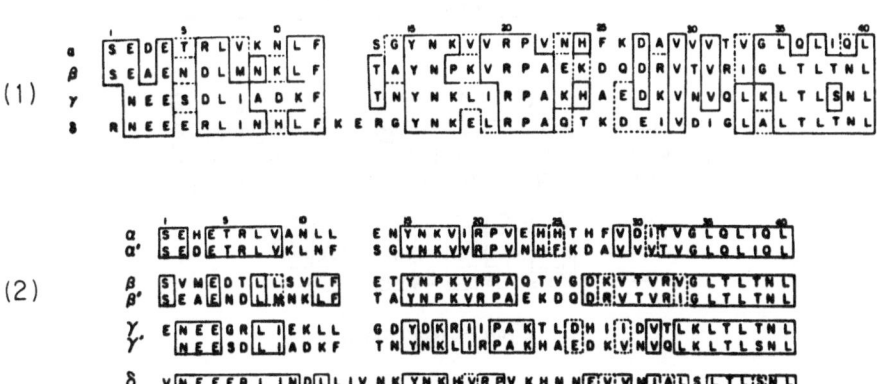

Figure 4. Amino-terminal sequences of the four subunits of *Electrophorus* (1) and fetal calf (4) AcChR and (2) corresponding subunits of *Torpedo* (α, β, γ, δ) and *Electrophorus* (α', β', γ', δ') are compared. In 5, the α and β subunits from fetal calf, *Torpedo*, and *Electrophorus* AcChR are compared. In 3, a phylogenetic tree generated from the amino-terminal sequence data on the four AcChR subunit types from *T. californica* (α_1, β_1, γ_1, δ_1) and *E. electricus* (α_2, β_2, γ_2, δ_2) by using the best-fit matrix method. Each branch length represents the "accepted point mutations" (PAMs) per 100 amino acid residues that occurred in generating the contemporary subunits of both *Torpedo* and *Electrophorus* AcChR.

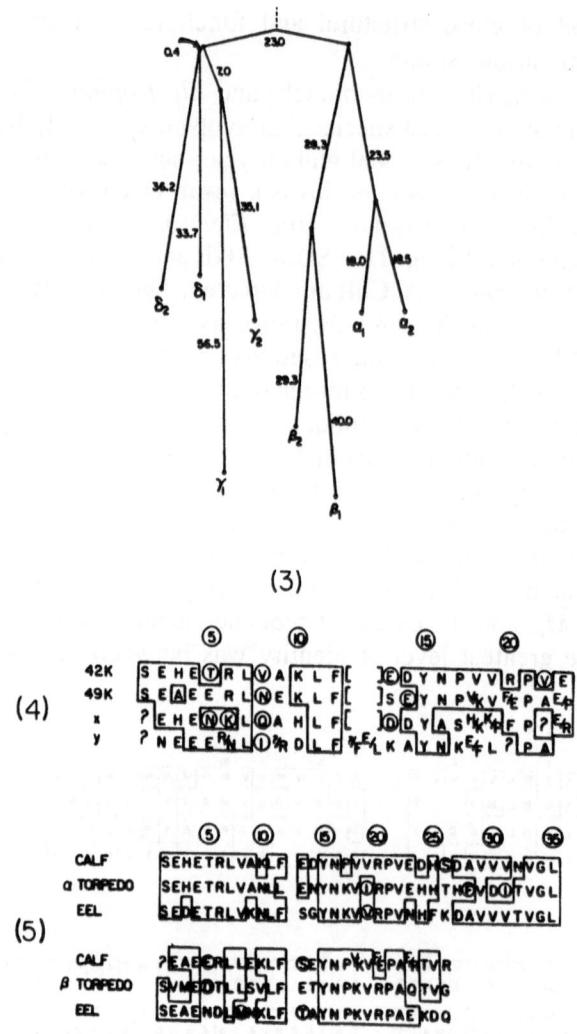

Figure 4. (continued)

subunits of each species, i.e., α to α₁, etc. Simultaneous quantitative sequencing of the peptides in preparations of intact *Electrophorus* AcChR allowed determination of the subunit stoichiometry (Table I), and molar ratios of 2:1:1:1 were obtained for the subunits of M_r 41,000 (α_1), 50,000 (β_1), 55,000 (γ_1), and 62,000 (δ_1), respectively. *Electrophorus* AcChR is therefore also a pseudosymmetric pentameric complex formed from four different polypeptides, and from their 2:1:1:1 stoichiometry a M_r of

Table I. Acetylcholine Receptor Subunit Stoichiometry[a]

Subunit	Residues			Triton-solubilized AcChR		Membrane-bound AcChR		Average
				Preparation 1	Preparation 2	Preparation 3	Preparation 4	
Torpedo californica electric organ								
α	Ala-9	Asn-10	Asn-14	1.93 ± 0.13	1.92 ± 0.14	1.96 ± 0.04	2.05 ± 0.16	1.97 ± 0.12
β	Ser-9	Val-10	Thr-14	1.02 ± 0.08	1.07 ± 0.09	1.03 ± 0.04	1.02 ± 0.01	1.03 ± 0.06
γ	Glu-9	Lys-10	Asp-14	1.00 ± 0.10	1.02 ± 0.21	1.01 ± 0.03	1.00 ± 0.07	1.01 ± 0.10
δ	Asn-9	Asp-10	Val-14	1.04 ± 0.07	1.00 ± 0.13	1.01 ± 0.08	0.93 ± 0.08	0.99 ± 0.09
Electrophorus electricus electric organ								
α_1	Val-8, Gly-14					1.90 ± 0.19	1.96 ± 0.18	1.93 ± 0.10
β_1	Met-8, Ala-14					1.02 ± 0.02	0.99 ± 0.01	1.01 ± 0.02
γ_1	Ala-8, Tyr-14					1.10 ± 0.27	1.04 ± 0.12	1.07 ± 0.20
δ_1	Ile-8, Glu-14					1.02 ± 0.07	1.00 ± 0.05	1.04 ± 0.06
Mammalian muscle (fetal calf)								
α	Val-8							2.16
β	Leu-8							0.95
x	Gln-8							0.92
y	Ile-8							0.98

[a] Values are means ± S.E.M.

249,000 can be calculated, which fits with experimental determinations (Conti-Tronconi and Raftery, 1982). This value is also consistent with the size of *Electrophorus* AcChR as determined by electron microscopy (Conti-Tronconi and Raftery, 1982).

A genealogical tree suggesting the evolutionary pathway by which the four contemporary subunits of both *Electrophorus* and *Torpedo* AcChRs can be generated from a single ancestral sequence via minimum nucleotide substitution is shown in Fig. 4C. The tree demonstrates that the two gene duplications that produced the four subunits present in the AcChR occurred before the divergence of *Electrophorus* from *Torpedo* at the beginning of vertebrate evolution if not before then.

7. Structure of Mammalian Muscle AcChR

Purified solubilized AcChR from fetal calf muscle has physical properties comparable with *Torpedo* AcChR monomers (Gotti *et al.*, 1982), and negatively stained preparations contained the same rosettelike structures with a diameter of ~95 Å and an electron-dense central pit as those found in *Torpedo* and *Electrophorus* AcChR preparations (Fig. 2).

On SDS gel electrophoresis, the purified mammalian AcChR was resolved into five major polypeptides having molecular weights (M_r) of 42, 44, 49, 55, and 58K (Fig. 2). The peptide of M_r 44K was identified as actin, since it contains 3-methylhistidine and binds antiactin antibodies, and the peptide of M_r 42K is labeled by [^{3}H]-bromoacetylcholine (Conti-Tronconi *et al.*, 1982c), suggesting close similarity to the α chains of *Torpedo* and *Electrophorus* AcChR.

The primary structure of each polypeptide of fetal calf AcChR was investigated by amino-terminal microsequence analysis (Conti-Tronconi *et al.*, 1982c). Three subunits (M_r 42K, 49K, and 53K) yielded distinct but homologous sequences. Because of the lack of identifiable sequences associated with the 55K and 58K polypeptides (most likely a result of blockade of their amino terminals during isolation), intact AcChR preparations were analyzed. Four homologous sequences were obtained (Fig. 4). Three of these were identical with those independently determined, and the fourth sequence could therefore be deduced. The four sequences were present in a stoichiometry of 2:1:1:1 (Table I), which demonstrates that mammalian muscle nicotinic receptor is a pentameric complex composed of two equivalent and three pseudoequivalent subunits, as are the AcChRs of *Torpedo* and *Electrophorus*.

In Fig. 4, comparison is made among the amino-terminal sequences of the two lighter subunits (α, β) of AcChR from fetal calf muscle, *Tor-*

pedo, and *Electrophorus*. Among the α-subunits, 51% of the residues were identical, and in an additional 36% of the positions two of the three polypeptides had the same residue. In the case of β-subunits, 38% of the residues were identical, and a comparison of all six polypeptide sequences shows that 23% of all positions were identical for the first 26 residues and an additional 19% identity was observed in five out of the six polypeptides at other positions.

8. The AcChR as a Cation Channel

Correlation of the subunit structure of the AcChR with its physiological function necessitates the use of quantitative methods to evaluate the efficiency of receptor-mediated cation transport *in vitro*. We developed a rapid kinetic method for this purpose that allows spectroscopic detection of monovalent cation transport on a millisecond time scale (Moore and Raftery, 1980). A water-soluble fluorophore (8-aminonaphthalene-1,3,6,-trisulfonate, ANTS, or pyrenetetrasulfonate) is trapped within AcChR-enriched vesicles, and the fluorescence quench caused by agonist-mediated inward transport of thallium (I) ion can be monitored by stopped-flow spectroscopy. The rate of fluorescence decay is dependent on the number of activated AcChRs, and this can be used to determine the ion transport efficiency of single AcChR molecules. In Fig. 5A, the effect of carbamylcholine (Carb) on the Tl^+-mediated fluorescence decay rate of ANTS is shown. The dose–response curve obtained is plotted in Fig. 5B, demonstrating saturation of the rate ($\sim 1500 \ sec^{-1}$) at high agonist concentration. This was shown (Moore and Raftery, 1980) to correspond to 7×10^6 ions per second per receptor in *Torpedo* Ringers, a value very close to that ($\sim 10^7$) estimated for the AcChR at the neuromuscular junction *in vivo* (see Lester, 1977). The experiments demonstrate that the receptor in isolated vesicles is fully functional.

Preparations of membrane-bound AcChR composed only of the four homologous polypeptides previously discussed transported Tl^+ as efficiently (Fig. 5C,D) as membranes containing other protein components such as the 43K protein discovered by Sobel *et al.* (1977) and promoted as a separate ionophore (Sobel *et al.*, 1978) associated with the AcCh binding protein ("AcCh modulator"). Such erroneous notions are eliminated by the above results, which show that the pentameric complex $\alpha_2\beta\gamma\delta$ constitutes a complete fully functional receptor. Our conclusions in this regard were confirmed by reconstitution studies using detergent-solubilized AcChR preparations (Wu *et al.*, 1981) in which the quantitative Tl^+ flux assay was used to analyze the efficiency of cation transport by

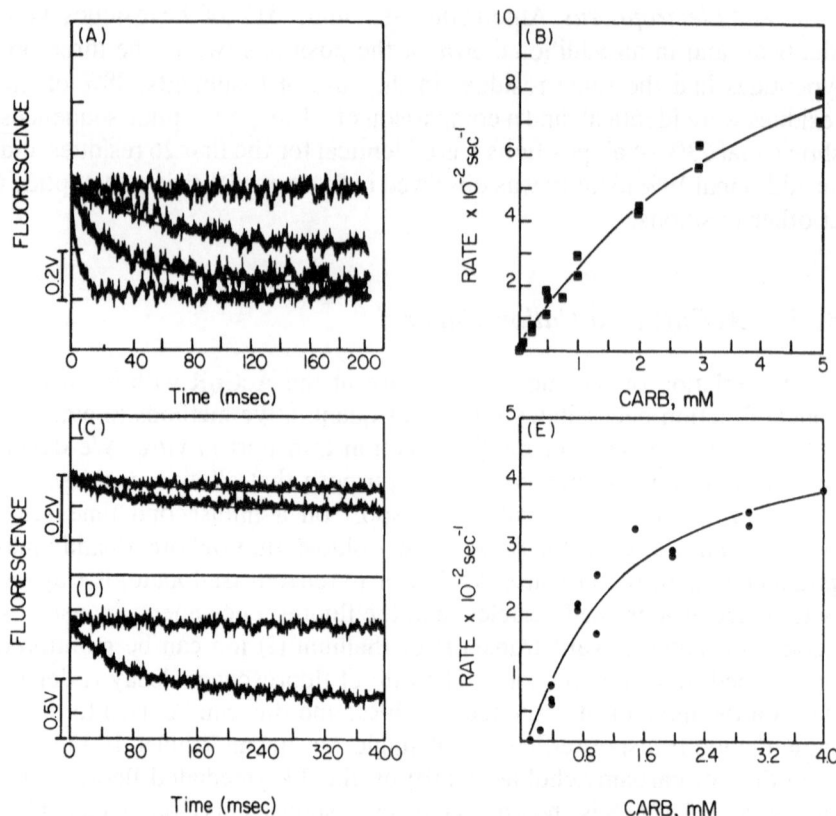

Figure 5. (A) Kinetics of T1$^+$ influx into ANTS-loaded vesicles induced by Carb concentrations (top to bottom trace) of 0, 50, 100, and 500 μM Carb. (B) Effect of Carb concentration on flux rate constant. Solid line is best fit to two-ligand binding model $k_{app} = k_{max}/(1 + K_d/[L])^2$ using $k_{max} = 1100$ sec^{-1} and $K_d = 1$ mM. (C) Comparison of T1$^+$ influx observed using AcChR-enriched vesicles before and (D) after alkali extraction of nonreceptor proteins. Upper trace in each panel is leakage in absence of agonist, and lower is in presence of 100 μM Carb. (D) Effect of Carb concentration on flux rate using reconstituted vesicles. Best fit to two-ligand binding model with $k_{max} = 490$ sec^{-1} and $K_d = 500$ μM.

AcChR monomers (9 S) and dimers (13.7 S) (Wu and Raftery, 1981b). The specific cation transport per AcChR molecule (monomer or dimer) in such reconstituted systems was within a factor of two of those discussed above for native membrane preparations (Moore and Raftery, 1980).

   The midpoints of the dose–response curves for Carb-mediated flux in native membranes (Fig. 5B) or in reconstitued vesicles (Fig. 5E) are in the millimolar range, in agreement with physiological studies (see Adams, 1981), as were values obtained for AcCh (~100 μM) and other

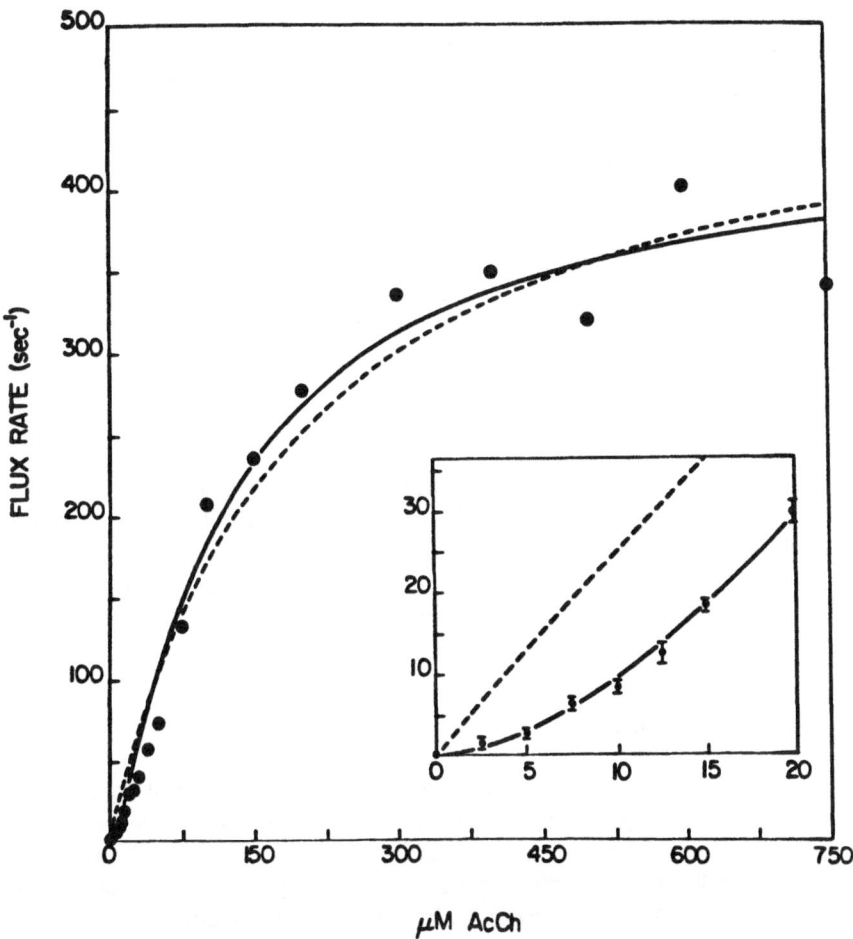

Figure 6. Effect of acetylcholine on kinetics of thallium influx (50 mM after mixing) monitored by fluorescence changes of pyrene tetrasulfonic acid (17 mM) trapped within AcChR-enriched membrane vesicles. Solid line is fit to two-ligand binding model (see legend to Fig. 5B) using $k_{max} = 440$ sec^{-1} and $K_d = 57$ μM. Dashed line is fit to single-ligand binding model $k_{app} = k_{max}[L]/(K_d + [L])$ with $k_{max} = 486$ sec^{-1} and $K_d = 186$ μM. Insert is expansion of low-concentration region and lines obtained from best-fit parameters above.

agonists (see Table II). Such high concentrations of agonist for activation are at variance with many direct studies of ligand binding to AcChR in either the resting or desensitized state and emphasize the difficulty of relating binding studies conducted *in vitro* to relevant physiological effects. Recent results (Fig. 6) that we have obtained using pyrenetetrasulfonic acid as an entrapped fluorophore (suggested by Haugland, 1982)

Table II. Effects of Agonists on Fluorescence of NBD-Labeled AcChR Before and After Solubilization and Comparison with K_d Values from Thallium Flux Data

| Agonist | K_d(Fluorescence)(mM) | | K_d(Thallium flux)(mM) | |
	Membrane bound	Solubilized	$(A)^a$	$(B)^b$
AcCh	0.09 ± 0.02	1.5 ± 0.2	0.16	0.057
Carb	0.96 ± 0.17	7.9 ± 1.2	5.0	1.0
Nicotine	0.60 ± 0.21	0.24 ± 0.17	0.26	0.13
PTA	0.47 ± 0.09	—	0.29	0.11
choline	35	—	—	—

^{a} Data fitted to single ligand binding model: $k_{app} = k_{max}[L]/(K_d + [L])$.
^{b} Data fitted to model in which two ligand binding sites must be occupied for channel to open: $k_{app} = k_{max}[L]^2/(K_d + [L])^2$.

in place of ANTS clearly demonstrate that the flux response is a co-operative phenomenon with a Hill coefficient of 1.7. Therefore, a major question with respect to AcChR mechanism is whether this cooperative flux response reflects multiple ligand binding in a positively cooperative manner (see below) or some other mechanism.

9. Agonist Binding to Torpedo AcChR

Measurement of the binding of agonists to the AcChR at equilibrium has demonstrated that under these conditions the receptor is desensitized, the ion channel is closed, and the affinity for agonists is high ($K_d \sim 10$ nM for AcCh). It is therefore necessary to study conformational transitions of the AcChR occurring under preequilibrium conditions (Conti-Tronconi and Raftery, 1982) using stopped-flow methods.

One study in this laboratory of agonist-binding kinetics to the membrane-bound AcChR covalently labeled by a fluorescent probe, 5-io-doacetamidosalicyclic acid (IAS), revealed that the fluorescence of the probe was enhanced on binding of agonists (Dunn et al., 1980), and at least three kinetic processes were observed, which could be described by the following mechanism (equation 1):

$$R + L \underset{}{\overset{K_1\ 14.6\ \mu M}{\rightleftharpoons}} RL \underset{k_{-2}\ 0.12\ sec^{-1}}{\overset{k_2\ 13.1\ sec^{-1}}{\rightleftharpoons}} R'L \overset{K_4\ 17.4\ \mu M}{\longrightarrow} R'L_2$$

$$k_3 \left\| k_{-3} \right. \qquad\qquad k_5 \left\| k_{-5} \right.$$
$$0.2\ sec^{-1}\ \ 0.006\ sec^{-1} \qquad 2\ sec^{-1}\ \ 0.6\ sec^{-1}$$

$$C_1 \qquad\qquad\qquad C_2 \tag{1}$$

where the kinetic parameters are for acetylcholine as ligand L (Blanchard *et al.*, 1982). This mechanism is in good agreement with one previously proposed to account for the kinetics observed when ethidium was used as an extrinsic probe (Quast *et al.*, 1979) except for the inclusion of an addition conformational transition (RL $\leftrightharpoons$ R'L). The resting state of the AcChR has an initial low affinity for agonists, but two sequential conformational changes result in a tightly bound C_1 complex. This C_1 complex can be correlated with the high-affinity binding site(s) observed under equilibrium conditions, since the overall dissociation constant of 4 nM ($K_1 k_{-2} k_{-3}/k_2 k_3$) for its formation is in agreement with K_d values obtained in equilibrium experiments. Binding of the second ligand occurs only at higher ligand concentrations, when the formation of C_2, which has lower affinity for agonist, becomes the predominant kinetic process. From the parameters above, an effective dissociation constant of 174 μM for the second ligand binding can be obtained from the ratio of C_1 to C_2 at equilibrium ($\overline{C}_1/\overline{C}_2 = K_4 k_{-5} k_3/k_5 k_{-3}$) (Quast *et al.*, 1979).

There are two major problems in correlating any observed receptor conformation with the open channel state: (1) apparent dissociation constants for agonist binding to the resting state of the AcChR are lower than those obtained from the concentration dependence of the permeability response; (2) no observed conformational change is fast enough to be correlated with channel opening, which must occur on a millisecond time scale. It is therefore likely that the slow processes observed in most studies (the fastest being 60 sec^{-1}; see Conti-Tronconi and Raftery, 1982) are related to desensitization or other inactivation mechanisms, which in electrophysiological experiments have been shown to occur on similar time scales (see Adams, 1981).

A high-affinity binding site for agonists has been assigned to each subunit of $M_r \sim 40K$ since, following reduction with DTT, they can be labeled by affinity alkylating agents such as bromoacetylcholine (see Conti-Tronconi and Raftery, 1982). The fluorescent probe IAS was shown to react with the same reduced disulfide bond near these sites, since after such labeling covalent binding of [³H]-bromoacetylcholine was much reduced (Dunn *et al.*, 1980). In view of the proximity of the fluorophore, agonist binding to these sites on the α (40K) subunit was most likely measured, and it is notable that all observed conformational changes were too slow to represent primary events in channel activation. On the other hand, as discussed above, the affinities measured related quite closely to independently determined values under equilibrium conditions, i.e., values typical for desensitized AcChR.

The general assumption that occupancy of a single class of sites leads to both functional responses of channel opening and desensitization has resulted in proposals of complex kinetic schemes that include both a low-affinity state in which the ion channel is open and a high-affinity desensitized state reached by sequential conformational transitions of the AcChR–agonist complex (Hess et al., 1979; Neubig and Cohen, 1980; Heidmann et al., 1983). We have recently obtained evidence for the existence of a low-affinity site(s) specific for agonist binding that is present under both initial and equilibrium conditions and is distinct from the high-affinity sites that are labeled by bromoacetylcholine (Conti-Tronconi et al., 1982b; Dunn and Raftery, 1982a,b; Dunn et al., 1983). In these experiments, the membrane-bound AcChR was first covalently labeled by the fluorescent probe IANBD, and the fluorescence of this probe was enhanced in a saturable manner by the binding of agonists (Dunn and Raftery, 1982a,b) with the same dose dependence as the flux response (see Table II). A typical fluorescence titration curve for AcCh binding in this system is shown in Fig. 7A. The fluorescence enhancement has a simple hyperbolic dependence on agonist concentration with a K_d of 75 μM and a Hill coefficient of approximately 1. Kinetic experiments have demonstrated that the fluorescence change is a monophasic process occurring on a rapid time scale. The rate and amplitude each had a hyperbolic dependence on agonist concentration (Fig. 7B), and the observed transition must therefore be a conformational change of the receptor–ligand complex.

This low-affinity site(s) is distinct from those sites of high affinity on the α-subunits since even when there latter sites were maximally labeled by BrAcCh, the fluorescence enhancement (and therefore agonist binding) to the low-affinity site was unaltered (Conti-Tronconi et al., 1982b; Dunn et al., 1983). In the presence of covalently bound BrAcCh, no ion flux was observed, presumably because of desensitization caused by occupancy of the binding sites on the α-subunits. The conformational change occurring on agonist binding to the low-affinity site is therefore independent of such other transitions, which likely inhibit channel opening.

The binding site revealed by the NBD-fluorescence experiments is present after solubilization of the AcChR in cholate or Triton X-100, and the affinity of solubilized receptor for AcCh and Carb is about tenfold lower, although the affinity for nicotine is unaltered (Table II). Reconstitution of cholate-solubilized AcChR with asolectin by the method of Wu and Raftery (1981a) had little effect on the K_d for Carb binding, which was 5.5 ± 2.7 mM in the solubilized state and 3.7 ± 1.6 mM after reconstitution, both measured for the NBD-labeled AcChR.

Several lines of evidence implicate the low-affinity site in transitions related to channel opening and suggest that activation and desensitization are parallel pathways that are mediated by agonist binding to different sites: (1) the fluorescence enhancement is specific for agonists, is abol-

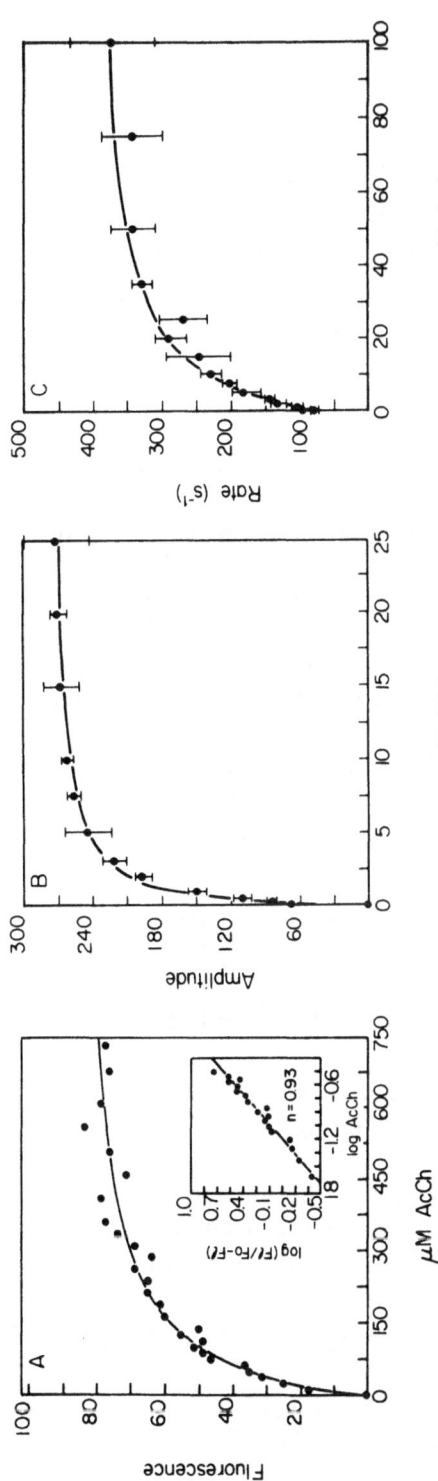

Figure 7. (A) Equilibrium fluorescence titration of NBD-labeled AcChR by acetylcholine. Solid line is fit to $F1 = F_0[L]/(K_d \pm [L])$, where $K_d = 75 \pm 2$ μM. F1 is observed fluorescence, and F_0 is maximum fluorescence change. Insert is Hill plot of the same data. Effect of Carb concentration on (B) amplitude and (C) on rate of binding to NBD-labeled AcChR measured in stopped-flow fluorescence experiments. Solid lines are calculated from best-fit parameters obtained from fit of mechanism 2 as described in the text.

ished by prior incubation with α-bungarotoxin, and reflects a conformational transition of the receptor–agonist complex; (2) K_d values for agonist binding correspond to those for activation (Table II); (3) the conformational change is rapid, reaching $\sim$400 sec^{-1} for Carb and $\sim$600 sec^{-1} for AcCh; (4) Q_{10} is $\sim$2.5, in agreement with electrophysiological measurements for channel opening; (5) the binding is unaffected by desensitization, covalently bound bromoacetylcholine, or prior incubation with physiologically active concentrations of curare, HTX, or local anesthetics.

10. The Mechanism of Agonist Binding to the Low-Affinity Site

The simplest model that predicts a hyperbolic dependence of both rate and amplitude on agonist concentration is one in which rapid binding is followed by a conformational change of the receptor–agonist complex $(R + L \xrightarrow{K_1} RL \underset{k_{-2}}{\overset{k_2}{\rightleftharpoons}} R^*L)$. This mechanism gave adequate fits to the Carb binding data shown in Fig. 7B, but there was a discrepancy between the overall K_d values obtained from the amplitude data ($\sim$0.73 mM) and those calculated from the best-fit rate parameters (Dunn and Raftery, 1982b). An extension of this model to include a preequilibrium between R and R* in the absence of ligand removes this discrepancy, and the following model adequately describes the data.

$$
\begin{array}{ccc}
R + L & \xrightarrow{\;K_1\;} & RL \\[2pt]
{\scriptstyle k_0}\big\updownarrow{\scriptstyle k_{-0}} & & {\scriptstyle k_2}\big\updownarrow{\scriptstyle k_{-2}} \\[2pt]
R^* + L & \xrightarrow{\;K_3\;} & R^*L
\end{array}
\qquad (2)
$$

In this model it is assumed that the ligand-binding steps are unobservably fast and that the fluorescence change occurs in the R$\rightleftharpoons$R* and RL$\rightleftharpoons$R*L transitions.

Under these circumstances, the fraction of subunits in the R* conformation in the presence of ligand concentration [L] is estimated from the amplitude of the signal change and is described by

$$
(R^* + R^*L)/R_0 = 1 \Big/ \left[1 + K_0 \left(\frac{1 + [L]/K_1}{1 + [L]/K_3} \right) \right]
\qquad (3)
$$

which may be approximated by $[L]/([L] + K_0 K_3)$ when the equilibrium lies in favor of R in the absence of ligand and agonist binds more tightly to R* than to R, i.e., $K_0(= k_{-0}/k_0) \gg \gg K_0K_3/K_1$ (see Janin, 1973). Changes in fluorescence of the NBD chromophore would then reflect the transition of R to R*, and the fluorescence would have a simple hyperbolic dependence on ligand concentration with an apparent dissociation constant of K_0K_3.

Equations describing the behavior of the apparent rate and procedures used for fitting this model have been described previously (Dunn and Raftery, 1982b), and the best-fit parameters for the rate data for Carb binding (Fig. 7B) are given by $k_{-0} = 130$ sec^{-1}, $k_0 = 18$ sec^{-1}, $k_{-2} = 18.3$ sec^{-1}, $k_2 = 385$ sec^{-1}, $K_1 = 9.1$ mM, and $K_3 = 0.11$ mM, giving an overall equilibrium constant of approximately 0.78 mM in good agreement with the concentration dependence of the amplitude data (0.73 mM).

The conformational transition detected by changes in fluorescence of bound NBD has many characteristics that suggest its involvement in the channel-opening event. However, it is not possible in terms of mechanism 2 to equate, in a simplistic manner, the R* conformation with the major conducting form of the AcChR. In the absence of ligand, from the kinetic parameters given above, the fraction of subunits in the R* state $\frac{1}{(1 + K_0)}$, may be calculated as 0.12, and this is clearly inconsistent with electrophysiological and *in vitro* ion flux data, which show that in the absence of agonist the major fraction of AcChR ion channels are closed and little ion transport occurs. A possible reason for this descrepancy is discussed below.

11. Conformational Coupling between Agonist Binding and Channel Opening

The observed binding of agonists to the low-affinity site revealed by NBD fluorescence showed no evidence of cooperativity, and Hill coefficients not significantly different from 1 were obtained (Fig. 7A). However, the concentration dependence of the flux response was clearly sigmoidal (Fig. 6) and had a Hill coefficient of 1.7, suggesting that two ligand molecules must bind for channel opening to occur. The simplest explanation for this discrepancy is that not one but two subunits of the AcChR must undergo the conformational change before channel opening occurs. In terms of mechanism 2 above, using the same assumptions and parameters as before, the fraction of subunits in the R* conformation in the absence of ligand was 0.12. However, the probability of two subunits of the AcChR being simultaneously in the R* conformation, i.e., the number of open channels is $p(\text{open}) = 1/(1 + K_0)^2$, i.e., 0.014, which is consistent with the poor conductivity observed in the absence of agonist. This model predicts that ligand binding to individual subunits follows a simple Langmuir isotherm and displays no cooperativity, but the rate of flux, since it is dependent on the number of open channels, has a sigmoidal dependence on agonist concentration:

$$k_{app}/k_{max} \sim p(\text{open}) \sim \left(\frac{[L]}{[L] + K_0 K_3}\right)^2$$

This model is illustrated schematically in Fig. 8A. The location of the low-affinity agonist binding site(s) has not been elucidated, and although in Fig. 8A both subunits are depicted as being identical, it is possible that two binding sites may exist on nonidentical but homologous subunits that have similar affinities for agonist.

12. Independent Pathways for Channel Activation and Desensitization

Torpedo AcChR appears to have two classes of agonist binding sites—those of high affinity on the 40K subunits, which may be labeled by bromoacetylcholine, and those of low affinity, which are revealed by NBD fluorescence changes. The conformational change occurring on agonist binding to the low-affinity site(s) is unaffected by desensitization or by covalent labeling by bromoacetylcholine, and the two pathways must therefore be independent. Such a model is illustrated pictorially in Fig. 8B. In the resting state, the "activation gate" is closed, but in the presence of high concentrations of agonists, the low-affinity sites are occupied, and the AcChR undergoes a rapid conformational transition to an open-channel state. Over longer time scales, slow conformational transitions in another part of the molecule mediated by agonist binding to other sites (possibly those of high equilibrium affinity on the α-subunits) cause the channel to close. Alternatively, at lower concentrations of agonist, occupancy of these latter sites causes slow conformational changes that close an "inactivation gate." Under these circumstances, the same conformational change may be induced by agonist binding to the low-affinity sites, but this transition cannot now lead to opening of the ion channel. Such a model is sufficient to explain the NBD fluorescence data.

13. Conclusion

The structure of the nicotinic AcChR has been highly conserved during animal evolution, and in all the species and tissues studied so far,

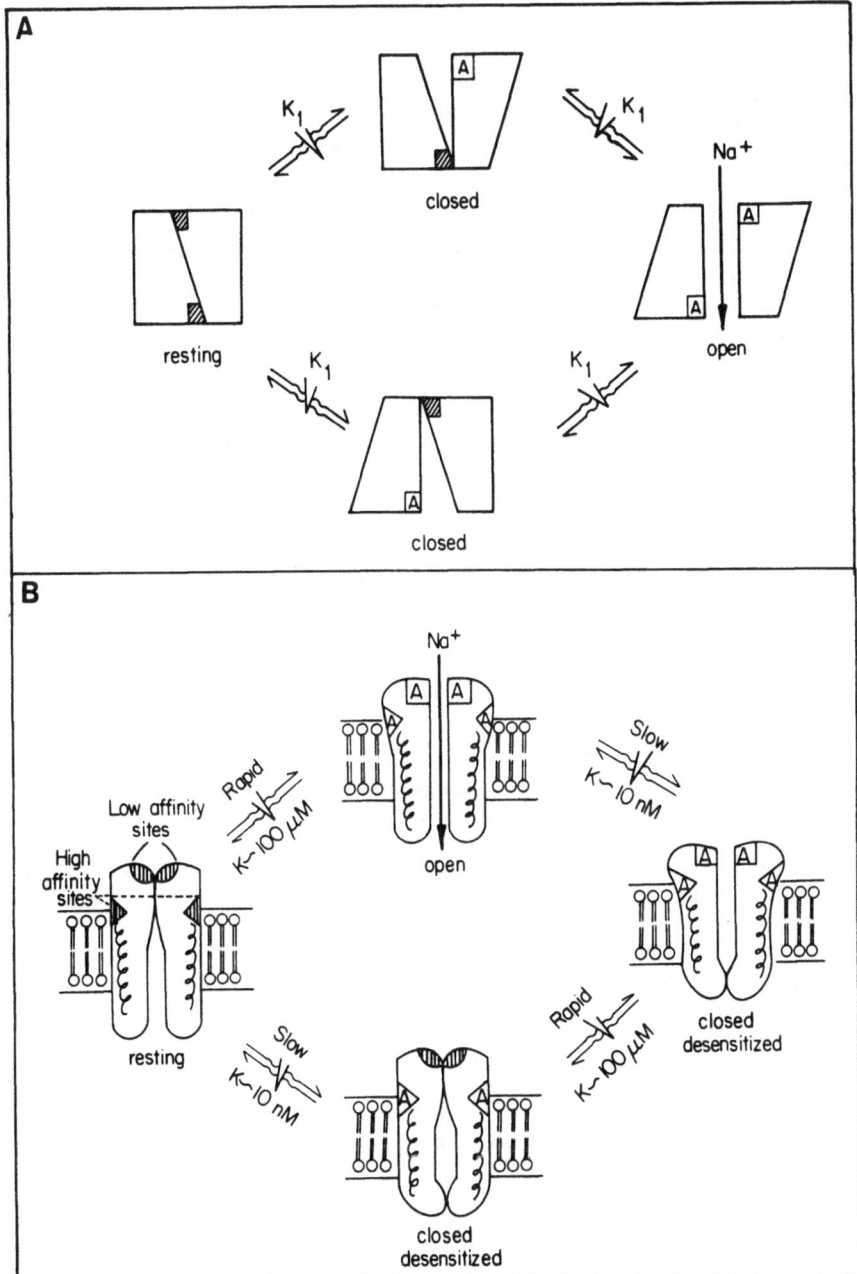

Figure 8. Schematic representation of ligand binding mechanisms. (A) Model in which two independent subunits must undergo identical conformational changes before channel opens. (B) Model in which activation and desensitization are parallel and independent mechanisms mediated by agonist binding to different sites.

including mammals, it is a pseudosymmetric pentameric complex of related subunits with very similar physical properties. All subunits of these nicotinic receptors were derived from a common ancestral gene, probably by way of gene duplication occurring early in animal evolution. The likely existence of such a unique acetylcholine-binding ancestral protein raises the possibility of a shared ancestry for many or even all proteins able to bind acetylcholine, such as the muscarinic receptor, acetylcholinesterase, etc., whose genes would have diverged very early to perform different receptor-mediated or enzymatic functions. The relatively early divergence of the genes coding for the subunits of the nicotinic AcChR, as well as their highly conserved primary structure and molecular weight, supports the possibility that the AcChR subunits themselves evolved separately to perform discrete functions such as activation, inactivation, and desensitization. The demonstration of multiple binding sites for agonists in *Torpedo* AcChR provides strong support for this hypothesis.

ACKNOWLEDGMENTS. This work was supported by USPHS Grant NS 10294, by ARO Contract No. DAMD17-82-C-2175, and by a grant from the Myasthenia Gravis Foundation, Los Angeles Chapter.

References

Adams, P. R., 1981, Acetylcholine receptor kinetics, *J. Membr. Biol.* **58**:161.

Bayley, H., and Knowles, J. R., 1978, Photogenerated reagents for membrane labeling. 2. Phenylcarbene and adamantylidene formed within the lipid bilayer, *Biochemistry* **17**:2420.

Bayley, H., and Knowles, J. R., 1980, Photogenerated reagents for membranes: Selective labeling of intrinsic membrane proteins in the human erythrocyte membrane, *Biochemistry* **19**:3883.

Blanchard, S. G., Dunn, S. M. J., and Raftery, M. A., 1982, Effects of reduction and alkylation on ligand binding and cation transport by *Torpedo californica* acetylcholine receptor, *Biochemistry* **24**:6258.

Claudio, T., and Raftery, M. A., 1980, Inhibition of α-bungarotoxin binding to acetylcholine receptors by antisera from animals with experimental autoimmune myasthenia gravis, *J. Supramol. Struct.* **14**:267.

Claudio, T., Ballivet, M., Patrick, J., and Heinemann, S., 1983, Nucleotide and deduced amino acid sequences of *Torpedo californica* acetylcholine receptor γ subunit, *Proc. Natl. Acad. Sci. U.S.A.* **80**:1111.

Conti-Tronconi, B. M., and Raftery, M. A., 1982, The nicotinic cholinergic receptor: Correlation of molecular structure with functional properties, *Annu. Rev. Biochem.* **51**:491.

Conti-Tronconi, B. M., Dunn, S. M. J., Raftery, M. A., 1982a, Functional stability of *Torpedo* acetylcholine receptor: Effects of protease treatment, *Biochemistry* **21**:893.

Conti-Tronconi, B. M., Dunn, S. M. J., and Raftery, M. A., 1982b, Independent sites of low and high affinity for agonists on *Torpedo californica* acetylcholine receptor, *Biochem. Biophys. Res. Commun.* **107**:123.

Conti-Tronconi, B. M., Gotti, C., Hunkapiller, M., and Raftery, M. A., 1982c, Mammalian muscle acetylcholine receptor: A supramolecular structure formed by four related proteins, *Science* **218**:1227.

Conti-Tronconi, B. M., Hunkapiller, M. W., Lindstrom, J. M., and Raftery, M. A., 1982d. Subunit structure of the acetylcholine receptor from *Electrophorus electricus*, *Proc. Natl. Acad. Sci. U.S.A.* **79**:6489.

Devillers-Thiery, A., Giraudat, J., Bentaboulet, M., and Changeux, J.-P., 1983, Complete mRNA coding sequence of the acetylcholine binding α-subunit of *Torpedo marmorata* acetylcholine receptor: A model for the transmembrane organization of the polypeptide chain, *Proc. Natl. Acad. Sci. U.S.A.* **80**:2067.

Dunn, S. M. J., and Raftery, M. A., 1982a, Activation and desensitization of *Torpedo* acetylcholine receptor: Evidence of separate binding sites, *Proc. Natl. Acad. Sci. U.S.A.* **79**:6757.

Dunn, S. M. J., and Raftery, M. A., 1982b, Multiple binding sites for agonists on *Torpedo californica* AcChR, *Biochemistry* **21**:6264.

Dunn, S. M. J., Blanchard, S. G., and Raftery, M. A., 1980, Kinetics of carbamylcholine binding to membrane-bound acetylcholine receptor monitored by fluorescence changes of a covalently bound probe, *Biochemistry* **19**:5645.

Dunn, S. M. J., Conti-Tronconi, B. M., and Raftery, M. A., 1983, Separate sites of low and high affinity for agonists on *Torpedo californica* acetylcholine receptor, *Biochemistry* **22**:2512.

Elliott, J., Blancard, S. G., Wu, W. C.-S., Miller, J., Strader C., Hartig, P., Moore, H.-H., Racs, J., and Raftery, M. A., 1980, Purification of *Torpedo californica* postsynaptic membranes and fractionation of their constituent proteins, *Biochem. J.* **185**:667.

Gotti, C., Conti-Tronconi, B. M., and Raftery, M. A., 1982, Mammalian muscle acetylcholine receptor purification and characterization, *Biochemistry* **21**:3148.

Gullick, W., Tzartos, S., and Lindstrom, J., 1981, Monoclonal antibodies as probes of acetylcholine receptor structure. 1. Peptide mapping, *Biochemistry* **20**:2173.

Haugland, R. P., 1982, *Handbook of Fluorescent Probes*, Molecular Probes, Inc., Texas, p. 43.

Heidmann, T., Bernhardt, J., Neumann, E., and Changeux, J.-P., 1983, Rapid kinetics of agonist binding and permeability response analysed in parallel on acetylcholine receptor rich membranes from *Torpedo marmorata*, *Biochemistry* **22**:5452.

Hess, G. P., Cash, D. G., and Aoshima, H., 1979, Acetylcholine receptor-controlled ion fluxes in membrane vesicles investigated by fast reaction techniques, *Nature* **282**:329.

Janin, J., 1973, The study of allosteric proteins, *Prog. Biophys. Mol. Biol.* **27**:77.

Lester, H. A., 1977, The response to acetylcholine, *Sci. Am.* **236**:106.

Lindstrom, J., Gullick, W., Conti-Tronconi, B., and Ellisman, M., 1980, Proteolytic nicking of the acetylcholine receptor, *Biochemistry* **19**:4791.

Martinez-Carrion, M., Sator, V., and Raftery, M. A., 1975, The molecular weight of an acetylcholine receptor isolated from *Torpedo californica*, *Biochem. Biophys. Res. Commun.* **65**:129.

Moore, H.-P. H., and Raftery, M. A., 1980, Direct spectroscopic studies of cation translocation by *Torpedo* acetylcholine receptor on a time scale of physiological relevance, *Proc. Natl. Acad. Sci. U.S.A.* **77**:4509.

Neubig, R. R., and Cohen, J. B., 1980, Permeability control by cholinergic receptor in *Torpedo* postsynaptic membranes: Agonist dose–response relations measured at second and millisecond times, *Biochemistry* **19**:2770.

Noda, M., Takahashi, H., Tanabe, T., Toyosato, M., Furutani, Y., Hirose, T., Asai, M. Inayama, S., Miyata, T., and Numa, S., 1982, Primary structure of α-subunit precursor of *T. californica* acetylcholine receptor deduced from cDNA sequence, *Nature* **299**:793.

Noda, M., Takahashi, H., Tanabe, T., Toyosato, M., Kikyotani, S., Hirose, T., Asai, M., Takashima, H., Inayame, S., Miyata, T., and Numa, S., 1983a, Primary structures and β and δ subunit precursors of *Torpedo californica* acetylcholine receptor deduced from cDNA sequences, *Nature* **30**:251.

Noda, M., Takahashi, H., Tanabe, T., Toyosato, M., Kikyotani, S., Furotani, Y., Hirose, T., Takashima, H., Mayama, S., Miyata, T., and Numa, S., 1983b, Structural homology of *Torpedo californica* acetylcholine receptor subunits, *Nature* **30**:528.

Quast, U., Schimerlik, M. I., and Raftery, M. A., 1979, Ligand-induced changes in membrane-bound acetylcholine receptor observed by ethidium fluorescence. 2. Stopped-flow studies with agonists and antagonists, *Biochemistry* **18**:1891.

Raftery, M. A., Vandlen, R., Michaelson, D., Bode, J., Moody, T., Chao, Y., Reed, K., Deutsch, J., and Duguid, J., 1974, The biochemistry of an acetylcholine receptor, *J. Supramol. Struct.* **2**:582.

Raftery, M. A., Hunkapiller, M. W., Strader, C. D., and Hood, L. E., 1980, Acetylcholine receptor: Complex of homologous subunits, *Science* **208**:1454.

Reynolds, M. A., and Karlin, A., 1978, Molecular weight in detergent solution of acetylcholine receptor from *Torpedo californica*, *Biochemistry* **17**:2035.

Sobel A., Weber, M., and Changeux, J.-P. 1977, Large-scale purification of the acetylcholine-receptor protein in its membrane-bound and detergent-extracted forms from *Torpedo marmorata* electric organ, *Eur. J. Biochem.* **80**:215.

Sobel, A., Heidmann, T., Hofler, J., and Changeux, J.-P., 1978, Distinct protein components from *Torpedo marmorata* membranes carry the acetylcholine receptor site and the binding site for local anesthetics and histrionicotoxin, *Proc. Natl. Acad. Sci. U.S.A.* **75**:510.

Sobel, A., Hofler, J., Heidmann, T., and Changeux, J.-P., 1979, Structural and functional properties of the acetylcholine regulator, *Adv. Cytopharmacol.* **3**:191.

Strader, C. D., and Raftery, M. A., 1980, Topographic studies of *Torpedo* acetylcholine receptor subunits as a transmembrane complex, *Proc. Natl. Acad. Sci. U.S.A.* **77**:5807.

Strader, C. D., Revel, J.-P., and Raftery, M. A., 1979, Demonstration of the transmembrane nature of the acetylcholine receptor by labeling with anti-receptor antibodies, *J. Cell Biol.* **83**:499.

Sumikawa, K., Houghton, M., Smith, M. C., Bell, L., Richards, B. M., and Barnard, E. A., 1982, The molecular cloning and characterization of cDNA coding for the α-subunit of the acetylcholine receptor, *Nucleic Acids Res.* **10**:5809.

Weill, C. L., McNamee, M. G., and Karlin, A., 1974, Affinity-labeling of purified acetylcholine receptor from *Torpedo californica*, *Biochem. Biophys. Res. Commun.* **61**:997.

Wu, W. C.-S., and Raftery, M. A., 1981a, Reconstitution of acetylcholine receptor function using purified receptor protein, *Biochemistry* **20**:694.

Wu, W. C.-S., and Raftery, M. A., 1981b, Functional properties of acetylcholine receptor monomeric and dimeric forms in reconstituted membranes, *Biochem. Biophys. Res. Commun.* **99**:436.

Wu, W. C.-S. Moore, H.-P. H., and Raftery, M. A., 1981, Quantitation of cation transport by reconstituted membrane vesicles containing purified acetylcholine receptor, *Proc. Natl. Acad. Sci. U.S.A.* **78**:775.

Adenylate-Cyclase-Coupled β-Adrenergic Receptors

Biochemical Mechanisms of Desensitization

BERTA STRULOVICI, JEFFREY M. STADEL,
and ROBERT J. LEFKOWITZ

1. Introduction

A wide variety of hormones initiate their effects on target cells by binding to specific cell surface receptors. In numerous systems, this interaction leads to the generation of the second messenger, cAMP, which initiates a cascade of events responsible for the ultimate response to that particular hormone. Catecholamines such as epinephrine and norepinephrine and various polypeptide hormones such as ACTH, glucagon, LH, and FSH are examples of hormones that stimulate responsive cells by binding to their specific receptors and by increasing the concentration of cAMP within the cells. However, prolonged exposure of target tissues to hormones results in a rapid rise followed by a decline in intracellular levels of cAMP. This phenomenon, termed "desensitization," "refractoriness," or "tachyphylaxis," appears to be quite a general mechanism for regulation of cellular sensitivity to hormonal stimulation (for review see Lefkowitz *et al.*, 1980; Perkins, 1983; Harden, 1983).

Of the adenylate-cyclase-coupled systems, few have attracted more attention than the adenylate-cyclase-coupled β-adrenergic receptors.

BERTA STRULOVICI, JEFFREY M. STADEL, and ROBERT J. LEFKOWITZ • Howard Hughes Medical Institute and Departments of Medicine and Biochemistry, Duke University Medical Center, Durham, North Carolina 27710. *Present address of B.S.:* Syntex Corporation, Palo Alto, California 94304. *Present address of J.M.S.:* Smith Kline & French Laboratories, Philadelphia, Pennsylvania 19101.

They have become one of the most thoroughly studied model systems for regulation of receptors and for the mechanisms by which they accomplish transmission of biological signals across the plasma membrane. In recent years, significant progress has been made towards understanding the mechanisms of catecholamine-induced desensitization. This progress has been made possible by advances in understanding of the molecular basis of hormone-mediated activation of adenylate cyclase. Thus, although the topic of this chapter is the mechanism of catecholamine-induced desensitization, we begin with a brief discussion of the components of the adenylate cyclase system and their normal interactions.

An enormous amount of information about the β-Adrenergic receptors for catecholamines has become available since their successful identification by ligand-binding studies 10 years ago, which permitted direct assay of the receptors under a wide variety of circumstances (Lefkowitz et al. 1974; Aurbach et al., 1974; Atlas et al., 1974). Thus, by using radiolabeled antagonists and agonists, it was possible to characterize pharmacologically the receptors (R) in whole cells and membrane preparations and to study their number and dynamics during desensitization or other physiological or pathological states. The ligand-binding approach also facilitated the ultimate purification to apparent homogeneity of the receptor by affinity chromatography (Shorr et al., 1981; 1982; Homcy et al., 1983; Benovic et al., 1984; for review see Lefkowitz et al., 1983).

The development of photoaffinity probes for the receptors with extremely high affinity (50–100 pM) has made possible the visualization of the β-adrenergic receptor macromolecules by SDS-PAGE followed by autoradiography. The method of photoaffinity labeling has been applied to a wide variety of mammalian and nonmammalian β_1- and β_2-adrenergic receptors (Lavin et al., 1982; Rashidbaigi and Ruoho, 1982; Benovic et al., 1983). Thus, the frog erythrocyte β_2-adrenergic receptor seems to reside exclusively on one peptide with a molecular weight of 58,000 (Lavin et al., 1981; Rashidbaigi and Ruoho, 1982), whereas the avian erythrocyte β_1-adrenergic receptor appears to reside on several peptides with molecular weights in the 40,000–50,000 range (Lavin et al., 1982; Rashidbaigi and Ruoho, 1982). In mammalian tissues, both β_1- and β_2-adrenergic receptors appear to reside on peptides of $M_r = 60,000$–65,000 (Benovic et al., 1983; Homcy et al., 1983).

The guanine nucleotide regulatory protein (N) to which the receptors are coupled in the plasma membrane has also been identified. The N stimulatory protein (N_s) has been purified to homogeneity, and it has been shown to be a heterodimer of 45,000 (α) and 35,000 (β) subunits (Northup et al., 1980; Sternweis et al., 1981; Hanski et al., 1981; for review see Gilman, 1984). The α subunit contains a site for NAD-dependent ADP-

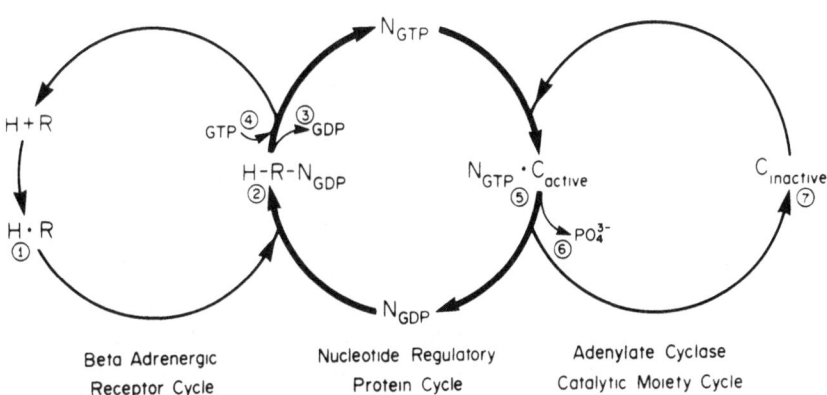

Figure 1. Model of the interactions among a catecholamine (H), the β-adrenergic receptor (R), the stimulatory guanine nucleotide regulatory protein (N), and the catalyst (C).

ribosylation catalyzed by cholera toxin, and it also contains a high-affinity guanine-nucleotide-binding site. Resolution of its subunits was achieved by high-performance gel filtration, and it was determined that the α subunit of the N protein, when bound to GTPγS, is the activator of C. The β subunit seems to act as an inhibitor by promoting the formation of the dimer, which in this form is inactive (Northup *et al.*, 1980, 1983a,b). Much less information is available about the catalyst (C), which has not yet been purified because of its hydrophobicity and lability but has only resolved from N_s (Ross, 1981). However, recently, progress in this direction has been reported with the use of forskolin affinity columns (Pfeuffer and Metzger, 1982, 1984).

Studies in intact membrane systems have shed light on the mechanisms by which agonist–receptor occupancy leads to an increased generation of cAMP. Based on such studies, the interactions of the β-adrenergic receptor with N have been suggested to result in an increase in the affinity of the receptor for agonists (De Lean *et al.*, 1980) The binding of guanine nucleotides to the ternary complex agonist–receptor–N (HRN) appears to result in the dissociation of a modified N–GTP complex and a return of the receptor to a low-affinity state for agonists (De Lean *et al.*, 1980) (Fig. 1). The actual stimulation of C seems to reflect the direct interaction between the enzyme and N–GTP, with the activation cycle being terminated by the hydrolysis of GTP to GDP by a GTPase present on N (Cassel and Selinger, 1976, 1978). The recent development of reconstitution techniques using resolved or purified preparations of R and N_s have shown that the interaction of these components in phospholipid vesicles results in a hormone-responsive GTPase activity (Brandt et al., 1983; Cerione et al., 1984).

2. Mechanisms of Desensitization of the Adenylate Cyclase Response to Catecholamines

The mechanisms of desensitization of β-adrenergic-receptor-coupled adenylate cyclase appear to be multiple. In some cases, incubation with agonists leads to a general dampening of the reponsiveness of the adenylate cyclase to further stimulation by a variety of different hormones as well as guanine nucleotides and NaF (Terasaki et al., 1978; Kassis and Fishman, 1982; Garrity et al., 1983). This phenomenon is termed "heterologous" desensitization. By contrast, in some cell types "homologous" desensitization is seen. This process is characterized by an attenuated responsiveness of the adenylate cyclase only to the desensitizing hormone without affecting the enzyme's responsiveness to other hormones, guanine nucleotides, or NaF (Mukherjee et al., 1975; Mickey et al., 1976; Su et al., 1979).

Here we describe two model systems in which we have probed the phenomenon of desensitization, the frog and the turkey erythrocyte. Two mechanisms of desensitization are suggested by our recent work: (1) the physical sequestration of the receptors within the cell, away from the other elements of the adenylate cyclase (frog erythrocyte), and (2) covalent modification—phosphorylation—of the receptors (turkey erythrocyte), both being reflected in the state of desensitization of that particular system.

2.1. Physical Sequestration of β-Adrenergic Receptors: The Frog Erythrocyte Model System

The first direct evidence for an alteration in hormone binding as an explanation for desensitization of adenylate cyclase came from studies done in vivo by Mukherjee et al. (1975). They demonstrated that chronic administation of β-adrenergic catecholamines to frogs led to a 68% decrease in [^{3}H]-DHA binding to the erythrocyte membranes as well as a 77% fall in adenylate cyclase response to β-adrenergic stimulation. Mickey et al. (1976) showed similar decrements in [^{3}H]-DHA binding after in vitro desensitization of frog erythrocytes with β-adrenergic agonists. Detailed binding studies showed that the diminshed [^{3}H]-DHA binding was the result of decreased receptor number in the plasma membranes—

"down-regulation"—and not a change in the affinity of the ligand for the receptor binding sites (Mickey *et al.*, 1976). Since other hormones such as PGE_1 and nonspecific stimulators such as NaF are able to stimulate the adenylate cyclase fully and only the ability of β-adrenergic catecholamines to stimulate the enzyme is diminished after exposure to β-adrenergic agonists, the frog erythrocyte represents a model system for the homologous type of desensitization.

Su *et al.* (1979, 1980) have demonstrated, using various cell lines such as 1321N human astrocytoma cells and S49 mouse lymphoma cells, that the desensitization process occurs in several identifiable steps. In these systems, there is an initial uncoupling of the β-adrenergic receptor from the other elements of the adenylate cyclase, which results in a diminished ability of β-adrenergic agonists to stimulate the enzyme, followed by loss of assayable β-adrenergic receptors from the cell surface. In the frog erythrocyte, the two alterations in receptor properties, namely, the "uncoupling" and the "down-regulation," cannot be separated temporally (Mickey *et al.*, 1976).

Receptor uncoupling has been studied by radioligand-binding techniques and is apparent as a decreased ability of an agonist to stabilize the high-affinity complex with the receptor and the guanine nucleotide regulatory protein (HRN). In plasma membranes prepared, for example, from control frog erythrocytes, agonists are able to promote the formation of a high-affinity, nucleotide-sensitive state of the receptor (the "coupled" form of the receptor) as assessed by competition binding of the agonist isoproterenol with radiolabeled antagonists (Kent *et al.*, 1980; Stadel *et al.*, 1983a). When the guanine nucleotide analogue Gpp(NH)p is present during the binding assay, the competition curve for the agonist is shifted to the right, indicating an interconversion of the high-affinity state of the receptor to a form that recognizes agonist with only low affinity. The competition curve for agonist in plasma membranes from desensitized cells (which show a reduction of ~50% of assayable β-adrenergic receptors) is partially shifted to the right compared to the control, indicating an impaired ability of agonist to induce the coupled state of the receptor (HRN) (Stadel et al., 1983a). The agonist–receptor complex in this preparation, however, is still sensitive to guanine nucleotides. Thus, in this form of desensitization, one observes some "uncoupling" and some "down-regulation" of the receptors.

Until recently, relatively little has been known about the fate of the "down-regulated" β-adrenergic receptors after their agonist-promoted loss from the cell surface. Using the frog erythrocyte model system, Chuang and Costa (1979) were able to recover ~20% of the "lost" receptors in the cytosol of lysates prepared from desensitized cells (see also

Chuang *et al.*, 1980). They have referred to these as "soluble" receptors. Subcellular fractionation of the erythrocytes subsequent to incubation with isoproterenol revealed that the internalized receptors were associated with lysosomal markers (Chuang, 1981, 1982). The authors contended that the very small recovery of "lost" receptors was attributable to the fact that the rest of them were processed within the cell. The fraction of the recovered down-regulated receptors was increased by pretreatment of cells with the lysosomotropic agent chloroquine before desensitization (Chuang, 1982).

Harden *et al.* (1980; for review see Perkins, 1983; Harden, 1983) also reported that in 1321N1 astrocytoma cells, uncoupled receptors produced during desensitization accumulate in a subpopulation of "light" membranes that show altered sedimentation properties on sucrose gradients. The rapid change in the membrane form of the β-adrenergic receptor during catecholamine-induced desensitization has been subsequently demonstrated to occur in many other systems in which "down-regulation" occurs (Hertel *et al.*, 1983a,b: Frederick *et al.*, 1983).

Using the frog erythrocyte model system, we have been able to recover ~85% of the "lost" desensitized β-adrenergic receptors in a light vesicular fraction that was obtained by centrifugation of the cytosol derived from the desensitized cells at $105,000 \times g$ for 1 hr. (Stadel *et al.*, 1983a). We have demonstrated that these vesicles are present in an intracellular compartment, sequestered away from the other elements of the adenylate cyclase system and from other plasma membrane markers. Thus, catecholamine-induced desensitization in the frog erythrocyte promotes the movement of the β-adrenergic receptors from the plasma membrane into a distinctly different membrane environment. Further characterization of the "down-regulated" β-adrenergic receptors by photoaffinity labeling and SDS-PAGE demonstrated that these receptors were not apparently altered or processed within the cell (Fig. 2).

As a means of characterizing the functionality of the vesicular "lost" receptors, we assessed their ability to form the high-affinity "coupled" state in ligand-binding studies. These studies revealed that these receptors are totally "uncoupled" in the sense that they are incapable of forming the high-affinity coupled ternary complex HRN. These data could be explained by two alternative hypotheses: (1) the vesicular receptors might have been modified in such a way that, although they bind ligands, they cannot couple to the other elements of the adenylate cyclase, or (2) they might be functionally intact, but the "uncoupling" could have resulted from their physical sequestration away from the effector components necessary to form this state (i.e., the N regulatory protein).

To test these hypotheses we fused the vesicles containing the inter-

MEMBRANES : VESICLES

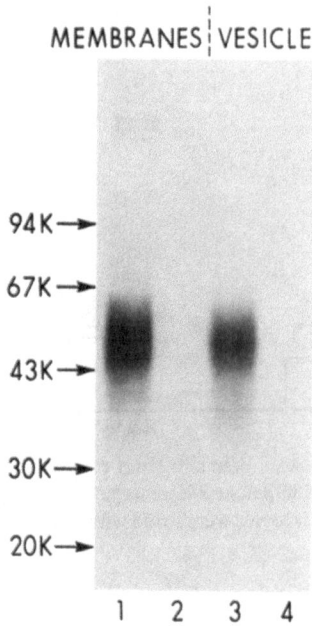

Figure 2. Sodium dodecylsulfate-polyacrylamide gel electrophoresis of [^{125}I]-pABC-labeled β-adrenergic receptors in plasma membranes from control frog erythrocytes and vesicles prepared from desensitized frog erythrocytes. The samples are: 1, [^{125}I]-pABC-labeled plasma membranes from control cells; 2, plasma membranes from control cells labeled with [^{125}I]-pABC in the presence of 10^{-5} M (−)alprenolol; 3, [^{125}I]-pABC-labeled vesicles from desensitized cells; 4, vesicles from desensitized cells labeled with [^{125}I]-pABC in the presence of 10^{-5} M (−)alprenolol (Stadel *et al.*, 1983a).

nalized "desensitized" receptors (but no N and C subunits) with an acceptor cell that had the effector component of the adenylate cyclase system but lacked the β-adrenergic receptors, namely, the *Xenopus laevis* erythrocyte (Strulovici *et al.*, 1983). The results of such an experiment are shown in Fig. 3. It can be seen that the receptors in these vesicles are highly active in establishing a catecholamine-responsive adenylate cyclse in the *X. laevis* erythrocytes. Thus, although these internalized β-adrenergic receptors appeared "uncoupled" as assessed by binding studies, they are quite active when assessed in this fusion system. These data suggest that in the case of the frog erythrocyte model system, an agonist-promoted physical sequestration of the receptors away from their effector components may be a major mechanism responsible for the desensitization of the cell to catecholamine stimulation.

The recovery from the desensitized state and the fate of the internalized receptors once the agonist agent is removed differs with the system and, in the case of the cultured cell systems, with the state of confluency of the cell line (Doss *et al.*, 1981). Thus, the recovery from catecholamine-induced desensitization (as assessed by recovery of assayable β-adrenergic receptors in the plasma membranes of the previously desensitized cells) is sometimes dependent on protein synthesis (Doss *et*

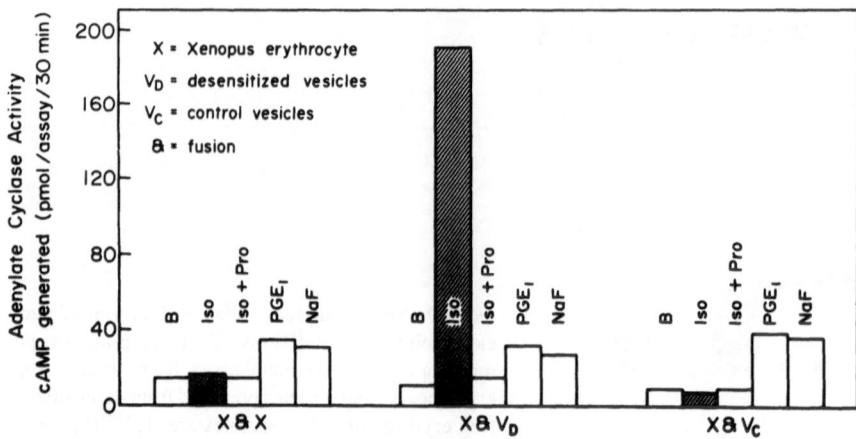

Figure 3. Fusion of *Xenopus laevis* erythrocytes with vesicles prepared from control and desensitized frog erythrocytes. Vesicles from control (0.180 pmole β-adrenergic receptor) and desensitized (2 pmole β-adrenergic receptor) frog erythrocytes were fused with *X. laevis* erythrocytes as described (Strulovici *et al.*, 1983).

al., 1981; Franklin *et al.*, 1975) and sometimes not (Su *et al.*, 1976). The receptors that appear at the cell surface once agonist is removed can be synthesized *de novo*, may reappear from an internal pool, or might represent the "down-regulated" receptors, which recycle back to the cell surface.

Doss *et al.* (1981) have shown that in preconfluent 1321N1 cell cultures the loss of β-adrenergic receptors during desensitization was completely reversed on removal of isoproterenol and was uninfluenced by protein synthesis inhibition. In contrast, in confluent cultures of the same cell type, the recovery of receptors was only 60%, and this was completely blocked by cycloheximide.

In the frog erythrocyte model system, earlier work *in vivo* (Mukherjee *et al.*, 1975) and *in vitro* (Mickey *et al.*, 1976) has shown that the cells can recover their full complement of receptors in the plasma membrane once the agonist is removed, even when protein synthesis is blocked. These data and the observations that the internalized β-adrenergic receptors did not seem structurally and functionally altered after agonist-promoted desensitization (Stadel *et al.*, 1983a; Strulovici *et al.*, 1983) suggested indirectly that the "down-regulated" receptors recycle back to the cell surface on removal of the agonist. Recently, we were able to develop more direct evidence in favor of this mechanism. By measuring the receptors simultaneously in the plasma membrane and cytosolic vesicular compartments of the cell, we found a reciprocal relationship during both desensitization and resensitization in the presence of cycloheximide (Strulovici and Lefkowitz, 1984). Thus, as the receptors disappeared from the plasma membrane during desensitization, they appeared in the vesicles. Conversely, as the receptors reappeared in the plasma membrane during resensitization, they were depleted from the vesicular compartment. These observations strongly support a recycling mechanism. The nature of the agonist-promoted alterations in the receptors that initially triggers their uncoupling and subsequent internalization into sequestered vesicles away from the other components of the adenylate cyclase system, as well as the physical location of these internalized receptors within the cell, remain to be elucidated.

2.2. *Covalent Modification of the Receptors: The Turkey Erythrocyte Model System*

In contrast to the situation observed during desensitization in the frog erythrocyte, changes in receptor number are not involved in the alteration in adenylate cyclase responsiveness during desensitization of the turkey erythrocyte adenylate cyclase system (Hoffman *et al.*, 1979). In avian erythrocytes, although β-adrenergic receptors are coupled to adenylate cyclase, there is also a small but significant decrease in both guanine-nucleotide- and NaF-stimulated enzyme activities as a result of desensitization (Hoffman *et al.*, 1979; Simpson and Pfeuffer, 1980). In the turkey and pigeon erythrocyte, this phenomenon is mimicked by nucleotide analogues such as 8-bromo cAMP of dibutyryl cAMP (Stadel *et al.*, 1981; Simpson *et al.*, 1980). Both agonist- and cyclic-nucleotide-induced desensitization are associated with a functional "uncoupling" of the β-adrenergic receptor, which is evidenced by an impaired ability of the receptors to form a high-affinity guanine-nucleotide-sensitive complex with agonist as assessed by radioligand-binding studies (Stadel *et al.*, 1981).

To determine the locus of alteration that is responsible for the "uncoupling" of the receptors and ultimately for the "desensitized" state of the adenylate cyclase system of the turkey erythrocyte, we first considered the receptors themselves. The β-adrenergic receptors of turkey erythrocyte membranes prepared from control and desensitized cells were covalently labeled with the photoaffinity probe [^{125}I]-pABC. Figure 4, left, demonstrates a difference in the patterns of migration in SDS-PAGE of the β-adrenergic receptor polypeptides derived from control and desensitized cells (Stadel *et al.*, 1982). the time course of alteration in β-adrenergic receptor mobility correlates with that for desensitization of isoproterenol-stimulated adenylate cyclase activity. Moreover, 8-bromo

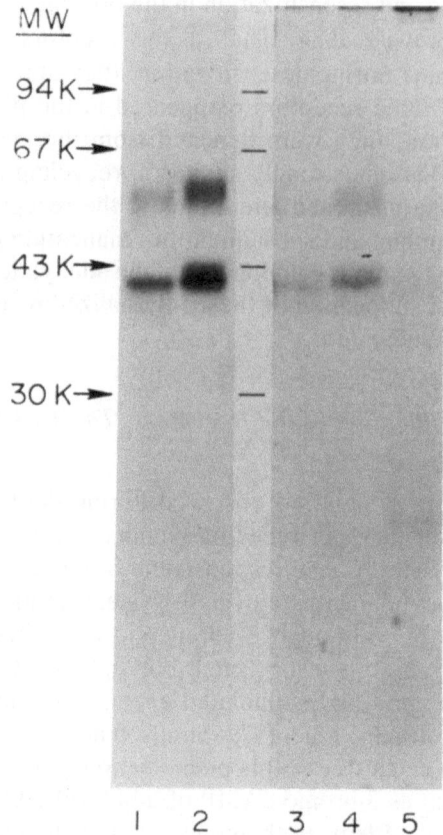

Figure 4. Sodium dodecylsulfate-polyacrylamide gel electrophoresis of [^{32}P]-labeled β-adrenergic receptor peptides partially purified from control and isoproterenol-desensitized turkey erythrocytes. The [^{32}P$_i$]-labeled cells were incubated for 4 hr in buffer alone (lane 3), with 10^{-6} M isoproterenol (lane 4), or with 10^{-6} M isoproterenol + 10^{-5} M propranolol (lane 5). Included in this panel for comparison are β-adrenergic receptor peptides labeled with [^{125}I]-pABC in membranes prepared from control (lane 1) and isoproterenol-desensitized (lane 2) erythrocytes. (From Stadel *et al.*, 1983b.)

cAMP can also produce this change in mobility to the same extent that it can mimic the catecholamine-induced desensitization (about 50% maximum desensitization) (Stadel *et al.*, 1983b). Thus, the agonist-induced desensitization in the turkey erythrocyte is associated with an alteration in the structure of the receptors that can be visualized as an altered mobility of the photoaffinity-labeled receptors on SDS-PAGE.

More recently, Stadel *et al.* (1983b) were able to demonstrate, by first incubating turkey erythrocytes with [^{32}P]-orthophosphate in order to label the endogenous ATP pool, that after desensitization there is a two- to threefold stimulation of phosphate incorporation into the receptors. The phosphorylated and desensitized receptors also show the altered mobility pattern on SDS-PAGE (Fig. 4, right). By including the antagonist propranolol along with the agonist during the desensitization procedure, it was possible to prevent the diminution in adenylate cyclase activation, the altered mobility of the β-adrenergic receptors on SDS-PAGE, and also the phosphorylation of the receptor polypeptides (Stadel *et al.*, 1983b). Other studies with this system have shown that the characteristics of the desensitization process and of the receptor phosphorylation, such as time course, agonist concentration relationships, etc., are highly correlated (Sibley *et al.*, 1984).

We wished to directly assess the functionality of these "desensitized" β-adrenergic receptors and to establish whether their covalent modification during desensitization is associated with an altered ability to couple to the other elements of the adenylate cyclase. Recently, we were able to accomplish this by partially purifying the receptors from control and desensitized turkey erythrocytes by affinity chromatography, reconstituting them into phospholipid vesicles, and fusing them with *X. laevis* erythrocytes (Strulovici *et al.*, 1984). Desensitized β-adrenergic receptors showed a 40–50% reduction in their ability to couple to the heterologous adenylate cyclase system (Fig. 5), comparable to the reduction in their functionality observed in their original membrane environment. The inset to Fig. 5 shows an autoradiogram of an SDS-PAGE of the [^{125}I]-pABC-labeled affinity-purified "control" and "desensitized" β-adrenergic receptors that were used in these reconstitution–fusion experiments. The altered mobility of the photoaffinity-labeled purified receptor on gel electrophoresis is quite comparable to that observed with receptors in crude membranes. The stability of this structural modification of the receptor induced during desensitization presumably underlies our ability to observe the diminished functionality of the desensitized receptor even after purification, reconstitution and fusion. Thus, in the turkey erythrocyte model system, there seems to be a direct relationship between a stable modification of the receptor (phosphorylation) and its impaired functionality, which presumably is reflected in the "desensitized" state of the adenylate cyclase system of the cell.

Prolonged exposure of turkey (Hoffman *et al.*, 1979) or pigeon erythrocytes (Simpson and Pfeuffer, 1980) to catecholamines and cyclic nucleotides also results in a small impairment of adenylate cyclase stimulation by guanine nucleotides and NaF. The desensitization to these nonspecific effectors is less than that seen with isoproterenol. This ob-

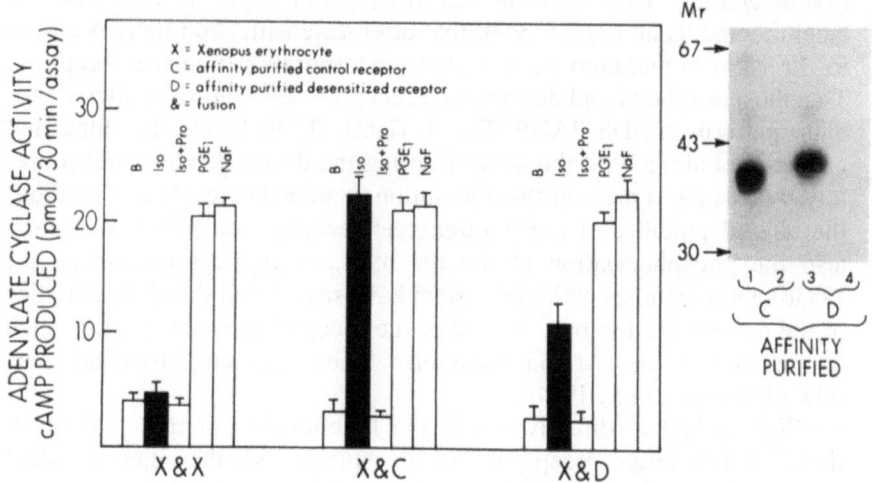

Figure 5. Fusion of *X. laevis* erythrocytes with affinity-purified and reconstituted β-adrenergic receptors from control and desensitized turkey erythrocytes. Inset: Autoradiogram of SDS-polyacrylamide gel electrophoresis of [^{125}I]-pABC-labeled purified β-adrenergic receptors from control and desensitized turkey erythrocytes used in these experiments. The samples are: 1, [^{125}I]-pABC-labeled purified "control" β-adrenergic receptor (C); 2, purified "control" β-adrenergic receptor labeled in the presence of 10^{-5} M (−)alprenolol; 3, [^{125}I]-pABC-labeled purified "desensitized" β-adrenergic receptor (D); 4, purified "desensitized" β-adrenergic receptor labeled in the presence of 10^{-5} M (−)alprenolol. (From Strulovici *et al.*, 1984.)

servation suggests that a cAMP-mediated modification of a component distal to the receptor also occurs in addition to receptor phosphorylation. Recently, Briggs *et al.* (1983) have reported a functional alteration in the turkey erythrocyte N protein subsequent to desensitization. Thus, it is likely that a second mechanism contributes to heterologous desensitization in the turkey erthrocyte in addition to a covalent modification of the receptor. This might involve a covalent modification of one or more of the subunits of the guanine nucleotide regulatory protein. Whether this phenomenon involves a cAMP-dependent phosphorylation or some other type of modification is not yet know.

3. Summary

In summary, desensitization of catecholamine-sensitive adenylate cyclase likely involves regulation of the function of various components of the system. Currently, most is known about regulation at the receptor level. At least two types of mechanisms of receptor regulation have thus far been identified: agonist-promoted sequestration of the receptors within the cell away from the effector adenylate cyclase and covalent modification as by phosphorylation.

References

Atlas, D., Steer, M. L., and Levitzki, A., 1974, Stereospecific binding of propranolol and catecholamines to the β-adrenergic receptor, *Proc. Natl. Acad. Sci. U.S.A.* **71**:4246–4248.

Aurbach, G. D., Fedak, S. A. Woodard, C. J., Palmer, J. S., Hauser, D., and Troxler, F., 1974, β-Adrenergic receptor: Stereospecific interaction of iodinated β-blocking agent with a high affinity size, *Science* **186**:1223–1225.

Benovic, J. L., Stiles, G. L., Lefkowitz, R. J., and Caron, M. G., 1983, Photoaffinity labeling of mammalian β-adrenergic receptors: Metal-dependent proteolysis explains apparent heterogeneity, *Biochem. Biophys. Res. Commun.* **110**:504–511.

Benovic, J. L., Shorr, R. G. L., Caron, M. G., and Lefkowitz, R. J., 1984, The mammalian β-adrenergic receptor: Purification and characterization, *Biochemistry* **23**:4510–4518

Brandt, D. R., Asano, T., Pedersen, S. E., and Ross, E. M., 1983, Reconstitution of catecholamine-stimulated guanosimethiphosphatase activity, *Biochemistry* **22**:4357–4362.

Briggs, M. M., Stadel, J. M., Iyengar, R., and Lefkowitz, R. J., 1983, Functional modification of the guanine nucleotide regulator protein after desensitization of turkey erythrocytes by catecholamines, *Arch. Biochem. Biophys.* **224**:142–151.

Cassel, D., and Selinger, Z., 1976, Catecholamine-stimulated GTP-ase activity in turkey erythrocyte membranes, *Biochim. Biophys. Acta* **452**:538–551.

Cassel, D., and Selinger, Z., 1978, Mechanism of adenylate cyclase activation through the β-adrenergic receptor: Catecholamine-induced displacement of bound GDP by GTP, *Proc. Natl. Acad. Sci. U.S.A.* **75**:4155–4159.

Cerione, R. A., Codina, J., Benovic, J. L., Lefkowitz, R. J., Birnbaumer, L., and Caron, M. G., 1984, The mammalian β$_2$-adrenergic receptor: Reconstitution of the functional interactions between the pure receptor and the pure stimulatory nucleotide binding protein (N$_s$) of the adenylate cyclase system, *Biochemistry* **23**:4519-4525.

Chuang, D.-M., 1981, Inhibition of transglutaninose prevents agonist-mediated internalization of β-adrenergic receptors, *J. Biol. Chem.* **256**:8291–8293.

Chuang, D.-M., 1982, Internalization of β-adrenergic receptor binding sites: Involvement of lysosomal enzymes, *Biochem. Biophys. Res. Commun.* **105**:1466–1472.

Chuang, D.-M., and Costa, E., 1979, Evidence for internalization of the recognition site of β-adrenergic receptors during receptor subsensitivity induced by isoproterenol, *Proc. Natl. Acad. Sci. U.S.A.* **76**:3024–3028.

Chuang, D.-M., Kinnier, W. J., Farber, L., and Costa, E., 1980, A biochemical study of receptor internalization during β-adrenergic receptor desensitization in frog erythrocytes, *Mol. Pharmacol.* **18**:348–355.

De Lean, A., Stadel, J. M., and Lefkowitz, R. J., 1980, A ternary complex model explains the agonist-specific binding properties of the adenylate cyclase-coupled β-adrenergic receptor, *J. Biol. Chem.* **255**:7108–7117.

Doss, R. C., Perkins, J. P., and Harden, T. K., 1981, Recovery of β-adrenergic receptors following long term exposure of astrocytoma cells to catecholamine: Role of protein synthesis, *J. Biol. Chem.* **256**:12281–12286.

Franklin, T. J., Morris, W. P., and Truose, P. A., 1975, Desensitization of β-adrenergic receptors in human fibroblasts in tissue culture, *Mol. Pharmacol.* **11**:485–491.

Frederick, R. C., Jr., Waldo, G. L., Harden, T. K., and Perkins, J. P., 1983, Characterization of agonist-induced β-adrenergic receptor-specific desensitization in C62B glioma cells, *J. Cyclic Nucleotide Res.* **9**:103–118.

Garrity, M. J., Andreasen, T. J., Storm, D. R., and Robertson, R. P., 1983, Prostaglandin E-induced heterologous desensitization of hepatic adenylate cyclase: Consequences on the guanyl nucleotide regulatory protein, *J. Biol. Chem.* **285**:8692–8697.

Gilman, A. G., 1984, Guanine nucleotide-binding regulatory proteins and dual control of adenylate cyclase, *J. Clin. Invest.* **73**:1–4.

Hanski, E., Sternweis, P. C., Northup, J. K., Dromerick, A. W., and Gilman, A. G., 1981, The regulatory component of adenylate cyclase—purification and properties of the turkey erythrocyte protein, *J. Biol. Chem.* **256**:12911–12919.

Harden, T. K., 1983, Agonist-induced desensitization of the β-adrenergic receptor-linked adenylate cyclase, *Pharmacol. Rev.* **35**:5–32.

Harden, T. K., Cotton, C. V., Waldo, G. L., Lutton, J. K., and Perkins, J. P., 1980, Catecholamine-induced alteration in sedimentation behavior of membrane-bound β-adrenergic receptors, *Science* **210**:441–443.

Hertel, C., Muller, P., Portenier, M., and Staehelin, M., 1983a, Determination of the desensitization of β-adrenergic receptors by [^{3}H]CGP-12177, *Biochem. J.* **216**:2–6.

Hertel, C., Staehelin, M., and Perkins, J. P., 1983b, Evidence for intravesicular β-adrenergic receptors in membrane fractions from desensitized cells: Binding of the hydrophobic ligand [^{3}H]CGP-12177 in the presence of alamethicin, *J. Cyclic Nucleotide Res.* **9**:119–28.

Hoffman, B. B., Mullikin-Kilpatrick, D., and Lefkowitz, R. J., 1979, Desensitization of β-adrenergic stimulated adenylate cyclase in turkey erythrocytes, *J. Cyclic Nucleotide Res.* **5**:355–366.

Homcy, C. J., Rockson, S. G., Countaway, J. R., and Egan, D. A., 1983, Purification and characterization of the mammalian β$_2$-adrenergic receptor, *Biochemistry* **22**:660–668.

Kassis, S., and Fishman, P. H., 1982, Different mechanisms of desensitization of adenylate cyclase by isoproterenol and prostaglandin E$_1$ in human fibroblasts: Role of regulatory components in desensitization, *J. Biol. Chem.* **257**:5312–5318.

Kent, R. S., De Lean, A., and Lefkowitz, R. J., 1980, A quantitative analysis of β-adrenergic receptor interactions: Resolution of high affinity and low affinity states of the receptor by computer modeling of ligand binding data, *Mol. Pharmacol.* **17**:14–23.

Lavin, T. N., Heald, S. L., Jeffs, P. W., Shorr, R. G. L., Lefkowitz, R. J., and Caron, M. G., 1981, Photoaffinity labeling of the β-adrenergic receptor, *J. Biol. Chem.* **256**:11944–11950.

Lavin, T. N., Nambi, P., Heald, S. L., Jeffs, P. W., Lefkowitz, R. J., and Caron, M. G., 1982, ^{125}I-Labeled *p*-azidobenzylcarazolol, a photoaffinity label for the β-adrenergic receptor, *J. Biol. Chem.* **257**:12332–12340.

Lefkowitz, R. J., Mukherjee, C., Coverstone, M., and Caron, M. G., 1974, Stereospecific [^{3}H] (−)alprenolol binding sites, β-adrenergic receptors and adenylate cyclase, *Biochem. Biophys. Res. Commun.* **60**:703–709.

Lefkowitz, R. J., Wessels, M. R., and Stadel, J. M., 1980, Hormones, receptors and cyclic AMP: Their role in target cell refractoriness, *Curr. Top. Cell. Regul.* **17**:205–230.

Lefkowitz, R. J., Stadel, J. M., and Caron, M. G., 1983, Adenylate cyclase-coupled β-adrenergic receptors: Structure and mechanisms of activation and desensitization, *Annu. Rev. Biochem.* **52**:159–186.

Mickey, J. V., Tate, R., Mullikin, D., and Lefkowitz, R. J., 1976, Regulation of adenylate cyclase-coupled β-adrenergic receptor binding sites by β-adrenergic catecholamines *in vitro, Mol. Pharmacol.* **12**:409–419.

Mukherjee, C., Caron, M. G., and Lefkowitz, R. J., 1975, Catecholamine-induced subsensitivity of adenylate cyclase associated with loss of β-adrenergic receptor binding sites, *Proc. Natl. Acad. Sci. U.S.A.* **72**:1945–1949.

Northup, J. K., Sternweis, P. C., Smigel, M. D., Schleifer, L. S., Ross, E. M., and Gilman, A. G., 1980, Purification of the regulatory component of adenylate cyclase, *Proc. Natl. Acad. Sci. U.S.A.* **77**:6516–6520.

Northup, J. K., Sternweis, P. C., and Gilman, A. G., 1983a, The subunits of the stimulatory regulatory component of adenylate cyclase—resolution, activity and properties of the 35,000 dalton (β) subunit, *J. Biol. Chem.* **258**:11361–11368

Northup, J. K., Smigel, M. D., Sternweis, P. C., and Gilman, A. G., 1983b, The subunits of the stimulatory regulatory component of adenylate cyclase—resolution of the activated 45,000 dalton (ga) subunit, *J. Biol. Chem.* **258**:11369–11376.

Perkins, J. P., 1983, Desensitization of the response of adenylate cyclase to catecholamines, *Curr. Top. Membr. Transp.* **18**:85–108.

Pfeuffer, T., and Metzger, H., 1982, 7-O-Hemisuccinyldeacetyl forskolin-Sepharose: A novel affinity support for purification of adenylate cyclase, *FEBS Lett.* **146**:369–375.

Pfeuffer, T., and Metzger, H., 1984, Isolation of homologous and heterologous complexes between catalytic and regulatory components of adenylate cyclase by forskolin-Sepharose, *FEBS Lett.* **164**:154–160.

Rashidbaigi, A., and Ruoho, A. E., 1982, Photoaffinity labeling of β-adrenergic receptors: Identification of the β-receptor binding site(s) from turkey, pigeon and frog erythrocytes, *Biochem. Biophys. Res. Commun.* **106**:139–148.

Ross, E. M., 1981 Physical separation of the catalytic and regulatory proteins of hepatic adenylate cyclase, *J. Biol. Chem.* **256**:1949–1953.

Shorr, R. G. L., Lefkowitz, R. J., and Caron, M. G., 1981, Purification of the β-adrenergic receptor: Identification of the hormone binding subunit, *J. Biol. Chem.* **256**:5820–5826.

Shorr, R. G. L., Strohsacker, M. W., Lavin, T. N., Lefkowitz, R. J., and Caron, M. G., 1982, The β₁-adrenergic receptor of the turkey erythrocyte: Molecular heterogeneity revealed by purification and photoaffinity labeling, *J. Biol. Chem.* **257**:12341–12350.

Sibley, D. R., Peters, J. R., Nambi, P., Caron, M. G., and Lefkowitz, R. J., 1984, Desensitization of turkey erythrocyte adenylate cyclase: β-Adrenergic receptor phosphorylation is correlated with attenuation of adenylate cyclase activity, *J. Biol. Chem.* **259**:9742-9749.

Simpson, I. A., and Pfeuffer, T., 1980, Functional desensitization of β-adrenergic receptors of avian erythrocytes by catecholamines and adenosine 3′,5′phosphate, *Eur. J. Biochem.* **111**:111–116.

Stadfel, J. M., De Lean, A., Mullikin-Kilpatrick, D., Sawyer, D. D., and Lefkowitz, R. J., 1981, Catecholamine-induced desensitization in turkey erythrocytes: cAMP mediated impairment of high affinity agonist binding without alteration in receptor number, *J. Cyclic Nucleotide Res.* **7**:37–47.

Stadel, J. M., Nambi, P., Lavin, T. N., Heald, S. L., Caron, M. G., and Lefkowitz, R. J., 1982, Catecholamine–induced desensitization of turkey erythrocyte adenylate cyclase, *J. Biol. Chem.* **257**:9242–9245.

Stadel, J. M. Strulovici, B., Nambi, P., Lavin, T. N., Briggs, M. M., Caron, M. G., and Lefkowitz, R. J., 1983a, Desensitization of the β-adrenergic receptor of frog erythrocytes: Recovery and characterization of the down-regulated receptors in sequestered vesicles, *J. Biol. Chem.* **258**:3032–3038.

Stadel, J. M., Nambi, P., Shorr, R. G. L., Sawyer, D. F., Caron, M. G., and Lefkowitz, R. J., 1983b, Catecholamine-induced desensitization of turkey erythrocyte adenylate cyclase is associated with phosphorylation of the β-adrenergic receptor, *Proc. Natl. Acad. Sci. U.S.A.* **80**:3173–3177.

Sternweis, P. C., Northup, J. K., Smigel, M. D., and Gilman, A. G., 1981, The regulatory component of adenylate cyclase—purification and properties, *J. Biol. Chem.* **256**:11517–11526.

Strulovici, B., and Lefkowitz, R. J., 1984, Activation, desensitization and recycling of frog erythrocyte β-adrenergic receptors: Differential perturbation by in situ trypsinization, *J. Biol. Chem.* **259**:4389–4395.

Strulovici, B., Stadel, J. M., and Lefkowitz, R. J., 1983, Functional integrity of desensitized β-adrenergic receptors: Internalized receptors reconstitute catecholamine-stimulated adenylate cyclase activity, *J. Biol. Chem.* **258**:6410–6414.

Strulovici, B., Cerione, R. A., Kilpatrick, B. F., Caron, M. G., and Lefkowitz, R. J., 1984, Direct demonstration of altered functionality of a purified desensitized receptor in a reconstituted system, *Science* **225**:837–840.

Su, Y.-F., Cubeddu, X. L., and Perkins, J. P., 1976, Regulation of adenosine 3'-5'monophosphate content of human astrocytoma cells: Desensitization to catecholamines and prostaglandins, *J. Cyclic Nucleotide Res.* **2**:257–270.

Su, Y.-F., Harden, T. K., and Perkins, J. P., 1979, Isoproterenol-induced desensitization of adenylate cyclase in human astrocytoma cells, *J. Biol. Chem.* **254**:38–41.

Su, Y.-F., Harden, T. K., and Perkins, J. P., 1980, Catecholamine-specific desensitization of adenylate cyclase: Evidence for a multistep process, *J. Biol. Chem.* **255**:7410–7419.

Terasaki, W. L., Brooker, G., de Vellis, J., Inglish, D., Hsu, C.-Y., and Moylan, R. D., 1978, Involvement of cyclic AMP and protein synthesis in catecholamine refractoriness, *Adv. Cyclic Nucleotide Res.* **9**:33–52.

Control of Receptor Function by Homologous and Heterologous Ligands

MORLEY D. HOLLENBERG

1. Introduction

1.1. Defining Receptors as Heterogeneous Pharmacological Entities

It is largely to the credit of Ehrlich (1908) that early in the development of the receptor concept it was realized that agents that can cause the same biological effect (e.g., trypanocidal dyes) could bind to receptors that display a strict chemical specificity to react with one family of compounds but not another. This fundamental concept formed the basis of work that led to the clear-cut distinction among receptors for agents such as histamine, acetylcholine, and angiotensin, all of which cause contraction in smooth muscle. It was quickly appreciated, however, that a single compound, e.g., acetylcholine, could react with quite different receptors in different tissues (e.g., nicotinic receptors in striated muscle and muscarinic receptors in smooth muscle). Further, it was realized that histamine, which, at comparatively low concentrations (e.g., 10^{-6} M), can activate its own specific receptors, can, at comparatively high concentrations (e.g., $>10^{-4}$ M), activate other receptors specific for chemically distantly related agents such as acetylcholine.

Thus, as outlined in Fig. 1, on the one hand, a single receptor can be activated by two compounds that are chemically quite distinct, and on the other hand, a single compound can activate two quite distinct specific receptor systems. The classical example of the distinction of two receptor systems reacting with a single agonist can be seen in the work of Ahlquist

MORLEY D. HOLLENBERG • Endocrine Research Group, Department of Pharmacology and Therapeutics, Faculty of Medicine, University of Calgary, Calgary, Alberta, Canada T2N 4N1.

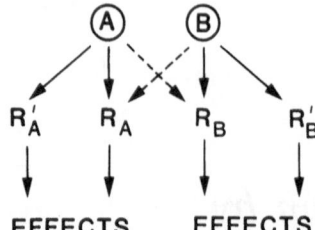

Figure 1. Receptor heterogeneity and receptor cross-over. The figure illustrates the interactions of two ligands, A and B, with their multiple receptor systems, R_A, R'_A and R_B, and R'_B. In addition, the figure indicates the possible cross reactivity of ligand A with one of the receptors for ligand B (R_B) and vice versa.

(1949), who was the first to clarify the interaction of catecholamines with α- and β-adrenergic receptors. The subsequent distinction of multiple receptors of the β-adrenergic class ($β_1$, $β_2$, etc.) or of multiple receptors for histamine (H_1, H_2, etc.) and dopamine (D_1, D_2, etc.) further emphasizes the generality of this situation. In essence, the code for receptor recognition (the ligand recognizing its receptor, or vice versa) is degenerate. It is the control by a given ligand of the function of its own receptor (homologous control) or of the function of a receptor intended for a second ligand (heterologous control) that forms the focus of this chapter. Attention is paid primarily to membrane-localized hormone receptors such as those for peptides and amines. These pharmacological receptors, as discussed elsewhere (Hollenberg and Goren, Chapter 18, this volume), exhibit a dual recognition–action property that is distinct from other cell surface recognition moieties.

1.2. Receptors as Dynamic Cell Surface Entities

As dealt with elsewhere in this volume, it is now realized that hormone receptors are dynamic elements both in the cell membrane and in the cytoplasmic compartment, as summarized by Fig. 2. The ligand-occupied receptor can migrate in the plane of the membrane, cluster, and aggregate in supramolecular patches at an internalization site; and the hormone–receptor complex can then undergo endocytosis either with subsequent proteolytic processing of the receptor and/or the ligand or with a recycling of the receptor and/or the ligand to the cell surface. These ligand-triggered dynamic receptor processes (sites 6 to 9, Fig. 2) are superimposed on the events that proceed in the absence of ligand and are concerned with steady-state receptor biosynthesis, membrane insertion, and turnover. A variety of factors other than the occupation of a receptor by its ligand can alter receptor dynamics (Hollenberg, 1981; Hollenberg and Goren, Chapter 18, this volume). For instance, tissue denervation, cell differentiation, changes in cellular growth rate, viral transformation, and nonspecific chemical stimuli (e.g., extracellular matrix) can all lead

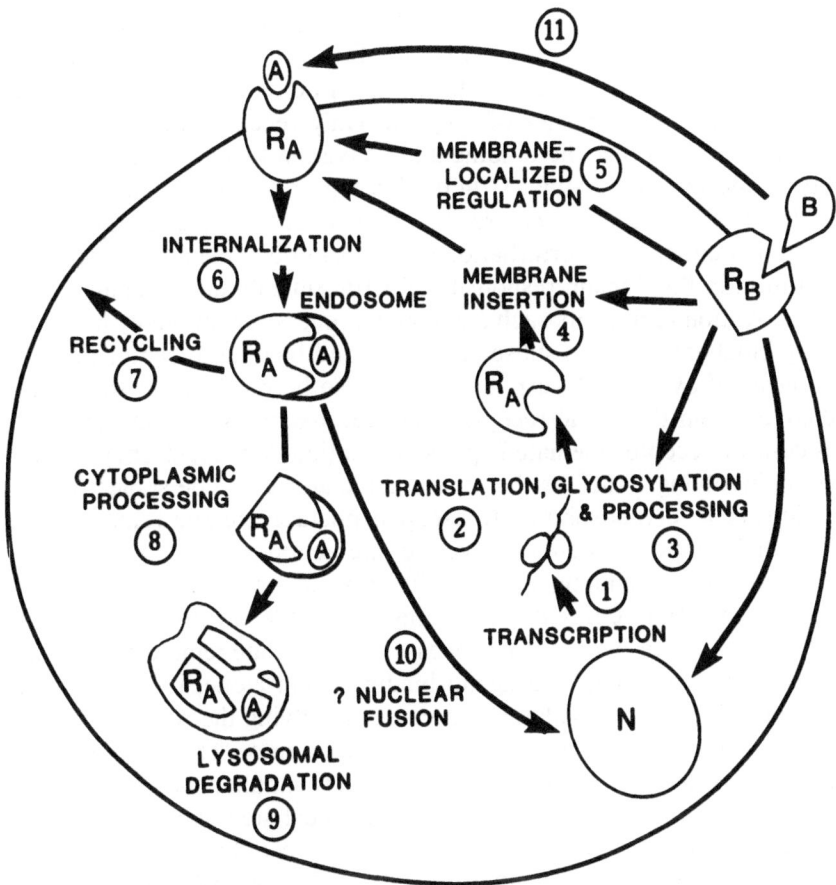

Figure 2. Levels of control of receptor regulation. The various possibilities of the control of receptor biosynthesis, turnover, and function, as discussed in the text, are indicated numerically.

to changes in receptor number and/or function. Added to these nonspecific receptor-regulatory factors, there are a variety of receptor-mediated events that can regulate receptor dynamics; these processes are dealt with in the following sections.

2. Ligand-Mediated Receptor Control

2.1. Levels of Control of Receptor Function

The overall function of a pharmacological receptor requires a dual recognition–activation process that leads ultimately to a cell response.

Some time ago, in a formal manner, the pharmacological concepts of intrinsic activity (Ariens *et al.*, 1954) or efficacy (Stephenson and Barlow, 1970) were developed to deal with the separate receptor-binding and receptor-activating properties of a ligand. Only recently, however, has it been appreciated that the binding and activation functions of a receptor may reside in quite distinct structural receptor domains. Thus, even at the level of the receptor *per se*, receptor function could, in theory, be controlled either by a perturbation of the ligand-binding domain or by an alteration in the domain responsible for transmembrane signaling. Since cell activation represents such a complex process, undoubtedly involving many biochemical steps subsequent to ligand binding, receptor function (in terms of overall cell response) could also be regulated at steps quite removed from the initial receptor-triggered reactions. For instance, the process of receptor-mediated, phosphorylation-dependent enzyme activation involving cAMP as the kinase modulator could be subject to regulation at a level (phosphodiesterase, phosphatase, etc.) far removed from the initial hormone–adenylate cyclase interaction.

Thus, by examining receptor function in terms of cell activation alone, it may not be possible to distinguish regulatory processes that involve the receptor *per se* from processes that regulate remote biochemical steps; and by studying the ligand binding event alone, it would be impossible to evaluate regulatory processes involving the receptor that influence only the "activation" domain of the receptor. Thus, the challenge for future studies of receptor control is to study simultaneously both ligand binding and, where possible, a key biochemical step likely to be tied to cell activation; optimally, this step should involve the receptor as one of the reactants (e.g., ion channel function, kinase activity, adenylate cyclase activation). Only by this dual approach will it be possible to evaluate cell regulation that occurs truly at the level of the receptor.

2.2. Levels of Control of Receptor Number

As indicated above, many nonspecific factors can affect the number of hormone receptors on responsive cells. As is elaborated on below, agents acting via receptor-mediated processes can also regulate receptor number. In terms of analyzing receptor-mediated control of receptor number, it is useful to distinguish among the several levels at which these effects may occur. As summarized in Fig. 2, for instance, one site of control would be at the level of gene transcription; the increase by steroid hormones of the numbers of cell surface receptors for peptide hormones might occur at this level. Regulation of the rates of mRNA translation or of the stability of cytoplasmic mRNA would provide other levels at which

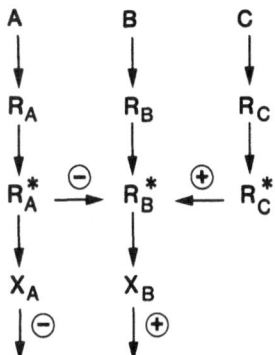

Figure 3. Levels of control of receptor function. The receptor-mediated control of a hypothetical regulatory enzyme is depicted. Increased enzyme activity in response to ligand B is shown to be caused by a mediator, X_B; decreased enzyme activity resulting from ligand A is shown to be caused by mediator X_A. The possible receptor-mediated control by ligands A and C of the process generating mediator X_B, at the level of the activated receptor R_B^*, is also illustrated. Thus, the function of ligand B could be regulated either at the level of cytoplasmic mediators or at the level of the membrane receptor, R_B.

hormones could alter cellular receptor number. An example of the complexities of multihormonal control in such a process can be seen in the regulation of casein synthesis by three hormones: insulin, prolactin, and hydrocortisone (Topper and Freeman, 1980). There is every reason to suspect that similar complex multihormonal receptor-mediated processes that regulate cell differentiation are involved in the control of receptor biosynthesis. The regulation of hormone receptors and cell responsiveness during cell differentiation has been discussed for several cultured cell systems by Lin and Beckner (1983).

Aside from these levels of biosynthetic control that receptors would have in common with other cellular proteins (sites 1 and 2, Fig. 2), it now appears that receptors may, in addition, be subject to rather specialized controls. One control point relates to posttranslational proteolytic processing and glycosylation reactions (site 3, Fig. 2), which undoubtedly affect receptors as well as other glycoproteins that eventually function in the plasma membrane. In addition, the apparently rapid process that can recruit previously synthesized receptors from the cytoplasmic compartment to the cell surface (site 4, Fig. 2) merits further attention. Such a process may, for instance, play a role in the recruitment of cell surface adipocyte IGF receptors caused by insulin (King et al., 1980; Bhaumick et al., 1981; Czech et al., Chapter 20, this volume).

A most important level of receptor regulation which is focused on below, involves processes localized in the plasma membrane (sites 5 and 6, Fig. 2). At this level, receptor number and/or ligand affinity can be regulated either as a consequence of receptor occupation (e.g., ligand A binding to receptor A, Fig. 3) or as a result of a separate receptor-mediated process (e.g., site 5, Fig. 2, wherein ligand B activates receptor B to regulate receptor A). The process of ligand-triggered receptor internalization has been documented for a variety of cell surface constituents;

however, comparatively little is known about factors (other than the ligand *per se*) that may regulate receptor number at this level. Once a receptor is internalized, further levels of control are possible along the complex routes that the receptor may take (Ciechanover *et al.*, Chapter 14) either via recycling (site 7, Fig. 2) back to the cell surface or via cytoplasmic routing (possibly including proteolytic processing; site 8, Fig. 2) to eventual lysosomal fusion and proteolytic degradation (site 9, Fig. 2). In summary, all of sites 4 to 9 depicted in Fig. 2 provide for potential levels of receptor regulation via receptor-mediated processes.

Two questions can be asked with regard to ligand-mediated receptor regulation that may help to sort out the level at which regulation might be occurring. (1) What is the time course of the regulation? (2) Can regulation of the kind observed (e.g., change in receptor number and/or affinity) be detected in membrane preparations as well as in intact cell systems? For instance, a time course of hours to tens of hours has been observed for the appearance in the plasma membrane of newly synthesized receptors such as the one for acetylcholine (nicotonic receptor) (Devreotes and Fambrough, 1975, 1976) or for epidermal growth factor–urogastrone (EGF–URO) (Krupp *et al.*, 1982). A similar time frame might be anticipated for receptor-mediated regulation that occurs at the level of receptor biosynthesis (e.g., sites 1 to 3, Fig. 2).

In contrast, the recruitment of IGF receptors to the adipocyte cell surface alluded to above takes place within tens of minutes up to an hour (King *et al.*, 1980; Bhaumick *et al.*, 1981; Czech *et al.*, Chapter 20, this volume); for this process, control at sites 4, 5, or 7 might thus be anticipated. Additionally, it would be expected that receptor recruitment from an intracellular pool (site 4 or 7) would take place only in intact cells. Even a process of receptor regulation occurring rapidly at the level of the membrane (site 5) might require the participation of cytoplasmic factors. For instance, as discussed below, the regulation of EGF receptor binding by phorbol esters, which takes place over a 20-min time course and appears to involve a membrane-localized process, occurs in intact cells but not in membrane preparations (Hollenberg *et al.*, 1981; Shoyab *et al.*, 1979). In contrast, the rapid effect of insulin on the hydrodynamic forms of the insulin receptor can take place either in membranes or in purified soluble receptor preparations (Maturo *et al.*, 1983). Thus, experiments designed to evaluate both the time course and the possible participation of cytoplasmic factors involved in receptor regulation may help to define the level at which ligand-mediated control may be exerted.

2.3. Homospecific versus Heterospecific Receptor Regulation

2.3.1. Homospecific Regulation

As summarized by Raff (1976), it was recognized some time ago that the binding of a ligand could induce the disappearance of a cell surface molecule. Although many of the initial experiments in this area were focused on the modulation of cell surface antigenic sites by antibodies, it was quickly realized that, via an analogous mechanism, hormones could also regulate the concentration and/or binding affinities of their own receptors (so-called "homospecific" receptor regulation). The initial studies by Gavin *et al.* (1974) were the first to suggest that the number of insulin binding sites in cultured IM-9 lymphocytes could be reduced by preincubating cells with insulin, albeit at concentrations far above those that could be considered physiological. Perhaps the best data related to receptor regulation and insulin action have come from studies with rat adipocytes, wherein preincubation of cells with elevated insulin concentrations within the physiological range leads to a diminution of receptor binding and a predicted shift to the right of the dose–response curve (Livingston *et al.*, 1978). An analogous insulin-dependent concentration-related reduction in insulin binding can also be observed within 4 hr at 37°C in cultures of human fibroblasts (Mott *et al.*, 1979).

Similar observations have been made in a number of cell systems containing receptors for EGF–URO and for β-adrenergic agonists (Carpenter and Cohen, 1976; Lefkowitz, 1978). A key feature of such studies relates to the ligand specificity of the so-called "down-regulatory" process. For instance, in studies of the regulation of insulin receptors in cultured fibroblasts, it was important to demonstrate that the order of potency for the effect was: porcine insulin > guinea pig insulin > proinsulin (Mott *et al.*, 1979); similarly, in studies with the β-adrenergic receptor, the receptor-related nature of the "down-regulation" phenomenon was indicated by the order of potency: isoproterenol > epinephrine > norepinephrine (Lefkowitz, 1978). This process, whereby a ligand can regulate the number of its own receptors, can be termed "homospecific" receptor regulation.

Homospecific receptor regulation can be in the positive as well as the negative direction. For instance, insulin, in addition to causing down-regulation of its receptor in isolated rat adipocytes and in cultured human fibroblasts, can cause an increase of its own receptors in cultured rat chondrosarcoma cells (Stevens *et al.*, 1983). Thus, even in cells from the same animal, insulin may cause down-regulation of its receptor in one tissue (fat) and up-regulation of its rceptor in another (cartilage). The administration *in vivo* of prolactin (Posner *et al.*, 1975) as well as angiotensin II (Hauger *et al.*, 1978) and pentagastrin (Takeuchi *et al.*, 1980) leads to an up-regulation of their respective receptors. In the studies with prolactin, gastrin, and angiotensin II, it remains to be determined if the effects are caused by a direct action of each hormone on the target cells or occur via an indirect process, possibly mediated by a second hormone signal generated *in vivo*. The ability of gonadotropin-releasing hormone to increase the number of its receptors in cultured pituitary cells appears to be a direct effect (Loumaye and Catt, 1982); however, the biphasic nature of the effect (low concentrations stimulate and higher concentrations reduce binding) and the long time course (six to nine hr) suggest a complex process. A number of interesting examples of more rapid (minutes to tens of minutes) homospecific receptor recruitment have been summarized by Triggle (1982).

Quite remarkable aspects of the regulation of gonadotropin receptors in Leydig cells of the testis and luteal cells of the ovary have been summarized by Catt and co-workers (1979). Three distinct features of this regulatory process were observed after the administration of gonadotropins to animals *in vivo:* (1) the production of a high-affinity ligand–receptor complex, causing a masking of luteinizing hormone receptor sites for periods up to 24 hr; (2) a concurrent desensitization of the effector system (in this case, adenylate cyclase) in step with this persistent receptor occupancy; and (3) an eventual depletion of receptors from the plasma membrane to an extent even greater than would have been anticipated on the basis of the initial degree of receptor occupancy. The results with the gonadotropins indicate the complex nature of the receptor-regulatory process and emphasize the importance of examining events over both a short and comparatively long time course.

2.3.2. Heterospecific Regulation

The regulation of the binding of one hormone to its receptor by the binding of a second hormone to its receptor (so-called "heterospecific" receptor regulation) has now been observed in a variety of instances. As with homospecific receptor regulation, heterospecific regulation can lead to either an increase or a decrease in ligand binding. An intriguing example of this kind or regulation relates to the action of the gonadotropins. In cultured rat ovarian fragments, follicle-stimulating hormone (FSH) causes the appearance of binding sites for [^{125}I]-labeled human chorionic gonadotropin (Nimrod *et al.*, 1977); these sites are presumably intended for luteinizing hormone (LH). In contrast, the hypothalamic modulatory factor that stimulates LH release (LHRH) causes a reduction in testicular gonadotropin receptors, seemingly via a direct effect on the Leydig cell (Lefebvre *et al.*, 1980; Labrie *et al.*, 1980). This interesting example of heterospecific receptor regulation merits further investigation to determine the mechanism(s) involved.

Steroid hormones acting via their specific receptors can also regulate the numbers of plasma membrane receptors for polypeptides. For instance, estrogen can elevate the number of oxytocin receptors in the rat uterus (Soloff, 1975) and can increase lactogenic receptors in rat liver (Posner *et al.*, 1974, 1975; Shiu and Friesen, 1981). Similarly, corticosteroids have been observed to elevate the binding of EGF–URO to cultured fibroblasts (Baker et al., 1978), and thyroid hormone has been observed to regulate rat hepatic EGF–URO receptors *in vivo* (Hayden and Severson, 1983). In a cultured preadipocyte system, glucocorticoids not only regulate the numbers of β-adrenergic receptors, but can also cause a switch in subtype from β_1 to β_2 (Lai *et al.*, 1982). As will be mentioned below, steroids (progesterone, β-estradiol) can affect muscarinic receptor binding in homogenates of rat hypothalamus and anterior pituitary (Sokolovsky *et al.*, 1981). To date, this kind of heterospecific receptor regulation has not been sufficiently explored vis à vis the influence of polypeptide hormones on the numbers and functions of receptors for steroid hormones.

The analysis of the effect of estrogens on the level of prolactin receptors is particularly instructive (Shiu and Friesen, 1981). Not only is it evident that estrogen acts indirectly *in vivo* by a mechanism involving the pituitary, but the data suggest that the pituitary mediator of the rise in lactogenic receptors, thought to be prolactin itself, may possibly be an as yet unidentified pituitary factor. The ability of insulin to regulate lactogenic binding sites in liver (Baxter and Turtle, 1978) further complicates this picture. The complexities of analyzing *in vivo* the effects of estrogens on lactogenic receptors emphasize the need for appropriate tissue culture systems in which receptor regulation can be analyzed.

The examples of heterospecific receptor regulation discussed above very likely involve mechanisms that relate to the rates of receptor biosynthesis and turnover. Recent studies using cell culture systems have revealed examples of very rapid (minutes to tens of minutes) heterospecific receptor regulation that appear to involve mechanisms located at or within the plasma membrane. This kind of regulation is well illustrated by the effects of a variety of agents on the binding of EGF–URO to its receptor. Studies with the phorbol ester tumor promotor 12-O-tetradecanoylphorbol-13-acetate (TPA) were the first to demonstrate a rapid, temperature-dependent reduction of EFG–URO binding via a heterospecific receptor mechanism (Lee and Weinstein, 1978; Brown *et al.*, 1979; Shoyab *et al.*, 1979; Hollenberg *et al.*, 1981). A similar reduction in EGF–URO binding is caused by another tumor promoter, teleocidin (Umezawa *et al.*, 1981). It was apparent in such studies that TPA acted via a binding site distinct from the EGF–URO receptor (Hollenberg *et al.*, 1979, 1981). Recent evidence suggests that the cellular "receptor" for TPA is the calcium and phospholipid-dependent protein kinase, kinase C (Kikkawa *et al.*, 1982; Castagna *et al.*, 1982; Niedel *et al.*, 1983).

Thus, TPA, interacting with its own "receptor," can regulate the affinity and/or number of receptors for EGF–URO. Schematically, this process is depicted in Fig. 3, where ligand A might represent TPA and ligand B might represent EGF–URO. The effect of TPA appears to be somewhat specific in that the receptors for insulin and transcobalamin II are unaffected (Hollenberg et al., 1979, 1981).

An analogous regulation of the EGF–URO receptor has also been observed in response to two other ligands, platelet-derived growth factor (PDGF) (Wrann et al., 1980; Wharton et al., 1982; Bowen-Pope and Ross, 1983; Collins et al., 1983) and vasopressin (Rozengurt et al., 1981). Each of these polypeptides acts via its own distinct receptor. Interestingly, both PDGF and vasopressin, like TPA, can cause a decrease in EGF–URO receptor affinity in certain cells (Collins et al., 1983; Rozengurt et al., 1981), and both PDGF and vasopressin appear to act on the EGF–URO receptor via a time- (tens of minutes to several hours) and temperature-dependent process that occurs in intact cells. It has been observed that membranes isolated from TPA-treated cells do not exhibit the same reduction in EGF–URO binding that can be documented in the intact cell system (Hollenberg et al., 1981). Thus, for TPA and possibly for PDGF and vasopressin, the regulation of the EGF–URO receptor very likely involves the participation of cytoplasmic factors.

Heterospecific mechanisms can also lead to an increase in ligand binding via a mechanism that appears to be localized at or near the plasma membrane. The ability of insulin to increase the binding of insulinlike growth factor II (IGF-II) in rat adipocytes can be seen as a counterpart to TPA-mediated reduction inEGF–URO binding described above. In adipocytes, insulin causes a rapid (minutes) increase in the binding of IGF-II by raising the IGF-II receptor affinity without changing the number of cell surface binding sites (King et al., 1982; Oppenheimer et al., 1983). As with the effect of TPA on EGF–URO binding, the increase of IGF-II binding by insulin requires an intact cell and does not occur at or below 15°C. It is of interest that antimicrotubule or antimicrofilament agents inhibit the insulin-mediated effect. Further, isolated membranes form insulin-treated adipocytes exhibit the same IGF-II binding properties as do untreated cells. Thus, in many respects (temperature dependence, requirement for intact cell, etc.), the effects of insulin on the binding of IGF-II mirror the effects of TPA on EGF–URO binding.

A variety of other examples of heterospecific receptor regulation have been observed to occur in isolated membrane preparations in contrast with the examples of EGF–URO and IGF-II receptor regulation, which occur only in intact cells. The data point to the participation of membrane-localized reactions in which adsorbed cytoplasmic factors or peripheral membrane proteins may or may not play a role. For instance, vasoactive intestinal polypeptide (VIP) very rapidly (3 to 5 min) enhances the binding of a muscarinic ligand to membranes isolated from the cat salivary gland (Lundberg *et al.*, 1982). The effect of VIP on the membranes, which is maximal at 5 min and disappears at 10 min, is to increase the association rate constant. Muscarinic receptor binding in homogenates of rat hypothalamus and adenohypophysis can also be affected (increased affinity; reduced proportion of high-affinity sites) by steroids (progesterone, β-estradiol) (Sokolovsky *et al.*, 1981). Muscarinic agents, in turn, are able under certain conditions *in vitro* to increase the binding affinity of an α_1-adrenergic receptor probe in isolated membranes from rat heart ventricle (Yamada *et al.*, 1980). In a like vein, antidepressant drugs such as imipramine can rapidly (10 min at 37°C) increase the affinity of [³H]-labeled serotonin binding to synaptic membranes from rat brain (Fillion and Fillion, 1981).

A particularly intriguing example of this kind of heterospecific receptor regulation relates to the ability of γ-aminobutyric acid (GABA) to modulate the binding of benzodiazepines and vice versa (Guidotti *et al.*, 1979; Costa, 1979; Braestrup and Nielsen, 1980). In isolated rat brain membranes, GABA and other related agonists can increase the binding of [³H]-labeled diazepam by increasing receptor affinity. The GABA receptor involved in regulating diazepam binding has the stereoselective characteristics of the *in vivo* GABA receptor. The reciprocal effect of benzodiazepines in increasing the binding of GABA to its receptor in similarly prepared brain membranes (both the affinity and binding capacity for GABA are increased) appears to involve a 15,000-mol.-wt. heat-stable peripheral membrane protein, GABA-modulin, which blocks GABA binding via an allosteric mechanism (Costa, 1979). It is thought that the benzodiazepines release GABA-modulin from the membranes, thereby enhancing the binding of GABA. The close association of the GABA and benzodiazepine receptors, possibly in the same macromolecular complex, is further emphasized by the copurification of both receptors from solubilized brain membrane preparations (Gavish and Snyder, 1981); the stimulation of flunitrazepam binding by GABA can be observed even in the affinity-purified receptor preparation.

As summarized recently (Fuxe and Agnati, 1985), the kind of heterospecific regulation heralded by the observations on GABA-benzodiazepine interactions may be widespread for receptors in the central nervous system. Considerable data (see summary by Fuxe and Agnati, 1985) now point to interactions between receptors for dopamine and those for cholecystokinin (CCK), glutamate and neurotensin; between receptors for 5-hydroxytryptamine (5H-T or serotonin) and those for CCK, vaso-

Table I. Heterospecific Receptor Regulation[a]

Receptor	Regulator(s)	Comments
EGF–URO	TPA PDGF Vasopressin	Time course: minutes to tens of minutes Reduction in binding ($K_D \uparrow$ or $B_{max} \downarrow$) Effect is temperature dependent Requires intact cell Effect not seen in membranes isolated from treated cells
IGF–II	Insulin	Time course: minutes Increase in binding ($K_D \downarrow$, B_{max} unchanged) Effect is temperature dependent Requires intact cell Effect not seen in membranes from treated cells
Muscarinic	VIP	Time course: minutes, transient Transient increase in binding ($K_D \downarrow$, $k_1 \uparrow$) Temperature dependence not reported Effect seen in membranes
	Progesterone, β-estradiol	Increase in binding affinity; decrease in proportion of high-affinity binding sites
α$_1$-adrenergic	Carbachol via muscarinic receptor	Time course: minutes at 24°C Increases WB4101 binding in the presence of Gpp(NH)p Temperature dependence not reported Effect seen in membranes
α$_2$-Adrenergic	Neuropeptide-Y	$K_D \uparrow$, $B_{max} \uparrow$
GABA	Benzodiazepine	Time course: minutes Increase in binding ($K_D \downarrow$, ? $\uparrow$ B_{max}) Temperature dependence not reported Effect seen in membranes Effect depends on method of membrane preparation Mediated via GABA-modulin
Benzodiazepine	GABA	Time course: minutes Increased binding ($K_D \downarrow$; B_{max} unchanged) Temperature dependence not reported Effect seen in membranes Effect seen in soluble purified receptor preparations
Dopamine	Cholecystokinin	$K_D \uparrow$; effects vary from one brain region to another; binding of agonist probes primarily affected
	Neurotensin	$K_D \uparrow$; $B_{max} \uparrow$; variable effects in different brain regions
	Glutamate	$K_D \uparrow$; B_{max} unchanged; agonist binding primarily affected
Neurotensin	Dopamine	$K_D \uparrow$; $B_{max} \uparrow$

Table I. (continued)

Receptor	Regulator(s)	Comments
Serotonin (5HT-1)	Substance P	K_D ↑ ; B_{max} ↑
	VIP (Vasoactive Intestinal Polypeptide)	K_D ↑ ; B_{max} ↑
Serotonin (5HT-2)	CCK	K_D ↑ ; B_{max} ↑ ; effect depends on CCK analogue used
Glutamate	CCK	K_D ↑ ; B_{max} ↓
	Kainic acid	K_D ↓ ; B_{max} ↓
Kainic acid	Glutamate	K_D ↑ ; B_{max} ↑

[a] The table summarizes the properties of the examples of heterospecific receptor regulation, as discussed in the text. References for the examples are summarized in the text.

active intestinal polypeptide (VIP) and substance (P) (SP); between α_2-adrenergic receptors and those for neuropeptide-Y; and between glutamate receptors and those for kainic acid. Thus, overall, heterospecific regulation of the kind described above and summarized in Table I can be observed in intact cells, in isolated membranes, and in purified receptor preparations.

3. Mechanisms of Receptor Regulation

As indicated in the previous sections, receptor activation can lead to either an increase or a decrease of ligand binding, which can involve changes in receptor number, receptor affinity, or both parameters. As pointed out above, those effects that occur only in intact cells over a comparatively long time frame (hours to tens of hours) very likely involve mechanisms that regulate the synthesis and turnover of many other cellular constituents (sites 1 through 4 of Fig. 2). These mechanisms are not dealt with here.

Recently, attention has begun to focus on the possible membrane-localized mechanisms that might be involved in the events (sites 5 and 6, Fig. 2) that occur over a relatively short time frame (minutes to tens of minutes). In principle, it is possible to categorize the potential mechanisms that might be involved in terms of what is known about the manner whereby many enzymes are regulated. For instance, evidence has already accumulated to indicate that receptors can be modified by reactions leading either to covalent or noncovalent bond formation. In terms of covalent modifications, one can point to reactions involving receptor phosphorylation, disulfide–sulfhydryl exchange, and receptor proteolysis as possible mechanisms involved in changes in receptor function. Interactions that do not involve the formation or breaking of covalent bonds and that could alter receptor function include changes in membrane potential, changes in receptor distribution (patching, capping), allosteric interactions involving either protein–protein (e.g., mobile receptor model) or small ligand (cations, anions, nucleotides, etc.)–receptor interactions, and changes in the membrane lipid environment (e.g., lipid methylation or hydrolysis). Examples related to all of these possible mechanisms are discussed in the sections that follow.

3.1. Receptor Phosphorylation

An enlarging body of evidence, as summarized briefly elsewhere (Hollenberg, 1982b), points to the importance of receptor phosphorylation as a key event in the function of many receptors, including the receptor for light (rhodopsin), the nicotinic cholinergic receptor, and the receptors for EGF–URO, insulin, and PDGF. More recently, the β-adrenergic receptor has joined the list of receptors known to be subject to phosphorylation reactions (Stadel et al., 1983). Two separate but possibly connected mechanisms appear to be involved. First, a ligand-directed conformational change in the receptor may cause the receptor to become a substrate for distinct membrane-associated kinases or phosphatases. This situation appears to be the case for rhodopsin and the nicotinic receptor. Second, the receptor itself, possessing intrinsic kinase activity, may be activated on binding its ligand and then phosphorylate itself as well as other membrane-associated substrates. This second situation is in keeping with data obtained for the insulin, EGF–URO, and PDGF receptors.

Interestingly, phosphorylation per se appears to be involved in regulating the kinase activity of the insulin receptor (Rosen et al., 1983). In addition, both the insulin receptor and the IGF-I (somatomedin C or basic somatomedin) receptor can be phosphorylated in intact cells in response to phorbol ester (TPA), presumably via reactions involving kinase-C (Jacobs et al., 1983). Phosphorylation can occur at a number of serine, threonine, and tyrosine residues in the receptor structure. Parallel work has demonstrated that phosphorylated receptors such as the one for EGF–URO can be dephosphorylated rapidly by endogenous membrane-associated phosphatases (Carpenter et al., 1979). Thus, like other enzyme systems, receptors can be subject to a variety of phosphorylation–dephosphorylation reactions with the receptor, in many instances, serving as both the acceptor and the donor of a phosphate residue. Future work with many receptors will undoubtedly focus on the protein kinase as well as protein phosphatase reactions in which receptors participate.

With the data now available concerning receptor-associated phosphorylation reactions, it is easy to envision how the activation of one receptor might lead to self-phosphorylation, either via a direct (e.g., receptor–kinase) or indirect (e.g., as a consequence of elevated cellular cAMP or kinase-C activation) process, and it can be seen that the activation of one receptor could lead to the phosphorylation (or dephosphorylation) of a second receptor. These kinds of homospecific or heterospecific phosphorylation–dephosphorylation reactions may relate directly to the homospecific and heterospecific receptor regulation events discussed above. However, there are two key questions that remain unanswered at present. (1) Does receptor phosphorylation *per se* relate to the regulation of receptor affinity and/or number? (2) What role(s) if any does receptor phosphorylation play in the process of cell activation? Partly in answer to the first question, it is tempting to speculate that the effect of phorbol ester (TPA) on EGF–URO binding as discussed above is mediated via kinase-C-induced receptor phosphorylation. It is clear that much work remains to be done to address these two important questions.

3.2. Disulfide–Sulfhydryl Exchange Reactions

The possible involvement of disulfides in receptor function was heralded by studies in the mid-1960s of the effects of reducing agents on the response of electroplax cells to cholinergic agents (summarized by Karlin *et al.*, 1973). On reduction with dithiothreitol (DTT), the depolarizing responses to monoquaternary activators and succinylcholine are decreased, whereas the response to decamethonium is increased (Karlin *et al.*, 1973). Reduction leads to a fourfold increase in the ED_{50} for carbachol. On reoxidation with disulfide reagents such as 5,5'-dithio-bis(2-nitrobenzoate) (DTNB), the effects of DTT can be reversed, but if a thiol-alkylating reagent is applied after reduction but before reoxidation, the reversal is prevented. There do not appear to be free receptor SH groups related to receptor function, and the sensitive disulfide does not appear to be reduced and oxidized cyclically during receptor activation. Evidently, the sensitive disulfide plays an important structural function related to receptor activity. Possibly, this sensitive nicotinic receptor disulfide could play a role *in vivo* in terms of receptor regulation via a heterospecific mechanism.

Disulfide–sulfhydryl exchange reactions may also play a role in insulin receptor function. For instance, it has been observed that in adipocyte membranes, there is a reversible sulfhydryl-mediated change in state of the insulin receptor (Schweitzer *et al.*, 1980); in placenta membranes, reduction (DTT) of the insulin receptor leads to a decrease in binding and a disappearance of the high-affinity portion of the curvilinear Scatchard plot usually observed for insulin (Jacobs and Cuatrecasas, 1980). The effects of DTT appear to vary from tissue to tissue (in adipocytes and liver membranes, binding increases, whereas in placenta membranes, binding decreases); however, reduction of the so-called class I receptor disulfides does not impair receptor function except to shift the dose–response curve slightly to the right (Massagué and Czech, 1982). In crude solubilized or purified insulin receptor preparations, exposure of the receptor to insulin unmasks a receptor sulfhydryl that can react with N-ethylmaleimide (NEM) (Maturo *et al.*, 1983); in the presence of insulin, the receptor is converted from one hydrodynamic form (R_I) to another R_{II}), and there is a reduction in the receptor's affinity for insulin (Maturo *et al.*, 1983; Maturo and Hollenberg, 1978). On reoxidation of the insulin-treated receptor with oxidized glutathione, the receptor is converted back to the higher-affinity R_I form. A reversal of the insulin-mediated conversion to the R_{II} low-affinity receptor form can also be brought about by an endogenous membrane-associated glycoprotein (Maturo and Hollenberg, 1978). It is important to note that a critical NEM-reactive sulfhydryl in the insulin receptor as well as in the EGF–URO receptor appears to be involved in the tyrosine kinase activities of these receptors (K. A. Valentine and M. D. Hollenberg, unpublished data; J. M. Maturo III and M. D. Hollenberg, unpublished data). Thus, homospecific regulation of the insulin receptor *in vivo* may in part involve a cyclic disulfide–sulfhydryl interchange reaction.

A final intriguing example of the possible control of receptor function by sulfhydryl–disulfide exchange reactions relates to the β-adrenergic receptor–cyclase system. As summarized by Korner *et al.* (1982), a number of studies have indicated that the functions of both the catalytic moiety and guanine nucleotide regulatory components of the cyclase system are associated with specific essential sulfhydryl groups. Further, it has been observed that in the simultaneous presence of NEM, the binding of isoproterenol leads to an inactivation of about 50% of the adrenergic binding sites in a variety of membrane preparations (Vauquelin *et al.*, 1980). It appears that the addition of NEM locks the ligand onto the receptor in a high-affinity state (Korner *et al.*, 1982); a spontaneous reappearance of the receptor binding sites facilitated by divalent cation and guanine nucleotide can be observed over a 20-min period at 37°C. This reversible affinity change resulting from agonist (but not antagonist) binding is thought to involve a sulfhydryl present on the guanine nucleotide regulatory component (Korner *et al.*, 1982). Since analogous regulatory components may be involved in the action of a variety of hormones (Rodbell, 1980), this sulfhydryl-mediated mechanism may play a role in the desensitization phenomena associated with many agents. This phenomenon can lead to a homospecific reduction in ligand binding and a reversible ligand-specific decrease in cell sensitivity.

3.3. Receptor Proteolysis

It is well recognized that receptors for many hormones are sensitive to proteolytic enzymes. Nonetheless, treatment with a proteolytic agent need not abolish receptor activity. For example, mild proteolysis of the insulin receptor merely shifts both the binding isotherm and the insulin dose–response curve to the right (Cuatrecasas, 1971a). In contrast, with the EGF–URO receptor, even comparatively extensive proteolysis does not abrogate ligand binding (Armstrong and Hollenberg, 1985). Nonetheless, proteolytic cleavage of a comparatively small fragment of the EGF–URO receptor (about 25 kilodaltons) appears to abolish the kinase activity of the EGF–URO receptor (data summarized by O'Connor-McCourt and Hollenberg, 1983). Since cells are known to secrete proteases in response to various agents, and since cytoplasmic proteases can be activated by calcium, it is not unlikely that receptor proteolysis may provide yet another means of controlling receptor function via specific ligand-activated processes.

3.4. Change in Membrane Potential

It has been repeatedly pointed out by Zierler (Zierler and Rogus, 1981; Zierler, 1985) that changes in membrane potential can potentially have a profound effect on the orientation of membrane proteins. Presumably, such a mechanism is partly involved in control of the voltage-sensitive sodium channel (Catterall, 1982). Quite possibly, such a mechanism might be responsible for either homospecific or heterospecific receptor regulation.

3.5. Changes in Receptor Distribution

As summarized elsewhere in this volume (Hollenberg and Goren, Chapter 18), a change in distribution of a receptor on the cell surface *per se* (microclustering, aggregation) can lead to a reduction in receptor affinity. Although the magnitude of these effects may not be large enough to measure experimentally, the reduction of receptor affinity by 30% or so, which can theoretically occur when receptors dimerize (Goldstein and

Wiegel, 1983), may have an important impact on the biological response of a system. Although the factors controlling receptor microclustering and aggregation are poorly understood, such mechanisms may well play a role in the homologous or heterologous control of receptor function.

3.6. Allosteric Interactions

As postulated by the mobile receptor paradigm, dealt with elsewhere in this volume (Hollenberg and Goren, Chapter 18), ligand-modulated protein–protein interactions can potentially regulate receptor affinity. The action of GABA-modulin and the high-affinity state of the β-adrenergic receptor, as outlined in Section 2.3.2, can be seen as two examples of this kind of receptor regulation.

In addition to protein–protein interactions, it is evident that low-molecular-weight substances can also serve as allosteric regulators of receptor function. For instance, sodium ion is known to alter the binding characteristics of receptors of opiates and α_2-adrenergic agents (Pert et al., 1973; Michel et al., 1980), and chloride ion is thought to regulate the binding of GABA (Costa, 1979). Guanine nucleotides may turn out to be perhaps the most important regulators of ligand binding. As summarized by Rodbell (1980), guanine nucleotides play a critical role in coupling receptor occupation to the regulation (either stimulatory or inhibitory) of adenylate cyclase. Coupling is mediated via a stimulatory (N_s or G_s) or inhibitory (N_i or G_i) oligomeric guanine-nucleotide-binding substituent (Gilman, 1984; Rodbell, 1980). Likewise, guanine nucleotides can regulate the affinity of a variety of receptors for their specific ligands, including agents such as glucagon, angiotensin, catecholamines (α and β receptors), dopamine, opiates, acetycholine (muscarinic), and prostaglandin E; in all cases, guanine nucleotides decrease the affinity of the agonist–receptor complex (summarized by Rodbell, 1980). It is believed that the regulation of ligand affinity by guanine nucleotides is mediated via the regulatory subunit, N_x or G_x, and not via a direct binding of nucleotide to the receptor, R. Since structures of the type RN_x (or RG_x) may regulate a variety of activities apart from adenylate cyclase (Rodbell, 1980), it is possible that guanine nucleotide regulation of receptor affinity may play a widespread role in the control of cell function by homospecific or heterospecific mechanisms. Undoubtedly, the list of low-molecular-weight allosteric regulators of receptor function will enlarge with time.

3.7. Changes in Lipid Environment

In certain instances, membrane receptors can exist in a cryptic state and can be increased by agents that perturb membrane lipids. For instance, adipocyte receptors for insulin or placental receptors for EGF–URO can be unmasked by phospholipases (Cuatrecasas, 1971b; Hock and Hollenberg, 1980). Possibly, the lysophospholipids and arachidonic acid generated by such treatments lead to increased membrane fluidity, which exposes the cryptic receptors. Phospholipid methylation leading to the formation of phosphatidylcholine, which would also increase membrane fluidity, has also been shown to reveal cryptic β-adrenergic receptors in rat reticulocyte membranes (Strittmatter *et al.*, 1979). Curiously, β-adrenergic stimulation of the same cells, which led to a decrease in membrane viscosity and an increase in isoproterenol-stimulated adenylate cyclase, did not cause an increase in receptor number (Hirata *et al.*, 1979); possibly, these results can be accounted for by offsetting receptor regulatory events. The likely activation of phospholipid methylation by a variety of hormones (Hirata and Axelrod, 1980) could, via a similar mechanism, possibly lead to the rapid recruitment of membrane receptors either by a homospecific or a heterospecific receptor mechanism. This methylation process would provide for one of the few mechanisms that can lead to a rapid up-regulation of receptors in the plane of the membrane.

4. Consequences of Ligand-Modulated Receptor Regulation

Since receptors play such a critical rate-limiting role in the activation of cells by hormones, it is quite likely that the physiological function of receptor regulation of the kind discussed in this chapter is to modulate overall cell sensitivity. Thus, receptors, as a class of biological regulators, can readily take their place along with other key rate-limiting proteins such as certain cytoplasmic enzymes that regulate metabolic pathways. However, because receptors represent the interface between the external and internal cellular milieux, receptor regulation may have even more far-reaching pathophysiological and therapeutic implications than does the regulation of certain specific metabolic enzymes. In essence, receptors may be slightly higher up in the hierarchical biological network than are cytoplasmic enzymes.

4.1. Pathophysiological Implications

4.1.1. Receptor Crossover and Receptor Heterogeneity

Receptor crossover, as outline in Fig. 1, has both pathophysiological and therapeutic implications. For instance, in disease states in which elevated hormone levels may occur (e.g., tumors of the anterior pituitary), the crossover of one hormone to activate a second receptor system (heterologous control) may contribute to the disease state. The melanophore-stimulating effects of elevated levels of adrenocorticotrophic hormone (pituitary adenomas; Addison's disease) can be seen as one example of this crossover situation. As discussed in Section 2.3.1, elevated hormone levels may down-regulate or up-regulate receptors via either homospecific or heterospecific mechanisms. Thus, in a situation like that of pseudo-hypoparathyroidism in which there is a defect in the guanine nucleotide regulatory subunit that is involved in PTH-stimulated adenylate cyclase (Farfel *et al.*, 1980; Levine *et al.*, 1980) resulting in elevated levels of parathyroid hormone (PTH), the possible crossover of PTH with another receptor system could have pathological consequences beyond those localized in the kidney and bone.

A further complication of a pathological state that results in an abnormally high hormone level relates to the possible action of the hormone on several homologous receptor systems. It appears to be the rule rather than the exception that, as outlined in Fig. 1, one hormone can react with several receptor systems; this situation very likely applies to polypeptide hormones as well as to the amines dopamine, histamine, and epinephrine. Thus, in an individual with a pheochromocytoma, the pathology may relate not only to the activation of α- and β-adrenergic receptors but also to the activation of the multiple subtypes of these two major receptor classes. Furthermore, all of the receptor regulatory mechanisms for such receptors will be called into play by the abnormally high hormone levels. Thus, continuing with the example of the pheochromocytoma, it can be seen that in such patients, heterospecific receptor regulatory mechanisms might have an impact on a much wider spectrum of receptor systems than just the ones intended for catecholamines. A concrete possibility of a complication that may be present in pheochromocytoma patients can be seen in the ability of adrenergic agents to regulate insulin receptor binding and function (Czech *et al.*, Chapter 20, this volume). In summary, the potential for receptor crossover and for the activation or regulation of multiple receptors with a single ligand adds a new complexity to the interpretation of various pathological states.

4.1.2. Homologous Receptor Domains and Common Regulatory Subunits

The ability of ligands to bind to several distinct receptors and the ability of several receptors to interact with common regulatory moieties such as the guanine nucleotide regulatory subunit of the cyclase system imply that receptors possess common domains, as discussed briefly elsewhere (Hollenberg, 1982a). Thus, it is not unlikely that a disease state affecting a common regulatory component will involve multiple receptor systems. An intriguing example of this situation can be seen in the action of pertussis toxin, which by inactivating the N_i (or G_i) regulatory subunit of the cyclase system affects multiple hormone systems (summarized by Gilman, 1984). Further, in a complex multisystem disease such as type II diabetes, it is possible that the same defect (possibly in a receptor regulator moiety) that leads to insulin resistance may also lead to "resistance" in the action of other polypeptides with "insulinlike" activities [e.g., basic somatomedin (IGF-I), IGF-II, and oxytocin].

The structural relationships between the receptors themselves may also play a role in disease states in which antireceptor antibodies are present. For instance, the cross reactivity of anti-insulin-receptor antibodies with receptors for the somatomedins (IGF-I) (Armstrong *et al.*, 1983; Jacobs *et al.*, 1983) may potentially play a pathophysiological role in antireceptor-antibody-bearing patients with acanthosis nigricans and insulin resistance. In summary, an understanding of the implications of homologous receptor domains and common receptor regulatory subunits adds a new dimension to the understanding of various pathological states.

4.2. Therapeutic Implications

The control of receptor function by homologous and heterologous ligands has important implications for the therapeutic use of many drugs. Since many drugs (pressors, antihypertensives, insulin) may be administered at levels that are "supraphysiological," many of the homospecific and heterospecific receptor mechanisms discussed above may be activated. A consideration of these mechanisms may, in certain instances, provide a more rational approach for drug use as well as a new approach for drug design.

4.2.1. Dose–Response Curves

Receptor down-regulation or receptor recruitment can alter the dose–response curve in a predictable way, as discussed elsewhere in this volume (Hollenberg and Goren, Chapter 18). In brief, an increase in receptor density will lead to a leftward shift in the dose–response curve, whereas a decrease in receptor density will lead to a rightward shift. Since for many agents, prolonged cellular exposure leads to a down-regulation of receptors, it may be anticipated that with time, the dose–response curve for a continuously administered drug will shift to the right.

4.2.2. Desensitization and Hypersensitivity

It is well known that cellular desensitization following the repeated administration of a drug can be manifest not only by a shift to the right in the dose–response curve (via receptor down-regulation, as outlined above) but also by a reduction in the maximal obtainable response. The reduction in maximum responsiveness is often glibly attributed to a "post-receptor" defect. As elaborated on elsewhere in this volume (Strulovici et al., Chapter 16), new insights have been gained concerning the mechanism of desensitization for cyclase-associated hormones such as epinephrine. These mechanisms involve the coupling factors, the N_s or N_i subunits (Rodbell, 1980), that are integrally involved in hormone responsiveness. During the recovery of a system from a "ligand-overload" situation leading to desensitization, there need not be simultaneous return of ligand binding and effector coupling. An example of this kind can be seen in the regulation of muscarinic acetylcholine receptors by carbachol in embryonic chick heart (Halvorsen and Nathanson, 1981). Considerations like the ones discussed above may apply to antagonists as well as agonists, for just as agonists may lead to receptor down-regulation and cellular desensitization, so may antagonists result in receptor recruitment and a hypersensitization of cells.

In view of the homospecific and heterospecific mechanisms discussed in this chapter, it is evident that the administration of a single drug could potentially lead to the codesensitization of many receptor systems (or, for nonselective antagonists, hypersensitivity in several receptor systems). The time course with which the receptor numbers and effector systems might recover from high ambient drug levels could potentially call for a change in drug dosing schedules. Thus, the triggering of receptor regulatory events that might be caused by high drug levels may lead to the paradoxical situation wherein less drug is required to be more effective.

4.2.3. Drugs Acting Distal to the Ligand-Binding Event

It is important to point out that drugs acting on receptors at a site distinct from the ligand-binding region may also lead to receptor down-regulation, receptor recruitment, and a change in cellular sensitivity (either increased or decreased) at the postreceptor level. Examples of this kind that can be cited relate to the sensitizing effects of sulfonylureas on insulin action, the down-regulation of insulin receptors caused by agents such as vitamin K_5 and peroxide (Caro and Amatruda, 1980), and the recruitment of insulin receptor by oral hypoglycemic agents. Thus, an understanding of the membrane-localized events that regulate receptor function may point to new targets for the design of drugs intended to activate certain receptor systems.

5. Summary

Since the elucidation of the receptor concept at the turn of the century, enormous strides have been made in characterizing receptors by pharmacological and biological methods. We are now beginning to understand the mechanisms whereby receptor function can be controlled by homologous and heterologous ligands. These mechanisms have important implications both for understanding the pathophysiology of various disease states and, in a practical sense, for providing an additional dimension for the rational use of drugs.

ACKNOWLEDGMENTS. Work in the author's laboratory is supported in part by the Medical Research Council of Canada and the Alberta Heritage Foundation for Medical Research.

References

Ahlquist, R. P., 1948, A study of adrenotropic receptors, *Am. J. Physiol.* **153**:586–600.

Ariens, E. J., 1954, Affinity and intrinsic activity in the theory of competitive inhibition, *Arch. Int. Pharmacodyn.* **99**:32–49.

Armstrong, G. D., and Hollenberg, M. D., 1985, Epidermal growth factor–urogastrone and its receptor, in: *Polypeptide Hormone Receptors* (B. Posner, ed.), Marcel Dekker, New York, pp. 201–226.

Armstrong, G. D., Hollenberg, M. D., Bhaumick, B., Bala, R. M., and Maturo, J. M. III, 1983, Receptors for insulin and basic somatomedin: Immunological and affinity-chromatographic cross-reactivity, *Can. J. Biochem.* **61**:650–656.

Baker, J. B., Barsh, G. S., Carney, D. H., and Cunningham, D. D., 1978, Dexamethasone modulates binding and action of epidermal growth factor in serum-free cell culture, *Proc. Natl. Acad. Sci. U.S.A.* **75**:1882–1886.

Baxter, R. C., and Turtle, J. R., 1978, Regulation of hepatic growth hormone receptors by insulin, *Biochem. Biophys. Res. Commun.* **84**:350–357.

Bhaumick, B., Goren, H. J., and Bala, R. M., 1981, Further characterization of human basic-somatomedin: Comparison with insulin-like growth factors I and II, *Horm. Metab. Res.* **13**:515–518.

Bowen-Pope, D. F., and Ross, R., 1983, Is epidermal growth factor present in human blood? Interference with the radioreceptor assay for epidermal growth factor, *Biochem. Biophys. Res. Commun.* **114**:1036–1041.

Braestrup, C., and Nielsen, M., 1980, Searching for endogenous benzodiazepine receptor ligands, *Trends Pharmacol. Sci.* **2**:424–427.

Brown, K. D., Dicker, P., and Rozengurt, E., 1979, Inhibition of epidermal growth factor binding to surface receptors by tumor promotors, *Biochem. Biophys. Res. Commun.* **86:**1037–1043.

Caro, J. F., and Amatruda, J. M., 1980, Insulin receptors in hepatocytes: Postreceptor events mediate down regulation, *Science* **210:**1029–1031.

Carpenter, G., and Cohen, S., 1976, ^{125}I-Labeled human epidermal growth factor. Binding, internalization and degradation in human fibroblasts, *J. Cell Biol.* **71:**159–171.

Carpenter, G., King, L., Jr., and Cohen, S., 1979, Rapid enhancement of protein phosphorylation A-431 cell membrane preparations by epidermal growth factor, *J. Biol. Chem.* **254:**4884–4891.

Castagna, M., Takai, Y., Kaibuchi, K., Sano, K., Kikkawa, U., and Nishizuka, Y., 1982, Direct activation of calcium-activated, phospholipid-dependent protein kinase by tumor-promoting phorbol esters, *J. Biol.Chem.* **257:**7847–7851.

Catt, K. J., Harwood, J. P., Aguilera, G., and Dufau, M. L., 1979, Hormonal regulation of peptide receptors and target cell responses, *Nature* **280:**109–116.

Catterall, W. A., 1982, The emerging molecular view of the sodium channel, *Trends Neurosci.* **5:**303–306.

Collins, M. K. L., Sinnett-Smith, J. W., and Rozengurt, E., 1983, Platelet-derived growth factor treatment decreases the affinity of the epidermal growth factor receptors of Swiss 3T3 cells, *J. Biol. Chem.* **258:**11689–11693.

Costa, E., 1979, The role of gamma-aminobutyric acid in the action of 1,4-benzodiazepines, *Trends Pharmacol. Sci.* **1:**41–44.

Cuatrecasas, P., 1971a, Perturbation of the insulin receptor of isolated fat cells with proteolytic enzymes, *J. Biol. Chem.* **246:**6522–6531.

Cuatrecasas, P., 1971b, Unmasking of insulin receptors in fat cells and fat cell membranes, *J. Biol. Chem.* **246:**6532–6542.

Devreotes, P. N., and Fambrough, D. M., 1975, Acetylcholine receptor turnover in membranes of developing muscle fibres, *J. Cell Biol.* **65:**335–358.

Devreotes, P. N., and Fambrough, D. M., 1976, Synthesis of the acetylcholine receptors by cultured chick myotubes and denervated mouse extensor digitorum longus muscles, *Proc. Natl. Acad. Sci. U.S.A.* **73:**161–164.

Ehrlich, P., 1956, Nobel lecture (1908) on partial functions of the cell, in: *Himmelweit, Marquardt, Dale. The Collected Papers of P. Ehrlich,* Vol. III, Pergamon Press, Oxford, p. 183.

Farfel, Z., Brickman, A. S., Kaslow, H. R., Brothers, V. M., and Bourne, H. R., 1980, Defect of receptor–cyclase coupling protein in pseudohypoparathyroidism, *N. Engl. J. Med.* **303:**237–242.

Fillion, G., and Fillion, M. P., 1981, Modulation of affinity of postsynaptic serotonin receptors by antidepressant drugs, *Nature* **292:**349–351.

Fuxe, K., and Agnati, L. F., 1985, Receptor-receptor interactions in the central nervous system. A new integrative mechanism in synapses, *Medicinal Research Reviews* 5 (in press).

Gavin, J. R. III, Roth, J., Neville, D. M., Jr., De Meyts, P., and Buell, D. N., 1974, Insulin-dependent regulation of insulin receptor concentration. A direct demonstration in cell culture, *Proc. Natl. Acad. Sci. U.S.A.* **71:**84–88.

Gavish, M., and Synder, S. H., 1981, γ-Aminobutyric acid and benzodiazepine receptors: copurification and characterization, *Proc. Natl. Acad. Sci. U.S.A.* **78:**1939–1942.

Gilman, A. G., 1984, Guanine nucleotide-binding regulatory proteins and dual control of adenylate cyclase, *J. Clin. Invest.* **73:**1–4.

Goldstein, B., and Wiegel, F. W., 1983, The effect of receptor clustering on diffusion-limited forward rate constants, *Biophys. J.* **43:**121–125.

Guidotti, A., Baraldi, M., and Costa, E., 1979, 1,4-Benzodiazepines and gamma-amino-butyric acid: Pharmacological and biochemical correlates, *Pharmacology* 19:267–277.

Halvorsen, S. W., and Nathanson, N. M., 1981, *In vivo* regulation of muscarinic acetylcholine receptor number and function in embryonic chick heart, *J. Biol.Chem.* 256:7941–7948.

Hauger, R. L., Aguilera, G., and Catt, K. J., 1978, Angiotensin-II regulates its receptors in the adrenal glomerulosa zone, *Nature* 271:176–178.

Hayden, L. J., and Severson, D. L., 1983, Correlation of membrane phosphorylation and epidermal growth factor binding to hepatic membranes isolated from triiodothyronine-treated rats, *Biochim. Biophys. Acta* 730:226–230.

Hirata, F., and Axelrod, J., 1980, Phospholipid methylation and biological signal transmission, *Science* 209:1082–1090.

Hirata, F., Strittmatter, W. J., and Axelrod, J., 1979, β-Adrenergic receptor agonists increase phospholipid methylation, membrane fluidity, and β-adrenergic receptor-adenylate cyclase coupling, *Proc. Natl. Acad. Sci. U.S.A.* 76:368–372.

Hock, R. A., and Hollenberg, M. D., 1980, Characterization of the receptor for epidermal growth factor–urogastrone in human placenta membranes, *J. Biol. Chem.* 255:10731–10736.

Hollenberg, M. D., 1981, Membrane receptors and hormone action I: New trends related to receptor structure and receptor regulation, *Trends Pharmacol. Sci.* 2:320–323.

Hollenberg, M. D., 1982a, Membrane receptors and hormone action II: New perspectives for receptor-modulated cell function, *Trends Pharmacol. Sci.* 3:25–28.

Hollenberg, M. D., 1982b, Receptor mediated phosphorylation reactions, *Trends Pharmacol. Sci.* 3:271–273.

Hollenberg, M. D., Nexø, E., Hock, R. A., and Berhanu, P., 1979, Phorbol tumor promoter causes a selective reduction of epidermal growth factor-urogastrone receptors via a separate ligand recognition site, *Clin. Res.* 27:387A.

Hollenberg, M. D., Nexø, E., Berhanu, P., and Hock, R., 1981, Phorbol ester and the selective modulation of receptors for epidermal growth factor-urogastrone, in: *Receptor-Mediated Binding and Internalization of Toxins and Hormones* (J. L. Middlebrook and L. D. Kohn, eds.), Academic Press, New York, pp. 181–195.

Jacobs, S., and Cuatrecasas, P., 1980, Disulfide reduction converts the insulin receptor of human placenta to a low affinity form, *J. Clin. Invest.* 66:1424–1427.

Jacobs, S., Sahyoun, N. E., Saltiel, A. R., and Cuatrecasas, P., 1983, Phorbol esters stimulate the phosphorylation of receptors for insulin and somatomedin C, *Proc. Natl. Acad. Sci. U.S.A.* 80:6211–6213.

Karlin, A., Cowburn, D. A., and Reiter, M. J., 1973, Molecular properties of the acetylcholine receptor, in: *Drug Receptors* (H. P. Rang, ed.), University Park Press, Baltimore, pp. 193–209.

Kikkawa, U., Takai, Y., Minakuchi, R., Inohara, S., and Nishizuka, Y., 1982, Calcium-activated, phospholipid-dependent protein kinase from rat brain, *J. Biol. Chem.* 257:13341–13348.

King, G. L., Kahn, C. R., Rechler, M. M., and Nissley, S. P., 1980, Direct demonstration of separate receptors for growth and metabolic activities of insulin and multiplication-stimulating activity (an insulinlike growth factor) using antibodies to the insulin receptor, *J. Clin. Invest.* 66:130–140.

King, G. L., Rechler, M. M., and Kahn, C. R., 1982, Interactions between the receptors for insulin and the insulin-like growth factors on adipocytes, *J. Biol. Chem.* 257:10001–10006.

Korner, M. K., Gilon, C., and Schramm, M., 1982, Locking of hormone in the β-adrenergic receptor by attack on a sulfhydryl in an associated component, *J. Biol. Chem.* **257**:3389–3396.

Krupp, M. N., Connolly, D. T., and Lane, M. D., 1982, Synthesis, turnover, and down-regulation of epidermal growth factor receptors in human A431 epidermoid carcinoma cells skin fibroblasts, *J. Biol. Chem.* **257**:11489–11496.

Labrie, F., Belanger, A., Cusan, L., Seguin, C., Pelletier, G., Kelly, P. A., Reeves, J. J., Lefebvre, F. A., Lemay, A., Gordeau, Y., and Raynaud, J.-P., 1980, Antifertility effects of LHRH agonists in the male, *J. Androl.* **1**:209–228.

Lai, E., Rosen, O. M., and Rubin, C. S., 1982, Dexamethasone regulates the β-adrenergic receptor subtype expressed by 3T3-L1 preadipocytes and adipocytes, *J. Biol. Chem.* **257**:6691–6696.

Lee, L.-S., and Weinstein, I. B., Tumor-promoting phorbol esters inhibit binding of epidermal growth factor to cellular receptors, *Science* **202**:313–314.

Lefebvre, F. A., Reeves, J. J., Seguin, C., Massicote, J., and Labrie, F., 1980, Specific binding of a potent LHRH agonist in rat testis, *Mol. Cell. Endocrinol.* **20**:127–134.

Lefkowitz, R., 1978, Identification and regulation of alpha and beta-adrenergic receptors, *Fed. Proc.* **37**:123–129.

Levine, M. A., Downs, R. W., Singer, M., Marx, S. J., Aurbach, G. D., and Spiegel, A. M., 1980, Deficient activity of guanine nucleotide regulatory protein in erythrocytes from patients with pseudohypoparathyroidism, *Biochem. Biophys. Res. Commun.* **94**:1319–1324.

Lin, M. C., and Beckner, S. K., 1983, Induction of hormone receptors and responsiveness during cellular differentiation, in: *Current Topics in Membranes and Transport*, Vol. IX (A. Kleinzeller, ed.), Academic Press, New York, pp. 287–315.

Livingston, J. N., Purvis, B. J., and Lockwood, D. H., 1978, Insulin-dependent regulation of the insulin-sensitivity of adipocytes, *Nature* **273**:394–396.

Loumaye, E., and Catt, K. J., 1982, Homologous regulation of gonadotropin-releasing hormone receptors in cultured pituitary cells, *Science* **215**:983–985.

Lundberg, J. M., Hedlund, B., and Bartfai, T., 1982, Vasoactive intestinal polypeptide enhances muscarinic ligand binding in cat submandibular salivary gland, *Nature* **295**:147–149.

Massagué, J., and Czech, M. P., 1982, Role of disulfides in the subunit structure of the insulin receptor, *J. Biol. Chem.* **257**:6729–6738.

Maturo, J. M. III, and Hollenberg, M. D., 1978, Insulin receptor: Interaction with nonreceptor glycoprotein from liver cell membranes, *Proc. Natl. Acad. Sci. U.S.A.* **75**:3070–3074.

Maturo, J. M. III, Hollenberg, M. D., and Aglio, L. S., 1983, Insulin receptor: Insulin-modulated interconversion between distinct molecular forms involving disulfide–sulfhydryl exchange, *Biochemistry* **22**:2579–2586.

Michel, T., Hoffman, B. B., and Lefkowitz, R. J., 1980, Differential regulation of the α_2-adrenergic receptor by Na^+ and guanine nucleotides, *Nature* **288**:709–711.

Mott, D. M., Howard, B. V., and Bennett, P. H., 1979, Stoichiometric binding and regulation of insulin receptors on human diploid fibroblasts using physiologic insulin levels, *J. Biol. Chem.* **254**:8762–8767.

Niedel, J. E., Kuhn, L. J., and Vandenbark, G. R., 1983, Phorbol diester receptor copurified with protein kinase C, *Proc. Natl. Acad. Sci. U.S.A.* **80**:36–40.

Nimrod, A., Tsafriri, A., and Linder, H. R., 1977, *In vitro* induction of binding sites for hCG in rat granulosa cells by FSH, *Nature* **267**:632–633.

O'Connor-McCourt, M., and Hollenberg, M. D., 1983, Receptors, acceptors, and the action of polypeptide hormones: Illustrative studies with epidermal growth factor (urogastrone), *Can. J. Biochem. Cell Biol.* **61**:670–682.

Oppenheimer, C. L., Pessin, J. E., Massagué, J., Gitomer, W., and Czech, M. P., 1983, Insulin action rapidly modulates the apparent affinity of the insulin-like growth factor II receptor, *J. Biol. Chem.* **258**:4824–4830.

Pert, C. B., Pasternak, G. W., and Snyder, S. H., 1973, Opiate agonists and antagonists discriminated by receptor binding in brain, *Science* **182**:1359–1361.

Posner, B. I., Kelly, P. A., and Friesen, H. G., 1974, Induction of a lactogenic receptor in rat liver: Influence of estrogen and the pituitary, *Proc. Natl. Acad. Sci. U.S.A.* **71**:2407–2410.

Posner, B. I., Kelly, P. A., and Friesen, H. G., 1975, Prolactin receptors in rat liver: Possible induction by prolactin, *Science* **188**:57–59.

Raff, M., 1976, Self regulation of membrane receptors, *Nature* **259**:265–266.

Rodbell, M., l1980, The role of hormone receptors and GTP-regulatory proteins in membrane transduction, *Nature* **284**:17–22.

Rosen, O. M., Herrera, R., Olowe, Y., Petruzzelli, L. M., and Cobb, M. H., 1983, Phosphorylation activates the insulin receptor tyrosine protein kinase, *Proc. Natl. Acad. Sci. U.S.A.* **80**:3237–3240.

Rozengurt, E., Brown, K. D., and Pettican, P., 1981, Vasopressin inhibition of epidermal growth factor binding to cultured mouse cells, *J. Biol. Chem.* **256**:716–722.

Schweitzer, J. B., Smith, R. M., and Jarett, L., 1980, Differences in organizational structure of insulin receptor on rat adipocyte and liver plasma membranes: Role of disulfide bonds, *Proc. Natl. Acad. Sci. U.S.A.* **77**:4692–4696.

Shiu, R. P. C., and Friesen, H. G., 1981, Regulation of prolactin receptors in target cells, in: *Receptors and Recognition,* Series B, Vol. 13 (R. J. Lefkowitz, ed.), Chapman and Hall, London, pp. 69–81.

Shoyab, M., De Larco, J. E., and Todaro, G. J., 1979, Biologically active phorbol esters specifically alter affinity of epidermal growth factor membrane receptors, *Nature* **279**:387–391.

Sokolovsky, M., Egozi, Y., and Avissar, S., 1981, Molecular regulation of receptors: Interaction of β-estradiol and progesterone with the muscarinic system, *Proc. Natl. Acad. Sci. U.S.A.* **78**:5554–5558.

Soloff, M., 1975, Uterine receptor for oxytocin: Effects of estrogen, *Biochem. Biophys. Res. Commun.* **65**:205–212.

Stadel, J. M., Nambi, P., Shorr, R. G. L., Sawyer, D. F., Caron, M. G., and Lefkowitz, R. J., 1983, Catecholamine-induced desensitization of turkey erythrocyte adenylate cyclase is associated with phosphorylation of the β-adrenergic receptor, *Proc. Natl. Acad. Sci. U.S.A.* **80**:3173–3177.

Stephenson, R. P., and Barlow, R. B., 1970, Concepts of drug action, quantitative pharmacology and biological assay, in: *A Companion to Medical Studies,* Vol. 2 (R. Passmore and J. S. Robson, eds.), Blackwell, London, pp. 3.1–3.19.

Stevens, R. L., Austen, K. F., and Nissley, S. P., 1983, Insulin-induced increase in insulin binding to cultured chondrosarcoma chondrocytes, *J. Biol. Chem.* **258**:2940–2944.

Strittmatter, W. J., Hirata, F., and Axelrod, J., 1979, Phospholipid methylation unmasks cryptic β-adrenergic receptors in rat reticulocytes, *Science* **204**:1205–1207.

Takeuchi, K., Speir, G. R., and Johnson, L. R., 1980, Mucosal gastrin receptor. III. Regulation by gastrin, *Am. J. Physiol.* **238**:G135–G140.

Taylor, P., Brown, R. D., and Johnson, D. A., 1983, The linkage between ligand occupation and response of the nicotinic acetylcholine receptor, in: *Current Topics in Membranes and Transport,* Vol. 18 (A. Kleinzeller, ed.), Academic Press, New York, pp. 407–444.

Topper, Y. J., and Freeman, C. S., 1980, Multiple hormone interactions in the developmental biology of the mammary gland, *Physiol. Rev.* **60**:1049–1106.

Triggle, D. J., 1982, Receptor recruitment and cryptic signals, *Trends Pharmacol. Sci.* **3**:273–274.

Umezawa, K., Weinstein, I. B., Horowitz, A., Fujiki, H., Matsushima, T., and Sugimura, T., 1981, Similarity of teleocidin B and phorbol ester tumour promoters in effects on membrane receptors, *Nature* **290**:411–413.

Vauquelin, G., Bottari, S., and Strosberg, A. D., 1980, Inactivation of β-adrenergic receptors by N-ethylmaleimide: Permissive role of β-adrenergic agents in relation to adenylate cyclase activation, *Mol. Pharmacol.* **17**:163–171.

Wharton, W., Leof, E., Pledger, W. J., and O'Keefe, E. J., 1982, Modulation of the epidermal growth factor receptor by platelet-derived growth factor and choleragen: Effects of mitogenesis, *Proc. Natl. Acad. Sci. U.S.A.* **79**:5567–5571.

Wrann, M., Fox, C. F., and Ross, R., 1980, Modulation of epidermal growth factor receptors on 3T3 cells by platelet-derived growth factor, *Science* **210**:1363–1365.

Yamada, S., Yamamura, H. I., and Roeske, W. R., 1980, The regulation of cardiac α_1-adrenergic receptors by guanine nucleotides and by muscarinic cholinergic agonists, *Eur. J. Pharmacol.* **63**:239–241.

Zierler, K., 1985, Membrane polarization and insulin action, in: *Insulin, Its Receptor and Diabetes* (M. D. Hollenberg, ed.), Marcel Dekker, New York, pp. 141–179.

Zierler, K., and Rogus, E. M., 1981, Effects of peptide hormones and adrenergic agents on membrane potentials of target cells, *Fed. Proc.* **40**:121–124.

18

Ligand–Receptor Interactions at the Cell Surface

MORLEY D. HOLLENBERG and H. JOSEPH GOREN

1. Introduction

Over the past 20 years, studies of pharmacological receptors have undergone a veritable biochemical metamorphosis. The first thorough study of the binding of a pharmacologically active compound (atropine) to its receptor (muscarinic cholinergic: Paton and Rang, 1965) was soon followed by a multitude of studies of the binding of a variety of ligands to their putative receptors. Now, it is realized that receptor structures are dynamic cellular elements that display considerable structural and functional complexities. Thus, in order to measure the interaction of a ligand with its cell surface receptor, it is essential to have a grasp of the many variables that can affect such measurements. In this chapter, we attempt to provide a suitable context in which the measurements of ligand–receptor interactions can be interpreted, and we provide several models of receptor structure and function that have emerged over the past 10 years.

1.1. Defining a Receptor

On examining the effects of nicotine and curare in nerve–muscle preparations, Langley (1906) was led to the conclusion that there was a "receptive substance" in the muscle that was responsible for triggering

MORLEY D. HOLLENBERG • Endocrine Research Group, Department of Pharmacology and Therapeutics, Faculty of Medicine, University of Calgary, Calgary, Alberta T2N 4N1, Canada. *H. JOSEPH GOREN* • Endocrine Research Group, Department of Medical Biochemistry, Faculty of Medicine, University of Calgary, Calgary, Alberta T2N 4N1, Canada.

contraction. At about the same time, Ehrlich (1908), investigating the action of tetanus toxin, postulated that the toxin must unite with "poison receptors," or simply "receptors" in the cells in order for the toxin to have its effect. Drawing on his studies with antitrypanosomal agents, Ehrlich added to the receptor concept the property of chemical selectivity: receptors could be defined biologically by their ability to react with one series of chemical agents but not another. Thus, almost simultaneously at the turn of the century, two individuals crystallized the receptor concept that has proved so useful in analyzing the action of a wide variety of drugs and hormones.

It was Clark (1926a,b; 1933) who, in studying the action of acetylcholine in cardiac tissue, realized that receptors must be present in vanishingly small numbers on responsive cells. His estimate of the small number of acetylcholine molecules (about 20,000) that need to be bound by a cardiac cell to produce a response stands as a landmark observation in terms of the progress of receptor studies. Thus, it was appreciated early that receptors for a variety of active agents displayed high ligand affinities coupled with remarkable stereochemical specificity, and it was fully realized that receptors are present in or on cells at very low concentrations. It is the binding of ligands to their putative cell surface pharmacological receptors and the dynamic aspects of this process occurring in the plasma membrane that form the focus of this chapter.

It is important to remember that the concept of a receptor for a particular ligand is inextricably linked to the response of a biological system to the ligand. The properties (e.g., high affinity, reversibility, stereospecificity, and tissue specificity) that are usually thought of as characterizing the interactions of active ligands with their putative receptors derive largely from the analyses of dose–response data in biologically responsive systems. Thus, in examining the molecular properties of a pharmacological receptor, it is essential to consider not only the recognition function of the receptor but also its ability to trigger the series of biochemical reactions that lead to a cellular response. It is this dual recognition–activation property that distinguishes the pharmacological receptors to be dealt with in this chapter from other macromolecules (e.g., enzymes, metabolite transporters, HLA antigens) that in another context might be thought of as "receptors."

It was the realization that the binding property of a receptor could be formally separated from its activation function that led Ariens, Stephenson, and their co-workers to develop the concepts of intrinsic activity (Ariens, 1954) or efficacy (Stephenson and Barlow, 1970), which terms relate to the ability of an agonist to activate receptors subsequent to binding. Accordingly, drug antagonists, which bind to but do not activate receptors, have intrinsic activities (or efficacies) of zero, whereas agonists have efficacies of varying magnitude. The mechanisms whereby agonists perturb receptor structure to initiate a cell response are presently under intensive study. It is the change in receptor properties subsequent to the binding of agonists that can distinguish pharmacological receptors from other cell surface constituents that may function as recognition moieties; this change in the properties of a receptor is accommodated by the "mobile" or "floating" receptor paradigm discussed below.

1.2. Receptors versus Acceptors

Ironically, the "poison receptors" that were the focus of Ehrlich's attention would not really qualify as "receptors" in the pharmacological sense intended in this chapter, for it is the tetanus toxin itself that embodies the biological activity and not the cell surface constituents to which the toxin binds. On the other hand, the nicotinic-cholinergic receptor, with which Langley was concerned and which possesses both the ligand recognition site(s) and the cell activation moiety (ion channel) in a unique oligomeric structure, does typify the kind of cell surface receptor that is dealt with in this chapter. Thus, there is a functional difference between the cell surface recognition sites responsible for the selective cellular uptake of toxins and the surface-localized pharmacological receptors with their dual recognition/activation properties.

In addition to moieties that serve as binding sites for a variety of toxins, there are also specialized membrane constituents that, apart from the familiar ion or metabolite transport structures, play a vital role in the communication of chemical information from the cell exterior to the cytoplasm. These sites participate in the highly selective adsorptive pinocytosis of important serum-borne regulatory ligands. Compounds such as cholesterol, cobalamin (vitamin B_{12}), and iron are internalized by such a selective adsorptive process. It has proved convenient to refer to the membrane-localized substituents involved in this internalization process as "acceptors" rather than receptors (Hollenberg, 1979; O'Connor-McCourt and Hollenberg, 1983); others have referred to such constituents as receptors of the class II type (Kaplan, 1981).

The role of acceptors is well illustrated by the functional properties of transcobalamin II (TCII) (Nexø, 1978; Mahoney and Rosenberg, 1975), a protein that serves as a transporter of cobalamin in the circulation. The membrane-localized TCII acceptor is able to recognize the TCII–cobalamin complex, resulting in a translocation of the complex and the subsequent intracellular release of cobalamin for further metabolic processes (Mahoney and Rosenberg, 1975). In this instance, cobalamin can be thought of as the pharmacologically active agent that, on release from TCII, binds to an appropriate intracellular (enzyme) receptor; the membrane-localized constituent that recognizes the TCII–cobalamin complex in a highly specific manner clearly functions in a manner different from the one required of pharmacological receptors and is thus more accurately termed an "acceptor" [the TCII–cobalamin complex but neither free TCII nor free cobalamin binds to the recognition site with high affinity (Seligman and Allan, 1978; Nexø et al., 1979)].

The cellular binding site for low-density lipoprotein (LDL) (Goldstein and Brown, 1975) can be thought of in similar terms, since it is the feedback regulator cholesterol and not the recognition moiety that is the pharmacologically active agent in the cell interior subsequent to internalization via the LDL acceptor. The transferrin acceptor represents a third interesting example of a highly selective cell recognition site responsible for the cellular uptake of an important cytoplasmic constituent (in this case, iron) (Lamb et al., 1983).

For acceptors as well as pharmacological receptors, it is to be expected that a strict chemical specificity for ligand binding will be observed, along with other properties that are consistent with a high-affinity recognition function. Nonetheless, because acceptors do not exhibit a ligand-triggered "activation state," it is very likely that the cellular processes (e.g., mobility, clustering, internalization, recycling) undergone by acceptors following ligand binding may differ in a number of important respects from the analogous processes in which pharmacological receptors participate.

Despite the functional differences between cell-surface acceptors and pharmacological receptors, it is now becoming increasingly evident that subsequent to ligand binding both acceptors and receptors can participate in a very similar cascade of interactions beginning in the plane of the membrane and proceeding from there to the cytoplasmic space. This similarity is illustrated by studies of the surface mobility and internalization of the receptors for insulin and EGF–URO (Maxfield et al., 1978) and of the acceptors for α_2-macroglobulin and vesicular stomatitis virus (Dickson et al., 1981). The complicated processes that follow receptor or acceptor occupation are dealt with below as well as by other chapters in this volume; in general, there is a redistribution of the ligand–receptor or ligand–acceptor complexes in the plane of the membrane followed by cellular internalization, possible recycling, and by lysosomal degradation of the receptor or acceptor constituents.

It is a challenge to determine which of the events involved in redistribution and internalization involve both receptors and acceptors and which events (if any) are unique for receptors. The above emphasis on distinguishing acceptors from receptors is of importance, because it is likely that the cell-surface redistribution events in which both receptors and acceptors participate will have an impact on the kinetic properties of the ligand-binding characteristics. Thus, differences in the dynamic processes in which receptors and acceptors participate may be reflected in part by differences in their interaction with ligands at the cell surface.

1.3. Distinguishing Receptor from Nonreceptor Interactions

Major advances in the study of receptors have come from the development of methods for the direct analysis of the interaction of radiolabeled ligands with their receptors. Although much work has been done using [^{3}H]-labeled ligand probes, it has been the adaptation of immunoassay technology that has permitted the explosive advance in the studies of receptors for many hormones. The radioimmunoassay technology provided for the preparation of high-specific-activity [^{125}I]-radiolabeled ligands that retain both biological and immunologic activity. In addition, the immunoassay methods were directly applicable to the rapid separation of receptor-bound from unbound labeled ligand (for the methodologies involved, see Cuatrecasas and Hollenberg, 1976; Hollenberg and Cuatrecasas, 1979; Hollenberg and Nexø, 1981).

Because of the technological advances, in many instances, the binding of a ligand to a cell surface site can be readily measured. However, it is now well recognized that just about anything binds to anything; it is simply a question of affinity, capacity, and specificity. Thus, it can often be a problem to sort out the pharmacologically relevant binding of a ligand from the "nonspecific" or nonrelevant binding. This issue has been discussed at some length elsewhere (Hollenberg and Cuatrecasas, 1979). The key to the interpretation of any binding data lies in the ability to match the ligand-binding properties with the bioassay data available for the ligand of interest. In essence, the same structural specificity, high potency (or affinity), reversibility, and tissue specificity that have been observed in bioassay systems must be reflected by the binding data.

Perhaps one of the most instructive examples of the interpretation of binding data can be seen in the elegant study of the binding of [^{3}H]-labeled atropine to smooth muscle (Paton and Rang, 1965). In that study, a complete binding isotherm was determined for interaction of [^{3}H]-atropine with strips of smooth muscle, and the binding curve was factored mathematically into a number of saturable and unsaturable components. Simultaneously, bioassay measurements were done to estimate, using the rate theory, the affinity of atropine for its muscarinic receptor; this affinity had also been estimated by the dose-ratio method (Arunlakshana and

Schild, 1959). In the final analysis, only the high-affinity saturable atropine-binding component, with a K_D of about 1 nM, was consistent with the bioassay data: an appropriate match had thus been made between the binding and bioassay data.

In more recent studies, it has been customary to use a variety of ligand analogues to evaluate by competitive binding curves the relative affinities of various analogues for a putative receptor site. When the relative affinities of the analogues are in agreement with the relative potencies of such analogues in a bioassay system, it is usually safe to assume that the binding site measured does indeed comprise the pharmacological receptor. Since, however, binding measurements often reveal two or more saturable binding sites of different affinities in membrane preparations, it is usually necessary to examine critically the ligand specificity of both the high- and the relatively low-affinity site(s) to distinguish receptor from nonreceptor binding. As can be seen below, it is probably the rule rather than the exception that receptors situated in the plasma membrane may exhibit more than one affinity state. Thus, the analysis of receptor binding data often represents a difficult challenge, particularly in cases in which multiple saturable high-affinity sites can be detected.

2. Receptor Dynamics and Hormone Action

2.1. The Mobile or Floating Receptor Paradigm

Perhaps because many of the early studies designed to characterize receptors employed nerve–muscle preparations, receptors were originally thought of (either explicitly or implicitly) as comparatively static entities localized at specific anatomic sites, as are the nicotinic–cholinergic receptors at the neuromuscular junction. However, progress in understanding the organization of the cell membrane has radically changed this "static" view of receptor function. The change in thinking about pharmacological receptors was stimulated by two series of observations. (1) Experiments using fluorescently labeled antibodies directed at cell-surface constituents demonstrated convincingly that membrane proteins could diffuse freely in the plane of the membrane (Frye and Edidin, 1970; Edidin et al., 1976). (2) Experiments dealing with the activation of adipocyte adenylate cyclase by a number of receptor-specific agonists yielded data suggesting that a unique adenylate cyclase enzyme was responding in a complex way to many distinct receptors. Taken together, the above observations led to the development of a "floating" or "mobile" receptor model of hormone action, conceived independently by a

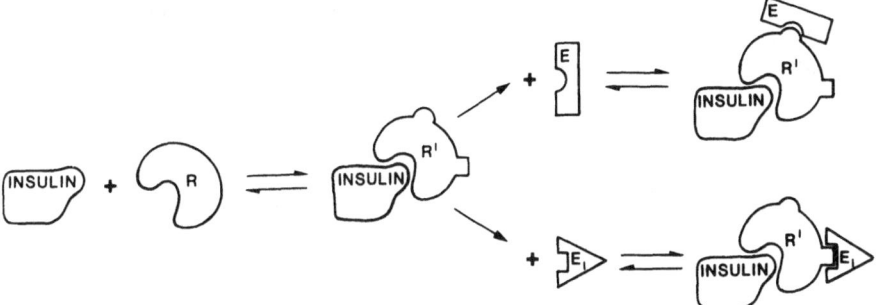

Figure 1. Interaction of mobile hormone receptors with multiple membrane effectors. In terms of the "mobile" or "floating" receptor paradigm of hormone action, a hormone receptor (e.g., for insulin) on binding its ligand, is depicted as changing its molecular conformation. The conformation of the hormone–receptor complex is shown to be capable of interacting in the plane of the membrane with two separate membrane effector molecules (E and E_1). Neither insulin nor the receptor and effector molecules are drawn to scale.

number of individuals (Cuatrecasas, 1974; Bennett *et al.*, 1975; Jacobs and Cuatrecasas, 1976, 1977; DeHaen, 1976; Boeynaems and Dumont, 1977, 1980).

In terms of hormone action, receptor mobility in the plane of the membrane is now viewed as a critical factor related to cell activation. It is a fundamental hypothesis of the "mobile" or "floating" receptor paradigm of hormone action (summarized in Cuatrecasas and Hollenberg, 1976; Valentine and Hollenberg, 1984; Hollenberg, Chapter 17, this volume) that the ability of a receptor to migrate in the plane of the membrane and to interact with other membrane constituents is dramatically altered when the receptor is occupied by its specific ligand (e.g., insulin). This situation is illustrated schematically in Fig. 1, where the formation of the insulin–receptor complex (IR') is depicted as causing a conformational change in the receptor, permitting the subsequent interaction of the receptor with hypothetical "effector" molecules in the plane of the membrane to form ternary complexes like those depicted as IR'E or IR'E₁. In principle, there is no limit to the number of membrane-localized "effectors" (e.g., ion channels, glucose transporters, adenylate cyclase) with which the ligand-occupied receptor might interact. As opposed to agonists, receptor antagonists would presumably be unable to cause the necessary conformation change in the receptor to permit the secondary interactions with membrane effectors.

Thus, the mobile receptor model introduces enormous flexibility into the possible functions of receptors whereby a single hormone–receptor complex may simultaneously activate a variety of cell processes. As is shown seen below, this model also has important implications related to the affinities with which ligands may bind to a receptor depending on the secondary interactions undergone by the receptor subsequent to ligand binding.

In a simplified version, the equilibria involved in the mobile receptor model can be expressed:

$$mH + R \underset{k_{-1}}{\overset{k_1}{\rightleftharpoons}} RH_m \qquad K_1 = k_1/k_{-1} \qquad (1)$$

$$n(RH_m) + nE \underset{k_{-2}}{\overset{k_2}{\rightleftharpoons}} E_n(RH_m)_n \qquad K_2 = k_2/k_{-2} \qquad (2)$$

where H, R, and E represent hormone, receptor, and effector (e.g., membrane-bound adenylate cyclase), respectively, and where m and n are integers that may account for either multiple hormone-binding sites (m) on the receptor or for aggregating hormone–receptor complexes. The equations also indicate an interaction between multiple hormone–receptor complexes (n) and the effector. It is evident that the equilibria could readily be made more complex to account for cooperative phenomena by varying the stoichiometry of the reacting species (e.g., either the receptor or effector may well represent oligomeric macromolecular species), as suggested by the integers m and n. The values, K_1 and K_2 represent the microscopic equilibrium association constants for the reactions with overall forward and reverse rate constants k_1, k_{-1}, k_2, and k_{-2}. In the simplest case, both m and n will be unity.

It is a fundamental hypothesis of the mobile receptor model that the affinity of the hormone–receptor complex for the effector, as expressed by the equilibrium constant K_2, is greater than the affinity of the uncomplexed receptor for the effector, as given by the following equation:

$$nR + nE \underset{k_{-3}}{\overset{k_3}{\rightleftharpoons}} E_n(R)_n \qquad K_3 = k_3/k_{-3} \qquad (3)$$

and expressed by the equilibrium constant K_3. It is thus proposed that for a hormone inhibitor, I, it would be expected that the ternary complex (IRE) would be biologically inactive and that for the inhibitor, K_2 would equal K_3. A fourth equilibrium that can occur is the dissociation of the hormone–receptor–effector complex according to the equation

$$(m \cdot n)H + (R)_n E_n \underset{k_{-4}}{\overset{k_4}{\rightleftharpoons}} (H_m R)_n E_n \qquad K_4 = k_4/k_{-4} \qquad (4)$$

An even more generalized scheme than the one described by the above four equations has been described by DeHaen (1976) for the action of hormones that stimulate adenylate cyclase.

The above equilibria lead to an interesting conclusion regarding the affinity of a ligand for its receptor in either the absence or the presence of an effector. If it is assumed that $n = m = 1$, it can be readily derived from equations 1 to 4 that

$$K_3 \cdot K_4 = K_1 \cdot K_2 \tag{5}$$

According to the main tenet of the mobile receptor hypothesis, the complex HR will have a greater affinity for the species E than will the uncomplexed receptor R; i.e., $K_2 > K_3$. Thus, from equation 5, K_4 must be greater than K_1, and therefore the same receptor will bind the ligand with a different affinity depending on whether or not the receptor is complexed with an effector.

The mobile receptor model thus predicts that a homogeneous population of receptors can, depending on receptor–effector stoichiometry, exhibit complicated binding kinetics. The detailed mathematical analyses of the mobile receptor model, developed by a number of investigators (DeHaen, 1976; Jacobs and Cuatrecasas, 1976; Boeynaems and Dumont, 1977, 1980), indicate that the binding of a homogeneous ligand with a unique receptor molecule can exhibit nonlinear Scatchard plots, Hill plots consistent with negative cooperativity, and ligand dissociation rates that depend on the concentration of unbound ligand. In a situation in which there is an excess of receptors (e.g., 10- to 20-fold) relative to the number of effectors (i.e., there are "spare" but equivalent receptors), the model predicts (Jacobs and Cuatrecasas, 1976) that at least two "affinity states" for the ligand would be detected by binding studies; only one of the affinity states would appear to correlate with the ED_{50} for a biological response curve. All of the receptors would, nonetheless, be equivalent, and all of the receptors would contribute equally to the responsiveness of the system.

Variations of the mobile receptor paradigm can be developed to accommodate many of the previously described models of hormone action. For instance, the "efficacy" term of Stephenson and co-workers may be viewed in terms of the ability of an agonist to perturb receptor structure $(H + R \rightarrow HR')$ and to permit a rapid turnover of the complex, HR'E. Alternatively, the kinetic models of drug action can be accommodated by assigning kinetic lifetimes to the various hormone–receptor–effector complexes depicted by equations 1 to 4. For instance, a very short half-life of the hormone–receptor–effector complex is proposed for the "collision coupling" model of adenylate cyclase activation developed by Tolkovsky, Levitzki, and colleagues (Tolkovsky and Levitzki, 1978). In essence, it is suggested that a very brief collision between the occupied receptor and the cyclase system results in enzyme activation and that the complex HRE does not accumulate. On the other hand, if there were a rapid interaction of the species HR with an effector E to yield altered forms both of E (let us say E*) and of R (for instance R → R*), such that HR* could no longer activate E, one could accommodate the "rate theory" as developed by Paton and co-workers (Paton, 1961). In such a situation, the initial binding of the ligand would yield a brief activation signal that could only occur again after recycling of the receptor back from the inactive R* form to the active R receptor form.

An important aspect of the mobile receptor model not dealt with by previous receptor models concerns both the levels of effectors and receptors present in a given cell and the receptor/effector ratio that may well vary under different physiological conditions or different states of cell differentiation. For example, denervation supersensitivity might be rationalized in terms of variations in the receptor-to-effector ratio. Alternatively, the ratio of adenylate cyclase effector molecules relative to the total number of receptors for distinct hormone activators would determine whether or not "additivity" of enzymatic activation might be observed for separate hormones. Thus, the mobile receptor model puts into focus potential effects not only of variations in receptor number but also of changes in the membrane content of effectors. The above discussion should serve to illustrate the many implications the mobile receptor model has in terms of interpreting the interactions of a variety of ligands with cell surface receptors.

2.2. Receptor Microclustering, Aggregation, Ligand Internalization, and Hormone Action

2.2.1. Surface Binding and Receptor Redistribution

Largely stimulated by the mobile receptor model of hormone action, recent work with receptors has turned from simple ligand-binding measurements to studies of the surface mobility of receptors using fluorescently labeled hormone probes. The fluorescence photobleaching recovery technique (Koppel et al., 1976) has been used to obtain estimates of receptor lateral diffusion coefficients in the range of 5×10^{-10} cm^2/sec (Hillman and Schlessinger, 1982). This value, typical of a number of membrane proteins, indicates that the occupied receptor can diffuse rapidly in the lipid bilayer.

The surface behavior of receptors has also been studied using video intensification microscopy, which allows the continuous observation of a small number of fluorescent ligands on the living cell surface (Willingham and Pastan, 1978). With this approach, it was observed that at

4°C, fluorescent insulin and EGF–URO derivatives bind to mobile receptors that are initially diffusely distributed on the surface of mouse 3T3 fibroblasts; however, when the temperature was raised to 23°C or 37°C, the ligands were observed to coalesce and form visible aggregates (Schlessinger *et al.*, 1978).

Recently, measurements of phosphorescence emission and anisotropy have been used to estimate the rotational motion of labeled molecules on the surface of living cells. This approach has provided more direct evidence for the existence of EGF–URO receptor microclusters (Hillman and Schlessinger, 1982; data summarized by O'Connor-McCourt and Hollenberg, 1983). The dramatic finding of the work of Hillman and Schlessinger (1982) was that the rotation of the EGF–URO–receptor complex becomes slower as the temperature increases even though membrane fluidity is increasing. This result was interpreted to reflect the progressive formation of receptor microclusters. These measurements, together with the earlier studies on EGF–URO internalization, have enabled Hillman and Schlessinger (1982) to draw a comprehensive picture of EGF receptor redistribution and internalization; that is, (1) at 4°C EGF–URO receptor-ligand complexes diffuse and rotate rapidly within the plane of the membrane; (2) at 37°C the receptors form microclusters of 10–50 molecules; and (3) these microclusters aggregate and are then internalized, some through coated pits. It was also proposed (Hillman and Schlessinger, 1982) that the visible patches of labeled receptor observed using fluorescence microscopy may represent endocytotic vesicles formed by the fusion of smaller vesicles containing microclusters of EGF–URO–receptor complexes.

In our own work (O'Connor-McCourt and Hollenberg, 1983), we have used a somewhat different approach to monitor EGF–URO receptor aggregation in a choriocarcinoma cell line (BeWo) (Friedman and Skehan, 1979) that provides a useful counterpart for our previous work with the human placenta membrane receptor. We have been following the metabolic fate of the receptor, employing affinity cross-linking methodology with the bifunctional reagent disuccinimidyl suberate (DSS) (Armstrong and Hollenberg, 1982) and with a monovalent photoaffinity probe (Hock *et al.*, 1979). When the binding, cross-linking, and solubilization procedures were done at 0°C, a temperature that inhibits patch formation and endocytosis, DSS cross linking led to the formation of a single specifically labeled band of apparent molecular weight 180,000 (Fig. 2A). This value is in good agreement with the molecular weights previously estimated for the EGF–URO receptor found in a variety of cell types as well as in human placenta membranes (Hock *et al.*, 1979). However, when the temperature was raised to 22°C, after binding but prior to cross linking, the

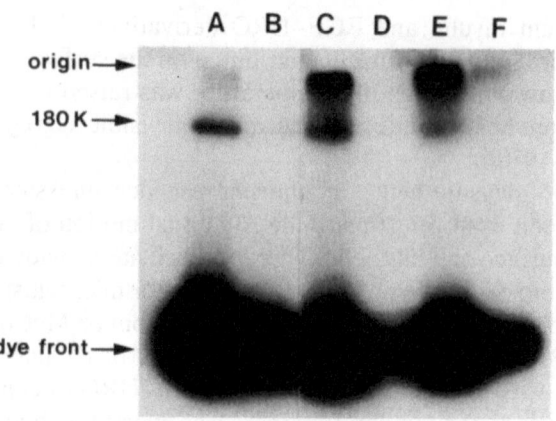

Figure 2. Cross linking of clustered EGF–URO receptors on cultured BeWo cells at 22°C. All samples were incubated for 50 min at 0°C with iodinated EGF–URO (10 ng in 0.5 ml binding medium). Excess (1 μg) unlabeled EGF–URO was added where noted. The cells were washed four times with 1 ml ice-cold binding medium to remove unbound labeled EGF–URO, and then receptor aggregation was initiated by incubating the cells in 0.5 ml of binding medium at 22°C for the indicated times. The [^{125}I]-labeled EGF–URO–receptor complexes on the cells were cross linked at 0°C using disuccinimidyl suberate (DSS) (pH 7.4). The cells were solubilized, and aliquots of each sample were analyzed by polyacrylamide gel electrophoresis (7.5% gel) and autoradiography. The preparations were as follows: (A) visualization of the 180,000-molecular-weight radiolabeled EGF–URO receptor present on cells at 0°C (no incubation at 22°C prior to cross linking with DSS); (B) as in A, but with excess unlabeled EGF–URO present during [^{125}I]-labeled EGF–URO binding (the disappearance of the 180,000 band indicates that it is specific for EGF–URO); (C and D) same as A and B, but with a 1-hr incubation at 22°C prior to cross linking; (E and F) same as A and B, but with a 2-hr incubation at 22°C prior to cross linking (data from O'Connor-McCourt and Hollenberg, 1983).

subsequent addition of DSS led to the formation of specifically labeled, high-molecular-weight material that migrated at the interface between the stacking and resolving gels (Fig. 2C). The amount of radioactivity in this band increased substantially when the incubation time was increased from 1 to 2 hr (Fig. 2E).

These results indicate that even at room temperature, the receptor complexes are aggregating, so that the DSS not only cross links the ligand to its receptor but also cross links receptor aggregates. It is not yet certain whether these cross-linked high-molecular-weight complexes are composed solely of receptor molecules or whether the cross-linked aggregates also include other membrane proteins that become associated with the receptor during the clustering process. In parallel experiments with the BeWo cells, using the monovalent EGF–URO photoprobe, no evidence

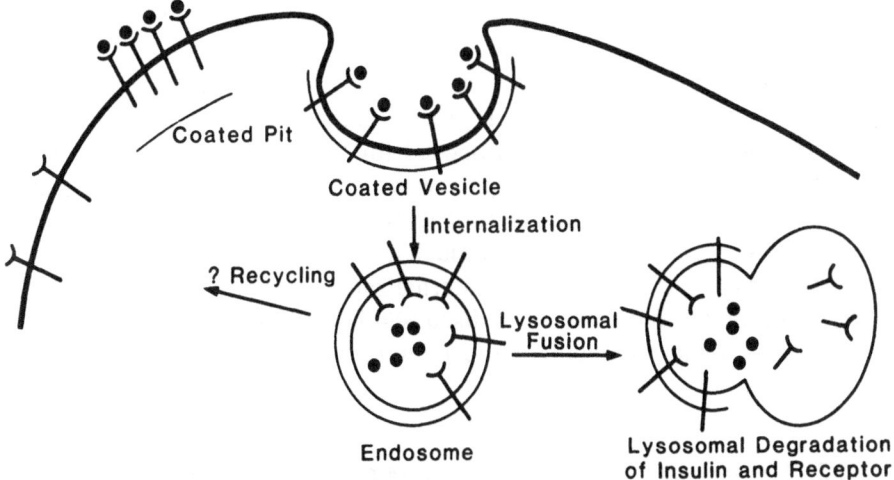

Figure 3. Formation and migration of receptor-bearing vesicles. It is thought the aggregated receptors become trapped in an endocytic vesicle that buds inward to form an intracellular vesicle, the endosome (or receptosome). Evidence suggests that the internal environment of the endosome becomes acidified, favoring the dissociation of many ligands including insulin from their receptors. After the initial internalization event, the receptor-bearing vesicles are thought to change their shape (possibly by fusing with other nonlysosomal intracellular constituents) and to migrate to a variety of cellular locations. Two further possibilities are depicted: (1) fusion with lysosomal structures and (2) recycling of the receptor to the cell surface.

was obtained to indicate the endogenous formation of covalently linked receptor aggregates during the process of receptor patching (O'Connor-McCourt and Hollenberg, 1983).

Taken together, the observations documenting the surface mobility of receptors and acceptors point to a complex series of reactions initiated by ligand binding. As alluded to briefly above and as summarized in Fig. 3, the processes believed in general to follow receptor or acceptor occupation include (1) redistribution of the ligand–receptor or ligand–acceptor complexes in the plane of the membrane (both microclustering and aggregation), (2) interaction of the complex with specialized membrane regions ("coated pits" in certain cell types), (3) internalization of the ligand–receptor or ligand–acceptor complexes from both coated and noncoated cell surface regions into a discrete nonlysosomal compartment [this has been termed the endosome or "receptosome" (Pastan and Willingham, 1981)], (4) either recycling to the cell surface or fusion with lysosomes of the internalized receptor-bearing membrane structure, and (5) lysosomal processing of the ligand and/or receptor with either further degradation or, possibly, further intracellular action of the degraded receptor fragments.

Although this scheme may not apply in its entirety for all receptors or acceptors, it is clear that such processes must be taken into account in terms of the mechanism(s) proposed for the action of specific polypeptide hormones. Thus, although few would dispute the importance of the cell-surface receptor as the critical initial point of contact between a cell and its regulator, there is now much active discussion concerning the potential role(s) of the receptor internalization process in terms of the regulation of cell function. As is discussed in Section 3, these processes also bear on measurement of the interaction of ligands with the cell surface.

2.2.2. Microclustering, Aggregation, and Cell Stimulation

Studies with antibodies were the first to point out the importance of multivalency both for stimulating patch and cap formation in lymphocytes and for initiating a cell response. As indicated in the above discussion, it is now realized that the binding of hormones also leads to receptor microclustering (groups of two or more, up to 50) followed by the appearance of discrete aggregates. In certain cells, such as adipocytes, receptors may be present in groups of two or more even before the addition of a ligand such as insulin (Smith and Jarett, 1985). Data from several studies now indicate that the formation of receptor dimers or microclusters is very likely a critical event for cell activation. The key observations have been made using three independent approaches: (1) the use of bivalent and monovalent antireceptor antibodies; (2) the use of a derivative of EGF–URO that can bind to receptors but cannot stimulate DNA synthesis in cells; and (3) the use of luteinizing hormone-releasing hormone (LHRH, also called gonadotropin releasing hormone or GnRH) and specific LHRH antagonists that can bind to LHRH receptors but cannot activate LH release from pituitary gonadotrophs.

The results obtained with antireceptor antibodies are summarized in Fig. 4. In brief, bivalent polyclonal antiinsulin receptor antibodies have been observed to mimic insulin action (Kahn et al., 1981; Jacobs et al., 1978). In contrast, monovalent F_{ab} antireceptor antibody fragments that were competitive inhibitors of insulin binding were unable to stimulate cells; the biological activity of the F_{ab} fragments was restored with bivalent anti-F_{ab} immunoglobulin (Kahn et al., 1981). Analogous results have been obtained with antibodies directed against the receptor for EGF–URO (Schreiber et al., 1983). These data have been taken to in-

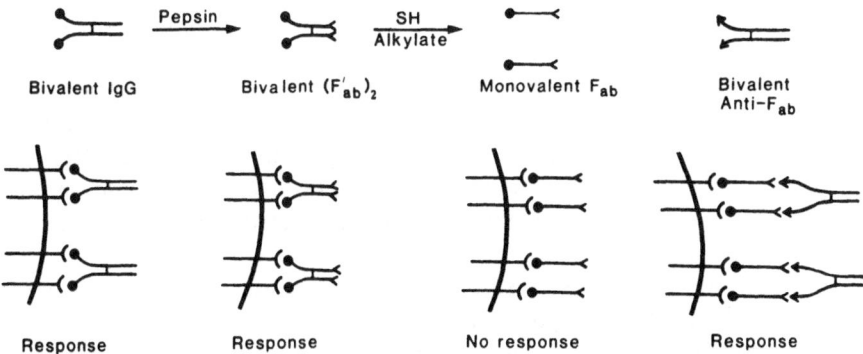

Figure 4. Effects of divalent and monovalent antireceptor antibodies. The scheme illustrates results obtained with polyclonal anti-insulin-receptor antibodies (Kahn *et al.*, 1981). Intact antibody (IgG) and the bivalent antibody derivative generated by pepsin cleavage (Fab)₂ were capable of stimulating cells. However, the monovalent antibody derivative (Fab) produced by reduction (SH) and alkylation of the (Fab)₂ species did not generate a cellular response. Nonetheless, cross linking the receptor-bound Fab fragments by bivalent anti-Fab molecules yielded a cellular response. Results akin to these have been obtained via a totally different approach using derivatives of epidermal growth factor–urogastrone and of leuteinizing hormone-releasing hormone (see text). A role for receptor microclustering in generating a cell response has thus been postulated. (Adapted from Kahn *et al.*, 1981.)

dicate that receptor microclustering is a key event in the process of cell activation.

It is important to point out that receptor microclustering *per se*, although necessary, may not be sufficient to cause cell activation. For instance, in adipocytes, insulin receptors have been observed to be clustered in groups of two or more prior to the addition of insulin (Smith and Jarett, 1985), yet the addition of insulin (or antireceptor antibody) is required for cell activation. Further, a new preparation has been described (Roth *et al.*, 1982) of monoclonal anti-insulin-receptor antibodies that can bind to the receptor to block insulin binding and action but do not exhibit the insulinlike activity observed for the polyclonal antibodies previously studied. There is no reason to suspect that the bivalent monoclonal antibodies could not cross link two insulin receptors as is thought to happen with the bivalent polyclonal antibodies. Further work will be required to resolve the apparent discrepancies between the results with the monoclonal and polyclonal antibodies. The results suggest that receptor microclustering may have to be accompanied by an additional conformational perturbation of the receptor to result in cell activation.

Evidence implicating receptor microclustering as a key event for cell activation has also come from work with derivatives of EGF–URO

(Shechter *et al.*, 1979; Yarden *et al.*, 1982) and LHRH (Conn *et al.*, 1982; Gregory *et al.*, 1982; Hopkins *et al.*, 1981). In these studies, inactive derivatives of LHRH and EGF–URO were made active by the addition of bivalent antiligand antibodies. Further, the ability of ferritin derivatives of LHRH to cause the microclustering and internalization of LHRH receptors, which in the absence of ligand were diffusely distributed on the cell surface, has also been observed at the electron microscopic level (Hopkins *et al.*, 1981).

Taken together, these results have led to the following conclusions: (1) evidently it is the receptor and not the ligand *per se* that embodies the information that directs cellular responsiveness; (2) receptor microclustering appears to be an important factor in initiating a cellular response; and (3) perturbation of a receptor site other than the ligand-binding region can cause a cellular response.

In the context of these conclusions, the trigger function of the activating ligand is brought into focus. For full agonists, it appears that a number of receptor domains interacting with multiple sites on the bound ligand must be involved in causing full receptor activation. This multisubsite interaction, which has been discussed in some detail (DeLean *et al.*, 1979), may explain the very high affinity many ligands have for their receptors and may rationalize some of the cooperative (both negative and positive) ligand–receptor interactions that have been discussed previously (DeLean and Rodbard, 1979; Burgen, 1981; Burgen *et al.*, 1975; DeMeyts *et al.*, 1978). The results with bivalent antibodies also raise the possibility that the hormonal ligands themselves may, at least in some instances, act as bivalent cross-linking agents to cause cell activation; according to a recently developed paradigm, the multivalency of an active ligand might not only be involved in receptor–receptor microclustering but might also participate in a receptor–effector microaggregation process (Minton, 1981). In summary, the comparatively recent observations and conclusions related to receptor microclustering, stimulated by the mobile receptor model, have led to new insights concerning the processes that lead to cell activation subsequent to the binding of a ligand.

2.2.3. Receptor Internalization and Hormone Action: The Pac-Man® Model

For many hormones, the rapidly modulated cell activities (ion flux, exocytosis) probably occur as a result of rapid cell-surface reactions such as the microclustering events described above. In particular, for those agents active via the adenylate cyclase system, all of the biological effects may stem from reactions localized in the plasma membrane. However, for certain neuromodulators, e.g., nerve growth factor (NGF), there are delayed biological effects that occur long after the initial receptor binding event; in the case of NGF, it is believed that the NGF–receptor complex may act, in part, by finding its way to the nucleus via retrograde axonal transport (reviewed by Yankner and Shooter, 1982). It is also recognized that nerve cells possess reuptake mechanisms for neurotransmitters. Thus, the processes that begin at the cell surface may continue in the cytoplasmic environment, and a role for an internalized ligand or for its internalized receptor must be considered.

As indicated by the scheme in Fig. 3, it is now realized that in a number of cell types, appreciable amounts of a ligand such as insulin can be taken up along with its receptor into an intracellular compartment *en route* to lysosomal degradation or, via recycling, back to the cell surface. In view of the insulinlike activity of the anti-insulin-receptor antibodies, a role for internalized insulin has all but been ruled out in terms of the cellular activation process. However, there is considerable conjecture about a possible role for the internalized receptor or its degradation products (Das, 1980), in terms of the delayed effects (enzyme synthesis, RNA or DNA synthesis) of hormones such as insulin or EGF–URO. In terms of hormone action, one might, for instance, envision a role for internalized receptor (possibly a presynaptic neurotransmitter receptor) in terms of the regulatory mechanisms that modulate the neuronal enzymes (both enzyme levels and catalytic activity) involved in neurotransmitter synthesis.

In connection with the possible role of internalized receptor, two possibilities are now being considered: (1) an active fragment of the receptor may be liberated at an intracellular site to regulate such biological processes as DNA synthesis, or (2) the receptor itself, via an intrinsic kinase activity, may be able to regulate enzymatic pathways via phosphorylation–dephosphorylation reactions. This second possibility has arisen from observations demonstrating that the receptors for EGF–URO, insulin, platelet-derived growth factor, and possibly for tumor-derived growth factor possess intrinsic tyrosine kinase activity (reviewed briefly by Hollenberg, 1982b). Thus, a single hormone receptor that does not act via the adenylate cyclase system may, on the one hand, regulate rapid cellular responses (milliseconds to tens of minutes) via membrane-localized reactions and, on the other hand, may modulate delayed responses (tens of minutes to tens of hours) via intracellular receptor-mediated processes. These two time frames of receptor-mediated cell activation are illustrated in Fig. 5.

How, one may ask, could the internalized receptor reach its intracellular sites of action? It is possible that the receptor-bearing endosome

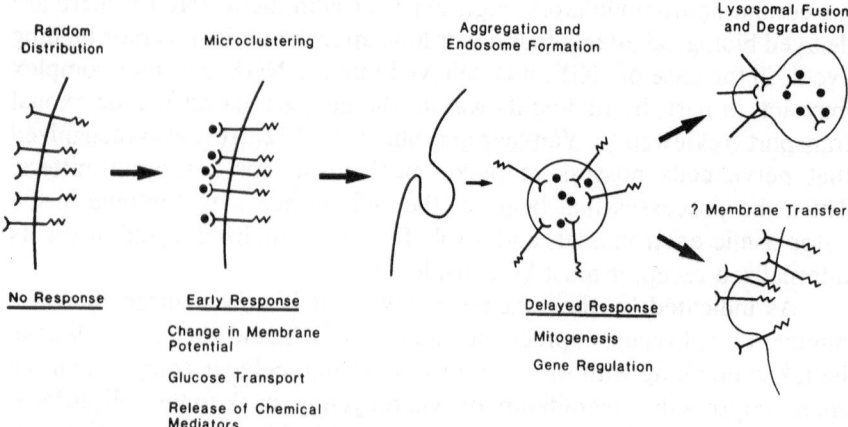

Figure 5. Receptor dynamics and cell activation. A general scheme for cell activation is depicted, pointing out the receptor dynamics thought to be involved in the process of cell activation. Very likely, cell responses that are rapidly regulated by neurotransmitters or hormones involve the initial microclustering event. Delayed effects of the ligand may be caused by receptor that is internalized in the endosomal organelle. The topography of the endosome would permit the intracellular portion of the receptor (designated ᷭ) to interact with a variety of intracellular constituents located at considerable distances from the plasma membrane; the endosomal form of the receptor represents an ideal vehicle to carry the receptor rapidly to selected sites of intracellular action. Possibly, the intracellular portion of some neurotransmitter receptors will be found to contain a kinase domain like the ones present in receptors for epidermal growth factor–urogastrone and insulin. The kinase may regulate a variety of enzymatic processes via phosphorylation reactions. In the course of its intracellular migration, the receptor-bearing endosome could ultimately fuse either with lysosomes or with other membrane structures (e.g., Golgi elements, nuclear membranes), resulting in a further relocation of the receptor.

(see Chapter 2 by Kaplan in this volume) may represent an ideal vehicle for directing such intracellular processes. As hypothesized in Fig. 5, the endosome (or "receptosome") may do more than simply function as a waystation for the receptor *en route* either to the lysosome or, via recycling, back to the cell surface. Rather, the endosome may travel like a Pac-Man®, conveying the receptor efficiently and rapidly to intracellular sites (including the nuclear membrane) where phosphorylation reactions directed by the receptor kinase (or, alternatively, receptor fragments released by proteolysis) could regulate processes far removed from the cell surface. In its intracellular location, the receptor would no longer be exposed to variations in ligand concentrations that might occur in the extracellular milieu. Thus, the time course of reactions in which the internalized receptor might participate could differ considerably from the time frame of those reactions occurring in the plasma membrane. The essence of the above discussion is that the temporally distinct actions of certain hormones directed by the same receptor interaction at the cell surface may relate directly to the topographically distinct receptor events that, subsequent to ligand binding, occur over quite different time frames.

2.3. Receptor Regulation

As the previous discussion has indicated, hormone receptors can no longer be thought of as static elements in the cell membrane. On combination with a hormone, receptors can cluster, aggregate in supramolecular patches at an internalization site, and can then undergo endocytosis with subsequent proteolytic receptor processing. Added to this dynamic state of receptor mobility caused by specific hormone–receptor interactions are a number of other factors that must be considered with respect to the control of receptor number at the cell surface.

2.3.1. Receptor Synthesis and Turnover

The rates of synthesis and turnover of hormone receptors can be measured by a combination of ligand-binding methods and with the heavy-isotope technique originally developed by Devreotes and Fambrough for the study of nicotinic acetylcholine receptors in cultured chick skeletal muscle cells (Devreotes and Fambrough, 1975, 1976; Devreotes *et al.*, 1977; Hartzell and Fambrough, 1973). More recently, this method has been used to study the insulin receptor in 3T3-L1 preadipocyte cells (Reed and Lane, 1980; Reed *et al.*, 1981). The technique depends on the synthesis of new, "heavy" receptor produced in the presence of $[^2H]$-, $[^{13}C]$-, and $[^{15}N]$-labeled "heavy" amino acids and on the subsequent separation of heavy and light receptors by density gradient ultracentrifugation.

In cultured chick skeletal cells, it was observed that the newly synthesized, "heavy" nicotinic receptors appeared in the membrane within 3 to 3.5 hr after exposure of the cells to heavy amino acids (Devreotes and Fambrough, 1975, 1976; Devreotes *et al.*, 1977; Hartzell and Fambrough, 1973). The precursor–product relationship between the intracellular pool of receptors and surface receptors was found to be somewhere between a strictly linear assembly-line process and a random selection of intracellular receptors to be inserted into the cell surface.

Similar experiments using the heavy-isotope technique in 3T3-L1 differentiating preadipocytes (Reed and Lane, 1980; Reed *et al.*, 1981) have shown that "heavy" insulin receptors appear in the plasma membrane of differentiated 3T3-L1 cells within 2 to 3 hr of exposure to the heavy-isotope-labeled amino acids. It was further noted that within 24 hr the heavy receptor replaces the light receptor completely in the plasma membrane. The half-life of the insulin receptor in differentiated 3T3-L1 cells was estimated to be 6.7 hr (Reed and Lane, 1980). The differentiation process in these cells involves a 10- to 20-fold increase in the numbers of insulin receptors present on the cell surface. On the basis of their data with the heavy-receptor technique, Reed et al. (1981) concluded that the increase in receptor resulted from an increase in the synthesis of the receptor rather than from decreased degradation of receptors or the unmasking of cryptic membrane receptors.

This same approach has been used to estimate the rates of synthesis and turnover of the EGF–URO receptor in human fibroblasts and tumor-derived A431 cells (Krupp et al., 1982). More recently, the use of anti-receptor antibodies has permitted the isolation of biosynthetically labeled [^{35}S-Met]-receptors for insulin and EGF–URO from cultured cells. Work with these two receptors (Kasuga et al., 1982a; Deutsch et al., 1983; Jacobs et al., 1983; Mayes and Waterfield, 1984) indicates a complex biosynthetic process involving both proteolytic cleavage and glycosylation of receptor precursors prior to the insertion of the mature receptors into the plasma membrane. Very likely the dynamics of biosynthesis, membrane insertion, and turnover of the insulin, acetylcholine, and EGF–URO receptors are representative of the dynamics of other macromolecular glycoprotein receptors as well.

In addition to changes in the rates of receptor synthesis or turnover, one must consider the *de novo* synthesis of a specific receptor, as has been reported for insulin receptors in lymphocytes activated by a number of natural and artificial stimuli (Helderman and Strom, 1978a,b; Hollenberg and Cuatrecasas, 1974; Krug et al., 1972). Neither cell-surface nor cytoplasmic binding sites for insulin can be detected in subpopulations of T- or B-cell lymphocytes before stimulation. After activation of the cells either with plant lectins or by immunologic stimuli, binding sites for insulin can be detected. The appearance of insulin binding sites is independent of DNA synthesis and cell division but can be inhibited by inhibitors of RNA synthesis. The evidence suggests that the receptors appear *de novo*, dependent on the state of differentiation of the lymphocyte induced by the mitogenic stimulus.

2.3.2. Developmental Aspects of Receptor Regulation

Regulation of hormone action can be accomplished in developing systems by regulation of the appearance of receptors as well as by regulating the receptor–effector coupling mechanisms. Receptors for epidermal growth factor–urogastrone (EGF–URO) have been observed to increase with age in mouse embryos, with pronounced changes evident in a target tissue such as the maxilla (Nexø *et al.*, 1980). Similarly, chick dorsal root ganglia have been observed to be highly responsive to β-nerve growth factor (β-NGF), as indicated by the neurite outgrowth bioassay and by binding studies with [^{125}I]-β-NGF during early development (8–14 days). However after 18–21 days, both the response to β-NGF and [^{125}I]-β-NGF binding are reduced markedly (Herrup and Shooter, 1975).

Hormone binding and hormone responsiveness need not parallel each other. For example, the maturation of rat reticulocytes into erythrocytes results in a marked reduction of response (activation of adenylate cyclase) to the adrenergic agonist isoproterenol. However, the reduction in response is not paralleled by a concomitant decrease in the number of β-adrenergic receptors on the plasma membrane (Charness *et al.*, 1976). In this case, the reduction of response is attributed to a change in the coupling between receptor occupation and adenylate cyclase activation (Beckman and Hollenberg, 1979; Bilezikian *et al.*, 1977a,b). Some examples of the effects of cell differentiation on hormone receptors and cellular responsiveness in cultured cell systems have been summarized by Lin and Beckner (1983).

2.3.3. Growth Control and Receptor Regulation

Cells grown in culture exhibit a variety of changes in receptor number depending on the stage of the cell cycle. For example, BSC-1 cells in low-density culture possess eight times as many receptors for EGF–URO as do cells at confluency (Holley *et al.*, 1977). Conversely, BALB/3T3 fibroblasts exhibit greater insulin binding either at confluency or under conditions of reduced serum concentration (i.e., cells that have stopped growing) compared with cells that are rapidly growing (Thomopoulous *et al.*, 1977). Another example of receptor regulation during cell growth has been observed in melanoma cell cultures (Varga *et al.*, 1974). In this case, receptors for melanocyte-stimulating hormone (MSH) have been detected only in the G$_2$ phase of the cell cycle of a melanoma cell line. The biological response to MSH parallels the appearance and disappearance of MSH receptors.

2.3.4. Hormonal Regulation of Receptors

Early studies by Gavin *et al.* (1974) indicated that the number of insulin binding sites in cultured IM-9 lymphocytes could be reduced by preincubating the cells with insulin, albeit at concentrations markedly above those that could be considered physiological. The original data with insulin in lymphocytes were not entirely confirmed in studies with fibro-blasts (Huang and Cuatrecasas, 1975). However, data obtained with short-term adipocyte cultures indicate that elevated insulin concentrations within the physiological range can indeed cause a diminution in receptor binding and a predictable shift in the dose–response curve to the right (Livingston et al., 1978). As summarized elsewhere in this volume (Hollenberg, Chapter 17), similar observations have now been made in a number of systems containing receptors for EGF–URO and for β-adrenergic agonists (Carpenter and Cohen, 1976; Lefkowitz, 1978). Some of the best data demonstrating the hormone-specific nature of this down-regulatory process have come from studies of these latter two receptors. As with any receptor-related process, it is essential to demonstrate an appropriate ligand specificity (e.g., potency of isoproterenol > epinephrine > nor-epinephrine for the β_2-adrenergic system) in order to characterize completely the receptor-related nature of the "down-regulation" phenomenon. This process whereby a ligand can regulate the numbers of its own receptors has been termed "homospecific" receptor regulation.

Very interesting aspects of the regulation of gonadotropin receptors in Leydig cells of the testis and luteal cells of the ovary have been pointed out by Catt and co-workers (results summarized by Catt et al., 1979). In essence, three distinct aspects of the regulatory process were illustrated after the administration of gonadotropins to animals in vivo: (1) the production of a high-affinity ligand–recepotor complex, causing a masking of luteinizing hormone receptor sites for periods up to 24 hr; (2) a concurrent desensitization of the effector system (in this case, adenylate cyclase) in step with this persistent receptor occupancy; and (3) an eventual depletion of receptors from the plasma membrane to an extent even greater than would have been anticipated on the basis of the initial degree of receptor occupancy. The results with the gonadotropins indicate the complex nature of the down-regulatory process.

Homospecific receptor regulation can be in the positive as well as negative direction. The phenomenon of receptor recruitment, as summarized by Triggle (1982) and as dealt with elsewhere in this volume (Hollenberg, Chapter 17), has been observed for a number of receptors. For instance, prolactin administration results in the appearance of its own receptors in the liver of hypophysectomized rats (Posner et al., 1974, 1975). In addition, angiotensin II administration in rats has been observed to cause initially (within 24 hr) an increase in both receptor number and affinity and subsequently (36 hr) an increase only in adrenal receptor number (Hauger et al., 1978). In parallel with the increase in receptor number, an increase in adrenal cell responsiveness was observed (Aguilera *et al.*, 1978). In the studies with prolactin and angiotensin, it remains to be determined if the effects are caused by a direct action of each hormone on the receptor-bearing cells or via an indirect process, possibly mediated by a second hormone signal generated *in vivo*; presumably, studies with cultured cells will be able to resolve this issue. Insulin has also been observed to cause an elevation of its own receptors in cultured chondrosarcoma cells (Stevens *et al.*, 1983). Thus, even a single ligand may cause down-regulation of its own receptor in one cell type and up-regulation in another.

The regulation of the numbers of receptors for one hormone by a second hormone (so-called "heterospecific" receptor regulation) has been observed in a number of instances for both steroid and polypeptide hormones. A particularly intriguing example of this kind of regulation relates to the action of the gonadotropins. In cultured rat ovarian fragments (Nimrod *et al.*, 1977), follicle-stimulating hormone (FSH) causes the appearance of binding sites for $[^{125}I]$-labeled human chorionic gonadotropin (these sites are presumably intended for leuteinizing hormone, LH). In contrast, the hypothalamic modulatory factor that stimulates LH release (LHRH) causes a reduction in gonadotropin receptors, seemingly via a direct effect on the Leydig cells (Lefebvre *et al.*, 1980; Labrie *et al.*, 1980). This intriguing example of heterospecific receptor regulation merits further investigation to determine the mechanism involved.

The influence of steroid hormones on the numbers of receptors for polypeptide hormones has been examined in a number of instances. For example, estrogens can cause an increase in the number of oxytocin receptors in rat uterus (Soloff, 1975) and can, via an effect on the pituitary, augment receptors for prolactin and growth hormone in rat liver (Posner *et al.*, 1974, 1975). Similarly, corticosteroids can increase the binding of EGF–URO to cultured fibroblasts (Baker *et al.*, 1978). To date, this kind of heterospecific receptor regulation has not been sufficiently explored *vis à vis* the influence of polypeptide hormones on the numbers of receptors for steroid hormones; such studies appear more than warranted.

The above examples of heterospecific receptor regulation presumably involve mechanisms that relate to the rates of synthesis and turnover of receptor. Recent data have brought to light examples of very rapid (within tenths of minutes) heterospecific receptor regulation that very likely involve mechanisms located solely in the plasma membrane. For instance, the phorbol ester tumor promoter 12-O-tetradecanoylphorbol-13-acetate (TPA), platelet-derived growth factor, and fibroblast growth factor can, on interacting with their own distinct membrane receptors, rapidly alter the number (and possibly affinity) of receptors for EGF–URO (Wrann *et al.*, 1980; see summarized references in Hollenberg *et al.*, 1981). An intriguing aspect of this kind of heterospecific receptor regulation relates to the requirement in some instances (e.g., for the TPA effect) of an intact cell for the process to occur. Possibly, a novel membrane-localized mechanism, involving the participation of cytoplasmic factors (?kinases/phosphatases) is involved in this process.

Complex heterospecific receptor regulation has also been documented in a variety of other receptor systems, including the following interactions: muscarinic/α_1-adrenergic (Yamada *et al.*, 1980); insulin/multiplication-stimulating activity (Bhaumick *et al.*, 1981; King *et al.*, 1980, 1982; Oppenheimer *et al.*, 1983); and benzodiazepine/γ-aminobutyric acid (Braestrup and Nielsen, 1980; Gavish *et al.*, 1979; Guidotti *et al.*, 1979; Muller, 1981). Very likely this kind of regulation will be found in a wide variety of receptor systems.

2.3.5. Nonspecific Factors Affecting Receptors

In addition to being hormonally regulated, receptors have also been shown to be regulated by a wide variety of compounds and agents acting via nonreceptor mechanisms. For example, butyric acid can induce the appearance of β-adrenergic receptors in cultured HeLa cells (Tallman *et al.*, 1977), and cAMP increases the numbers of insulin receptors in cultured fibroblasts and lymphocytes (Thomopoulos *et al.*, 1977). Murine or feline sarcoma virus-mediated transformation of cells causes a marked reduction of EGF–URO receptors (Todaro *et al.*, 1976), as does the transformation of cultured hamster fibroblasts by chemicals (Hollenberg *et al.*, 1979) or by DNA tumor virus (Berhanu and Hollenberg, 1980).

In the case of the RNA tumor virus-mediated transformation, the reduction in EGF–URO binding can be attributed to the production of substances (so-called tumor growth factors) that are distinct from EGF–URO but are able to occupy the EGF–URO receptors and mimic the effects of EGF–URO in cells (Todaro *et al.*, 1976; Hollenberg *et al.*, 1979; Berhanu and Hollenberg, 1980; Roberts *et al.*, 1983). In addition to changes in receptor numbers, changes in the specificity of receptors have also been noted. For example, simian virus-40 transformation of mouse 3T3 cells causes a change in the catecholamine receptors from a β_1 specificity to a β_2 specificity (Sheppard, 1977).

It is evident from the above discussion that hormone receptors are regulated by a wide variety of factors related both to intracellular events such as turnover, cell cycle, and cell differentiation and to extracellular stimuli caused by hormones and other agents.

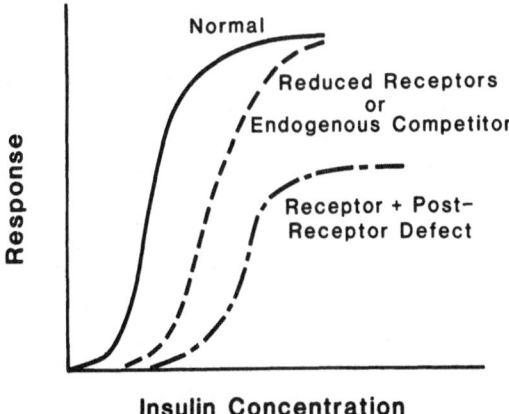

Insulin Concentration

Figure 6. Effects of receptor regulation and postreceptor defects on dose–response curves. A hypothetical situation is depicted for a cell that has insulin receptors in excess of the number required for a maximal cell response. In the normal situation, occupation of a fraction of the available receptors causes a dose–response curve shown as "normal." When the receptor number is decreased, or when endogenous substances (e.g., antibodies) block insulin binding in a competitive manner, there is a shift in the dose–response curve to the right; however, since in this situation there are still enough receptors present to cause maximum activation, raising the insulin concentration can still result in a "normal" maximal response (middle response curve). When the receptor number is reduced below the level necessary for a maximum response, or when postreceptor defects interfere with insulin action, there is a reduction in the maximum attainable response even at very high insulin concentrations (right-hand curve). A combination of reduced receptors and a postreceptor defect would lead to a rightward shift of the dose–response curve as well as a reduction in the maximum response (right-hand response curve).

2.3.6. Effects of Receptor Regulation on Cell Response

Since hormone action is initiated via an interaction at the cell surface, the regulation of the numbers of receptors provides a means of controlling cell responsiveness. The consequences of increases or decreases in receptor number can be predicted on the basis of two assumptions: (1) that the initial hormone–receptor interaction obeys the law of mass action, i.e., that for the reaction $H + R \rightleftharpoons HR$; $K = [HR]/[H][R]$, and (2) that the magnitude of the cellular response is in some way proportional to the number of occupied receptors (i.e., response $\propto [HR]$). From these assumptions it follows that, if the concentration of receptors decreases (e.g., down-regulation), there must be a compensatory increase in hormone concentration, $[H]$, to achieve the same concentration of hormone–receptor complexes and thereby the same cell response; thus the dose–response curve will be shifted to the right. Figure 6 illustrates this theoretical shift for the action of insulin, for which shifts in the dose–response curve have been documented both *in vivo* and *in vitro* (Livingston *et al.*, 1978; Kolterman *et al.*, 1980, 1981). Analogous shifts in the dose–response curve subsequent to receptor down-regulation have been observed for the action of gonadotropins (Catt *et al.*, 1979).

In many systems so far studied, there are more receptors present on cells then are required for a maximum response (i.e., "spare" but equivalent receptors are present). Thus, even in a "down-regulated" system, a sufficiently high concentration of hormone should still be able to occupy the number of receptors that must be activated for a maximum cell response. Therefore, in many receptor systems, a decrease in receptor number will result in a shift of the dose–response curve to the right without a change in the maximum response (middle curve, Fig. 6). Similarly, an increase in receptor number would result in a leftward shift in the dose–response curve. In the situation in which postreceptor mechanisms reduce cell responsiveness, the maximum response will not be achieved even if all receptors are occupied; thus, the combined effect of a reduction of receptors with a postreceptor defect would be to shift the dose–response curve to the right with a simultaneous reduction in the maximum achievable response (curve on the far right, Fig. 6). In practice, dose–response curves that have been measured for the action of insulin in obese and in diabetic subjects (Kolterman *et al.*, 1980, 1981) have been observed to be consistent with both of the curves shown on the right in Fig. 6.

It is evident that the situation with receptors differs somewhat from the situation with the action of enzymes on substrate molecules. In contrast to the situation with enzymes, where the velocity relative to V_{max} is proportional to the fraction of enzyme molecules occupied, in the cellular receptor system, it is the total number rather than the fraction of available receptor entities occupied that governs net cellular responsiveness. Thus, in enzyme systems, reducing the enzyme concentration does not alter the concentration of substrate at which reaction velocity is half-maximal, whereas in receptor systems, a reduction in receptor number causes a rightward shift in the dose–response curve. It is to the credit of the studies of ligand binding that it is possible to dissect the changes in responsiveness that can be attributed to changes in receptor number from the changes in responsiveness that can be attributed to postreceptor mechanisms.

3. Kinetics of Ligand Binding

As indicated previously in this chapter, a variety of methods are now available to measure the interaction of various ligands with their putative receptors (Cuatrecasas and Hollenberg, 1976; Hollenberg and Cuatre-casas, 1979; Hollenberg and Nexø, 1981). It is now clear that the dynamic aspects of the receptor processes (microclustering, aggregation, internalization) triggered by ligand binding introduce a complexity that represents a particular challenge to those analyzing ligand-binding data (Perelson and DeLisi, 1980; Goldstein and Wiegel, 1983; Beck and Goren, 1983, 1985). In the sections that follow, the approaches that have been used for the analysis of binding data are reviewed briefly. For detailed analyses, the reader is referred to more extensive treatises (e.g., DeLean and Rodbard, 1979; Schafer, 1983; and references therein).

3.1. Binding Kinetics for a Simplified Model

3.1.1. Measurements of Binding at Equilibrium

Most frequently, the analysis of binding data found in the literature dealing with hormone receptors is based on a simple mass action model. This model assumes (1) that there is a single class of binding sites, (2) that the receptor is monovalent, (3) that the receptor and the ligand have a unique affinity for each other, (4) that the receptor binding sites do not interact with each other, and (5) that the ligand and the receptor are entities free in solution with similar kinetic properties (e.g., brownian motion, diffusion). Many receptors appear to fulfill the first four criteria; however, the fifth assumption really only applies to solubilized receptors. Based on the above assumptions, the interaction of a receptor (R) with ligand (H) to form the hormone–receptor complex (HR) may be described in terms of the mass action equation discussed above and represented by a simplified version of equation 1 in which $m = 1$.

$$H + R \underset{k_{-1}}{\overset{k_1}{\rightleftharpoons}} HR \qquad (1A)$$

At any time, the rate of formation of the complex, $d[HR]/dt$, may be expressed in terms of the association (k_1) and dissociation (k_{-1}) rate contants:

$$d[HR]/dt = k_1[H][R] - k_{-1}[HR] \qquad (6A)$$

where [H], [R], and [HR] represent the concentrations of ligand, receptor, and ligand–receptor complex, respectively. Thus, at equilibrium, where $d[HR]/dt = 0$, equation 6A simplifies to:

$$[H][R]/[HR] = k_{-1}/k_1 = K_D \qquad (6B)$$

The affinity of a ligand for its receptor is reflected by the magnitude of the equilibrium constant (either the dissociation constant, K_D, or its reciprocal, the association constant, K_A). In essence, the larger K_A (or the smaller K_D), the larger is the energy of interaction between the receptor and the ligand (Goren *et al.*, 1977). Equation 6B is very frequently rearranged in the form discussed by Scatchard (1949):

$$B/F = R_T/K_D - (1/K_D) \cdot B \qquad (7)$$

where B represents the concentration of bound ligand, i.e., [HR], F is the concentration of free ligand, i.e., [H], R_T is the total receptor concentration (i.e., R_T or $[R_T] = [R] + [HR]$). A plot of B/F versus B is routinely used to estimate the equilibrium dissociation constant (reciprocal of the slope of the curve) and the total number of receptors present in the preparation (X-intercept $= R_T$). As with the analysis of enzyme kinetics, there are a variety of ways of rearranging equation 6B to yield equations equivalent to those denoted by the eponyms "Hanes plot," "Eadie–Hofstee plot," "Lineweaver–Burke," etc. (Boeynaems and Dumont, 1975; Schafer, 1983). The advantages or disadvantages of analyzing binding data according to one or other of the alternatives to the so-called "Scatchard plot" (equation 7) have been discussed at length elsewhere (Deranleau, 1969; Thompson and Klotz, 1971; Klotz, 1982; Munson *et al.*, 1983).

It is important to point out that only if the five assumptions outlined above hold and if the system is at equilibrium will the various analyses of the binding data yield simple linear plots. If data are obtained before equilibrium is reached, nonlinear plots will result, as illustrated by a variety of computer simulations (Boeynaems and Dumont, 1975; DeLean and Rodbard, 1979; Beck and Goren, 1983). As is discussed below, the deviation from linearity of the plots of binding data (e.g., according to Scatchard) are difficult to interpret unequivocally.

3.1.2. Measurements of the Rates of Association and Dissociation

As indicated by equation 6B, the association and dissociation rate constants for the receptor–ligand interaction can also be used to estimate the equilibrium dissociation constant. Experimentally, if conditions can be arranged such that [H], the concentration of hormone, is very small (thus, the product $k_1[H][R]$ will be negligible compared with the magnitude of $k_{-1}[HR]$, then equation 6B simplifies to:

$$d[HR]/dt = -k_{-1}[HR] \qquad (8A)$$

This equation can be rearranged to $d\ln[HR] = -k_{-1}dt$, to yield, on integration,

$$2.303 \log([HR]_t/[HR]_0) = k_{-1}t \qquad (8B)$$

In equation 8B, $[HR]_0$ is the concentration of hormone–receptor complex at $t = 0$, and $[HR]_t$ is the concentration of complex remaining at time t.

In practice, the dissociation rate constant is estimated by first allowing a ligand to equilibrate with its receptor and then rapidly washing the hormone–receptor complex free from unbound ligand (either by a 100-fold dilution with buffer or by filtration onto membrane filters). The amount of ligand–receptor complex is then measured as a function of time; a logarithmic plot of the fraction of complex remaining ($[HR]_t/[HR]_0$), versus time yields a straight line with a negative slope from which the dissociation rate constant is calculated. The half-life of the complex ($t_{1/2}$) can be estimated from such plots (i.e., the time at which $[HR]_t/[HR]_0 = 0.5$), and k_{-1} may be calculated:

$$k_{-1} = 0.693/t_{1/2} \qquad (8C)$$

From the simple relationship expressed by equation 8C, it is comparatively easy to see that the longer the half-life of the complex, the smaller will be the value of k_{-1} and, hence (from equation 6B), the lower the equilibrium dissociation constant, K_D (or the higher the equilibrium affinity constant K_A). Because of the nature of the species forming the hormone–receptor complex (e.g., small neurotransmitter molecules reacting with comparatively large receptors in isolated cells, membranes, or in detergent solution), it is believed (Weber, 1975; Koren and Hammes, 1976; Triggle and Triggle, 1976) that the value of the association rate constant (k_1, discussed below) is essentially diffusion limited and, therefore, that the differences in the dissociation rate constant will have the most profound impact on the value of the equilibrium constant. In view of these considerations, there has been a great deal of interest, in studies of receptor binding, on factors that may affect the dissociation rate (Rodbard, 1979).

In practice, the rate with which the hormone–receptor complex dissociates ($k_{-1}[HR]$) is slower than the rate of complex formation ($k_1[II][R]$). Also, experimentally, conditions can be arranged so that over the time period of measurement, the reverse reaction (i.e., complex dissociation) is negligible with respect to the forward reaction. Thus, equation 8 can be rewritten

$$d[HR]/dt = k_1[H][R] \qquad (8D)$$

Since $[H] = [H]_0 - [HR]$ and $[R] = [R_T] - [HR]$, where $[H]_0$ is the initial concentration of hormone, equation 8D becomes:

$$d[HR]/dt = k_1([H]_0 - [HR])([R] - [HR]) \qquad (9)$$

Integration of this equation yields the expression:

$$k_1 t = \{2.303/([H]_0 - [R_T])\}\cdot\log\{[R_T]$$

$$([H]_0 - [HR])/[H]_0([R_T] - [HR])\} \qquad (10)$$

Thus, a plot of the right-hand side of equation 10 versus time will yield the association rate constant as the slope.

An alternative integrated form of equation 9 can be written if experimental conditions are arranged so that only a small amount of receptor is complexed, i.e., $[R_T] \simeq [R_T] - [HR]$:

$$1 - [HR]/[H]_0 = \exp(-k_1\cdot t\cdot[R_T]) \qquad (11)$$

Using a Taylor expansion for the right-hand side of equation 11 and eliminating terms in $(k_1\,R_T\,t)^n$, where n = 2, 3 . . . and for which the magnitudes are negligible,* one obtains a useful expression for the association rate constant:

$$[HR]/[H]_0 = k_1\cdot[R_T]\cdot t \qquad (11A)$$

Thus, a simple plot of the fraction of ligand bound, ($[HR]/[H]_0$), versus time yields, as the limiting slope (at time close to zero), the value of the association rate constant multiplied by $[R_T]$.

Although experimentally it is relatively easy to estimate the values of $[H]_0$ and $[HR]$, it is often difficult to determine the precise value of $[R_T]$ for a given cell, membrane, or soluble receptor preparation. In practice, an estimate of $[R_T]$ can be measured from a simple equilibrium binding isotherm that yields the limiting amount of ligand–receptor complex formed at comparatively high concentrations of radioligand (Thompson and Klotz, 1971).

As an alternative approach to the estimation of the association rate constant k_1, binding conditions can be arranged so that even at equilibrium, only a small fraction of the total receptor is occupied; that is, [R]

---

* This assumption holds over a period up to 100 min for receptor concentrations in the range 10^{-12} M, with $k_1 \simeq 10^8$ M^{-1} min^{-1}.

= $[R_T]$ − $[HR]$ ≃ $[R_T]$. If the term $k_1[R]$, which by the above assumption equals $k_1[R_T]$, is redefined as a pseudo-first-order rate constant, $k_1' = k_1[R_T]$, equation 6A can be rewritten:

$$d[HR]/dt = k_1'[H] − k_{-1}[HR] \qquad (6C)$$

Since $[H] = [H]_0 − [HR]$, then integration of equation 6C yields the relationship between the concentration of hormone–receptor complex and time:

$$1 − \{(k_1' + k_{-1})/k_1'\}·[HR]/[H]_0 = exp − (k_1' + k_{-1})·t \qquad (12)$$

It follows that as t increases and as the system approaches equilibrium, the term exp-$(k_1' + k_{-1})t$ becomes negligible, and equation 12 may be rewritten:

$$[HR]_{eq} = \{k_1'/(k_1' + k_{-1})\}·[H]_0 \qquad (12A)$$

where $[HR]_{eq}$ represents the limiting amount of hormone–receptor complex formed at equilibrium. Thus, equation 12 may be rewritten, substituting for $[HR]_{eq}$ from equation 12A:

$$1 − [HR]/[HR]_{eq} = exp − (k_1' + k_{-1})·t \qquad (13)$$

Experimentally, log$([HR]_{eq} − [HR])/[HR]_{eq}$ is plotted as a function of time. The slope (multiplied by −2.303) yields the value of $(k_1' + k_{-1})$ or $(k_1[R_T] + k_{-1})$. To determine the value of the true association rate constant, k_1, by this method, it is necessary to measure independently k_{-1} (e.g., by equation 8B and $[R_T]$).

Equation 13 is also applicable in those association experiments in which the concentration of hormone–receptor complex is much smaller than the concentration of ligand present (that is, $[H] = [H]_0 − [HR] ≃ [H]_0$). In equation 13, the pesudo-first-order rate constant, k_1', can be redefined: $k_1' = k_1[H]_0$. Experimentally, the binding data are plotted as described above, but k_1 is calculated without requiring a knowledge of receptor concentration: $k_1 = [(−2.303·slope) − k_{-1}]/[H]_0$ (Pollet *et al.*, 1977).

The advantage of these alternative approaches for the estimate of k_1 are twofold. First, the absolute concentration of the hormone–receptor complex need not be determined; only the relative value $[HR]/[HR]_{eq}$ need be measured at any time t. Second, the precise value of R_T does not (as is the case for the use of equation 10) enter into the data plot used to estimate the magnitude of $(k_1' + k_{-1})$. Thus, overall, this approach introduces less experimental uncertainty into the estimate of k_1. As indicated, an example of the use of this approach for the estimate of the association rate constant can be seen in the paper by Pollett *et al.* (1977) dealing with the insulin receptor.

3.2. Departures from the Simple Model of Ligand Binding

The above sections have dealt with the simplest model of the binding of a ligand to its receptor. It is now realized that the state of the receptor in a cell membrane departs considerably from the idealized model of a soluble ligand-binding substance free to diffuse in solution. In the following sections, the impact of the state of the receptor (e.g., fixed in a membrane, clustered on the cell surface) on the kinetics of ligand binding is explored. Although it is not possible, within the context of this chapter, to deal in depth with all of the complexities that may be encountered, a number of issues are addressed, and the reader is directed to more detailed references that deal with the areas discussed.

3.2.1. Impact of Embedding the Receptor in a Membrane

3.2.1a. Cell Surface Localization. A simple change of location of the receptor from the soluble state in solution to the two-dimensional environment of a cell membrane (or liposome surface) will have an impact on the affinity of the receptor–ligand interaction. In essence, the receptor will lose one of its dimensional degrees of freedom, and the chances of the ligand finding the receptor fixed on a particle, as opposed to the receptor being free in solution, will be reduced. The consequence (DeLisi and Wiegel, 1981) will be a lowering of the association rate constant, which in turn (equation 6B) will raise the value of the equilibrium dissociation constant K_D. The smaller the surface area of the particle on which the receptor is embedded, the greater will be the effect on k_1. Thus, even in "solution," if the receptor were to reside in heterogeneous micelles of differing size, there could be a heterogeneity of binding affinities generated by a population of receptors with a homogeneous intrinsic ligand affinity; Scatchard plots of the data would be curvilinear. The same situation might arise (e.g., curvilinear Scatchard plots) for homogeneous receptors embedded in liposomes of markedly disparate size.

3.2.1b. Effects of Microclustering and Aggregation. In Section 3.2.1a, the receptors were assumed to be diffusely distributed in the membrane. However, as outlined in Section 2.2, it is now realized that ligand binding can lead to receptor microaggregation and clustering. These phenomena, as well, will have an impact on the interpretation of ligand-binding data. Even though the association rate constant (k_1) is largely diffusion limited, the aggregation of receptors into patches can lead to a decrease in k_1 compared to the situation in which receptors are diffusely distributed over the cell surface (Perelson and DeLisi, 1980; Goldstein and Wiegel, 1983). The reduction in k_1 caused by ligand-induced aggregation would add to the effect of membrane localization (Section 3.2.1a), further raising the value of K_D (or lowering the ligand affinity). Thus, membrane localization and ligand-induced aggregation alone, in the absence of any secondary interactions envisioned by the mobile receptor model, could lead to apparent negative cooperativity (concave-up Scatchard plot) for a homogeneous receptor population (DeLisi and Chabay, 1979).

3.2.1c. Effects of Secondary Receptor Interactions within the Plane of the Membrane (Mobile Receptor Paradigm). The mobile receptor paradigm, discussed at length above, highlights yet another consequence of localizing the receptor in the membrane. This theme is recapitulated here to emphasize the difference in the effects of these secondary receptor interactions, which can lead to multiple ligand affinities, from the membrane-localization effects *per se*, which, as has been discussed, can also lead to heterogeneity of ligand binding.

3.2.2. Heterogeneity of Binding Sites and/or Receptor Cooperativity

As opposed to the previous discussion, indicating how a homogeneous population of receptors might give rise to multiple ligand affinities, it is also important to consider the case in which the receptor binding sites themselves may be heterogeneous. Equation 7, which was derived for a single class of binding sites, can be generalized to accommodate any number (i) of independent sites:

$$B_i/F = R_{T,i}/K_{D,i} - (1/K_{D,i}) \cdot B_i \tag{7A}$$

where the subscript, i, refers to the designated values (B, K_D, R_T) for the ith site. The total binding of a ligand is given by the summation of equation 7A for each binding site:

$$B/F = \sum_{i=1}^{n} (R_{T,i}/K_{D,i} - (1/K_{D,i}) \cdot B_i) \tag{7B}$$

where n is the total number of different sites. Equation 7B indicates that for a heterogeneous population of binding sites, a plot of B/F versus B ("Scatchard plot") will be curvilinear, concave-up. The values of $K_{D,i}$ and $R_{T,i}$ may be calculated in some cases where $n = 2$ or 3 (Klotz and Hunston, 1971; Feldman, 1972; Thakur et al., 1980; DeLean et al., 1982; Peters and Pingoud, 1982).

Curvilinear plots of equilibrium binding data, according to equation 7 (i.e., B/F versus B), can also result from receptor cooperativity rather than from binding site heterogeneity. The plots can be curvilinear, concave-down (consistent with positive cooperativity) or concave-up (negative cooperativity) (DeLean and Rodbard, 1979). The cooperativity of the ligand–receptor interaction will be evidenced by a ligand-induced change in the rate constants: either or both of the association (k_1) and dissociation (k_{-1}) rate constants may be affected. DeMeyts and colleagues, studying the binding of insulin, demonstrated an increase in the dissociation rate of insulin from its receptor that was dependent on the concentration of insulin (DeMeyts et al., 1973; DeMeyts and Roth, 1975; DeMeyts, 1976). The data, although controversial (Pollet et al., 1977; Rodbard, 1979), are consistent with the concept that insulin receptor interactions may be negatively cooperative.

In principle, either the association or dissociation rate constant may vary independently with respect to ligand–receptor complex concentration to change the overall equilibrium constant at varying degrees of receptor occupancy (Rescigno et al., 1982). The dependence (either negative or positive cooperativity) of the association and dissociation rate constants on ligand concentration has been dealt with in considerable detail by DeLean and Rodbard (1979) using models in which either k_1 or k_{-1} was linearly dependent on receptor occupancy. The computer simulations developed by these authors and others (Jose and Larralde, 1982) for the association, dissociation, and equilibrium binding curves illustrate graphically the possible consequences of positive and negative cooperativity. In a model in which k_{-1} exhibits cooperativity whereas k_1 is constant, the simulations demonstrated that negative cooperativity of k_{-1} would yield Scatchard plots concave-up, whereas positive cooperativity would lead to curves that are concave-down. The simulations pointed out that dissociation studies would be the most sensitive approach to validate a cooperative model in which k_{-1} is a function of occupancy.

In contrast, if only k_1 exhibits cooperativity, dissociation experiments would not be informative; equilibrium binding data would be insufficient to characterize the model. Positive cooperativity (k_1 increases with increasing receptor occupancy) would yield bell-shaped Scatchard plots exhibiting a larger maximum value, the stronger the cooperativity (Dahlquist, 1974). In contrast, if k_1 decreases with increasing receptor occupancy, Scatchard plots would be concave-up at low degrees of cooperativity, and linear plots with reduced values of B_{max} would result from a strong negative dependence of k_1 on complex concentration. Appropriate tests for positive and negative cooperativity of either k_1 or k_{-1} are suggested by the simulations (DeLean and Rodbard, 1979); in addition, specific pitfalls in experimental design are pointed out. The interested reader is strongly encouraged to consult the chapter by DeLean and Rodbard (1979) for an in-depth evaluation of cooperative receptor binding.

3.2.3. Impact of Internalization and Degradation of Receptor and/or Ligand

In studies with intact cell systems, the compartmentalization of the reacting species (ligand, receptor, and ligand-receptor complex) that results from cellular sequestration adds an additional dimension to data interpretation. The situation for a binding experiment using intact cells can be illustrated by the following scheme:

$$H + R \underset{k_{-1}}{\overset{k_1}{\rightleftharpoons}} HR \begin{cases} \overset{k_a}{\longrightarrow} HR^* \\ \overset{k_b}{\longrightarrow} R^* + H \\ \overset{k_c}{\longrightarrow} H^* + R \end{cases} \tag{I}$$

Extracellular $\rightarrow$ Intracellular (*)

In the scheme, the reaction of hormone with the receptor at the cell surface is shown to be followed by three possible events, identified by the rate constants, k_a, k_b, k_c: k_a refers to the internalization (or surface sequestration) of the hormone receptor complex (HR^*), which remains unavailable for further binding reactions; k_b refers to internalization followed by retention of the receptor (R^*) but recycling of the hormone to the extracellular space, where it can participate again in the binding reaction; k_c refers to internalization of the hormone–receptor complex followed by recycling of the receptor (but not the ligand, H^*) for participation in the reversible reaction characterized by the rate constants k_1 and k_{-1}.

The degradation of the ligand or the receptor, illustrated by scheme II, is characterized by the rate constants k_H and k_R.

$$\begin{array}{ccc} H & + & R \underset{k_{-1}}{\overset{k_1}{\rightleftharpoons}} HR \\ k_H \downarrow & k_R \downarrow & \\ H^* & R^* & \end{array} \tag{II}$$

In scheme II, the asterisk denotes species that are degraded and so are unavailable for participation in a binding reaction.

In cell systems, ligand degradation and internalization can be minimized by performing experiments at 0°C. However, at this temperature, the time to reach binding equilibrium may be markedly prolonged. In experiments using intact cells, it is important to distinguish the amount of surface-bound ligand from ligand that is internalized. The amount of radioactivity that can be eluted from cells in isotonic buffers at acid pH (pH 3 to 5) usually reflects the amount of surface-bound ligand. Further, the amount of ligand degradation in the binding medium (expressed by k_L) can be monitored by evaluating the integrity of the radioligand probe (e.g., precipitability with trichloroacetic acid or ability to rebind to receptor) during the course of a binding experiment. The degradation of receptors is more difficult to monitor. For instance, proteolytic cleavage can yield receptor fragments that can still bind hormone, albeit with different affinities than the intact receptor.

The magnitudes of the rate constants depicted in schemes I and II can have a major impact on the results of ligand-binding experiments, as described in detail elsewhere (Beck and Goren, 1983, 1985). For instance, computer simulation of the ligand association rates and resulting Scatchard plots observed in the presence of ligand internalization alone ($k_a > 0$, but $k_b = k_c = 0$) reveals that, depending on the magnitude of k_a, Scatchard plots can be markedly curvilinear, concave-down, with a pronounced leftward hook (Fig. 7; Beck and Goren, 1983). The simulation predicts this abnormal Scatchard plot even when a distinction is made between internalized and cell-surface-bound ligand. In contrast, if receptor recycling occurs (e.g., with low-density lipoprotein receptor, thought to be characterized in part by a reaction depicted by k_c), then, depending on the magnitude of k_c, the slope of the Scatchard plot can decrease to yield an inappropriately low affinity for the binding reaction. In such a case ($k_c > 0$), the resulting Scatchard plot will be linear, and the X-intercept will still yield the correct estimate of R_T, the total receptor concentration.

Clearly, any combination of the reactions characterized by the rate constants illustrated in schemes I and II (k_H, k_R, k_a, etc.) would complicate further the interpretation of the binding data. In essence, the simulations (Beck and Goren, 1983) suggest (1) that in such experiments with intact cells, there may be serious errors in estimates of the magnitudes of the affinity constants from Scatchard plots, and (2) that the most common deviation from linearity of such plots will be a concave-downward curve. Surprisingly, however, the estimates of the total receptor concentration in such experiments can be obtained provided suitable precautions

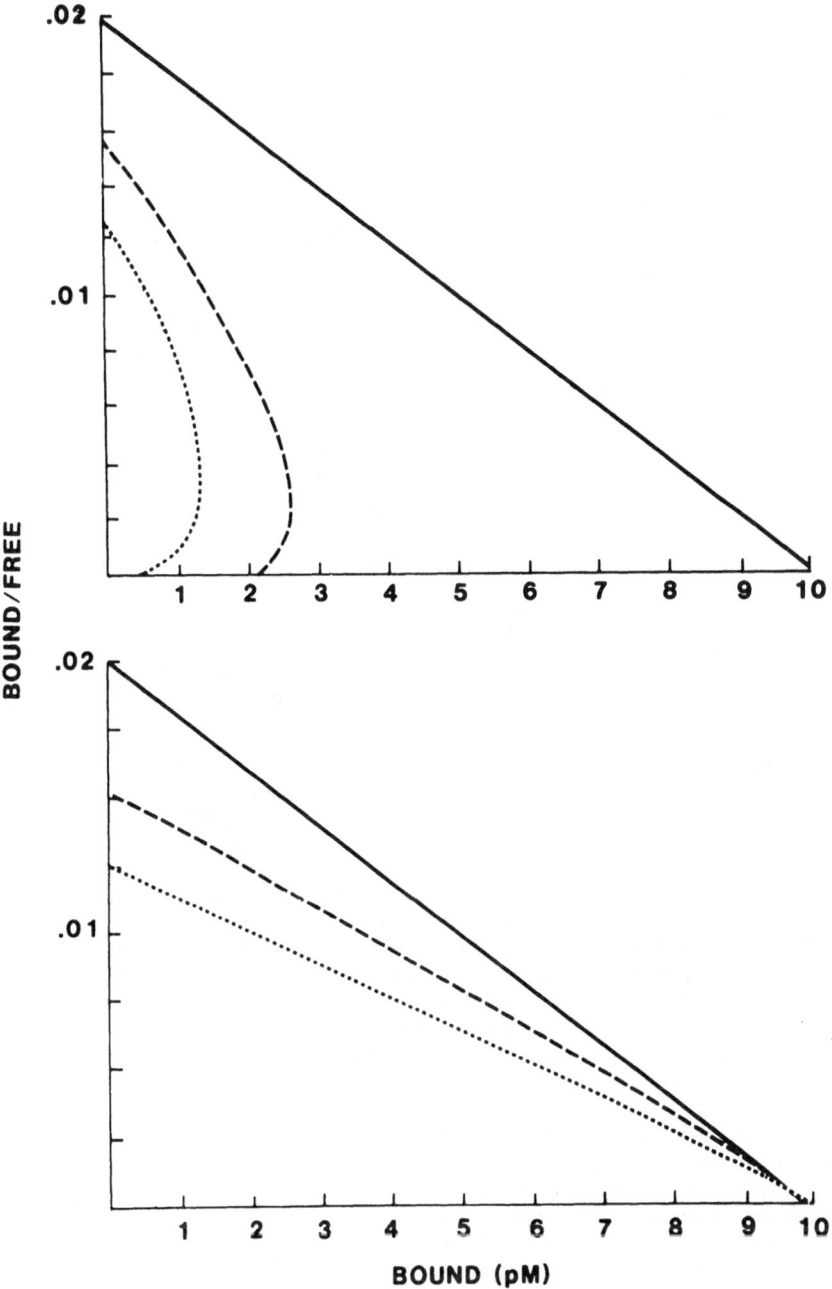

Figure 7. Effect of ligand–receptor (upper) or ligand (lower) internalization on Scatchard plot of ligand-binding assay. Curves were simulated as previously described (Beck and Goren, 1983), where $R_T = 1 \times 10^{-11}$ M, $[H]_0 = 1 \times 10^{-11}$ to 1×10^{-8} M, $k_1 = 1.2 \times 10^8$ M^{-1}·min^{-1}, $k_{-1} = 0.06$ min^{-1}, and k_a (scheme I) for upper figure and k_c (scheme I) for lower figure $= 0$ (——), 0.0167 (– –), and 0.0334 (···) min^{-1}. Bound ([HR]) and free ligand ([H]) concentrations were calculated at 90 min, a time when [HR] is at steady state.

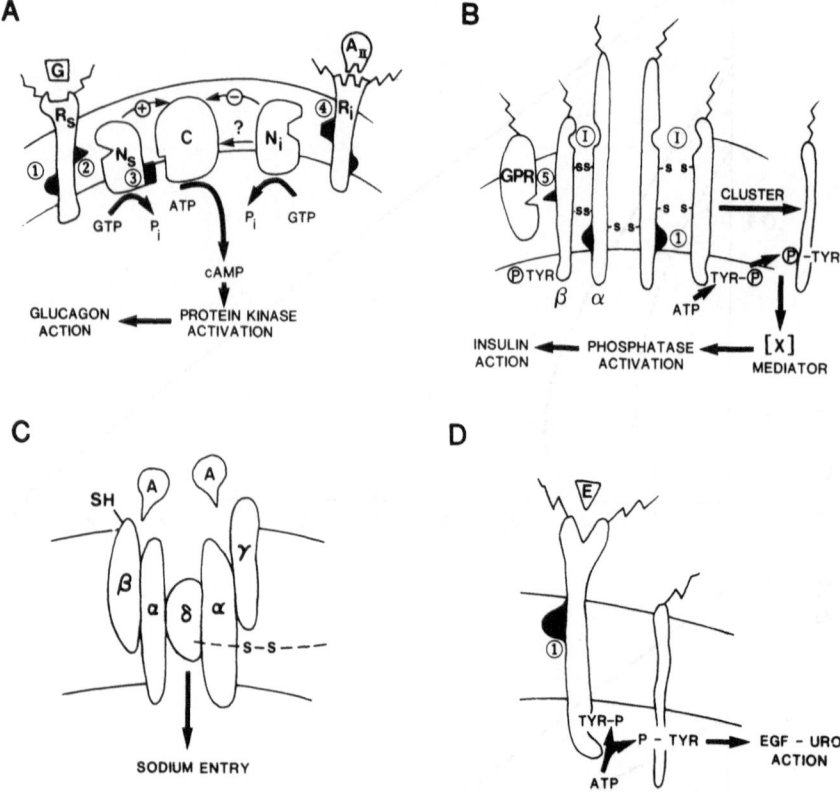

Figure 8. Hypothetical molecular models of receptor-mediated cell regulation. Receptors for glucagon (G, panel A), angiotensin II (A-II, panel A), acetylcholine (A, panel C), insulin (I, panel B), and epidermal growth factor–urogastrone (E, panel D) are shown (not drawn to scale) as integral membrane glycoproteins (oligosaccharide constituents ∿) that traverse the plasma membrane. Regions that may be homologous in different receptors and that relate to receptor function are depicted schematically: region 1 is involved in receptor internalization; regions 2 and 3 are related to the coupling of hormone receptor occupation to cyclase activation; region 4 is related to the coupling of receptor occupation to cyclase inhibition (it is not known if the inhibitory subunit (N_i) interacts directly with the catalytic subunit (C) as does the stimulatory subunit (N_s). Region 5 on the insulin receptor is shown to interact with an affinity regulator glycoprotein (GPR). The four distinct mechanisms are discussed in the text. Panel A: G, glucagon; R_s, glucagon stimulatory receptor; N_s, cyclase stimulatory regulatory subunit; C, cyclase catalytic subunit; A_{II}, angiotensin II; R_i, angiotensin II inhibitory receptor; N_i, cyclase inhibitory regulatory subunit; ATP, adenosine-5'-triphosphate; GTP, guanosine-5'-triphosphate; cAMP, cyclic adenosine 3'-5' monophosphate. Panel B: insulin; α, β, subunits of the insulin receptor; GPR, glycoprotein regulator of insulin receptor affinity; [X], chemical mediator of insulin action; tyr-Ⓟ, phosphotyrosine on putative kinase site of the insulin receptor. Panel C: A, acetylcholine; α, β, γ, δ, subunits of the nicotinic receptor forming the ion channel. Panel D: E, epidermal growth factor–urogastrone; P-tyr, phosphotyrosine. Disulfide bonds that link receptor oligomers are shown (—S—S—), as is a free sulfhydryl (SH) on the acetylcholine receptor.

are taken (Beck and Goren, 1983). The above considerations demonstrate that the interpretation of binding data obtained in intact cell systems, because of the various possible side reactions (aggregation, internalization, etc.), requires a focus on factors in addition to those that are usually considered for studies of membrane-bound or soluble receptor preparations (Beck and Goren, 1983, 1985).

3.2.4. Nonlinear Scatchard Plots

As discussed above, the causes of nonlinear Scatchard plots for ligand–receptor binding are many, and examples of a number of causes of nonlinear plots have already been observed in biological model systems. In addition, nonlinear Scatchard plots may be caused by mechanisms for which experimental examples have not yet been found (Parsons and Vollner, 1978a,b,c). It may be assumed that in time these and other mechanisms for the generation of nonlinear Scatchard plots will be found.

4. Receptor Structure and Molecular Models for Hormone-Mediated Cell Activation

Over the past several years, a great deal of information has become available related to the detailed chemistry and subunit composition of a variety of hormone receptors. The detailed structural and biochemical data permit the description of four distinct molecular models for receptor function (Hollenberg, this volume; and Fig. 8) in addition to the Pac-Man® model discussed above. Most impressive is the information available for the nicotinic cholinergic receptor, for which the detailed molecular structure can now be correlated with the functional properties of the receptor (Conti-Tronconi and Raftery, 1982; Karlin, 1980; Stevens, 1982; Taylor *et al.*, 1983). The receptor is a transmembrane glycosylated protein oligomer, situated normal to the plasma membrane, with about 50% of its 10-nm length extending above the membrane surface. The receptor oligomers form an annulus of about 9 nm outside diameter with a 1.5- to 2.5-nm hole that extends for an unknown distance. Intracellularly, the receptor extends 1 to 2 nm from the cytoplasmic face of the membrane; the diameter of the internal projection is not known. In terms of receptor mechanisms, the nicotinic cholinergic receptor can be taken as representative of an integrated structure that performs both the recognition function and action (ion channel modulation) as a single species. This distinct receptor mechanism is illustrated in Fig. 8C.

Although compared with the nicotinic receptor fewer precise details are known about the receptor systems that modulate adenylate cyclase, it is clear that the mechanisms are distinct from the manner whereby the nicotinic receptor regulates ion flux. Considerable information is now becoming available about the number of membrane components that regulate the cyclase catalytic subunit. As outlined in Fig. 8A, receptors (R_s) for those hormones such as glucagon (G) that stimulate adenylate cyclase interact via an oligomeric stimulatory nucleotide (GTP)-binding subunit (the so-called G_s or N_s) subunit; Rodbell, 1980) to modulate the activity of the catalytic subunit (C). In Fig. 8A, it is suggested that a receptor domain (region 2), which may be homologous for a variety of cyclase-related hormone receptors, interacts directly with the N_s subunit. In Fig. 8A, the catalytic subunit of the cyclase is shown to have a region (region 3) for interacting with a specific portion of the N_s regulatory subunit; evidence suggests a conservation of this region between cell types and between species.

Recently, it has been discovered that receptors for hormones such as angiotensin II (R_i) that cause an inhibition of the cyclase activity do so via interaction with an inhibitory GTP-binding subunit, designated G_i (or N_i) (Bokoch *et al.*, 1983; Gilman, 1984). It is not known whether or not the G_i subunit on interacting with an inhibitory receptor (R_i), interacts directly with the cyclase catalytic subunit. Thus, the cyclase catalytic activity can come under bidirectional control from the interaction of either stimulatory or inhibitory receptors and their respective stimulatory or inhibitory regulatory subunits.

Although at least part of the catalytic subunit must face the inner aspect of the plasma membrane, the precise transmembrane distribution of all the components (receptor, catalytic subunit, and regulatory subunits) of the cyclase systems has yet to be determined. Since guanosine triphosphate (GTP) promotes the interaction of the receptors with the G_s or G_i subunits, it is perhaps not surprising that GTP has an effect on the ligand–receptor affinities measured in binding experiments (Rodbell, 1980). It is to the credit of the mobile receptor model that it was able to foresee at least some of the complex interactions that regulate the adenylate cyclase system.

The receptor for insulin, like the receptor for insulinlike growth factor I (IGF-I), contains a disulfide-stabilized structure ($\alpha\beta$) formed from glycoproteins having molecular weights of about 135,000 (α chain) and 95,000 (β chain). Although the $\alpha\beta$ structure appears to form the principal ligand-binding region of the insulin receptor (Fig. 8B), it has been suggested that other closely associated membrane glycoprotein constituents may regulate the receptor's binding activity (reviewed briefly by Hollenberg, 1981, 1982a). Two models of the insulin receptor have been proposed. A generally accepted model (Jacobs and Cuatrecasas, 1983) envisions the receptor as a heterodimeric $(\alpha\beta)_2$ structure, as illustrated in Fig. 8B. An alternative model (Yip and Moule, 1983) proposes that the α and β subunits are associated with two other glycoproteins of about 40 kilodaltons and a disulfide-linked substituent of 85 kilodaltons. The cloning and sequencing of the human insulin receptor gene has provided the amino acid sequences of the key α- and β-receptor subunits (Ullrich *et al.*, 1985; Ebina *et al.*, 1985).

Two most intriguing series of results have been reported that relate directly to the possible mechanism whereby insulin activates its target cells. First, work from a number of laboratories (Jarett and Seals, 1979; Larner *et al.*, 1979; Saltiel *et al.*, 1982; Seals and Czech, 1980) indicates that on combining with target cell membranes, insulin causes the release of low-molecular-weight chemical mediators that, via modulation of phosphatases and/or cAMP phosphodiesterases, can regulate the activities of insulin-responsive enzymes such as pyruvate dehydrogenase and glycogen synthetase. Second, several groups have obtained data to indicate that the insulin receptor, like the receptor for EGF–URO, possesses intrinsic tyrosine kinase activity (Kasuga *et al.*, 1982b; Rosen *et al.*, 1983; Roth and Cassell, 1983; Van Obberghen and Kowalski, 1982). The β-subunit of the receptor appears to be the site of tyrosine phosphorylation. Thus, for receptors like the one for insulin, there may be a dual membrane-localized activation process, one related to the phosphorylation of intrinsic membrane proteins and a second mechanism related to the release intracellularly of low-molecular-weight chemical mediators. The dual activation process possible for the insulin receptor is illustrated in Fig. 8B.

In contrast with the receptors for insulin and acetylcholine (nicotinic), the EGF-URO receptor (Fig. 8D) is a single-chain glycoprotein that has a molecular weight of about 180,000 (reviewed by Carpenter, 1981; Hollenberg and Armstrong, 1985). Cloning and sequencing of the human placental EGF–URO receptor gene has yielded the complete sequence of the receptor (Ullrich *et al.*, 1984). The novel tyrosine kinase activity of this receptor, discovered by Cohen and collaborators (Carpenter, 1981; Cohen *et al.*, 1980), has served as a major stimulus for the discovery of the intrinsic kinase activity of the insulin receptor and for the observation of other receptor-mediated phosphorylation reactions (briefly reviewed by Hollenberg, 1982b). It is believed that the catalytic site of the EGF–URO receptor resides on the inner aspect of the plasma membrane (Linsley and Fox, 1980) (Fig. 8D). It is quite likely that the phosphorylation of membrane protein tyrosine residues represents a first step in at least some of the actions of EGF–URO (e.g., change in cellular morphology, stimulation of cation transport); the membrane kinase reaction itself, however, does not appear to be sufficient for the activation of the mitogenic response (Schreiber *et al.*, 1983).

Thus, as outlined in Fig. 8, four distinct membrane-localized reaction pathways (cyclase activation and cAMP-modulated protein kinase activity; activation of the membrane to liberate low-molecular-weight chemical mediators; activation of ion transport; and intrinsic receptor kinase activity leading to membrane protein phosphorylation) appear to lead to cell activation. It is important to reemphasize that, in keeping with the mobile receptor model outlined above, any particular receptor oligomer could possibly modulate a number of such reaction pathways. Therefore, it may be inappropriate to attempt to identify a single reaction pathway as the primary event in the action of a particular hormone. Rather, the challenge in the future is to identify the matrix of concurrent receptor-modulated membrane-localized reactions that are involved in cell activation. In view of the diversity of differentiation of the cells in which the various receptors reside, it is perhaps not surprising that a number of distinct mechanisms are used by different receptors to cause cell activation. Because of this diversity, it is also not unreasonable to suspect that the same receptor species situated in different cell types may behave quite differently both in terms of the cell activation processes and in terms of the detailed kinetics of ligand binding.

ACKNOWLEDGMENTS. Work in the authors' laboratories has been made possible by grants from the Canadian Medical Research Council, the Alberta Heritage Foundation for Medical Research, and the Canadian Diabetes Association.

References

Aguilera, C., Hauger, R. L., and Catt, K. J., 1978, Control of aldosterone secretion during sodium restriction: Adrenal receptor regulation and increased adrenal sensitivity to angiotensin II, *Proc. Natl. Acad. Sci. U.S.A.* **75**:959–979.

Ariens, E. J., 1954, Affinity and intrinsic activity in the theory of competitive inhibition, *Arch. Int. Pharmacodyn.* **99**:32–49.

Armstrong, G. D., and Hollenberg, M. D., 1982, Crosslink-labeling and proteolytic mapping of the human placental receptor for epidermal growth factor–urogastrone, *Proc. Can. Fed. Biol. Soc.* **25**:15.

Arunlakshana, O., and Schild, H. O., 1959, Some quantitative uses of drug antagonists, *Br. J. Pharmacol.* **14**:48–58.

Baker, J. B., Barsh, G. S., Carney, D. H., and Cunningham, D. D., 1978, Dexamethasone modulates binding and action of epidermal growth factor in serum-free cell culture, *Proc. Natl. Acad. Sci. U.S.A.* **75**:1882–1886.

Beck, J. S., and Goren, H. J., 1983, Simulation of association curves and "Scatchard" plots of binding reactions where ligand and receptor are degraded or internalized, *J. Receptor Res.* 3:561–577.

Beck, J. S., and Goren, H. J., 1985, Determination of binding parameters in the presence of coupled reactions, *Cell Biophys.* 7:31–42.

Beckman, B. S., and Hollenberg, M. D., 1979, Beta-adrenergic receptors and adenylate cyclase activity in rat reticulocytes and mature erythrocytes, *Biochem. Pharmacol.* 28:239–248.

Bennett, G. V., O'Keefe, E., and Cuatrecasas, P., 1975, The mechanism of action of cholera toxin and the mobile theory of hormone-receptor–adenylate cyclase interactions, *Proc. Natl. Acad. Sci. U.S.A.* 72:33–37.

Berhanu, P., and Hollenberg, M. D., 1980, Epidermal growth factor–urogastrone receptor: Selective alteration in simian virus 40 transformed mouse fibroblasts, *Arch. Biochem. Biophys.* 203:134–144.

Bhaumick, B., Goren, H. J., and Bala, R. M., 1981, Further characterization of human basic-somatomedin: Comparison with insulin-like growth factors I and II, *Horm. Metab. Res.* 13:515–518.

Bilezikian, J. P., Speigel, A. M., Brown, E. M., and Aurbach, G. D., 1977a, Identification and persistence of beta-adrenergic receptors during maturation of the rat reticulocyte, *Mol. Pharmacol.* 13:775–785.

Bilezikian, J. P., Speigel, A. M., Gammon, D. E., and Aurbach, G. D., 1977b, The role of guanyl nucleotides in the expression of catecholamine-responsive adenylate cyclase during maturation of the rat reticulocyte, *Mol. Pharmacol.* 13:786–795.

Boeynaems, J. M., and Dumont, J. E., 1975, Quantitative analysis of the binding of ligands to their receptors, *J. Cyclic Nucleotide Res.* 1:123–142.

Boeynaems, J. M., and Dumont, J. E., 1977, The two-step model of ligand–receptor interaction, *Mol. Cell. Endocrinol.* 7:33–47.

Boeynaems, J. M., and Dumont, J. E., 1980, *Outlines of Receptor Theory*, Elsevier/North-Holland Biomedical Press, Amsterdam, New York.

Bokoch, G. M., Katada, T., Northup, J. K., Hewlett, E. L., and Gilman, A. G., 1983, Identification of the predominant substrate for ADP-ribosylation by islet activating protein, *J. Biol. Chem.* 258:2072–2075.

Braestrup, C., and Nielsen, M., 1980, Benzodiazepine receptors, Drug Res. 30:852–857.

Burgen, A. S. V., 1981, Conformational changes and drug action, *Fed. Proc.* 40:2723–2728.

Burgen, A. S. V., Roberts, G. C. K., and Feeney, J., 1975, Binding of flexible ligands to macromolecules, *Nature* 253:753–755.

Carpenter, G., 1981, Epidermal growth factor, in: *Tissue Growth Factors, Handbook of Experimental Pharmacology* (R. Baserga, ed.), Springer-Verlag, New York, pp. 89–132.

Carpenter, G., and Cohen, S., 1976, ^{125}I-Labeled human epidermal growth factor. Binding, internalization and degradation in human fibroblasts, *J. Cell. Biol.* 71:159–171.

Catt, K. J., Harwood, J. P., Aguilera, G., and Dafau, M. L., 1979, Hormonal regulation of peptide receptors and target cell responses, *Nature* 280:109–116.

Charness, M. D., Bylund, D. B., Beckman, B. S., Hollenberg, M. D., and Snyder, S. H., 1976, Independent variation of β-adrenergic receptor binding and catecholamine-stimulated adenylate cyclase activity in rat erythrocytes, *Life Sci.* 19:243–250.

Clark, A. J., 1926a, The reaction between acetylcholine and muscle cells, *J. Physiol. (Lond.)* 61:530–546.

Clark, A. J., 1926b, The antagonism of acetylcholine by atropine, *J. Physiol. (Lond.)* 61:547–556.

Clark, A. J., 1933, *The Mode of Action of Drugs on Cells*, Edward Arnold, London.

Cohen, S., Carpenter, G., and King, L., Jr., 1980, Epidermal growth factor–receptor-protein kinase interactions. Copurification of receptor and epidermal growth factor-enhanced phosphorylation activity, *J. Biol. Chem.* **255**:4834–4842.

Conn, P. M., Rogers, D. C., Stewart, J. M., Neidel, J., and Sheffield, T., 1982, Conversion of a gonadotropin-releasing hormone antagonist to an agonist, *Nature* **296**:653–655.

Conti-Tronconi, B. M., and Raftery, M. A., 1982, The nicotinic cholinergic receptor: Correlation of molecular structure with functional properties, *Annu. Rev. Biochem.* **51**:491–530.

Cuatrecasas, P., 1974, Membrane receptors, *Annu. Rev. Biochem.* **43**:169–214.

Cuatrecasas, P., and Hollenberg, M. D., 1976, Membrane receptors and hormone action, *Adv. Protein Chem.* **30**:251–451.

Dahlquist, F. W., 1974, The quantitative interpretation of maximum Scatchard plots, *FEBS Lett.* **49**:267–268.

Das, M., 1980, Mitogenic hormone-induced intracellular message: Assay and partial characterization of an activator of DNA replication induced by epidermal growth factor, *Proc. Natl. Acad. Sci. U.S.A.* **77**:112–116.

DeHaen, C., 1976, The non-stoichiometric floating receptor model for hormone-sensitive adenylate cyclase, *J. Theor. Biol.* **58**:383–400.

DeLean, A., and Rodbard, D., 1979, Kinetics of co-operative binding, in: *The Receptors: A Comprehensive Treatise*, Vol. 1 (R. D. O'Brien, ed.), Plenum Press, New York, pp. 143–192.

DeLean, A., Munson, P. J., and Rodbard, D., 1979, Multisubsite receptors for multivalent ligands, *Mol. Pharmacol.* **15**:60–70.

DeLean, A., Hancock, A. A., and Lefkowitz, R. J., 1982, Validation and statistical analysis of a computer modeling method for quantitative analysis of radioligand binding data for mixtures of pharmacologic receptor subtypes, *Mol. Pharmacol.* **21**:5–16.

DeLisi, C., and Chabay, R., 1979, The influence of cell surface receptor clustering on the thermodynamics of ligand binding and the kinetics of its dissociation, *Cell. Biophys.* **1**:117–131.

DeLisi, C., and Wiegel, F. W., 1981, Effect of nonspecific forces and finite receptor number on rate constants of ligand-cell brand receptor interactions, *Proc. Natl. Acad. Sci. USA* **78**:5569–5572.

DeMeyts, P., 1976, Cooperative properties of hormone receptors in cell membranes, *J. Supramol. Struct.* **4**:201–218.

DeMeyts, P., and Roth, J., 1975, Cooperativity in ligand binding: A new graphic analysis, *Biochem. Biophys. Res. Commun.* **66**:1118–1125.

DeMeyts, P., Roth, J., Neville, D. M., Jr., Gavin, J. R. III, and Lesniak, M. A., 1973, Insulin interactions with its receptors: Experimental evidence for negative cooperativity, *Biochem. Biophys. Res. Commun.* **55**:154–161.

DeMeyts, P., van Obberghen, E., Roth, J., Wollmer, A., and Brandenburg, D., 1978, Mapping of the residues responsible for the negative cooperativity of the receptor binding region of insulin, *Nature* **273**:504–509.

Deranleu, D. A., 1969, Theory of the measurement of weak molecular complexes. I. General considerations, *J. Am. Chem. Soc.* **91**:4044–4049.

Deutsch, P. J., Wan, C. F., Rosen, O. M., and Rubin, C. S., 1983, Latent insulin receptors and possible receptor precursors in 3T3-L1 adipocytes, *Proc. Natl. Acad. Sci. U.S.A.* **80**:133–136.

Devreotes, P. N., and Fambrough, D. M., 1975, Acetylcholine receptor turnover in membranes of developing muscle fibers, *J. Cell Biol.* **65**:335–358.

Devreotes, P. N., and Fambrough, D. M., 1976, Synthesis of the acetylcholine receptor by cultured chick myotubes and denervated mouse extensor digitorum longus muscles, *Proc. Natl. Acad. Sci. USA* **73**:161–164.

Devreotes, P. N, Gardner, J. M., and Fambrough, D. M., 1977, Kinetics of biosynthesis of acetylcholine receptor and subsequent incorporation into plasma membrane of cultured chick skeletal muscle, *Cell* **10**:365–373.

Dickson, R. B., Willingham, M. C., and Pastan, I., 1981, α_2-macroglobulin adsorbed to colloidal gold: A new probe in the study of receptor-mediated endocytosis, *J. Cell Biol.* **89**:29–34.

Ebina, Y., Ellis, L., Jarnagin, K., Edery, M., Graf, L., Clauser, E., Ou, J.-h., Maslarz, F., Kan, Y. W., Goldfine, I. D., Roth, R. A, and Rutter, W. J., 1985, The human insulin receptor cDNA: the structural basis for hormone-activated transmembrane signalling, *Cell* **40**:747–758.

Edidin, M., Laganasky, Y., and Lardner, T. J., 1976, Measurement of membrane protein lateral diffusion in single cells, *Science* **191**:466–468.

Ehrlich, P., 1908, Nobel lecture on partial functions of the cell, in: *The Collected Papers of P. Ehrlich*, Vol. III (F. Himmelweit, M. Marquardt, H. Dale, eds.), Pergamon Press, Oxford, p. 183.

Feldman, H. A., 1972, Mathematical theory of complex ligand-binding systems at equilibrium: Some methods for parameter fitting, *Anal. Biochem.* **48**:317–338.

Friedman, S. J., and Skehan, P., 1979, Morphological differentiation of human choriocarcinoma cells induced by methotrexate, *Cancer Res.* **39**:1960–1967.

Frye, L. D., and Edidin, M., 1970, The rapid intermixing of cell surface antigens after formation of mouse–human heterokaryons, *J. Cell Sci.* **7**:319–335.

Gavin, J. R., III, Roth, J., Neville, D. M., Jr., DeMeyts, P., and Buell, D. N., 1974, Insulin-dependent regulation of insulin receptor concentration. A direct demonstration in cell culture, *Proc. Natl. Acad. Sci. U.S.A.* **71**:84–88.

Gavish, M., Chang. R. S. L., and Snyder, S. H., 1979, Solubilization of histamine H-1, GABA and benzodiazepine receptors, *Life Sci.* **25**:783–790.

Gilman, A. G., 1984, Guanine nucleotide-binding regulatory proteins and dual control of adenylate cyclase, *J. Clin. Invest.* **73**:1–4.

Goldstein, B., and Wiegel, F. W., 1983, The effect of receptor clustering on diffusion-limited forward rate constants, *Biophys. J.* **43**:121–125.

Goldstein, J. L., and Brown, M. S., 1975, Hyperlipidemia in coronary artery disease: A biochemical genetic appraoch, *J. Lab. Clin. Med.* **85**:15–28.

Goren, H. J., Bauce, L. G., and Vale, W., 1977, Forces and structural limitations of binding of thyrotrophin-releasing factor to the thyrotrophin-releasing receptor: The pyroglutamic acid moiety, *Mol. Pharmacol.* **16**:2265–2279.

Gregory, H., Taylor, C. L., and Hopkins, C. R., 1982, Leuteinizing hormone release from dissociated pituitary cells by dimerization of occupied LHRH receptors, *Nature* **300**:269–271.

Guidotti, A., Baraldi, M., and Costa, E., 1979, 1,4-Benzodiazepines and gamma-aminobutyric acid: Pharmacological and biochemical correlates, *Pharmacology* **19**:267–277.

Hartzell, H. C., and Fambrough, D. M., 1973, Acetylcholine receptor production and incorporation into membranes of developing muscle fibers, *Dev. Biol.* **30**:153–165.

Hauger, R. L., Aguilera, G., and Catt, K. J., 1978, Angiotensin II regulates its receptor sites in the adrenal glomerulosa zone, *Nature* **271**:176–177.

Helderman, J. H., and Strom, T. B., 1978a, Emergence of insulin receptors upon alloimmune T cells in the rat, *J. Clin. Invest.* **59**:334–338.

Helderman, H. J., and Strom, T. B., 1978b, Specific insulin binding site on T and B lymphocytes as a marker of cell activation, *Nature* **274**:62–63.

Herrup, K., and Shooter, E. M., 1975, Properties of the β-nerve growth factor receptor in development, *J. Cell. Biol.* **67:**118–125.

Hillman, G. M., and Schlessinger, J., 1982, Lateral diffusion of epidermal growth factor complexed to its surface receptors does not account for the thermal sensitivity of patch formation and endocytosis, *Biochemistry* **21:**1667–1672.

Hock, R. A., Nexø, E., and Hollenberg, M. D., 1979, Isolation of the human placenta receptor for epidermal growth factor–urogastrone, *Nature* **277:**403–405.

Hollenberg, M. D., 1979, Membrane receptors and hormone action, *Pharmacol. Rev.* **30:**393–410.

Hollenberg, M. D, 1981, Membrane receptors and hormone action. I. New trends related to receptor structure and receptor regulation, *Trends Pharmacol. Sci.* **2:**320–323.

Hollenberg, M. D., 1982a, Membrane receptors and hormone action. II. New perspectives for receptor-modulated cell function, *Trends Pharmacol. Sci.* **3:**25–28.

Hollenberg, M. D., 1982b, Receptor-mediated phosphorylation reactions, *Trends Pharmacol. Sci.* **3:**271–273.

Hollenberg, M. D., 1984, Receptor models and the action of neurotransmitters and hormones: Some new perspectives, in: *Neurotransmitter Receptors*, 2nd ed. (H. I. Yamamura, S. J. Enna, and M. J. Kuhar, eds.), Raven Press, New York, pp. 1–39.

Hollenberg, M. D., and Armstrong, G., 1985, Epidermal growth factor-urogastrone and its receptor, in: *Polypeptide Hormone Receptors* (B. Posner, ed.), Marcel Dekker, New York, pp. 201–206.

Hollenberg, M. D., and Cuatrecasas, P., 1974, Hormone receptors and membrane glycoproteins during *in vitro* transformation of lymphocytes, in: *Control of Proliferation of Animal cells* (B. Clarkson and R. Baserga, eds.), Cold Spring Harbor Laboratory, Cold Spring Harbor, New York, pp. 423–434.

Hollenberg, M. D., and Cuatrecasas, P., 1979, Distinction of receptor from non-receptor interaction in binding studies: Historical and practical perspectives, in: *The Receptors, A Comprehensive Treatise*, Vol. I (R. D. O'Brien, ed.), Plenum Press, New York, pp. 193–214.

Hollenberg, M. D, and Nexø, E., 1981, Receptor binding assays, in: *Receptors and Recognition*, Series B, Vol. II: *Membrane Receptors: Methods for Purification and Characterization* (S. Jacobs and P. Cuatrecasas, eds.), Chapman and Hall, London, pp. 1–31.

Hollenberg, M. D., Barrett, J. C., Ts'o, P. O. P., and Berhanu, P., 1979, Selective reduction in receptors for epidermal growth factor–urogastrone in chemically transformed tumorigenic Syrian hamster embryo fibroblasts, *Cancer Res* **39:**4166–4169.

Hollenberg, M. D., Nexø, E., Berhanu, P., and Hock, R. A., 1981, Phorbol ester and the selective modulation of receptors for epidermal growth factor-urogastrone, in: *Receptor-Mediated Binding and Internalization of Toxins and Hormones* (J. L. Middlebrook and L. D. Kohn, eds.), Academic Press, New York, pp. 181–195.

Holley, R. W., Armour, R., Baldwin, J. H., Brown, K. D., and Yeh, Y.-C., 1977, Density-dependent regulation of growth of BSC-1 cell culture: Control of growth by serum factors, *Proc. Natl. Acad. Sci. U.S.A.* **74:**5046–5050.

Hopkins, C. R., Semoff, S., and Gregory, H., 1981, Regulation of gonadotrophin secretion to the anterior pituitary, *Phil. Trans R. Soc. Lond. [Biol.]* **296:**73–81.

Huang, D., and Cuatrecasas, P., 1975, Insulin-induced reduction of membrane receptor concentrations in isolated fat cells and lymphocytes: Independence from receptor occupation and possible relation to proteolytic activity of insulin, *J. Biol. Chem.* **250:**8251–8259.

Jacobs, S., and Cuatrecasas, P., 1976, The mobile receptor hypothesis and "cooperativity" of hormone binding application to insulin, *Biochim. Biophys. Acta* **433:**482–495.

Jacobs, S., and Cuatrecasas, P., 1977, The mobile receptor hypothesis for cell membrane receptor action, *Trends Biochem. Sci.* 2:280–282.

Jacobs, S., and Cuatrecasas, P., 1983, The insulin receptor, *Annu. Rev. Pharmacol. Toxicol.* 23:461–479.

Jacobs, S., Chang, K.-J., and Cuatrecasas, P., 1978, Antibodies to purified insulin receptor have insulin-like activity, *Science* 200:1283–1284.

Jacobs, S., Kull, F. C., and Cuatrecasas, 1983, Monensin blocks the maturation of receptors for insulin and somatomedin C: Identification of receptor precursors, *Proc. Natl. Acad. Sci. U.S.A.* 80:1228–1231.

Jarett, L., and Seals, J. R., 1979, Pyruvate dehydrogenase activation in adipocyte mitochondria by an insulin-generated mediator from muscle, *Science* 206:1407–1408.

Jose, M. V., and Larralde, C., 1982, Alternative interpretation of unusual Scatchard plots: Contribution of interactions and heterogeneity, *Math. Biosci.* 58:159–170.

Kahn, C. R., Baird, K. L., Flier, J. S., Grunfeld, C., Harmon, J. T., Harrison, L. C., Karlsson, F. A., Kasuga, M., King, G. L., Lang, U. C., Podskalny, J. M., and van Obberghen, E., 1981, Insulin receptor, receptor antibodies and the mechanism of insulin action, *Recent Prog. Horm Res.* 37:477–538.

Kaplan, J., 1981, Polypeptide-binding membrane receptors: Analysis and classification, *Science* 212:14–20.

Karlin, A., 1980, Molecular properties of nicotinic acetylcholine receptors, in: *The Cell Surface and Neuronal Function* (G. Poste, G. L. Nicolson, and C. W. Cotman, eds.), Elsevier/North-Holland Biomedical Press, Amsterdam, pp. 191–260.

Kasuga, M., Hedo, J. A., Yamada, K. M., and Kahn, C. R., 1982a, The structure of insulin receptor and its subunits. Evidence for multiple nonreduced forms and a 210,000 possible proreceptor, *J. Biol. Chem.* 257:10392–10399.

Kasuga, M., Zick, Y., Blithe, D. L., Crettaz, M., and Kahn, C. R., 1982b, Insulin stimulates tyrosine phosphorylation of the insulin receptor in a cell-free system, *Nature* 298:667–669.

King, G. L., Kahn, C. R., Rechler, M. M., and Nissley, S. P., 1980, Direct demonstration of separate receptors for growth and metabolic activities of insulin and multiplication-stimulating activity (an insulin-like growth factor) using antibodies to the insulin receptor, *J. Clin. Invest.* 66:130–140.

King, G. L., Rechler, M. M., and Kahn, C. R., 1982, Interactions between the receptors for insulin and the insulin-like growth factors on adipocytes, *J. Biol. Chem.* 257:10001–10006.

Klotz, I. M., 1982, Number of receptor sites from Scatchard graphs: Facts and fantasies, *Science* 217:1247–1249.

Klotz, I. M., and Hunston, D. L., 1971, Properties of graphical representations of multiple classes of binding sites, *Biochemistry* 10:3065–3069.

Kolterman, O. G., Insel, J., Saekow, M., and Olefsky, J. M., 1980, Mechanisms of insulin resistance in human obesity. Evidence for receptor and postreceptor defects, *J. Clin. Invest.* 65:1271–1284.

Kolterman, O. G., Gray, R. S., Griffin, J., Burstein, P., Insel, J., Scarlett, J. A., and Olefsky, J. M., 1981, Receptor and postreceptor defects contribute to the insulin resistance in non-insulin-dependent diabetes mellitus, *J. Clin. Invest.* 68:957–969.

Koppel, D. F., Axelrod, D., Schlessinger, J., Elson, E. L., and Webb, W. W., 1976, Dynamics of fluorescence marker concentration as a probe of mobility, *Biophys. J.* 16:1315–1329.

Koren, R., and Hammes, G. G., 1976, A kinetic study of protein–protein interactions, *Biochemistry* 15:1165–1175.

Krug, U., Krug, F., and Cuatrecasas, P., 1972, Emergence of insulin receptors on human lymphocytes during *in vitro* transformation, *Proc. Natl. Acad. Sci. U.S.A.* **69:**2604–2608.

Krupp, M. N., Connolly, D. T., and Lane, M. D., 1982, Synthesis, turnover, and down-regulation of epidermal growth factor receptors in human A431 epidermoid carcinoma cells and skin fibroblasts, *J. Biol. Chem.* **257:**11489–11496.

Labrie, F., Belanger, A., Cusan, L., Seguin, C., Pelletier, G., Kelly, P. A., Reeves, J. J., Lefebvre, F. A., Lemay, A., Gourdeau, Y., and Raynaud, J.-P., 1980, Antifertility effects of LHRH agonists in the male, *J. Androlog.* **1:**209–228.

Lamb, J. E., Ray, F., Ward, J. H., Kushner, J. P., and Kaplan, J., 1983, Internalization and subcellular localization of transferrin and transferrin receptors in HeLa cells, *J. Biol. Chem.* **258:**8751–8758.

Langley, J. N., 1906, On nerve endings and on special excitable substances, *Proc. R. Soc. Lond. [Biol.]* **78:**170–194.

Larner, J., Galasko, G., Cheng, K., DePaoli-Roach, A. A., Huang, L., Daggy, P., and Kellogg, J., 1979, Generation by insulin of a chemical mediator that controls phosphorylation–dephosphorylation, *Science* **206:**1408–1410.

Lefebvre, F. A., Reeves, J. J., Seguin, C., Massicot, J., and Labrie, F., 1980, Specific binding of a potent LHRH agonist in rat testis, *Mol. Cell. Endocrinol.* **20:**127–134.

Lefkowitz, R., 1978, Identification and regulation of alpha and beta-adrenergic receptors, *Fed. Proc.* **37:**123–129.

Lin, M. C., and Beckner, S. K., 1983, Induction of hormone receptors and responsiveness during cellular differentiation, in: *Current Topics in Membranes and Transport*, Vol. 18 (A. Kleinzeller, ed.), Academic Press, New York, pp. 287–315.

Linsley, P. S., and Fox, C. F., 1980, Controlled proteolysis of EGF receptors: Evidence for transmembrane distribution of the EGF binding and phosphate acceptor sites, *J. Supramol. Struct.* **14:**461–471.

Livingston, J. N., Purvis, B. J., and Lockwood, D. H., 1978, Insulin-dependent regulation of the insulin-sensitivity of adipocytes, *Nature* **273:**394–396.

Mahoney, M. S., and Rosenberg, L. E., 1975, Inborn errors of cobalamine metabolism, in: *Cobalamine Biochemistry and Pathophysiology* (B. M. Babior, ed.), John Wiley & Sons, New York, pp. 369–402.

Maxfield, F. R., Schlessinger, J., Schechter, Y., Pastan, I., and Willingham, M. C., 1978, Insulin, epidermal growth factor and α_2-macroglobulin rapidly collect in the same patches on the surface of cultured fibroblasts and are internalized together, *Cell* **14:**805–810.

Mayes, E. L. V., and Waterfield, M. D., 1984, Biosynthesis of the epidermal growth factor receptor in A431 cells, *EMBO J.* **3:**531–537.

Minton, A. P., 1981, The bivalent ligand hypothesis: A quantitative model for hormone action, *Mol. Pharmacol.* **19:**1–14.

Muller, W. E., 1981, The benzodiazepine receptor, *Pharmacology* **22:**153–161.

Munson, P. J., Rodbard, D., and Klotz, I. M., 1983, Number of receptor sites from Scatchard and Klotz graphs: A constructive critique, *Science* **220:**979–981.

Nexø, E., 1978, Transcobalamin I and other receptor-binders: Purification, structural, spectral, and physiologic studies, *Scand. J. Haematol.* **20:**221–236.

Nexø, E., Hollenberg, M. D., and Oleson, H., 1979, Solubilization and characterization of the transcobalamin-II acceptor from human placenta and rabbit liver, in: *Vitamin B_{12}* (B. Zagalak and W. Friedrich, eds.), Walter de Gruyter & Co., New York, pp. 843–850.

Nexø, E., Hollenberg, M. D., Figueroa, A., and Pratt, A. M., 1980, Detection of epidermal growth factor–urogastrone and its receptor during fetal mouse development, *Proc. Natl. Acad. Sci. U.S.A.* **77**:2782–2785.

Nimrod, A., Tsafriri, A., and Linder, H. R., 1977, *In vitro* induction of binding sites for hCG in rat granulosa cells by FSH, *Nature* **267**:632–633.

O'Connor-McCourt, M., and Hollenberg, M. D., 1983, Receptors, acceptors, and the action of polypeptide hormones: Illustrative studies with epidermal growth factor (urogastrone), *Can. J. Biochem. Cell. Biol.* **61**:670–682.

Oppenheimer, C. L., Pessin, J. E., Massague, J., Gitomer, W., and Czech, M. P., 1983, Insulin action rapidly modulates the apparent affinity of the insulin-like growth factor II receptor, *J. Biol. Chem.* **258**:4824–4833.

Parsons, D. L., and Vollner, J. J., 1978a, Theoretical models for cooperative binding. I. One-site creator of binding sites. *Math. Biosci.* **41**:189–215.

Parsons, D. L., and Vollner, J. J., 1978b, Theoretical models for cooperative binding. II. Two-site creator of sites and destruction of pre-existing sites, *Math. Biosci.* **41**:217–230.

Parsons, D. L., and Vollner, J. J., 1978c, Theoretical models for cooperative binding. III. Positive and negative site–site cooperativity, *Math. Biosci.* **41**:231–240.

Pastan, I. H., and Willingham, M. C., 1981, Journey to the center of the cell: Role of the receptosome, *Science* **214**:504–509.

Paton, W. D. M., 1961, A theory of drug action based on the rate of drug–receptor combination, *Proc. R. Soc. Lond. [Biol.]* **154**:21–69.

Paton, W. D. M., and Rang, H. P., 1965, The uptake of atropine and related drugs by intestinal smooth muscle of the guinea pig in relation to acetylcholine receptors, *Proc. R. Soc. Lond [Biol]* **163**:1–44.

Perelson, A. S., and DeLisi, C., 1980, Receptor clustering on a cell surface. I. Theory of receptor cross-linking by ligands bearing two chemically identical functional groups, *Math. Biosci.* **48**:71–110.

Peters, F., and Pingoud, V. A., 1982, A critical interpretation of experiments on binding of peptide hormones to specific receptors by computer modelling, *Biochim. Biophys. Acta* **714**:442–444.

Pollet, R. J., Standaert, M. L., and Haase, B. A., 1977, Insulin binding to the human lymphocyte receptor: Evaluation of the negative cooperativity model, *J. Biol. Chem.* **252**:5828–5834.

Posner, B. I., Kelley, P. A., and Friesen, H. G., 1974, Induction of lactogenic receptor in rat liver: Influence of estrogen and the pituitary, *Proc. Natl. Acad. Sci. U.S.A.* **71**:2407–2410.

Posner, B. I., Kelley, P. A., and Friesen, H. J., 1975, Prolactin receptor in rat liver: Possible induction by prolactin, *Science* **188**:57–59.

Reed, B. C., and Lane, M. D., 1980, Insulin receptor synthesis and turnover in differentiating 3T3-L1 preadipocytes, *Proc. Natl. Acad. Sci. U.S.A.* **77**:285–289.

Reed, B. C., Ronnett, G. V., Clements, P. R., and Lane, M. D., 1981, Regulation of insulin receptor metabolism. Differentiation-induced alteration of receptor synthesis and degradation, *J. Biol. Chem.* **256**:3917–3925.

Rescigno, A., Beck, J. S., and Goren, H. J., 1982, Determination of dependence of binding parameters on receptor occupancy, *Bull. Math. Biol.* **44**:477–489.

Roberts, A. B., Frolik, C. A., Anzano, M. A., and Sporn, M. B., 1983, Transforming growth factors from neoplastic and non-neoplastic tissues. *Fed. Proc.* **42**:2621–2626.

Rodbard, D., 1979, Negative cooperativity: A positive finding? *Am. J. Physiol.* **237**:E203–E205.

Rodbard, D., and Feldman, H. A., 1975, Theory of protein–ligand interaction, *Methods Enzymol* **36**:3–16.

Rodbell, M., 1980, The role of hormone receptors and GTP-regulatory proteins in membrane transduction, *Nature* **284:**17–22.

Rosen, O. M., Herrera, R., Olowe, Y., Petruzzelli, L. M., and Cobb, M. H., 1983, Phosphorylation activates the insulin receptor tyrosine protein kinase, *Proc. Natl. Acad. Sci. U.S.A.* **80:**3237–3240.

Roth, R. A., and Cassell, D. J., 1983, Evidence that the insulin receptor is a protein kinase, *Science* **219:**299–301.

Roth, R. A., Cassell, D. J., Wong, K. Y., Maddux, B. A., and Goldfine, I. D., 1982, Monoclonal antibodies to the human insulin receptor block insulin binding and inhibit insulin action, *Proc. Natl. Acad. Sci. U.S.A.* **79:**7312–7316.

Saltiel, A. R., Siegel, M. I., Jacobs, S., and Cuatrecasas, P., 1982, Putative mediators of insulin action: Regulation of pyruvate dehydrogenase and adenylate cyclase activities, *Proc. Natl. Acad. Sci. U.S.A.* **79:**3513–3517.

Scatchard, G., 1949, The attraction of proteins for small molecules and ions, *Ann. N.Y. Acad. Sci.* **51:**660–672.

Schafer, D. E., 1983, Measurement of receptor–ligand binding: Theory and practice, in: *Tracer Kinetics and Physiological Modeling* (R. M. Lambrecht and A. Rescigno, eds.), Springer-Verlag, Berlin, p. 445–507.

Schlessinger, J., Schechter, Y., Willingham, M. C., and Pastan, I., 1978, Direct visualization of binding, aggregation and internalization of insulin and epidermal growth factor on living fibroblastic cells, *Proc. Natl. Acad. Sci. U.S.A.* **75:**2659–2663.

Schreiber, A. B., Liberman, T. A., Lax, I., Yarden, Y., and Schlessinger, J., 1983, Biological role of epidermal growth factor–receptor clustering, *J. Biol. Chem.* **258:**846–853.

Seals, J. R., and Czech, M. P., 1980, Evidence that insulin activates an intrinsic plasma membrane protease in generating a secondary chemical mediator, *J. Biol. Chem.* **255:**6529–6531.

Seligman, P. A., and Allan, R. H., 1978, Characterization of the receptor for transcobalamin II isolated from human placenta, *J. Biol. Chem.* **253:**1766–1772.

Schechter, Y., Hernaez, L., Schlessinger, J., and Cuatrecasas, P., 1979, Local aggregation of hormone–receptor complexes is required for activation by epidermal growth factor, *Nature* **278:**835–838.

Sheppard, J., 1977, Catecholamine hormone receptor differences identified on 3T3 and simian virus transformed 3T3 cells, *Proc. Natl. Acad. Sci. U.S.A.* **73:**1091–1094.

Smith, R. L., and Jarett, L., 1984, Tissue specific variations in insulin receptor dynamics: A high resolution ultrastrucutral and biochemical approach, in: *Insulin, Its Receptor and Diabetes* (M. D. Hollenberg, ed.), Marcel Dekker, New York, pp. 105–139.

Soloff, M., 1975, Uterine receptor for oxytocin: Effects of estrogen, *Biochem. Biophys. Res. Commun.* **65:**205–212.

Stephenson, R. P., and Barlow, R. B., 1970, Concepts of drug action, quantitative pharmacology and biological assay, in: *A Companion to Medical Studies*, Vol. 2 (R. Passmore and J. S. Robson, eds.), Blackwell, London, pp. 3.1–3.19.

Stevens, C. F., 1982, The acetylcholine receptor cloned east and west; and . . ., *Nature* **299:**776; **300:**110.

Stevens, R. L., Austen, K. F., and Nissley, S. P., 1983, Insulin-induced increase in insulin binding to cultured chondrosarcoma chondrocytes, *J. Biol. Chem.* **258:**2940–2944.

Tallman, J. F., Smith, C. C., and Henneberry, R. C., 1977, Induction of functional β-adrenergic receptors in HeLa cells, *Proc. Natl. Acad. Sci. U.S.A.* **73:**873–877.

Taylor, P., Brown, R. D., and Johnson, D. A., 1983, The linkage betweeen ligand occupation and response of the nicotinic acetylcholine receptor, *Curr. Top. Membr. Transp.* **18:**407–444.

Thakur, A. K., Jaffe, M. L., and Rodbard, D., 1980, Graphical analysis of ligand-binding systems: Evaluation by Monte Carlo studies, *Anal. Biochem.* **107**:279–295.

Thomopoulos, P., Kosmakos, F. C., Pastan, I., and Lovelace, E., 1977, Cyclic AMP increases the concentration of insulin receptors in cultured fibroblasts and lymphocytes, *Biochem. Biophys. Res. Commun.* **73**:246–252.

Thompson, C. J., and Klotz, I. M., 1971, Macromolecule–small molecule interactions: Analytical and graphical re-examination, *Arch. Biochem. Biophys.* **147**:178–185.

Todaro, G. J., Delarco, J. E., and Cohen, S., 1976, Transformation by murine and feline sarcoma viruses specifically blocks binding of epidermal growth factor to cells, *Nature* **264**:26–31.

Tolkovsky, A. M., and Levitzki, A., 1978, Mode of coupling between the β-adrenergic receptor and adenylate cyclase in turkey erythrocytes, *Biochemistry* **17**:3795–3810.

Triggle, D. J., 1982, Receptor recruitment and cryptic signals, *Trends Pharmacol. Sci.* **3**:273–274.

Triggle, D. J., and Triggle, C. R., 1976, *Chemical Pharmacology of the Synapse*, Chapter 2, Academic Press, New York.

Ullrich, A., Coussens, L., Hayflick, J. S., Dull, T. J., Gray, A., Tam, A. W., Lee, J., Yarden, Y., Libermann, T. A., Schlessinger, J., Downward, J., Mayes, E. L. V., Whittle, N., Waterfield, M. D., and Seeburg, P. H., 1984, Human epidermal growth factor receptor cDNA sequence and aberrant expression of the amplified gene in A431 epidermoid carcinoma cells, *Nature* **309**:418–425.

Ulrich, A., Bell, J. R., Chen, E. Y., Herrara, R., Petruzzelli, L. M., Dull, T. J., Gray, A., Coussens, L., Liao, Y.-C., Tsubokawa, M., Mason, A., Seeburg, P. H., Grunfeld, C., Rosen, O. M., and Ramachandran, J., 1985, Human insulin receptor and its relationship to the tyrosine kinase family of oncogenes, *Nature* **313**:756–761.

Valentine, K. A., and Hollenberg, M. D., 1984, Membrane receptors and hormone action, in: *Cell Biology of the Secretory Process* (M. Cantin, Ed.), S. Karger, New York, pp. 1–5.

Van Obberghen, E., and Kowalski, A., 1982, Phosphorylation of the hepatic insulin receptor. Stimulating effect of insulin on intact cells and in a cell-free system, *FEBS Lett.* **143**:179–182.

Varga, J. M., Dipasquale, A., Pawelek, J., McGuire, J. S., and Lerner, A. B., 1974, Regulation of melanocyte stimulating hormone action at the receptor level: Discontinuous binding of hormone to synchronized mouse melanoma cells during the cell cycle, *Proc. Natl. Acad. Sci. U.S.A.* **71**:1590–1593.

Weber, G., 1975, Energetics of ligand binding to proteins, *Adv. Protein Chem.* **29**:1.

Willingham, M. C., and Pastan, I., 1978, The visualization of fluorescent proteins in living cells by video intensification microscopy (VIM), *Cell* **13**:501–507.

Wrann, M., Fox, C. F., and Ross, R., 1980, Modulation of epidermal growth factor receptors on 3T3 cells by platelet-derived growth factor, *Science* **210**:1363–1365.

Yamada, S., Yamamura, H. I., and Roeske, W. R., 1980, The regulation of cardiac α_1-adrenergic receptors by guanine nucleotides and by muscarinic cholinergic agonists, *Eur. J. Pharmacol.* **63**:239–241.

Yankner, B. A., and Shooter, E. M., 1982, The biology of mechanism of action of nerve growth factor, *Annu. Rev. Biochem.* **51**:845–868.

Yarden, Y., Schreiber, A. B., and Schlessinger, J., 1982, A non-mitogenic analogue of epidermal growth factor induces early responses mediated by epidermal growth factor, *J. Cell. Biol.* **92**:687–693.

Yip, C. C., and Moule, M. I., 1983, Structure of the insulin receptor of rat adipocytes: The three interconvertable redox forms, *Diabetes* **32**:760–767.

Iozzo, A. J., Jaffe, M. J., and Radwell, D. (1990). Significant studies of plant-based biology systems. Prokaryotic Mono-Cells studies application form, 10, 172–96.

Thorunduss, P., Kostmann, G. H., Pagan, F., and Havelaus, K., 1977, Cyclooxide increasing concentration of biosynthesis amino in cultured Hippolaris and lymphotases. Russian Biophys. Arca Germany, 3, 28–3.

Thompson, G. A., and Kumar, J. M., 1978, Membrane-deoxy-based molecular interactions. Alterbyllen and epithelic re-calibration study. Plant Physiol. Biochem, 2(10):24–38.

Todaro, G. J., Delanco, J. L., and Cohen, S., 1977, Transformation by murine and feline sarcoma viruses specifically the technique of cell growth in factor to death. Nature.

Innocente, A. M., and Hill, R. D., 1978, Mode of coupling between the photosystem receptor and identified acid-binder in key of factor in the membrane. Biochemistry 1977, 1988. Effect of calcium on the events of analysis interaction. Biochemistry 3223.

Trigg, D. J., and Gruell, C. R. H., Characterization by acceptance. Chemistry 1, Academic Press, New York.

Ullrich, P., Coussens, L., Hayflies, J. S., Dull, T. J., Gray, A., Tam, A. W., Lee, J., Yarden, Y., Liberman, T. A., Schlessinger, J., Downward, J., Mayes, E. L. V., Whittle, P., Waterfield, M. D., and Seeburg, P. N., 1984, Human epidermal growth factor receptor cDNA sequence and aberrant expression of the amplified gene in A431 epidermolid carcinoma cells. Nature, 309, 418–425.

Ullrich, A., Bell, J. R., Chen, E. Y., Gendele, R., Petruzzelli, L. M., Dull, T. J., Gray, A., Coussens, L., Liao, Y. C., Tsubokawa, M., Mason, A., Seeburg, P. H., Grunfeld, C., Rosen, O. M., and Ramachandran, J., 1985, Human insulin receptor and its relation to the tyrosine protein kinase family of oncogenes. Nature, 313, 756–761.

Valentine, R. A., and Hohenberg, M. D., 1985, Membrane receptors and cellular regulation, in Cell Membrane in Science Physiology and Health, (A. S. Kliber, New York, pp.

Van Dyne, G., and Rowland, E., 1985, Chemromation of methyl cortical-related pep-Stimulating effect of sulonting in acid soil high plant-cell-grass system. PNAS, Acad. Sci. USA.

Vale, R. M., Danenmoto, A., Hutson, T., Hitchison, S. G., and Leppert, A., 1985, The regulation of mesen-cytic sympathetic hormones action in the recovery level. Reintroduction of binding to a surgical in a sulphated matrix single with. Subbing the cell cycle. Russ. Appl. Acad. Sci. USA, 27(1):1140–1144.

Wagner, D., 1979, Freeze-sweat lipoid binding to acrosides. Adv. Phys. Acid Chem. 27:1–24.

Waughman, M. C., and Pastan, L., 1979, The adsorption of adherescent proteins in living cells from membrane lattice receptors of cellular VLDL. Cell, 6:4441–8.

Wasen, H., Dulley, R., 1979, Microfilter study on cell biochemistry receptors on 3T3 cells by phorbol-related growth factors. J. Acta. 30(1):1633–1680.

Yamazaki, D., Yamamura, H. H., and Rozelle, K. K., 1985, The mediation of receptor adrenergic receptor by agonist-projecting and by muscarinic, cholinergic receptor. Eur. J. Pharmacol. 113:81–91.

Yankner, B. A., and Shooter, E. M., 1982, The biology and mechanism of action of nerve growth factor in cell-receptors. Annu. Rev. Biochem. 3:845–868.

Yarden, Y., Schreiber, A. B., and Schlessinger, J., 1982, A nonmitogenic analogue of epidermal growth factor induces early responses mediated by epidermal growth factor. J. Cell Biol. 22:687–693.

Yuspa, C., and Morille, M. L., 1982, Quantitative alteration in receptors of cell autocrine. The three interrelated biochemical forms. Nature 22:381–387.

19

Unique Tumor-Specific Antigens as Altered Cell-Surface Receptors

HANS SCHREIBER, CARTER VAN WAES,
and HANS JOSEF STAUSS

1. Introduction and Definition of Unique Tumor-Specific Antigens

Cancer is often viewed as a primary disturbance of cell division. One might, therefore, expect to find the critical alterations exclusively in the nucleus of the malignant cells. Interestingly, however, many of the products of cancer-inducing viral genes—so called oncogenes—have now been traced to the periphery of the cell, namely, the plasma membrane. Furthermore, several of these viral oncogenes appear to be derived from cellular genes that code for products active at the periphery of the cell: these include growth factors, their membrane-bound receptors, and membrane-associated enzymes (Newmark, 1982). Cell biologists and developmental biologists have, of course, argued for years that the cell surface is a major component in the control of growth, differentiation, and movement of cells (Moscona, 1974) and that alterations of cell surface molecules are likely to be critically important in the establishment of malignant behavior. Immunologic probes have an extreme sensitivity for discovering even very subtle alterations in a surface molecule that change a normal "self" molecule to an abnormal "altered-self" molecule. Immunologic analysis of the tumor cell surface might, therefore, be extremely useful for discovering cell surface changes that are important for causing malignant behavior of the cell.

HANS SCHREIBER, CARTER VAN WAES, and HANS JOSEF STAUSS • Department of Pathology, The University of Chicago, La Rabida–University of Chicago Institute, Chicago, Illinois 60649.

There have been major problems in the past in how immunologists have approached the analysis of surface changes on cancer cells. First, it has only recently been realized how rapidly cancers can lose the initial antigenic surface markers through immunoselection of variants by the host's immune response to the tumor (Urban *et al.*, 1982a,b). Therefore, many of the tumors that were carefully analyzed in the past may have already lost the surface alterations that would have been initially detectable. In Section 5 of this chapter, we demonstrate that it appears to be more fruitful to analyze tumors isolated under conditions that prevent the rapid outgrowth of antigen-loss variants. The second major problem is that for decades immunologists have focused their major interest on finding a single antigenic change on cancer cells that would be present on all cancer cells but absent from normal cells. Even though cancers induced by the same virus share antigens, an antigen that is the same on all cancers but absent from any normal cell has not yet been discovered. Nevertheless, the search for these "common" tumor-specific antigens led to the most important discoveries of transplantation and differentiation molecules on normal cells (for review see Old, 1982). During this search for these common antigens, it also became clear that cancers—especially those induced by physical or chemical carcinogens—expressed tumor antigens that were specific for a particular tumor, and these were therefore called "unique" tumor-specific antigens.

The first clear evidence that chemically induced tumors displayed unique tumor-specific antigens came from experiments in the 1950s, since earlier studies had not been conclusive (Woglom, 1929). The individual tumor specificity of the antigens was shown by transplantation assays (Gross, 1943; Foley, 1953; Prehn and Main, 1957; Klein *et al.*, 1960). Mice were first immunized with a particular tumor and then challenged with the same or several other tumors. The protection observed was uniquely specific for each particular immunizing tumor. These unique antigens were found to be specific for a particular tumor even when compared to other tumors of the same histological type induced in the same organ by the same carcinogen and in the same strain of mice.

2. Appearance of Unique Tumor Antigens as a Result of Carcinogen Exposure

Explaining the origins of the unique tumor-specific antigens has been a great challenge for research. Among several conceivable mechanisms, three have been tested experimentally. The first hypothesis (I: activation of repressed genes) suggests that the normal genome already contains the

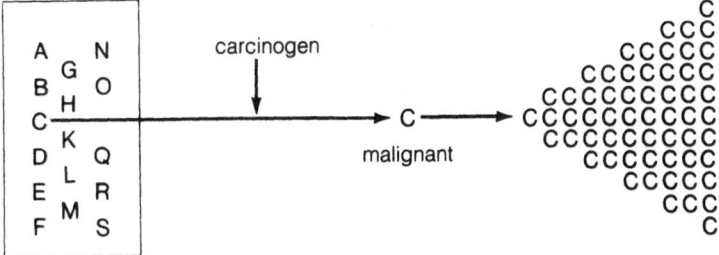

Figure 1. Scheme of the clonal amplification hypothesis. Each precursor cell contains different antigens that are insufficient to be recognized by the immune system until clonal amplification occurs as a result of malignant transformation.

information to produce the tumor antigens. If this mechanism operates, one would expect independently induced tumors to commonly express the same antigens. This seems to be the case in the aberrant expression of certain embryonic and differentiation antigens on malignant tumors (Abelev *et al.*, 1968; Gold and Freedman, 1965; Flaherty and Rinchik, 1978). However, an orderly search for antigenic similarity has failed to detect any reproducible cross reactivity among 90 tests for possible cross antigenicity (Basombrío, 1970). An alternative explanation (II: clonal amplification) is that the cell undergoing malignant transformation already possesses a cell- or clone-specific antigen (Burnett, 1970; Lampson and Levy, 1979). Analogous to idiotypes on B-cell malignancies, this clonal antigen would only be recognized as a unique tumor-specific antigen after

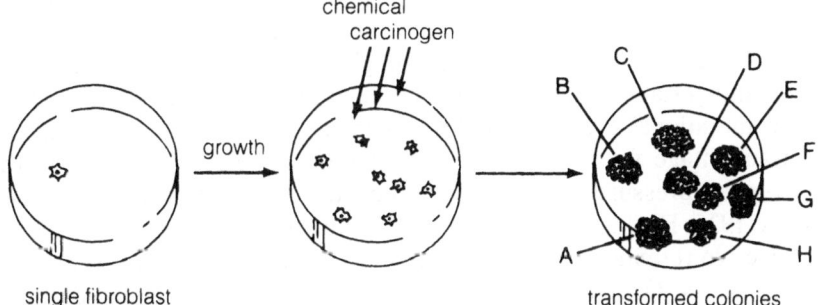

Figure 2. Scheme of experiment to test the clonal amplification hypothesis. The results show that progeny of a single cell, when exposed to carcinogen, give rise to tumors that have individually specific tumor antigens. Since the observed diversity could not be preexistent, it was (Embleton and Heidelbeger, 1972; Basombrío and Prehn, 1972) apparently induced as the result of carcinogen exposure.

the cell underwent clonal amplification, since it would then be in sufficient quantity to be recognized (Fig. 1). However, two studies (Embleton and Heidelberger, 1972; Basombrío and Prehn, 1972) have shown that independently transformed tumors have different antigens even when they are all derived from the same recent clone of a normal cell (Fig. 2). Thus, it appears that these antigens may arise as a consequence of the interaction of the target cell with the carcinogen and that carcinogens, usually being mutagens, might induce somatic mutation at the time of malignant transformation (III: neoantigen as a result of mutation).

3. Possible Relationship of Unique Tumor Antigens to the Malignant Phenotype

One of the major unanswered questions has been the relationship of the tumor antigens to transformation and expression of the malignant phenotype. Boyse (1970) was probably one of the first to suggest that tumor antigens may represent a stable mutation or misarrangement of cell–cell recognition sites. Such a heritable alteration ("mutation") could cause not only the appearance of a tumor antigen but also dysfunction of a recognition site resulting in abnormal invasive growth. Similarly, a complete loss of such cell–cell recognition sites could also cause malignant behavior with the added consequence that a tumor might not appear "antigenic." In view of the lack of selectivity of the mutational effects of chemical or physical carcinogens for particular genes (or even particular DNA sequences), one would expect that random mutation of multiple surface molecules occurs but only some of these changes would be important for the establishment of malignant behavior. Furthermore, we must consider the possibility that these mutational changes occurring during and after malignant transformation give the tumor cell a combination of advantages and disadvantages relative to the normal cells. Thus, many of the mutational events caused by chemical or physical carcinogens are a disadvantage for the cell and are, therefore, selected against by the Darwinian pressures that guide tumor development and progression (Rous, 1935; Foulds, 1954; Farber, 1973; Nowell, 1976). In contrast, there is selective retention of specific mutational changes that favor the malignant process. One such example might be the apparent selective tumor-specific mutation in a cellular protooncogene in tumors induced by physical or chemical carcinogens (Barbacid, 1984).

At present, it still is a major question whether the tumor-specific surface alterations described by immunologists as unique tumor-specific antigens are selective changes on the tumor with relevance to the malignant process. It has been argued in the past that the variety (i.e., the individual tumor specificity) of these antigens makes it unlikely that many or all of these antigens represent changes of the same receptor functions. However, as exemplified by the unique antigenicity of the idiotypes on immunoglobulin molecules present on different B-cell clones, enormous antigenic diversity can clearly be generated by genetic alteration of even a very small region of a single molecule. Finally, experimental studies on the origins of unique tumor antigens have implicated an alteration of only two major types of known cell–cell interaction molecules, namely, those coded for by the immunoglobulin and the major histocompatibility (MHC) gene clusters. Both of these gene clusters belong to a supergene family that originated from a common ancestral gene involved in basic cell–cell recognition functions. Thus, the unique tumor-specific antigens may represent alterations of molecules that have relevance to the malignant process. In addition, there is a third family of normal genes that has been proposed to play a critical role in differentiation and development (Weinberg, 1982; Müller *et al.*, 1982). These so-called protooncogenes are related to oncogenes that are pieces of modified cellular DNA found in certain highly oncogenic retroviruses. Mutational alteration of protooncogenes may not only lead to malignant transformation (Reddy *et al.*, 1983; Tabin *et al.*, 1983) but may also lead to the expression of unique antigenic specificities.

4. Three Gene Families Implicated in the Generation of Unique Tumor Antigens

At present, the evidence for the involvement of the three above-mentioned gene families (Ig, MHC, protooncogene) in the generation of unique tumor-specific antigens is not yet conclusive. The immunoglobulin gene clusters have been closely linked with the expression of a unique tumor-specific antigen defined by an absorbed antiserum on the Balb/c tumor Meth A (Pravtcheva *et al.*, 1981; Flood *et al.*, 1983); however, in the absence of monoclonal antibodies specific for this antigen, final biochemical purification and identification of the antigen are missing. There is a similar problem with the results that agree with the hypothesis that unique tumor-specific antigens may be related to products of mutated protooncogenes. Unique heritable differences among murine leukemia virus envelope proteins on the surface of the tumor cells have been suggested by some experiments with antisera (Lennox *et al.*, 1981; Roman *et al.*, 1981). However, even normal mouse sera contain ubiquitous antibodies to shared epitopes on MuLV-related antigens (Roman *et al.*, 1981;

Lennox *et al.*, 1981), and none of the gp70 determinants that have so far been defined by monoclonal antibodies had unique tumor specificity (Nowinski *et al.*, 1980; Pinter *et al.*, 1982).

A third, current hypothesis is that unique tumor-specific antigens represent altered (or abnormally expressed) histocompatibility antigens (AHA) (Invernizzi and Parmiani, 1975; Bowen and Baldwin, 1975; Garrido *et al.*, 1976). Although a large body of literature is now available on this subject (Bortin and Truitt, 1980, 1981; Festenstein and Schmidt, 1981), attempts at molecular characterizations are remarkably few (Schmidt and Festenstein, 1980; Evan *et al.*, 1983; Callahan *et al.*, 1983). When several trivial explanations such as cross contamination, spontaneous H-2 (MHC of mouse is called H-2) mutation in the mouse of tumor origin, or antigenic drift of the H-2 of the responder strain (Bailey, 1982) can be ruled out, there remain two hypotheses to explain the occurrence of such AHA on tumor cells. First, the AHA may be caused by mutations of the original H-2 genes, leading to epitopes that are cross reactive with allogenic H-2 antigens.

The major difficulty with this hypothesis is that such mutations should have caused only a partial alteration, with retention of some of the original H-2 specificities, in analogy to available examples of H-2 mutations in mice (Klein and Figueroa, 1981). However, up to now, only one report has suggested such partial alteration (Callahan *et al.*, 1983). Therefore, the second hypothesis, namely, that AHA may be caused by mere amplifications of previously hidden normal H-2 specificities or a depression of silent genes for H-2 in the tumor cell could most easily accommodate the published work on AHA (Parmiani *et al.*, 1979). This hypothesis was originally based on a model proposed by Bodmer, who in 1973 suggested that the MHC of each individual contained the genetic information for more than one allele and that allelism was maintained by a regulatory mechanism. This model is no longer tenable, since it is now known that each haplotype contains only a single set of unique structural genes encoding one allele per locus (Steinmetz *et al.*, 1981). But it is now clear that the MHC complex of a mouse contains over 30 distinct class I genes, most of which map into gene clusters outside the traditional boundaries of the H-2 complex in the Qa2,3/T1a region of mouse chromosome 17 (Steinmetz *et al.*, 1982; Hood *et al.*, 1983). Thus, derepression or amplified expression of these genes could generate a considerable diversity of tumor antigens, and some evidence for such derepression has been found (Flaherty and Rinchik, 1978). Nevertheless, it is quite possible that the variety of unique tumor-specific antigens on tumors is too large to be explained solely on the basis of derepression or amplification of these genes without implying an additional mutational mechanism.

5. Strong Unique Tumor-Specific Antigens on Potentially Malignant Cells

One of the major reasons for the slow progress in the dissection of the nature of the unique tumor-specific antigens during the past 30 years has been a total lack of probes that are specific, reliable, and permanent and thus exchangeable between laboratories involved in the analysis of such antigens. As mentioned in Section 1, these unique tumor-specific antigens were defined only by transplantation assays. Recently, however, we have succeeded in developing permanent cloned specific T-cell lines and monoclonal antibody-secreting hybridoma cell lines with high specificity for unique surface antigens on an ultraviolet-radiation (UV)-induced fibrosarcoma (see below).

The model of UV-induced tumors (Kripke 1981) seems to be ideal for the analysis of unique tumor-specific antigens for the following reasons: these tumors have been induced rather recently and express strong unique tumor-specific transplantation antigens; furthermore, the tumors have been induced in pathogen-free mice of defined genetic background, and large numbers of concurrently isolated tumors are available. Thus, the unique specificity for a particular tumor can easily be established, and any genetic drift of the responding inbred animal population (Bailey, 1982) can be detected were it to occur. Equally important is that many of these tumors were preserved as the original tumors without passage of the neoplastic cells *in vitro* or *in vivo*. In contrast, earlier investigators often studied tumors that had been transplanted repeatedly in normal individuals. One of the reasons for the difficulties of many earlier studies may have been that unique tumor-specific antigens were lost rather quickly during successive transplantations. We have in fact recently shown very clearly that this can readily happen (Urban *et al.*, 1982a,b).

In the case of UV-induced tumors, the preservation of the tumors in their original state was almost a necessity, since these tumors are so highly immunogenic that they are usually not transplantable into normal genetically identical mice. These tumors are therefore called regressor tumors, since they only grow regularly in immunodeficient or immunosuppressed mice (Fig. 3). These tumors are, therefore, quite different from other experimentally induced tumors, which can usually be transplanted into normal genetically identical animals. The reasons for this high immunogenicity of UV tumors are not yet fully understood, but it is rather clear that such immunogenic tumors could only arise because of the immunosuppression that is found in the aging UV-irradiated mice that develop the tumors. Thus, the unique tumor-specific antigens have been preserved

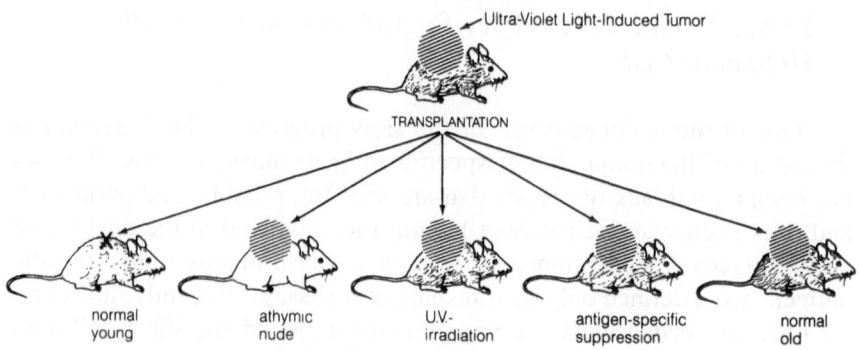

Figure 3. Growth characteristics of ultraviolet-light-induced tumors when transplanted into mice of the strain of tumor origin. Often such tumors are strongly antigenic (e.g., tumor 1591), and such tumors are then rejected regularly by normal mice without any prior immunization (Kripke, 1974). Only on rare occasions do progressor variants arise after tumor transplantation into normal mice (Urban *et al.*, 1982a). However, the tumors will grow regularly in athymic nude mice or mice immunosuppressed by UV irradiation (Kripke, 1974; Kripke and Fisher, 1976; Daynes and Spellman, 1977; Urban *et al.*, 1982b). Selective immune suppression of the tumor-specific T-cell response also appears to be sufficient to permit the outgrowth of these highly immunogenic tumors (Flood *et al.*, 1980, 1981; Schreiber, 1984). Finally, advanced age appears to be sufficient to make the normal mice accept transplanted UV-induced tumors (Spellman and Daynes, 1978; Flood *et al.*, 1981; Urban and Schreiber, 1984).

in these tumors, and we can develop immunologic probes to these tumor antigens in order to study them.

6. Multiplicity of Unique Tumor-Specific Antigens on a Single Tumor Cell

Tumor-specific antigens were classically defined only by transplantation rejection, which was known to be a T-cell-mediated phenomenon. Wortzel *et al.* (1983a,b) therefore decided to define such antigens on UV-induced tumors with cytolytic T-cell lines. (Such T cells kill target cells expressing the relevant antigen by specific binding and subsequent lysis, and each T-cell line was cloned to insure that it had only one specificity.) The particular UV-induced tumor chosen carried the number 1591, and the following approach was used (Fig. 4). A syngeneic (genetically identical) mouse was immunized with the 1591 regressor tumor, and cytolytic T-cell clones were generated that selectively reacted with the 1591 tumor cells. This unique tumor-specific antigen on the 1591 tumor was arbitrarily designated "A."

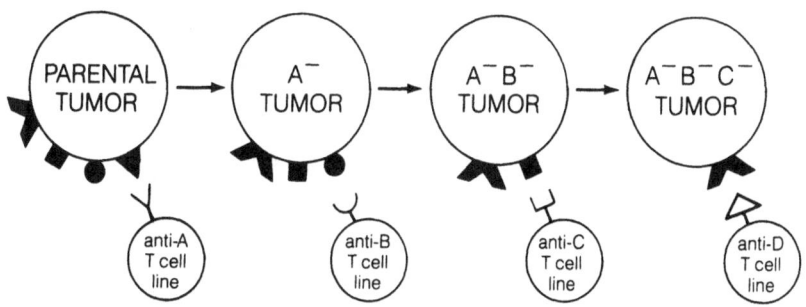

Figure 4. Schematic diagram of the principle of sequential selection whereby the complexity of a unique tumor-specific antigens can be dissected. For further explanation see text and Wortzel *et al.* (1983a,b, 1984).

Then, Wortzel *et al.* selected *in vitro* for an A^- tumor cell variant that was resistant to this anti-A T-cell line. Wortzel *et al.* then investigated whether, in fact, all of the unique tumor-specific antigen had been lost from the A^- tumor variant by using it as immunogen. Thus, another syngeneic mouse was immunized with the antigen-loss variant, and a second T-cell line was derived that reacted with an antigen designated as "B" that was present on the parental tumor as well as the A^- variant. Wortzel *et al.* then selected with the anti-B cytolytic T-cell line for a variant that had also lost the B antigen, and by continuing cycles of immunization, T-cell line generation, and selection for antigen loss variants, two more antigens, C and D, were also recognized. We have not examined how many more of such antigens can be defined on the 1591 tumor cell, and thus we actually do not know how large the antigenic repertoire of a single tumor cell may be. However, it is important to state that all of these antigens were uniquely tumor specific as shown by the selective destruction of the 1591 tumor with the anti-A, B, C, and D T-cell lines (Fig. 5). All four antigens were not only uniquely tumor specific but also independent of each other, since selection with one T-cell line never simultaneously selected for variants that had also lost a second antigen (Fig. 6). Using a similar approach Uyttenhove *et al.* (1983) showed that even a tumor much less immunogenic than the 1591 tumor can have multiple independent tumor antigens, although the unique tumor specificity of these antigens could not be established in the absence of simultaneously derived control tumors.

The above findings clearly demonstrate that the 1591 tumor expresses at least four antigens that are independent and uniquely tumor specific. If these four unique tumor-specific antigens were in fact relevant to the rejection process, it should be possible to convert the tumor *in vitro* from

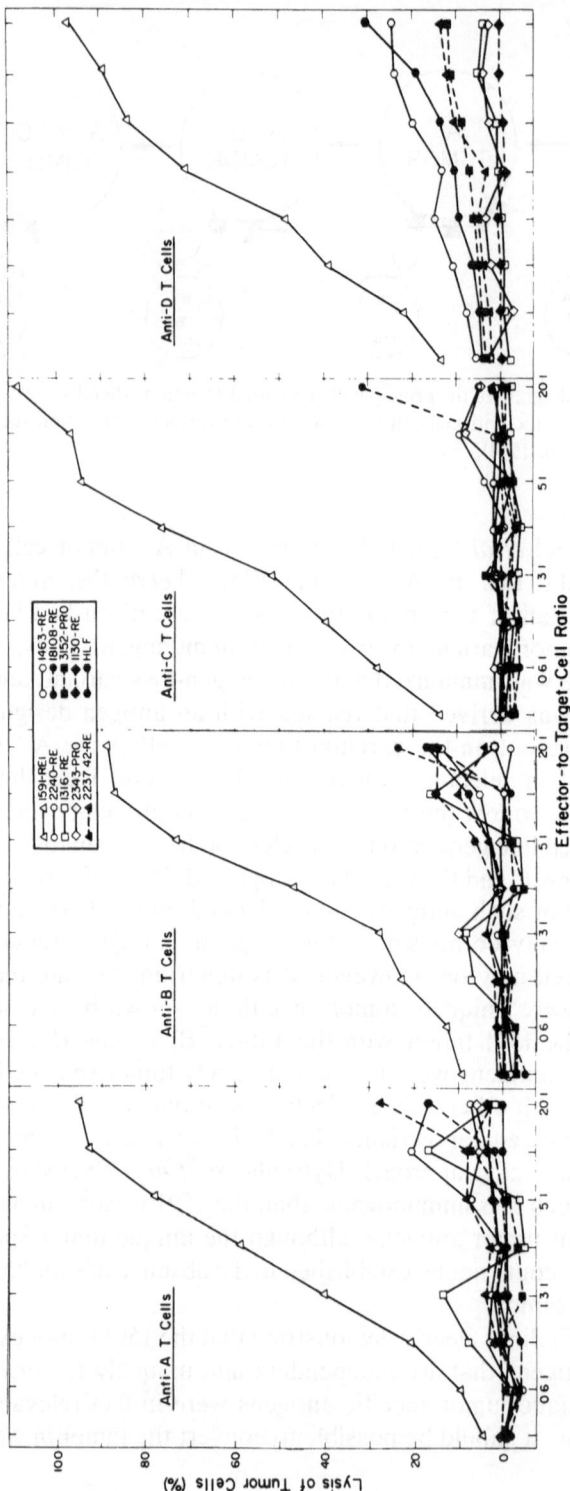

Figure 5. Unique specificity of tumor-reactive T-cell lines for 1591 tumor cells. The radioactive intracellular label ^{51}Cr is released from 1591 tumor cells but not from any other UV-induced fibrosarcoma cells tested; only some control tumors tested are shown here. (From Wortzel *et al.*, 1983b, with permission from Macmillan Journals Limited.)

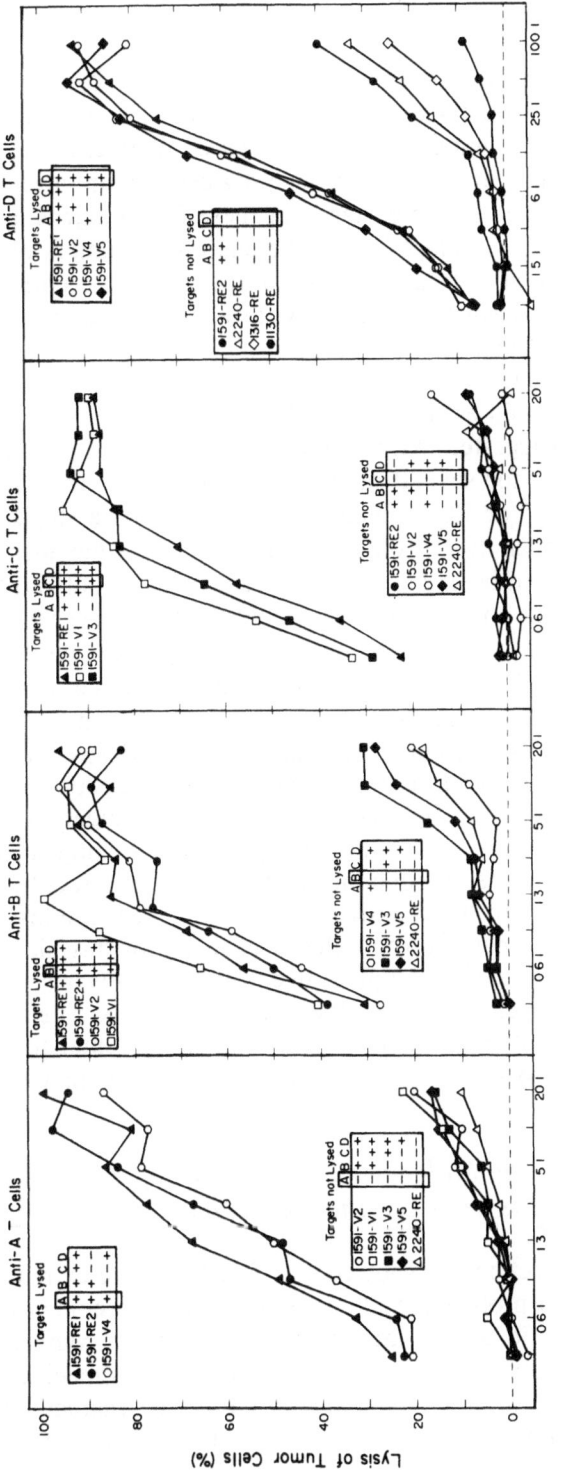

Figure 6. Multiplicity and independence of 1591-specific antigens as demonstrated by cytolytic T cells and immunoselected variants. The radioactive label ^{51}Cr is only released from tumor cells expressing the relevant unique tumor-specific antigen for which the T-cell line is specific. (From Wortzel *et al.*, 1983b, with permission from Macmillan Journals Limited.)

its normal regressor phenotype to a progressor tumor by selecting for antigen-loss tumor variants with the T-cell clones. This was in fact observed: by sequential immune selection *in vitro*, Wortzel *et al.* derived antigen loss variants that could grow progressively in normal mice (Wortzel *et al.*, 1984). The importance of these four antigens in the rejection process is also exemplified by the analysis of tumors that occasionally do develop after transplantation of the tumor into normal mice. These tumors when reisolated have always lost one or more unique tumor-specific antigens (Urban *et al.*, 1982a).

It was difficult to perceive how a tumor expressing multiple independent antigens could escape the host's immune system, since *in vitro* experiments had suggested that the frequency of antigen-loss variants in cell cultures grown for several months was always less than 10^{-4} and often less than 10^{-6} or 10^{-7}. Even if the frequency of a single antigen loss is high (less than 10^{-4}), only one tumor cell that has lost all four tumor antigens would be expected in 10^{16} cells, which would be a tumor larger than a human body. Thus, how could the tumor ever escape? In order to explore this question, the kinetics of antigen loss on the escape tumors were carefully analyzed at various times while the tumors were growing out in normal mice, and the results (J. L. Urban, unpublished data) clearly demonstrated that different unique antigens are lost sequentially. Why does the host allow such sequential selection to happen and not destroy the tumor by response to all of the antigens simultaneously? The answer comes from the finding that the A antigen—as long as it is present—will be immunodominant and thus will totally prevent an immune response to the B antigen until the A^+ cells are lost (Urban *et al.*, 1984). We do not yet know the mechanism for this hierarchy in the immune response to the multiple antigens, but we know it is not simple direct physical interference among the different antigens. The biological importance of this immunodominance is that the parental tumor protects—as long as it is present—the escape variants from immune attack.

7. Exploration of the Molecular Nature of Unique Tumor-Specific Antigens on a UV-Induced Tumor

In the above section, we showed that T-cell lines are extremely sensitive, well-defined probes that allow us to dissect the complexity of a unique tumor-specific antigen into multiple independent components by variant selection. However, monoclonal antibody probes specific for such antigens are needed in order to be able to purify and characterize these antigens. C. Philipps *et al.*, 1985 have recently isolated two syngeneic monoclonal antibodies that bind to the 1591 tumor very specifically and not to as many as 35 other UV-induced tumors tested by quantitative cytofluorometry. Both monoclonal antibodies do not react with other normal syngeneic adult or embryonic cells. Furthermore, several lines of evidence strongly suggested that the unique tumor-specific antigen recognized by the monoclonal antibody is identical or very similar to that recognized by our anti-A cytolytic T-cell line.

It has been shown on a number of occasions that the reactivity pattern of syngeneic monoclonal antibodies with lymphocytes from other strains can sometimes give a rather easily obtainable clue about the type of the molecule with which an antibody reacts (e.g., Flood *et al.*, 1983). Thus, the two syngeneic monoclonal tumor-specific antibodies were tested for possible reactivity with lymphocytes from several congenic strains, and the results suggested that the antibodies reacted with class I MHC antigens of lymphocytes from several other strains of mice (Philipps *et al.*, 1985); this cross reactivity was confirmed using class I gene-transacted L cells. Subsequent analysis using a battery of class I specific monoclonal antibodies showed that the antigens on the 1591 tumor shared only some antigenic determinants with alien class I antigens, suggesting similarity but nonidentity (Philipps *et al.*, 1985). By immune precipitation of radiolabeled antigen on the tumor with the syngeneic monoclonal tumor-specific antibodies, it was shown that the antigen was a 45K molecule associated with a 15K molecule (Philipps *et al.*, 1985). This was confirmed by Philipps *et al.* (1985), who further showed that the tumor-specific antigen was associated with β_2-microglobulin, and had the typical appearance of a class I molecule in two-dimensional gel electrophoresis. There was also clear evidence that the 1591 tumor expressed not just one but several molecularly independent altered class I antigens.

Although it is not yet clear how many of the T-cell-recognized unique tumor antigens belong to a family of related molecules, it is clear that several of them represent altered tumor-specific class I molecules. It therefore appears appropriate to discuss possible mechanisms for the origins of such molecules. A trivial explanation for the findings would be that the 1591 tumor had not originated in an inbred mouse and therefore contained class I genes of other strains. However, Southern blot analysis of Bam II1-digested 1591 tumor DNA using cDNA probes specific for class I genes (Steinmetz *et al.*, 1981) showed that the pattern for the class I genes of the 1591 tumor was characteristic for the genome of the C3H/HeN mouse strain from which the tumor had originated (J. Strauss, unpublished data). Therefore, it is more likely that the newly detected class I antigens on the 1591 tumor originated by some gene rearrangement or gene activation of the MHC genes during tumorigenesis.

It is important to mention in this regard that the MHC complex of a mouse contains over 30 distinct class I genes, most of which map outside the traditional boundaries of the H-2 complex into the Qa2,3/Tla region (Steinmetz *et al.*, 1982). If many or most of the tumor-specific antigens are, in fact, class I molecules, we do not know whether simple gene activation of silent class I MHC genes will account for sufficient pleomorphism of these antigens. It is also quite possible that some recombinant event leads to the activation of previously silent class I DNA sequences as well as to the expression of new unique antigenic sites on the expressed molecules. In this regard, it is relevant to mention that one of the altered MHC class I molecules expressed by the 1591 tumor has several of the antigenic determinants of a class I L molecule (Philipps *et al.*, 1985). A class I L molecule is normally never expressed in the C3H mouse strain from which the 1591 tumor originated. In fact, it had been shown in earlier studies (M. Steinmetz, H. Winoto, K. Mimrad, and L. Hood, 1982) that the flanking sequences for this gene are deleted in this mouse strain. Thus, one might assume that some recombinational mechanism activated L-like coding sequences in the genome.

Obviously, a large number of questions on the origins of the unique tumor-specific antigens remain to be answered by the gene-cloning and gene-transfection experiments presently in progress. This should determine whether all the multiple unique tumor-specific antigens belong to a family of related gene products. We will also determine the mechanisms for the genetic alterations (for example activation or recombination) and analyze the mechanism for antigen loss after variant selection, since antigen loss may not simply be loss of the entire molecule but rather loss of only the antigenic determinant because of some mutational alteration.

8. Possible Effects of Altered MHC Molecules on Various Receptor Functions at the Cancer Cell Surface

A powerful way to analyze the biological significance of the unique tumor-specific antigens may be to transfer the genes encoding them into normal cells, since they might alter the growth of differentiation of the normal cells. The results of such experiments are not yet available. We do, however, know that the 1591 tumor expresses multiple aberrant class I molecules. Furthermore, as mentioned earlier, there is a large body of other albeit only indirect evidence that expression of aberrant MHC class I molecules may occur in many or possibly all malignant tumors. Therefore, we would like to speculate on the significance of changes of MHC class I molecules, especially in the light of the notion that MHC class I

Table I. Immune and Nonimmune Functions of MHC Class I Molecules

1. Membrane-bound antigen associates with class I molecules for T cell recognition
2. Membrane-bound hormone receptors associate with class I molecules
3. Class I molecules may function as a cell–cell recognition system

molecules may have not only immune but also nonimmune functions (Table I).

First, it is well known that cytolytic T cells always need to recognize normal class I molecules in addition to the antigen on the surface of host cells, and the antigen may be either a viral antigen or an altered self molecule (Zinkernagel and Doherty, 1979). Any loss of class I molecules or β_2-microglobulin (Travers *et al.*, 1982) from the cell surface or quantitative reduction or certain mutational alterations of these molecules may lead to failure of the T cells to properly recognize antigens on the tumor cell surface and thus may lead to escape from immune surveillance (Schrier *et al.*, 1984; Bernards *et al.*, 1983; Sondel and Bach, 1980).

A second possible significance comes from the notion that certain growth factor receptors such as the epidermal growth factor receptor and the insulin receptor apparently associate with MHC class I molecules, since antibodies to some class I antigen can block their activity (A. B. Schreiber *et al.*, 1984; Simonson and Olsson, 1983; Edidin, 1983). Therefore, it is conceivable that changes in MHC class I antigen may make such growth factor receptors function abnormally. Qualitatively or quantitatively abnormal class I antigens may, for example, alter ligand binding by an allosteric effect. Abnormal class I molecules might also affect the events following ligand binding such as aggregation or internalization of the receptor. The importance of changes in such receptors for the development of malignancy is exemplified by the recent finding that certain oncogenes can encode defective growth factor receptors (Downward *et al.*, 1984).

The last and most speculative suggestion is that certain MHC class I molecules are used for normal cell–cell interactions and are needed for differentiation and orderly growth (Bartlett and Edidin, 1978; Curtis and Rooney, 1979; Zelený *et al.*, 1978; Curtis, 1979; Edidin, 1983). Although the importance of MHC molecules for cell–cell interaction is well documented in the immune system (for review see J. Klein, 1982), much less is known about the role of MHC molecules for cell interactions among nonimmune cells. Nevertheless, we might find that transfer of the altered class I MHC genes into normal cells alters their normal growth behavior, since the MHC class I molecules may have immune and nonimmune functions. In conclusion, it is possible that the presence of altered class I MHC molecules on a tumor has a dual significance, first as tumor-specific transplantation antigen (as long as the molecules are immunogenic) and second as surface structures that aid the escape of the cancer from immune surveillance, hormonal control, and contact inhibition.

ACKNOWLEDGMENTS. This work was supported by NCI grants RO1-CA-22677, PO1-CA-19266, and PHS 5 RO1 CA-37156. H.S. was supported by RCDA CA-00432, H.J.S. by a fellowship from the Deutsche Forschungsgemeinschaft, and C.V.W. by NIGMS grant PHS T32 GM-07281.

References

Abelev, G. I. 1968, Production of embryonal serum α globulin by hepatomas. Review of experimental and clinical data, *Cancer Res.* 28:1344.

Bailey, D. W., 1982, How pure are inbred strains of mice, *Immunol. Today* 3:210.

Barbacid, M., 1984, Molecular and cellular biology, in: *Genes and Cancer UCLA Symposia, New Series*, Vol. 17, Alan R. Liss, New York pp. 353–373.

Bartlett, P. F., and Edidin, M., 1978, Effect of the H-2 gene complex rates of fibroblast intracellular adhesion, *J. Cell Biol.* 77:377.

Basombrío, M. A., 1970, Search for common antigenicities among twenty-five sarcomas induced by methylcholanthrene, *Cancer Res.* 30:2458.

Basombrío, M. A., and Prehn, R. T., 1972, Studies on the basis for diversity and time of appearance of antigens in chemically induced tumors, *Natl. Cancer Inst. Monogr.* 35:117.

Bernards, R., Schrier, P. I., Houweling, A., Bos, J. L., Van der Eb, A. J., Zijstra, M., and Melief, C. J. M., 1983, Tumorigenicity of cells transformed by adenovirus type 12 by evasion of T cell immunity, *Nature* 305:776.

Bodmer, W. F., 1973, A new genetic model for allelism at histocompatibility and other complex loci. Polymorphism for control of gene expression, *Transplant. Proc.* 5:1471.

Bortin, M. M., and Truitt, R. L. (eds.), 1980, 1st International Symposium on Alien Histocompatibility Antigens on Cancer Cells, *Transplant. Proc.* 12:1.

Bortin, M. M., and Truitt, R. L. (eds.), 1981, 2nd International Symposium on Alien Histocompatibility Antigens on Cancer Cells, *Transplant. Proc.* 13:1751.

Bowen, J. G., and Baldwin, R. W., 1975, Tumor-specific antigens related to rat histocompatibility antigens, *Nature* 258:75.

Boyse, E. A., 1970, Cell membrane receptors, in: *Immunosurveillance* (R. T. Smith and M. Landy, eds.), Academic Press, New York, London, pp. 5–48.

Burnet, F. M., 1970, A certain symmetry: Histocompatibility antigens compared with immune receptors, *Nature* 226:124.

Callahan, G. N., Pardi, D., Giedlin, M. A., Allison, J. P., Morizot, D. M., and Martin, W. J., 1983, Biochemical evidence for the expression of a semiallogeneic, H-2 antigen by a murine adenocarcinoma, *J. Immunol.* 130:471.

Curtis, A. S. G., 1979, Histocompatibility systems, recognition and cell positioning, *Dev. Comp. Immunol.* 3:379.

Curtis, A. S. G., and Rooney, P. 1979, H-2 restriction of contact inhibition of epithelial cells, *Nature* **281**:222.

Daynes, R. A., and Spellman, C. W., 1977, Evidence for the generation of suppressor cells by ultraviolet radiation, *Cell. Immunol.* **31**:182.

Downward, J., Yarden, Y., Hayes, E., Scrace, G., Totty, N., Stockwell, P., Ullrich, A., Schlessinger, J., and Waterfield, M. D., 1984, Close similarity of epidermal growth factor receptor and v-*erb*-B oncogene protein sequences, *Nature* **307**:521.

Edidin, M., 1983, MHC antigens and nonimmune functions, *Immunol. Today* **4**:269.

Embleton, M. J., and Heidelberger, C., 1972, Antigenicity of clones of mouse prostate cells transformed *in vitro*, *Int. J. Cancer* **9**:8.

Evan, G. I., Lennox, E. S., Alderson, E. S., and Croft, L., 1983, A monoclonal anti-HLA antibody recognizes a mouse tumor associated antigen, *Eur. J. Immunol.* **13**:160.

Farber, E., 1973, Carcinogenesis cellular evolution as a unifying thread, *Cancer Res.* **33**:2537.

Festenstein, H., and Schmidt, W., 1981, Variation in MHC antigenic profiles of tumor cells and its biological effects, *Immunol. Rev.* **60**:85.

Fisher, M. S., and Kripke, M. L., 1978, Systemic alteration induced by mice by ultraviolet light irradiation and its relationship to ultraviolet carcinogenesis, *Proc. Natl. Acad. Sci. U.S.A.* **74**:1688.

Flaherty, L, and Rinchik, E., 1978, No evidence for foreign H-2 specificities on the EL4 mouse lymphoma, *Nature* **273**:52.

Flood, P. M., Kripke, M. L., Rowley, D. A., and Schreiber, H., 1980, Suppression of tumor rejection by antologous anti-idiotypic immunity, *Proc. Natl. Acad. Sci. U.S.A.* **77**:2209.

Flood, P. M., Urban, J. L., Kripke, M. L., and Schreiber, H., 1981, Loss of tumor-specific and anti-idiotypic immunity with age, *J. Exp. Med.* **154**:275.

Flood, P. M., DeLeo, A. B., Old, L. J., and Gershon, R. K., 1983, Relation of cell surface antigens on methylcholanthrene-induced fibrosarcomas to immunoglobulin heavy chain complex variable region-linked T cell interaction molecules, *Proc. Natl. Acad. Sci. U.S.A.* **80**:1683.

Foley, E. J., 1953, Antigenic properties of methylcholanthrene-induced tumors in mice of the strain of origin, *Cancer Res.* **13**:835.

Foulds, L., 1954, The experimental study of tumor progression: A review, *Cancer Res.* **14**:327.

Garrido, F., Festenstein, H., and Schirrmacher, V., 1976, Further evidence for derepression of H-2 and Ia like specificities of foreign haplotypes in mouse tumor cell lines, *Nature* **261**:705.

Gold, P., and Freedman, S. O., 1976, Specific carcinoembryonic antigens of the human digestive system, *J. Exp. Med.* **122**:467.

Gross, L., 1943, Intradermal immunization of C3H mice against a sarcoma that originated in an animal of the same line, *Cancer Res.* **3**:326.

Hood, L., Steinmetz, M., and Malissen, B., 1983, Genes of the major histocompatibility complex of the mouse, *Annu. Rev. Immunol.* **1**:529.

Invernizzi, G., and Parmiani, G., 1975, Tumor-associated transplantation antigens of chemically induced sarcoma cross-reactivity with allogeneic histocompatibility antigens, *Nature* **254**:713.

Klein, G., Sjögren, H. O., Klein, E., and Hellström, K. E., 1960, Demonstration of resistance against methylcholanthrene-induced sarcomas in the primary autochthonous host, *Cancer Res.* **20**:1561.

Klein, J., 1982, The science of self–nonself discrimination, in: *Immunology*, John Wiley & Sons, New York.

Klein, J., and Fiqueroa, F., 1981, Polymorphism of the mouse H-2 loci, *Immunol. Rev.* **60**:23.

Kripke, M. L., 1974, Antigenicity of murine skin tumors induced by ultraviolet light, *J. Natl. Cancer Inst* **53**:1333.

Kripke, M. L., 1981, Immunologic mechanisms in UV radiation carcinogenesis, *Adv. Cancer Res.* **34**:69.

Kripke, M. L., and Fisher, M. S., 1976, Immunologic parameters of ultraviolet carcinogenesis, *J. Natl. Cancer Inst.* **57**:211.

Lampson, L. A., and Levy, R., 1979, A role for clonal antigens in cancer diagnosis and therapy, *J. Natl. Cancer Inst.* **62**:217.

Lennox, E. S., Lowe, A. D., Cohn, J., and Evan, G., 1981, Specific antigens on methylcholanthrene-induced tumors of mice, *Transplant. Proc.* **13**:1759.

Moscona, A. A. (ed.), 1974, *The Cell Surface in Development*, John Wiley & Sons, New York.

Müller, R., Slamon, D. J., Tremblay, J. M., Cline, M. J., and Verma, I. M., 1982, Differential expression of cellular oncogenes during pre- and postnatal development of the mouse, *Nature* **299**:640.

Newmark, P., 1984, Cell and cancer biology meld, *Nature* **307**:499.

Nowell, P. C., 1976, The clonal evolution of tumor cell populations. Acquired genetic lability permits selection of variant sublines and underlies tumor progression, *Science* **194**:23.

Nowinski, R. C., Stone, M. R., Tam, M. R., Lostrom, M. E., Burnette, M. N., and O'-Donnell, P. V., 1980, Mapping of viral proteins with monoclonal antibodies, analysis of the envelope proteins of murine leukemia viruses, in: *Monoclonal Antibodies* (R. H. Kennett, T. J. McKearn, and K. B. Bechtol, eds.), Plenum Press, New York, London, pp. 295–315.

Old, L. J., 1982, Cancer immunology: The search for specificity, *Natl. Cancer Inst. Monogr.* **60**:193.

Parmiani, G., Carbone, G., Invernizzi, G., Pierotti, M. A., Sensi, M. L., Rogers, M. J., and Appella, E., 1979, Alien histocompatibility antigens on tumor cells, *Immunogenetics* **9**:1.

Philipps, C., McMillan, M., Flood, P. M., Murphy, D. B., Forman, J., Lancki, D., Womack, J. E., Goodenow, R. S., and Schreiber, H., 1985, Identification of a unique tumor-specific antigen as a novel class I major histocompatibility molecule, *Proc. Natl. Acad. Sci. U.S.A.* **82**:5140.

Pinter, A., Honnen, W. J., Tung, J. S., O'Donnell, P. V., and Hämmerling, U., 1982, Structural domains of endogenous murine leukemia virus gp 70s containing specific antigenic determinants defined by monoclonal antibodies, *Virology* **116**:499.

Pravtcheva, D. D., DeLeo, A. B., Ruddle, F. H., and Old, L. J., 1981, Chromosome assignment of the tumor-specific antigen of a 3-methylcholanthrene-induced mouse sarcoma, *J. Exp. Med.* **154**:964.

Prehn, R. T., and Main, J. M., 1957, Immunity to methylcholanthrene-induced sarcomas, *J. Natl. Cancer Inst.* **18**:769.

Reddy, E. P., Reynolds, R. K., Santos, E., and Barbacid, M., 1983, A point mutation is responsible for the acquisition of transforming properties by the T24 human bladder carcinoma oncogene, *Nature* **300**:149.

Roman, J. M., Hirsch, J., Readhead, C., Levy, D., DeOgny, L., and Dreyer, W. J., 1981, Heritable differences among gp 70-like molecules on C3H ultraviolet light-induced fibrosarcomas, *Transplant. Proc.* **13**:1782.

Rous, P., and Beard, J. W., 1935, The progression to carcinoma of virus-induced rabbit papillomas (Shope), *J. Exp. Med.* **62**:523.

Schmidt, W., and Festenstein, H., 1980, Serological and immunochemical studies on H-2 allospecificities on K36, a syngeneic tumor of AKR, *J. Immunogenet.* **7**:7.

Schreiber, A. B., Schlessinger, J., and Edidin, M., 1984, Interaction between major histocompatibility complex antigens and epidermal growth factor receptors on human cells, *J. Cell Biol.* **98**:725.

Schreiber, H., 1984, Idiotype-specific interactions in tumor immunity, *Adv. Cancer Res.* **41**:291–321.

Schrier, P. I., Bernards, R., Vaessen, R. T. M. J., Houweling, A., and Van der Eb, A. J., 1984, Expression of class I major histocompatibility antigens switched off by highly oncogenic adenovirus 12 in transformed rat cells, *Nature* **305**:771.

Simonson, M., and Olsson, L., 1983, Possible roles of compound membrane receptors in the immune system, *Ann. Immunol.* **134D**:85.

Sondel, P. M., and Bach, F. H., 1980, The alienation of tumor immunity. Alien-driven diversity and alien-selected escape, *Transplant. Proc.* **12**:211.

Spellman, C. W., and Daynes, R. A., 1978, Immunoregulation by ultraviolet light. III. Enhancement of suppressor cell activity in older animals, *Exp. Gerontol.* **13**:141.

Steinmetz, M., Moore, K. W., Frelinger, J. G., Sher, B. T., Shen, F. W., Boyse, E. A., and Hood, L., 1981, A pseudogene homologous to mouse transplantation antigens: Transplantation antigens are encoded by light exons that correlate with protein domains, *Cell* **25**:683.

Steinmetz, M., Winoto, A., Minard, K., and Hood, L., 1982, Clusters of genes encoding mouse transplantation antigens, *Cell* **28**:489.

Tabin, C. J., Bradley, S. M., Bargmann, C. I., Weinberg, R. A., Papageorge, A. G. Scolnick, E. M., Dhar, R., Lowly, D. R., and Chang, E. H., 1982, Mechanism of activation of a human oncogene, *Nature* **300**:143.

Travers, P. J., Arklie, J. L., Trowsdale, J., Patillo, R. A., and Bodmer, W. F., 1982, Lack of expression of HLA-ABC antigens in choriocarcinoma and other human tumor cell lines, *Natl. Cancer Inst. Monogr.* **60**:175.

Urban, J. L., and Schreiber, H., 1984, Rescue of the tumor-specific immune response of aged mice *in vitro*, *J. Immunol.* **133**:527.

Urban, J. L., Burton, R. C., Holland, J. M., Kripke, M. L., and Schreiber, H., 1982a, Mechanisms of syngeneic tumor rejection: Susceptibility of host-selected progressor variants to various immunological effector cells, *J. Exp. Med.* **155**:557.

Urban, J. L., Holland, J. M., Kripke, M. L., and Schreiber, H., 1982b, Immunoselection of tumor cell variants by mice suppressed with ultraviolet radiation, *J. Exp. Med.* **156**:1025.

Urban, J. L., Van Waes, C., and Schreiber, H., 1984, Pecking order among tumor-specific antigens, *Eur. J. Immunol.* **14**:18.

Uyttenhove, C., Maryanski, J., and Boon, T., 1983, Escape of mouse mastocytoma P815 after nearly complete rejection is due to antigen loss variants rather than immunosuppression, *J. Exp. Med.* **157**:1040.

Weinberg, R. A., 1982, Fewer and fewer oncogenes, *Cell*, **30**:3.

Woglom, W. H., 1929, Immunity to transplantable tumors, *Cancer Rev.* **4**:129.

Wortzel, R. D., Philipps, C., and Schreiber, H., 1983a, Multiplicity of unique tumor-specific antigens expressed on a single malignant cell, *Nature* **304**:165.

Wortzel, R. D., Philipps, C., Urban, J. L., Fitch, F. W., and Schreiber, H., 1983b, Independent immunodominant and immunorecessive tumor-specific determinants on a malignant tumor. Antigenic dissection with cytolytic T cell clones, *J. Immunol.* **130**:2461.

Wortzel, R. D., Urban, J. L., and Schreiber, H., 1984, Malignant growth in the normal host after variant selection *in vitro* with cytolytic T cell lines, *Proc. Natl. Acad. Sci. U.S.A.* **81**:2186.

Zelený, V., Matoušek, V., and Lengerová, A., 1978, Intercellular adhesiveness of H-2 identical and H-2 disparate cells, *J. Immunogenet.* **5**:41.

Zinkernagel, R. M., and Doherty, P. C., 1979, MHC-restricted cytotoxic cells: Studies on the biological role of the polymorphic major transplantation antigens determining T cell restriction specificity function and responsiveness, *Adv. Immunol.* **27**:51.

20

Mechanisms That Regulate Membrane Growth Factor Receptors

MICHAEL P. CZECH, ROGER J. DAVIS,
JEFFREY E. PESSIN, CRISTINA MOTTOLA, and
YOSHITOMO OKA

1. Introduction

Cell-surface receptors for hormones and other physiologically active agents play critically important roles in signaling and coordinating a multitude of cellular functions. The molecular mechanisms whereby receptor-mediated transmembrane signaling occurs have been intensively studied during the past two decades, and significant insight has been obtained about some of these mechanisms. For example, receptors linked to the generation of cAMP are known to interact with adenylate cyclase through the action of GTP binding (coupling) proteins, which have been purified to homogeneity (Gilman, 1984). On the other hand, many other receptor systems, notably the growth factor receptors, participate in signaling mechanisms that are as yet not understood. An important future objective is the elucidation of signaling mechanisms related to this latter category of receptor systems.

In this chapter, we consider an aspect of receptor biology that perhaps has received less attention than signaling mechanisms but is no less physiologically relevant—membrane receptors as targets of cellular signaling mechanisms initiated by other, distinct receptor systems. The concept that surface receptors are an important locus of regulatory control is consistent with general biochemical principles. Thus, metabolic pathways are regulated at enzyme steps located at initiation or branch points

MICHAEL P. CZECH, ROGER J. DAVIS, JEFFREY E. PESSIN, CRISTINA MOTTOLA, and YOSHITOMO OKA • Department of Biochemistry, University of Massachusetts Medical School, Worcester, Massachusetts 01605.

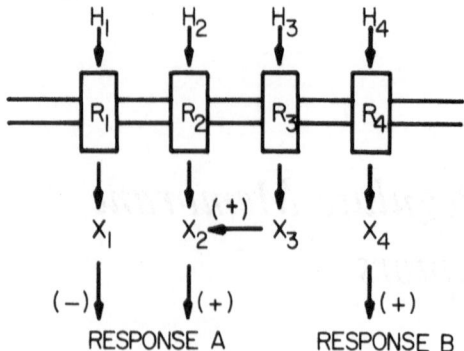

Figure 1. Schematic representation of several theoretical biological relationships among hormone receptor systems.

of the overall pathway. The fact that receptors initiate cellular signaling mechanisms and therefore may serve as sites of regulation is consistent with this concept. Receptor-mediated signaling mechanisms are also analogous to multistep metabolic pathways in that they probably involve many components and amplification steps. The conclusion from this line of reasoning is that we should expect that cell surface receptors serve as focal points for tight regulatory control.

There exist a number of biological relationships among cell-surface receptor systems that suggest the presence of mutual regulatory controls. Figure 1 depicts several relationships that may occur between two hormone receptor systems in respect to their biological responses. For example, receptor systems 1 and 2 modulate the same effect, response A, but in opposite directions. Thus, receptor 1 action inhibits the activity of target A, whereas signaling by receptor 2 stimulates this target activity. This antagonistic relationship is representative of several known hormonal interactions. Within this type of antagonistic relationship, one might predict that at least one contribution to the overall mechanisms that lead to the opposite responses is the action of one receptor to densensitize the signaling mechanism of the other. Thus, the signaling mechanism for receptor 1 might act on target A as well as on receptor 2 by inhibiting them both, perhaps by very similar mechanisms.

A second biological relationship depicted in Fig. 1 is that between receptor 2 and receptor 4. These two receptor systems hypothetically activate target A and target B, respectively. Thus, if these two target systems, A and B, are required to be highly coordinated in cell metabolism or in a cell process, it may be predicted that hormone signaling by receptor 2 leads to activation of the activity of receptor 4, or vice versa. Such receptor–receptor interaction would promote the coordinated responses of A and B. Still another example of biological linkages between receptor systems is the relationship exemplied by receptors 2 and 3 in Fig. 1.

I. STRUCTURAL MODULATION

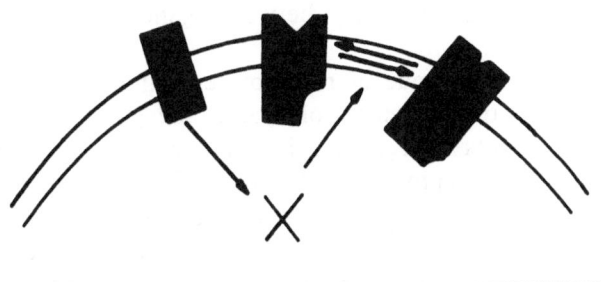

II. CHANGE IN CELLULAR LOCUS

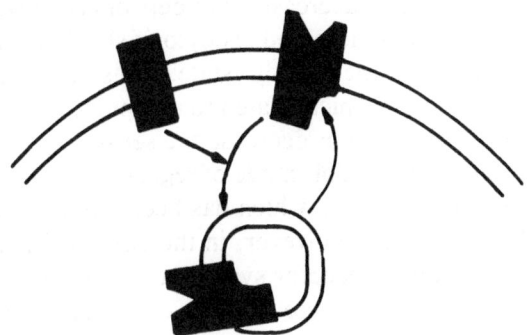

Figure 2. Two general mechanisms whereby a membrane receptor can be regulated by a second receptor system.

Receptor 3 has no action on target A or B through its own signaling mechanism but potentiates the action of receptor 2 to activate target A. Such potentiation of the action of receptor 2 could involve the direct modulation by receptor 3 of receptor 2 itself to enhance its signaling.

Prior to discussing specific examples of these types of receptor interactions, we should note that the cellular mechanisms involved in receptor regulation can be conveniently divided into two categories. Figure 2 illustrates in schematic form these two general classes of mechanisms. In the top panel is depicted a general mechanism whereby one receptor system modulates a second receptor system by direct alteration of one or more structural features of the target receptor. Thus, by covalent or noncovalent structural modifications, the target receptor responds by an alteration in its affinity for its hormone or in an alteration of some other activity on the receptor that leads to a change in its ability to participate in transmembrane signaling. This latter type of effect could involve changes in intrinsic enzyme activity associated with the target receptor or even an alteration in the linkage between ligand binding and receptor enzyme activation.

In contrast to this mode of receptor regulation, a second general possibility involves the movement of a target receptor from one cellular compartment to another. Because it is well appreciated that membrane components such as receptors have the ability to be internalized from the plasma membrane into intracellular compartments as well as to recycle back to the cell surface, both the endocytotic and exocytotic pathways are potential sites of regulation. As indicated by the bottom panel in Fig. 2, a target receptor might be modulated by a second receptor system with respect to its disposition within various cellular domains. Thus, increased internalization would lead to decreased numbers of cell-surface receptors in the steady state, whereas increased exocytosis should lead to increased cell-surface receptors in the steady state. This general mechanism whereby the membrane dynamics related to receptor movements are regulated can be seen to increase or decrease the sensitivity of cells to stimulation of the target receptor. This mode of regulation is not unlike ligand-induced receptor internalization, which has been described in many systems (Maxfield *et al.*, 1978). However, in the case in which a target receptor responds to another receptor system, there is greater diversity in ultimate potential effects, i.e., increased as well as decreased cell-surface receptor number.

The remaining sections of this chapter discuss specific biological systems in which target receptors are regulated by the above two general mechanisms. Studies in our laboratory have suggested that the insulin receptor is a target for structural modification by a cAMP-dependent mechanism that decreases its binding of [^{125}I]-insulin (Pessin *et al.*, 1983). In addition, insulin receptor tyrosine kinase activity appears to be uncoupled from insulin action in response to cAMP stimulation. In contrast, other studies we have recently performed suggest that the type II insulinlike growth factor (IGF) receptor is a target for insulin action such that its concentration in the plasma membrane is markedly stimulated (Oppenheimer *et al.*, 1983; Oka *et al.*, 1984). In addition, recent data we have obtained on a third target receptor system, the epidermal growth factor (EGF) receptor, suggest that it may undergo both general modes of regulation described in Fig. 2 in response to treatment of cells with tumor-promoting phorbol diesters (Davis and Czech, 1984). We now consider the experimental details that characterize each of these three target receptor systems.

2. Isoproterenol Action on the Insulin Receptor Kinase

The actions of insulin receptor and β-adrenergic receptors exemplify a classic antagonistic biological relationship (Czech, 1977). The β-adrenergic response, mediated by elevated cAMP levels, leads to a multitude of changes in enzyme activities, including activations of phosphorylase, hormone-sensitive lipase, and fructose 1,6-phosphatase and deactivations of glycogen synthase and phosphofructokinase. Insulin antagonizes all of these actions. Insulin itself promotes glucose and amino acid transport, systems that are little effected by β-catecholamines alone. On the other hand, β-adrenergic stimulation antagonizes insulin action on glucose transport (Pessin *et al.*, 1983; Kirsch *et al.*, 1983). The antagonism of β-catecholamine action by insulin is mediated in part by an effect of the latter hormone to reduce the β-adrenergically stimulated cAMP levels (Czech, 1977). However, it is generally thought that insulin probably antagonizes β-catecholamine action by additional mechanisms as well (Czech, 1977).

In light of the above biological facts, we tested the possibility that one contribution to this antagonistic relationship between insulin and β-catecholamines might be an effect of the latter hormones to desensitize directly insulin receptor activity. Isolated fat cells were treated with or without isoproterenol prior to addition of [^{125}I]-insulin and measurement of its binding to insulin receptors. In these experiments, [^{125}I]-EGF binding to fat cells was also monitored. Table I shows a dramatic inhibition of both [^{125}I]-insulin and [^{125}I]-EGF binding to the β-catecholamine-treated fat cells. Of particular interest was the discovery that all agents known to elevate cAMP levels produce the same inhibition of insulin receptor and EGF receptor activity (Pessin *et al.*, 1983). Such agents include phosphodiesterase inhibitors (e.g., 3-isobutyl-1-methylxanthine) as well as the cAMP analogue dibutryl cAMP (Table I). The effect of isoproterenol was blocked by propranolol but not by phentolamine, indicating that the effect resulted from β-adrenergic stimulation. α-Catecholamines such as phenylephrine were ineffective. Recently, similar evidence that a cAMP-mediated mechanism leads to decreased [^{125}I]-insulin binding to isolated adipocytes has been reported (Kirsch *et al.*, 1983; Lonnorth and Smith, 1983).

We recently completed four experimental approaches designed to probe the question of whether the decreased [^{125}I]-insulin binding noted above in response to isoproterenol results from decreased numbers of insulin receptors or decreased affinity of the same number of receptors (Pessin *et al.*, 1984). All four experimental protocols have led to results

Table I. The Effect of Various Agents on the Binding of Insulin and EGF to the Rat Adipocyte[a]

| | Percent inhibition of binding | |
Addition	Insulin	EGF
None	0	0
Isoproterenol (1 μM)	48 ± 15	46 ± 9
Norepinephrine (1 μM)	51 ± 17	48 ± 4
Epinephrine (1 μM)	47 ± 17	44 ± 2
Soterenol (1 μM)	53 ± 15	52 ± 10
Propranolol (10 μM)	0 ± 6	12 ± 5
Propranolol (10 μM) + isoproterenol (1 μM)	16 ± 12	16 ± 2
Alprenolol (10 μM)	0 ± 6	3 ± 3
Alprenolol (10 μM) + isoproterenol (1 μM)	2 ± 3	5 ± 4
Phenylephrine (10 μM)	6 ± 3	1 ± 1
Phentolamine (50 μM)	3 ± 3	7 ± 12
Phentolamine (50 μM) + isoproterenol (1 μM)	50 ± 12	54 ± 8
3-Isobutyl-1-methylxanthine (200 μM)	66 ± 3	66 ± 5
3-Isobutyl-1-methylxanthine (200 μM) + isoproterenol (1 μM)	72 ± 6	70 ± 6
Adenosine (1 mM)	0 ± 15	5 ± 7
Adenosine (1 mM) + isoproterenol (1 μM)	10 ± 5	15 ± 16
Bt₂cAMP (2 mM)	44 ± 8	25 ± 6
Bt₂cGMP (2 mM)	10 ± 1	9 ± 9
Sodium butyrate (1 mM)	1 ± 14	2 ± 12
Sodium oleate (1 mM)	0 ± 10	6 ± 6

[a] Isolated rat adipocytes were pretreated with various concentrations of agents for 30 min at 37°C. The percentage inhibition of binding at 1.0 nM [^{125}I]-EGF was determined. These results represent the average of two to four independent detreminations with their respective ranges of values. (Reprinted with permission from Pessin *et al.*, 1983.)

that support the conclusion that a decreased insulin receptor affinity is the cause of the β-adrenergic effect. Scatchard analysis of [^{125}I]-insulin binding to control and isoproterenol-treated cells showed markedly decreased binding at low [^{125}I]-insulin concentrations caused by the catecholamine, but no such difference was observed at high [^{125}I]-insulin concentrations. Acid extraction of fat cells treated with nothing or isoproterenol prior to addition of [^{125}I]-insulin indicated that the same relative amount of [^{125}I]-insulin was bound to cell surface versus intracellular membranes under both conditions. Similarly, when control and isoproterenol-treated fat cells were homogenized and fractionated into plasma membrane and low-density microsome membranes, plasma membranes derived from isoproterenol-treated cells exhibited a marked decrease in [^{125}I]-insulin binding. Low-density microsome fractions exhibited no difference in [^{125}I]-insulin binding when derived from control versus β-catecholamine-treated cells, indicating that the effect of β-adrenergic stimulation was localized to the plasma-membrane domain. Finally, anti-insulin-receptor Ig was employed to estimate directly the amount of insulin receptors on the intact fat cell surface. Under conditions in which β-catecholamine inhibited [^{125}I]-insulin binding to intact fat cells by 75%, no change was observed in the specific binding of anti-insulin-receptor Ig. These experiments strongly support the concept that one or more structural alterations in the insulin receptor led to a decreased affinity for hormone in response to elevated cAMP levels.

The above conclusion that insulin receptor structure is modified by β-adrenergic stimulation suggested the possibility that other receptor functions may be disrupted. It is now well established that a tyrosine kinase activity is associated with purified insulin receptor (Kasuga *et al.*, 1983). Insulin binding to the receptor complex stimulates the tyrosine kinase activity, which leads to autophosphorylation of the insulin receptor β subunit and also to the phosphorylation of exogeneous protein substrates (Kasuga *et al.*, 1983). In intact [^{32}P]-labeled cells, insulin binding to its receptor increases incorporation of ^{32}P into serine and threonine as well as tyrosine residues in the β subunit (Kahn *et al.*, 1985). Interestingly, Rosen and colleagues (1983) presented results indicating that phosphorylation of the insulin receptor β subunit leads to activation of the associated tyrosine kinase activity. Furthermore, results from that laboratory showed a dissociation of the insulin requirement for maintaining the receptor-associated tyrosine kinase activity in the phosphorylated state (Rosen *et al.*, 1983).

Studies in our laboratory extended these findings by demonstrating that [^{32}P]-phosphotyrosine content in a specific tryptic HPLC receptor phosphopeptide fraction correlates with kinase activation (Yu and Czech, 1984). Figure 3 depicts in schematic form the conclusions from the above studies. The concept developed in this figure that a specific phosphotyrosine (or phosphotyrosines) is involved in activation of insulin receptor kinase is supported by experiments in our laboratory showing specific dephosphorylation of insulin receptor phosphotyrosines and parallel deactivation of receptor kinase using alkaline phosphatase with purified phosphorylated insulin receptor (Yu and Czech, 1984). About 90% of the total phosphotyrosine dephosphorylated by alkaline phosphatase under these conditions is associated with receptor phosphopeptide fraction 2 on tryptic hydrolysis and reverse-phase HPLC. These data support the view that phosphorylation of one or more specific tyrosine residues on the receptor β subunit activates insulin receptor kinase and converts it to an insulin-dependent enzyme (Fig. 3).

With the above information about the regulation of insulin receptor

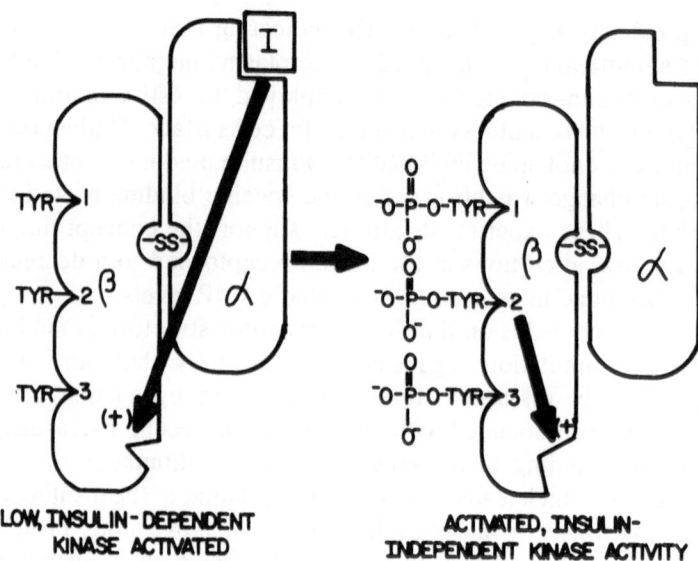

LOW, INSULIN-DEPENDENT
KINASE ACTIVATED

ACTIVATED, INSULIN-
INDEPENDENT KINASE ACTIVITY

Figure 3. Schematic representation of insulin receptor tyrosine kinase activation by autophosphorylation. For simplicity, only one α and one β subunit of the insulin receptor complex
are depicted. See text for explanation.

tyrosine kinase activity as a framework, we set out to test the hypothesis
that insulin receptor kinase activity may be a target of isoproterenol action
(Pessin *et al.*, 1984). The fact that the isoproterenol-induced affinity
change of the insulin receptor survives the preparation of plasma membranes suggested that possible effects on the tyrosine kinase activity might
also be stable to receptor purification. Isolated fat cells were therefore
treated with no addition or isoproterenol and homogenized for the preparation of plasma membranes. The plasma membranes were solubilized
in detergent, and insulin receptors were partially purified by adsorption
and specific elution from wheat germ agglutinin-Sepharose. Receptors
purified in this manner from control and isoproterenol-treated cells were
then incubated with $[\gamma\text{-}^{32}P]$-ATP in the presence and absence of a saturating concentration of insulin (200 nM). Under these conditions, the receptors from both control and isoproterenol-treated cells exhibit similar
binding because the high concentration of insulin overcomes the affinity
defect resulting from isoproterenol treatment. In order to visualize autophosphorylation of insulin receptors under these conditions, the receptors were immunoprecipitated with anti insulin-receptor Ig and electrophoresed on dodecylsulfate gels. The β subunit autophosphorylation was
used as an estimate of insulin receptor kinase activity.

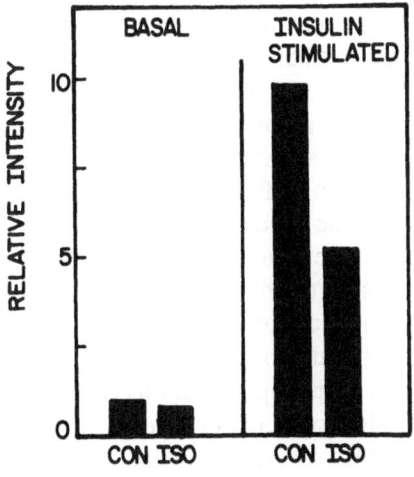

Figure 4. Uncoupling of insulin receptor tyrosine kinase activity following treatment of intact cells with isoproterenol. Crude plasma membranes prepared from control and isoproterenol-treated adipocytes were solubilized in 1.0% Triton X-100, and insulin receptors were partially purified by adsorption and elution from immobilized wheat germ agglutinin. The insulin receptors were first incubated for 30 min at 20°C with (right panel) or without (left panel) 200 nM insulin and then incubated for 2 min at 4°C in the presence of $[\gamma\text{-}^{32}\text{P}]$-ATP (5 μM, 50 μCi/ nmole). The insulin receptors were then immunoprecipitated using anti-insulin-receptor antibody (provided by Dr. C. R. Kahn), electrophoresed on a 7% SDS-polyacrylamide gel, and subjected to autoradiography. The relative amount of insulin receptor β subunit phosphorylation was determined from densitometric scans of the autoradiographs. CON, control; ISO, isoproterenol.

The results of the experiments described above are depicted in Fig. 4 and demonstrate a striking effect of isoproterenol on the insulin receptor tyrosine kinase. When no insulin was added to insulin receptors *in vitro*, autophosphorylation was similar for the partially purified insulin receptors derived from control and β-adrenergically stimulated cells. In contrast, on addition of a saturating concentration of insulin, insulin receptor autophosphorylation *in vitro* was significantly reduced in the receptor preparation derived from isoproterenol-treated cells (right panel, Fig. 4). Because the number of receptors is apparently similar between control and isoproterenol-treated cells in the immobilized wheat germ agglutinin eluates, these results strongly suggest that isoproterenol action mediates an uncoupling of the insulin-stimulated receptor tyrosine kinase activity. Similar experiments were performed in intact $[^{32}\text{P}]$-labeled adipocytes incubated with no addition, isoproterenol alone, insulin alone, or the combination of hormones. In these experiments, insulin receptors were immunoprecipitated, and the radioactivity incorporated into β subunit was determined by gel electrophoresis and autoradiography. The results demonstrated that isoproterenol blocks the increased insulin receptor phosphorylation caused by addition of insulin to intact adipocytes (Pessin *et al.*, 1984).

What is the mechanism whereby isoproterenol action modulates insulin receptors such that affinity for hormone and activation of tyrosine kinase activity are markedly inhibited? Because the results in Table I strongly suggest that cAMP mediates these effects of isoproterenol, it is reasonable to hypothesize that cAMP-dependent protein kinase is involved. This line of reasoning implies that serine phosphorylation on the insulin receptor itself or on a regulatory component is involved in these processes. We have not been able to detect in [^{32}P]-labeled fat cells significant direct effects of isoproterenol on the incorporation of ^{32}P into insulin receptor in the basal state. These experiments are extremely difficult technically, however, and the possibility that insulin receptor is phosphorylated in response to isoproterenol cannot yet be ruled out. Although an answer to this question remains unavailable at present, we suggest the hypothesis that one or more serine phosphorylations catalyzed directly or indirectly by the cAMP-dependent protein kinase regulates the insulin receptor such that its tyrosine kinase activity is uncoupled from insulin action.

Taken together, the data described in this section suggest the presence of a network of tyrosine and serine phosphorylations that tightly control insulin receptor activity. In concert with these data, we propose the hypothesis that tyrosine phosphorylations lead to activation of insulin receptor kinase activity whereas serine phosphorylations lead to deactivation of insulin-stimulated kinase. Because the insulin receptor tyrosine kinase activity is an excellent candidate for participating in transmembrane signaling of the insulin receptor, these hypothetical antagonisms between tyrosine and serine phosphorylations may well be critical regulatory components of the insulin signaling mechanism. This postulate would be consistent with the observations that insulin action is densensitized in isoproterenol-treated cells. This hypothesis also suggests that phosphotyrosine and phosphoserine phosphatases may play important roles in regulating insulin receptor activity. Further work is required to test these concepts rigorously in intact cells.

3. Insulin Stimulates IGF-II Receptor Recycling

The insulinlike growth factor (IGF-II) receptor structure was originally identified in our laboratory using affinity labeling techniques (Massague et al., 1981; Massague and Czech, 1982) and was found to consist of a 250,000-dalton protein component (for reviews, see Czech, 1982; Czech et al., 1984). This receptor structure binds IGF-II or multiple stimulating activity (MSA) with high affinity and binds IGF-I with lower affinity. This receptor structure has no affinity for insulin. This receptor

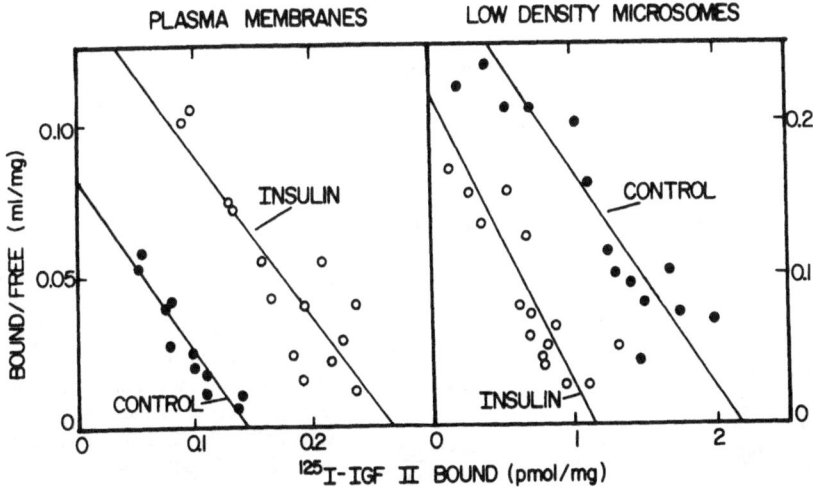

Figure 5. Scatchard analysis of [^{125}I]-IGF-II binding to isolated membrane fractions. Adipocyte plasma membranes and low-density microsome fractions were prepared and incubated with [^{125}I]-IGF-II for 90 min at 10°C. The binding was analyzed by the method of Scatchard. Left: Plasma membranes prepared from control (●) and insulin-treated (○) adipocytes. Right: Low-density microsomes from control (●) and insulin-treated (○) cells. The experiment was performed five times for plasma membrane and three times for low-density microsomes. (Reprinted with permission from Oppenheimer *et al.*, 1983).

structure presumably plays an important role in IGF action or other function, but that role has not been fully defined as yet. This receptor structure has been purified to homogeneity (Oppenheimer and Czech, 1983; August *et al.*, 1983), and potent rabbit antisera against this receptor have been prepared in our laboratory (Oka *et al.*, 1984).

Studies performed by Zapf *et al.* (1978) first suggested that the IGF-II receptor was a target of insulin action. These workers presented evidence that [^{125}I]-IGF-II binding increased in intact adipocytes in response to the addition of physiological concentrations of insulin. Subsequent studies confirmed this result (King *et al.*, 1982; Oppenheimer *et al.*, 1983). Interestingly, both subsequent studies presented data indicating that the modulation of IGF-II receptors by insulin represented an increase in affinity for [^{125}I]-IGF-II. Our suspicion that the mechanism of this effect might be more complicated was prompted by our results shown in Fig. 5. In these studies, intact adipocytes were incubated with or without insulin and then homogenized and fractionated into a plasma membrane and low-density microsome membrane fraction. Binding of [^{125}I]-IGF-II was then performed on the membrane fractions. As shown in Fig. 5, insulin caused an increase in the apparent number of IGF-II receptors in

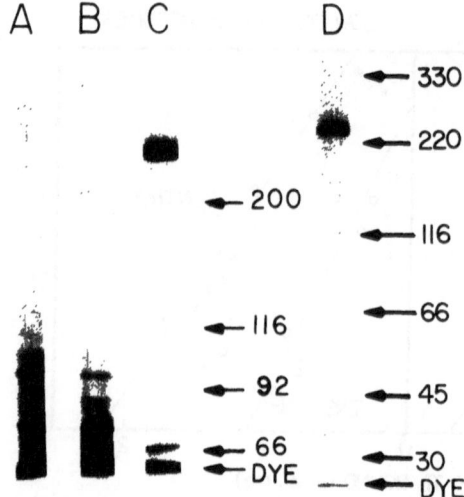

Figure 6. Sodium dodecylsulfate electrophoresis of the IGF-II receptor after purification. Samples were boiled in the presence of 1% sodium dodecylsulfate and 50 mM dithiothreitol and then electrophoresed on 5% polyacrylamide gels by the method of Laemmli and silver stained. Lanes A–C: 5% polyacrylamide gel. Lane A, rat placenta membranes (8 μg); lane B, Triton X-100 extract (9 μg); lane C, purified IGF-II receptor (0.9 μg). Arrows show the location of molecular weight standards: bovine serum albumin (66,200), phosphorylase B (92,500), β-galactosidase (116,250), and myosin (200,000). Lane D: 5% to 12% gradient polyacrylamide gel, purified IGF-II receptor (0.9 μg). Molecular weight standards: carbonic anhydrase (30,000), ovalbumin (45,000), bovine serum albumin (66,200), β-galactosidase (116,250), ferritin half-unit (220,000), and thryoglobulin (330,000). (Reprinted with permission from Oppenheimer and Czech, 1983.)

the plasma membrane fraction and a concomitant decrease in the apparent number of IGF-II receptors present in the low-density microsome fraction. No difference in apparent affinity was observed between IGF-II receptors in membranes derived from control and insulin-treated cells. These data are reminiscent of the findings reported for insulin action on hexose transporters, which also were found to redistribute among adipocyte membrane fractions in a similar manner (Cushman and Wardzala, 1980; Suzuki and Kono, 1980). Our data illustrated in Fig. 5 thus suggested the possibility that insulin action might initiate a mechanism that leads to increased movement of IGF-II receptors from an intracellular, nonexposed domain to the cell surface.

In order to test directly the hypotheses that insulin either increases IGF-II receptor affinity or increases IGF-II receptor numbers on the cell surface, we devised a strategy that allows estimation of IGF-II receptor numbers on the surface of intact adipocytes by a means that is independent of [^{125}I]-IGF-II binding (Oka et al., 1984). The reagent employed is specific anti-IGF-II-receptor Ig prepared in rabbits against purified IGF-II receptor. Figure 6 shows the dodecylsulfate gel record of our IGF-II receptor purification procedure. The gel is silver stained for visualization of proteins. Rat placenta membranes were solublized in Triton X-100, and the extracts passed over an immobilized rat IGF-II column for specific affinity purification of IGF-II receptor. Elution of receptor from the column is achieved by an acid wash. As is depicted in Fig. 6, purified IGF-II receptor migrates as a single band at approximately 250,000 daltons. This procedure results in an approximate 1100-fold purification and leads to a preparation that is near homogeneity (Oppenheimer and Czech, 1983). Antisera raised in rabbits injected with this receptor preparation are highly potent in binding to and immunoprecipitating IGF-II receptors in crude extracts of membranes. At 5000-fold dilution, it shows a positive reaction using the enzyme-linked immunosorbent assay (ELISA). Interestingly, these antisera as well as the Ig fraction prepared from the antisera are effective in inhibiting [^{125}I]-IGF-II binding to IGF-II receptors in intact cells as well as membrane preparations (Oka *et al.*, 1984).

The Ig fraction from anti-IGF-II-receptor antiserum was used to estimate IGF-II receptors on intact cell surfaces. Fat cells treated with no addition or with insulin were incubated with the anti-IGF-II-receptor Ig and washed. The cells were then incubated with [^{125}I]-goat-antirabbit IgG in order to assess the amount of anti-IGF-II-receptor Ig bound. Figure 7 depicts the results from such an experiment. As previously described, insulin action increases the binding of [^{125}I]-IGF-II to intact adipocytes (left panel of Fig. 7). Sigificantly, under the same experimental conditions, the binding of anti-IGF-II-receptor Ig is also stimulated by the addition of insulin to intact adipocytes. The data support our conclusion that insulin action leads to increased IGF-II receptors on the fat cell surface.

An insulin-induced exposure of IGF-II receptors in intact adipocytes could reflect a variety of possible cellular mechanisms. We believe that this phenomenon reflects a mechanism whereby the IGF-II receptor continuously recycles from plasma membrane to an internal domain or domains and then back to the plasma membrane. According to this hypothesis, insulin action could inhibit the internalization pathway or stimulate the movement of receptors to the cell surface. This proposed concept whereby IGF-II receptors undergo endocytotic and exocytotic movements is depicted in Fig. 8 in simplified form. Recent results (unpublished) from our laboratory suggesting various aspects of this hypothesis include the following. (1) Cell-surface IGF-II receptors affinity labeled at low temperature redistribute within minutes to the low-density microsome fraction on warming of the cells to 37°C. (2) Insulin action may not effect the initial rates of this internalization process. Thus, as depicted in Fig. 8, insulin action might act on the exocytotic aspect of the overall recycling pathway. (3) Under conditions in which rapid rates of internalization are proceeding, no change in the numbers

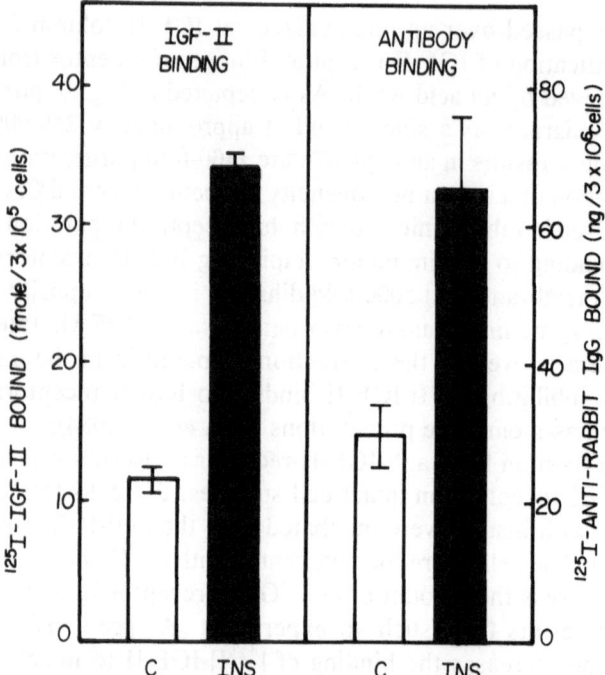

Figure 7. The effect of insulin on [^{125}I]-IGF-II binding (left) and antireceptor Ig binding (right) to intact adipocytes. Isolated adipocytes were incubated with or without 10 nM insulin for 15 min at 37°C. For measurement of [^{125}I]-IGF-II binding (left panel), cells were then incubated with 1 nM [^{125}I]-IGF-II for 30 min at 21°C, and bound [^{125}I]-IGF-II was measured by an oil floatation method. For measurement of antireceptor Ig binding (right panel), control and insulin-treated cells (3×10^6) were incubated with 5 μg/ml of antireceptor Ig for 30 min at 21°C. Cells were then washed and further incubated with 25 μg/ml of [^{125}I]-goat antirabbit IgG for 30 min at 21°C, and bound goat IgG was measured. The values presented are the mean ± standard deviations obtained in three separate experiments. C, control; INS, insulin. (Reprinted with permission from Oka *et al.*, 1984.)

of cell surface receptors is apparent as assessed by anti-IGF-II-receptor Ig binding to intact cells, thus demonstrating the recycling phenomenon.

Taken together, these data strongly suggest that two consequences result from insulin action on the IGF-II receptor. First, increased receptors are expressed on the cell surface at any given moment; secondly, the overall recycling rate of the IGF-II receptor is increased. The underlying physiological significance of these effects is not understood at present, but the data demonstrate that the IGF-II receptor is a striking target membrane component for the action of insulin.

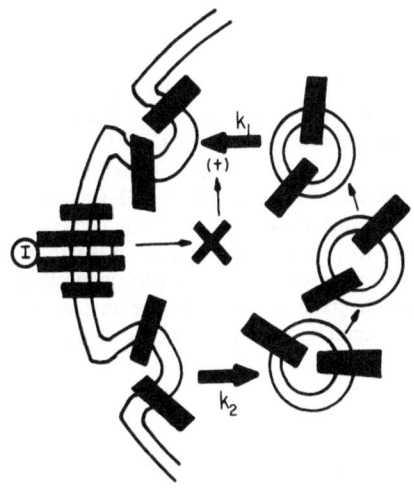

Figure 8. Schematic representation of a simplified hypothesis for the mechanism of insulin action on IGF-II binding to intact cells. Insulin may activate an exocytotic mechanism that increases the movement of IGF-II-receptor-rich vesicles to the cell surface. Other mechanisms are also possible. See text for details.

4. The Epidermal Growth Factor Receptor as a Target for Phorbol Diester Action

A number of reports have demonstrated that binding of $[^{125}I]$-EGF to its receptors in intact cells is rapidly and markedly inhibited by the addition of tumor-promoting phorbol diesters (Davis and Czech, 1984; Brown *et al.*, 1979; Lee and Weinstein, 1979; Shoyab *et al.*, 1979). Phorbol diesters are potent agents first recognized for the ability to promote skin tumors in mice that had been previously treated with low doses of carcinogens (Boutwell, 1974). At the cellular level, the phorbol esters exert a myriad of effects on cell growth and metabolism (for reviews, see Blumberg, 1980, 1981). Pertinent to the present discussion, phorbol esters have interesting effects on mitogenesis. Although these agents mediate a stimulation of $[^{3}H]$-thymidine incorporation into DNA and cell proliferation, perhaps more interesting is their ability to greatly potentiate the actions of epidermal growth factor on cell proliferation. This ability to potentiate the action of EGF on cell proliferation suggests the possibility that phorbol ester action may directly modulate the signaling pathway for EGF receptor.

Several laboratories have reported that the EGF receptor itself is a target of phorbol ester action (Davis and Czech, 1984; Cochet *et al.*, 1984; Iwashita and Fox, 1984). The tumor-promoting agents act rapidly to decrease $[^{125}I]$-EGF binding markedly in intact cells. Which of the two general mechanisms depicted in Fig. 2 applies to this EGF receptor regulation is unknown, although Scatchard analysis of $[^{125}I]$-EGF binding suggests

10 nM EGF   CONTROL

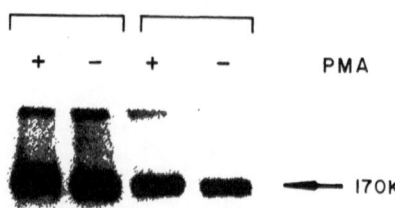

Figure 9. The effect of PMA on EGF receptor phosphorylation in intact A431 cells. A431 cells were incubated with 1 mCi/ml [^{32}P]-phosphate at 37°C. After 3 hr, 10 nM PMA was added, and after 3.5 hr, 10 nM EGF was added to some incubations. At 4 hr, the cells were harvested, and the EGF receptor was purified by affinity chromatography. The purified receptor was resolved from the phosphoproteins on a 7% polyacrylamide gel. The figure is an autoradiograph of the Coomassie-blue-stained dried gel using Kodak X-OMAT AR film with a Dupont Cronex lightning plus enhancing screen at −70°C.

that an affinity change is caused by the addition phorbol esters (Davis and Czech, 1984; Brown *et al.*, 1979; Lee and Weinstein, 1979; Shoyab *et al.*, 1979). However, there are also data suggesting that the EGF receptor redistributes to a trypsin-insensitive cellular compartment in response to phorbol esters (King and Cuatrecasas, 1982). This issue is an important one and requires further work for its resolution.

Whether the effect of phorbol esters on EGF receptors reflects a receptor affinity change or a decrease in cell surface receptor numbers, the molecular basis of this effect appears likely to involve a calcium- and phospholipid-activated protein kinase. This hypothesis is based on results demonstrating copurification of phorbol ester receptors and the C kinase (Niedel *et al.*, 1983; Aschendel *et al.*, 1983). Furthermore, phorbol diesters have been shown to activate C kinase, converting it to a form that is much less dependent on calcium (Castagna *et al.*, 1982; Kikkawa *et al.*, 1983). This protein kinase may play a central role in cellular regulation in view of its ability to respond to diacylglycerol, which is under hormonal control (for review, see Nishizuka, 1984). Taken together, these considerations suggested the hypothesis that phosphorylation of EGF receptors by a mechanism activated by phorbol esters might be directly involved in mediating the [^{125}I]-EGF binding inhibition by phorbol esters.

The results in Fig. 9 show that in intact cells EGF receptors are indeed the target of a phosphorylation reaction mediated by phorbol esters. A431 cells were incubated with ^{32}P for 3 hr followed by addition of EGF, 4β-phorbol-12β-myristate 13α-acetate (PMA), or EGF plus PMA. The EGF receptors were then purified from cell extracts using immobilized EGF and then electrophoresed on dodecylsulfate gels. Figure 9 depicts the autoradiograph from such an experiment: EGF stimulates incorporation of ^{32}P into EGF receptor, as has been previously reported. Addition of phorbol ester also leads to EGF receptor phosphorylation. Addition of

EGF to phorbol-diester-treated cells leads to phosphorylation of EGF receptor to an extent that is greater than either agent alone but to a level that is not fully additive.

Analysis of the phosphoamino acids comprising the [32P]-labeled EGF receptor derived from these incubations revealed significantly increased [32P]-phosphoserine content of EGF receptor derived from PMA-treated cells. Smaller amounts of phosphothreonine were observed, and PMA treatment increased this as well. Significantly, EGF receptor phosphorylation on serine and threonine residues in response to phorbol diesters parallels, with respect to time course and phorbol diester analogue potency, the effects on [125I]-EGF binding to receptors. These data suggest that phosphorylation may play a role in EGF receptor regulation with respect to either its affinity for [125I]-EGF or its internalization into the intracellular compartment.

Very recent results in our laboratory have shown that most of the PMA-induced phosphorylation of EGF receptor on phosphoserine is localized on a major phosphoserine-containing phosphopeptide following trypsin hydrolysis of the EGF receptor (Davis and Czech, 1985). Recent reports from other laboratories have indicated that unique phosphothreonine-containing receptor peptide fragments are phosphorylated in response to phorbol ester treatment of cells as well (Cochet *et al.*, 1984; Iwashita and Fox, 1984). Comparison of those results and our own indicate that the [32P]-phosphothreonine sites contain much smaller total amounts of 32P compared to the [32P]-phosphoserine sites of the EGF receptor. However, it is not yet clear which site or sites, if any, may be most important for EGF receptor regulation. A significant finding of our studies and those of other laboratories is that incorporation of 32P into EGF phosphotyrosine in response to EGF is markedly diminished in PMA-treated cells. The decreased autophosphorylation of tyrosine residues in response to PMA action is paralleled by decreased EGF-mediated [32P]-phosphotyrosine content in total cell constituents.

The results noted above are reminiscent of data presented in Section 3 indicating that isoproterenol action, presumably mediated through a serine phosphorylation site or sites, diminishes insulin receptor tyrosine kinase activity. These data also suggest that the EGF receptor may be modulated by phorbol diesters in a biomodal manner in which both general mechanisms illustrated in Fig. 2 are employed. That is, both EGF receptor internalization and inhibition of EGF receptor tyrosine kinase activity may result from PMA action. This concept is illustrated in Fig. 10. This working hypothesis suggests that phorbol diesters bind to and activate protein kinase C, which then either directly or indirectly catalyzes phosphorylation of EGF receptor on serine (and threonine, not shown) resi-

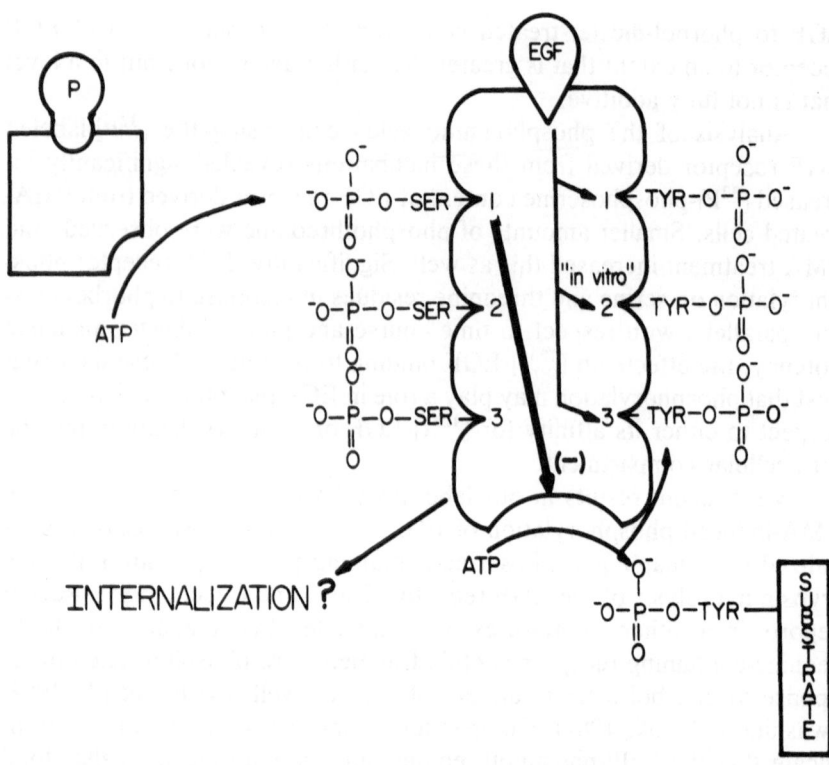

Figure 10. Schematic representation of the hypothesis that phorbol ester action modulates EGF receptor in intact cells by mediating one or more serine phosphorylations. This working hypothesis suggests that phorbol ester binding to kinase C in intact cells directly or indirectly activates serine (and threonine, not shown) phosphorylation of EGF receptor at a minimum of three sites. Such serine (or threonine) phosphorylation of EGF receptor is hypothesized to deactivate receptor tyrosine kinase or receptor internalization or both.

dues. Direct *in vitro* phosphorylation of EGF receptor by C kinase has indeed been demonstrated (Cochet *et al.*, 1984). We further hypothesize that a specific serine phospohorylation site or sites modulate EGF receptor structure such that tyrosine kinase activity is inhibited or uncoupled from activation by EGF. We cannot at present exclude the possibility that phosphothreonine sites may be involved in this hypothetical process. A second hypothetical process that may be initiated by EGF receptor phosphorylation is its internalization, leading to decreased cell-surface EGF receptors. Although much further work is required to test these possible interpretations, the postulates summarized in Fig. 10 do provide a framework for future studies.

5. Conclusions

We believe that a major conclusion derived from the experiments described in this chapter on the insulin, IGF-II, and EGF receptors is that regulation of a particular receptor system by the signaling mechanism of a second, distinct receptor system is an important and often used event in cell biology. Such receptor regulation most likely has many specific physiological roles in coordinating overall cellular functions. These specific roles may include desensitization of receptor function, potentiation of receptor action, modulation of receptor degradation, and others. Clearly, understanding the detailed mechanisms of receptor regulation will provide important insights into key regulatory pathways in cells.

A major aim of this chapter is to provide a summary of the methods that have been successfully utilized to date in investigating mechanisms of receptor regulation. It is apparent that we now have methods that can rigorously document whether specific examples of receptor modulation fall into one or the other of two general mechanistic categories, described in Fig. 2. As we learn more about the precise signaling mechanisms for the growth factor receptors, we ought to be able to more precisely define the mechanisms whereby such signaling is modified. It is also clear that we are now in a position to test directly whether receptor phosphorylation, mediated by a second receptor system, provides a specific signal for desensitizing receptor enzyme activity (e.g., tyrosine kinase) or possibly the initiation of an internalization pathway. That tyrosine phosphorylations and serine phosphorylations compete antagonistically for regulating receptor activities is another important hypothesis that has evolved from our recent studies. This topic should provide a fertile ground for future experimental thrusts.

References

Ascendel, C. L., Staller, J. M., and Boutwell, R. K., 1983, Protein kinase activity associated with a phorbol ester receptor purified from mouse brain, *Cancer Res.* **43**:4333.

August, G. P., Nissley, S. P., Kasuga, M., Lee, L., Greenstein, L., and Rechler, M. M., 1983, Purification of an insulin-like growth factor II receptor from rat chondrosarcoma cells, *J. Biol. Chem.* **258**:9033.

Blumberg, P. M., 1980, *In vitro* studies on the mode of action of the phorbol esters, potent tumor promoters: Part 1, *Crit. Rev. Toxicol.* **8**:153.

Blumberg, P. M., 1981, *In vitro* studies on the mode of action of the phorbol esters, potent tumor promotors: Part 2, *Crit. Rev. Toxicol.* **8**:199.

Boutwell, R. K., 1974, The function and mechanism of promoters of carcinogenesis, *Crit. Rev. Toxicol.* **2**:419.

Brown, K. D., Dicker, P., and Rozengurt, E., 1979, Inhibition of epidermal growth factor binding to surface receptors by tumor promoters, *Biochem. Biophys. Res. Commun.* **86:**1037.

Castagna, M., Takai, Y.,, Kaibuchi, K., Sano, K., Kikkawa, U., and Nishizuka, Y., 1982, Direct activation of calcium-activated, phospholopid-dependent protein kinase by tumor promoting phorbol esters, *J. Biol. Chem.* **257:**7847.

Cochet, C., Gill, G., Meisenhelder, J., Cooper, J., and Hunter, T., 1984, C-kinase phosphorylates the epidermal growth factor receptor and reduces its epidermal growth factor-stimulated tyrosine protein kinase activity, *J. Biol. Chem.* **259:**2553.

Cushman, S. W., and Wardzala, L. J., 1980, Potential mechanism of insulin action on glucose transport in the isolated rat adipose cell, *J. Biol. Chem.* **255:**4758.

Czech, M. P., 1977, Molecular basis of insulin action, *Ann. Rev. Biochem.* **46:**359.

Czech, M. P., 1982, Structural and functional homologies in the receptors for insulin and the insulin-like growth factors, *Cell* **31:**8.

Czech, M. P., Mottola, C., Yu, K.-T., and Oka, Y., 1984, The insulin-like growth factor receptors, in: *Human Growth Hormone* (S. Raiti, ed.), Plenum Press, New York (in press).

Davis, R. J., and Czech, M. P., 1984, Tumor-promoting phorbol diesters mediate phosphorylation of the epidermal growth factor receptor. *J. Biol.Chem.* **259:**8545–8549.

Davis, R. J., and Czech, M. P., 1985, Tumor-promoting phorbol diesters cause the phosphorylation of epidermal growth factor receptors in normal human fibroblasts at threonine-654, *Proc. Natl. Acad. Sci. U.S.A.*, **82:**1974–1978.

Gilman, A. G., 1984, G proteins and dual control of adenylate cyclase, *Cell* **37:**577.

Iwashita, S., and Fox, C. F., 1984, Epidermal growth factor and potent phorbol ester tumor promoters induce epidermal growth factor receptor phosphorylation in a similar but distinctly different manner in human epidermal carcinoma A431 cells, *J. Biol. Chem.* **259:**2559.

Kahn, C. R., White, M. F., Grigorescu, F., Takayama, S., Haring, H. U., and Crettaz, M., 1985, The insulin receptor protein kinase, in: *The Molecular Basis of Insulin Action* (M. P. Czech, ed.), Plenum Press, New York, pp. 67–93.

Kakkawa, U., Takai, Y., Tanaka, Y., Miyake, R., and Nishizuka, Y., 1983, Protein kinase C as a possible receptor protein of tumor promoting phorbol esters, *J. Biol. Chem.* **258:**11442.

Kasuga, M., Fujita-Yamaguchi, Y., Blith, D. L., and Kahn, C. R., 1983, Tyrosine-specific protein kinase activity is associated with the purified insulin receptor, *Proc. Natl. Acad. Sci. U.S.A.* **80:**2137.

King, A. C., and Cuatrecasas, P., 1982, Resolution of high and low affinity epidermal growth factor receptors, *J. Biol. Chem.* **257:**3053.

King, G. L., Rechler, M. M., and Kahn, C. R., 1982, Interactions between the receptors for insulin an the insulin-like growth factors on adipocytes, *J. Biol. Chem.* **257:**10001.

Kirsch, D. M., Baumgarten, M., Deufel, T., Rinninger, F., Kemmler, W., and Haring, H. U., 1983, Catecholamine-induced insulin resistance of glucose transport in isolated rat adipocytes, *Biochem. J.* **216:**737.

Lee, L. S., and Weinstein, I. B., 1979, Mechanism of tumor promotor inhibition of cellular binding of epidermal growth factor, *Proc. Natl. Acad. Sci. U.S.A.* **76:**5168.

Lonroth, P., and Smith, U., 1983, β-Adrenergic dependent down regulation of insulin binding in rat adipocytes, *Biochem. Biophys. Res. Commun.* **112:**972.

Massague, J., and Czech, M. P., 1982, The subunit structures of two distinct receptors for insulin-like growth factors I and II and their relationship to the insulin receptor, *J. Biol. Chem.* **257:**5038.

Massague, J., Guillette, B. J., and Czech, M. P., 1981, Affinity labeling of multiplication stimulating activity receptors in membranes from rat and human tissues, *J. Biol. Chem.* **256:**2122.

Maxfield, F. R., Schlessinger, J., Shecter, Y., Pastan, I., and Willingham, M. C., 1978, Collection of insulin, EGF, α_2-macroblogulin in the same patches on the surface of cultured fibroblasts and common internalization, *Cell* **14:**805.

Niedel, J. E., Kuhn, L. J., and Vandenbark, G. R., 1983, Phorbol diester receptor copurifies with protein kinase C, *Proc. Natl. Acad. Sci. U.S.A.* **80:**36.

Nishizuka, Y., 1984, The role of protein kinase C in cell surface signal transduction and tumor promotion, *Nature* **308:**693.

Oka, Y., Mottola, C., Oppenheimer, C. L., and Czech, M. P., 1984, Insulin activates the appearance of the IGF-II receptors on the adipocyte cell surface, *Proc. Natl. Acad. Sci. U.S.A.* **81:**4028–4032.

Oppenheimer, C. L., and Czech, M. P., 1983, Purification of the type II insulin-like growth factor receptor from rat placenta, *J. Biol. Chem.* **258:**8539.

Oppenheimer, C. L., Pessin, J. E., Massague, J., Gitomer, W., and Czech, M. P., 1983, Insulin action rapidly modulates the apparent affinity of the insulin-like growth factor II receptor, *J. Biol. Chem.* **258:**4824–4830.

Pessin, J. E., Gitomer, W., Oka, Y., Oppenheimer, C. L., and Czech, M. P., 1983, β-adrenergic regulation of insulin and epidermal growth factor receptors in rat adipocytes, *J. Biol. Chem.* **258:**7386–7394.

Pessin, J. E., Mottola, C., Yu, K.-T. and Czech, M. P., 1985, Subunit structure and regulation of the insulin receptor complex, in: *The Molecular Basis of Insulin Action* (M. P. Czech, ed.), Plenum Press, New York, pp. 3–29.

Rosen, O. M., Herrera, R., Olowe, U., Petruzzelli, L. M., and Cobb, M. H., 1983, Phosphorylation activates the insulin receptor tyrosine protein kinase, *Proc. Natl. Acad. Sci. U.S.A.* **80:**2327.

Shoyab, M., DeLarco, J. E., and Todaro, G. J., 1979, Biologically active phorbol esters specifically alters affinity of epidermal growth factor membrane receptors, *Nature* **279:**387.

Suzuki, K., and Kono, T., 1980, Evidence that insulin causes translocation of glucose transport activity to the plasma membrane from an intracellular storage site, *Proc. Natl. Acad. Sci. U.S.A.* **77:**2542.

Yu, K.-T., and Czech, M. P., 1984, Tyrosine phosphorylation of the insulin receptor β subunit activates the receptor-associated tyrosine kinase activity, *J. Biol. Chem.* **259:**5277.

Zapf, J., Schoenle, E., and Froesch, E. R., 1978, Insulin-like growth factors I and II: Some biological actions and receptor binding characteristics of two purified constituents of nonsuppressibe insulin-like activity of human serum, *Eur. J. Biochem.* **87:**285.

Index